Machine Learning with Amazon SageMaker Cookbook

80 proven recipes for data scientists and developers to perform machine learning experiments and deployments

Joshua Arvin Lat

BIRMINGHAM—MUMBAI

Machine Learning with Amazon SageMaker Cookbook

Publishing Product Manager: Sunith Shetty
Senior Editor: Mohammed Yusuf Imaratwale
Content Development Editor: Nazia Shaikh
Technical Editor: Arjun Varma
Copy Editor: Safis Editing
Project Coordinator: Aparna Ravikumar Nair
Proofreader: Safis Editing
Indexer: Tejal Daruwale Soni
Production Designer: Aparna Bhagat

First published: October 2021

Production reference: 2280921

Published by Packt Publishing Ltd.
Livery Place
35 Livery Street
Birmingham
B3 2PB, UK.

ISBN 978-1-80056-703-0

www.packt.com

Dear reader. Thank you for purchasing this book! Years ago, I relied on "cookbooks" to help me gain the hands-on skills needed to get the job done using tech frameworks, libraries, tools, and services. It is my turn to give back to the tech community and provide you a "cookbook" with practical and complete solutions to help you in your machine learning journey. I hope this book helps you achieve your goals and dreams as well.

First, I would like to acknowledge Sunith Shetty, Gebin George, Aparna Nair, Nazia Shaikh, Arjun Varma, Shifa Ansari, and everyone from the Packt team for making this book a success. I would also like to thank Raphael Jambalos, Mark Jimenez, and Lauren Yu for their patient support in helping significantly improve the quality of this book. Writing a book is a team game and I am grateful to everyone who has contributed to this book.

Next, I would also like to thank Ross Barich, Julien Simon, Cameron Peron, and everyone from the AWS team for the advice and support that helped me write this book. I would also like to give my sincere thanks to the AWS teams who have built, developed, and managed the different features and capabilities of Amazon SageMaker. I would also like to acknowledge and thank Raphael Quisumbing and the leaders of AWS User Group Philippines. Years ago, it was just me, Raphael Quisumbing, Diwa del Mundo, and Mike Rayco, leading and organizing these tech events. Now, the user group has grown significantly bigger and we now have more leaders and contributors trying to make the tech world a better place.

I would like to give my sincere thanks to my parents and my sister for their never-ending love and support. At the same time, I would like to thank my relatives, friends, and colleagues at work. I would not be able to list all your names here but this acknowledgment section would not be complete without giving credit to the support and advice you all have given me throughout the years.

Finally, I want to dedicate this book to Sophie Soliven, who has been very supportive in my career choices and decisions. It all started with the "commute adventure" years ago and we did not expect that to become a lifelong journey.

Contributors

About the author

Joshua Arvin Lat is the **Chief Technology Officer (CTO)** of **NuWorks Interactive Labs, Inc**. He previously served as the CTO of three Australian-owned companies and also served as the Director for Software Development and Engineering for multiple e-commerce start-ups in the past, which allowed him to be more effective as a leader. Years ago, he and his team won first place in a global cybersecurity competition with their published research paper. He is also an **AWS Machine Learning Hero** and has shared his knowledge at several international conferences, discussing practical strategies on machine learning, engineering, security, and management.

About the reviewers

Lauren Yu is a former software engineer currently pursuing a career in law. She previously worked at AWS on Amazon SageMaker, primarily focusing on the SageMaker Python SDK, as well as toolkits and Docker images for integrating deep learning frameworks into Amazon SageMaker. While at Amazon, she also helped co-found the Amazon Symphony Orchestra of Seattle. In her spare time, she enjoys playing viola and learning more about the intersection of law and technology.

Raphael Jambalos is a cloud-native developer with 8 years of experience developing in Ruby and Python. He currently leads the cloud-native development team of eCloudValley Philippines, focused on designing and implementing various solutions such as serverless applications, CI/CD pipelines, load testing, and web development. He also holds four AWS certifications, with all three Associate-level certs and a Specialty certification in security.

Mark Jimenez is a software developer with a decade of experience in the industry ranging from web development and mobile development to machine learning. He holds several AWS certifications, including the AWS Certified Machine Learning – Specialty, AWS Certified Developer – Associate, and AWS Certified Solutions Architect – Associate certifications.

Table of Contents

2
Building and Using Your Own Algorithm Container Image

3

Using Machine Learning and Deep Learning Frameworks with Amazon SageMaker

4
Preparing, Processing, and Analyzing the Data

5
Effectively Managing Machine Learning Experiments

6

Automated Machine Learning in Amazon SageMaker

8

Solving NLP, Image Classification, and Time-Series Forecasting Problems with Built-in Algorithms

9

Managing Machine Learning Workflows and Deployments

Other Books You May Enjoy

Index

Preface

Amazon SageMaker is a fully managed **machine learning** (**ML**) service that aims to help data scientists and ML practitioners manage ML experiments. In this book, you will use the different capabilities and features of Amazon SageMaker to solve relevant data science and ML requirements.

This step-by-step guide has 80 proven recipes designed to give you the hands-on experience needed to contribute to real-world ML experiments and projects. The book covers different algorithms and techniques for training and deploying NLP, time series forecasting, and computer vision models to solve various ML problems. You will explore various solutions when working with deep learning libraries and frameworks such as **TensorFlow**, **PyTorch**, and **Hugging Face Transformers** in Amazon SageMaker. In addition to these, you will learn how to use **SageMaker Clarify**, **SageMaker Model Monitor**, **SageMaker Debugger**, and **SageMaker Experiments** to debug, manage, and monitor multiple ML experiments and deployments. You will also have a better understanding of how **SageMaker Feature Store**, **SageMaker Autopilot**, and **SageMaker Pipelines** can solve the different needs of data science teams.

By the end of this book, you will be able to combine the different solutions you have learned as building blocks to solve real-world ML requirements.

Who this book is for

This book is for developers, data scientists, and ML practitioners interested in using Amazon SageMaker to build, analyze, and deploy ML models with 80 step-by-step recipes. All we need is an AWS account to get things running. Some knowledge of AWS, ML, and the Python programming language will help readers grasp the concepts in this book more effectively.

What this book covers

Chapter 1, Getting Started with Machine Learning Using Amazon SageMaker, focuses on getting our feet wet on training and deploying an ML model using Amazon SageMaker. You will perform a simplified end-to-end ML experiment using Amazon SageMaker and the Linear Learner built-in algorithm.

Chapter 2, Building and Using Your Own Algorithm Container Image, is devoted to helping us understand how model training and deployment with Amazon SageMaker works internally. You will work on creating and using your own algorithm container images and scripts to train and deploy a custom ML model in SageMaker.

Chapter 3, Using Machine Learning and Deep Learning Frameworks with Amazon SageMaker, teaches you how to define, train, and deploy your own models using several ML and deep learning frameworks with the Amazon SageMaker Python SDK. This will allow you to use any custom models you have prepared using libraries and frameworks such as TensorFlow, Keras, scikit-learn, and PyTorch, and port these to SageMaker.

Chapter 4, Preparing, Processing, and Analyzing the Data, explores the different techniques and solutions you can use to handle your different data processing and analysis requirements. You will work with recipes that make use of SageMaker Processing, Amazon Athena, and several unsupervised built-in SageMaker algorithms to perform various data preparation and processing tasks.

Chapter 5, Effectively Managing Machine Learning Experiments, provides practical solutions and examples on debugging and managing ML experiments. You will use SageMaker Debugger to detect issues in your training jobs. In addition to this, you will work with SageMaker Experiments to manage and track multiple experiments at the same time.

Chapter 6, Automated Machine Learning in Amazon SageMaker, reveals the capabilities and features of SageMaker that help us build, train, and tune ML models automatically. You will take a closer look at using AutoML in SageMaker using SageMaker Autopilot. In addition to this, you will use and configure the automatic model tuning capability to search for the optimal set of hyperparameter values for our model.

Chapter 7, Working with SageMaker Feature Store, SageMaker Clarify, and SageMaker Model Monitor, exposes a few more capabilities of SageMaker that have great integration with SageMaker Studio – SageMaker Feature Store, SageMaker Clarify, and SageMaker Model Monitor. These capabilities help data scientists and ML practitioners handle requirements that involve using online and offline feature stores, detecting bias in the data, enabling ML explainability, and monitoring a deployed model.

Chapter 8, Solving NLP, Image Classification, and Time-Series Forecasting Problems with Built-in Algorithms, is devoted to different solutions and recipes that make use of several built-in SageMaker algorithms to solve **natural language processing** (**NLP**), image classification, and time-series forecasting problems.

Chapter 9, Managing Machine Learning Workflows and Deployments, explores several intermediate solutions for real-time endpoint deployments and automated workflows. You will work on recipes that focus on deep learning model deployments for Hugging Face models, multi-model endpoint deployments, and ML workflows.

To get the most out of this book

You will need an AWS account and a stable internet connection to complete the recipes in this book. If you still do not have an AWS account, feel free to check the **AWS Free Tier** page and click **Create a Free Account**: `https://aws.amazon.com/free/`.

Software/hardware covered in the book	OS requirements
Chrome/Firefox/Safari/Edge/Opera (or alternative)	Windows/macOS/Linux

If you are using the digital version of this book, we advise you to type the code yourself or access the code via the GitHub repository (link available in the next section). Doing so will help you avoid any potential errors related to the copying and pasting of code.

Download the example code files

You can download the example code files for this book from **GitHub** at `https://github.com/PacktPublishing/Machine-Learning-with-Amazon-SageMaker-Cookbook`. In case there's an update to the code, it will be updated on the existing GitHub repository.

We also have other code bundles from our rich catalog of books and videos available at `https://github.com/PacktPublishing/`. Check them out!

Code in Action

Code in Action videos for this book can be viewed at `https://bit.ly/3DYHjoB`.

Download the color images

We also provide a PDF file that has color images of the screenshots/diagrams used in this book. You can download it here: `https://static.packt-cdn.com/downloads/9781800567030_ColorImages.pdf`.

Conventions used

There are a number of text conventions used throughout this book.

`Code in text`: Indicates code words in text, database table names, folder names, filenames, file extensions, pathnames, dummy URLs, user input, and Twitter handles. Here is an example: "Initialize the `HuggingFace` estimator object. Here, we specify `distilbert-base-uncased` for the `model_name` value."

A block of code is set as follows:

```
predictor = Predictor(
    endpoint_name=endpoint_name
)
```

When we wish to draw your attention to a particular part of a code block, the relevant lines or items are set in bold:

```
pipeline = Pipeline(
    name=pipeline_name,
    parameters=[
        processing_instance_type,
        training_instance_type,
        input_data,
    ],
    steps=[step_process, step_train],
)
```

Any command-line input or output is written as follows:

```
cd /home/ubuntu/environment/opt
mkdir -p ml-python ml-r
```

Bold: Indicates a new term, an important word, or words that you see onscreen. For example, words in menus or dialog boxes appear in the text like this. Here is an example: "Under **Network and storage**, select the existing default VPC and choose **Public internet Only**."

> **Tips or important notes**
> Appear like this.

Sections

In this book, you will find several headings that appear frequently (*Getting ready*, *How to do it...*, *How it works...*, *There's more...*, and *See also*).

To give clear instructions on how to complete a recipe, use these sections as follows:

Getting ready

This section tells you what to expect in the recipe and describes how to set up any software or any preliminary settings required for the recipe.

How to do it...

This section contains the steps required to follow the recipe.

How it works...

This section usually consists of a detailed explanation of what happened in the previous section.

There's more...

This section consists of additional information about the recipe in order to make you more knowledgeable about the recipe.

See also

This section provides helpful links to other useful information for the recipe.

Get in touch

Feedback from our readers is always welcome.

General feedback: If you have questions about any aspect of this book, mention the book title in the subject of your message and email us at customercare@packtpub.com.

Errata: Although we have taken every care to ensure the accuracy of our content, mistakes do happen. If you have found a mistake in this book, we would be grateful if you would report this to us. Please visit www.packtpub.com/support/errata, selecting your book, clicking on the Errata Submission Form link, and entering the details.

Piracy: If you come across any illegal copies of our works in any form on the Internet, we would be grateful if you would provide us with the location address or website name. Please contact us at copyright@packt.com with a link to the material.

If you are interested in becoming an author: If there is a topic that you have expertise in and you are interested in either writing or contributing to a book, please visit authors.packtpub.com.

Share Your Thoughts

Once you've read *Machine Learning with Amazon SageMaker Cookbook*, we'd love to hear your thoughts! Scan the QR code below to go straight to the Amazon review page for this book and share your feedback.

https://packt.link/r/1800567030

Your review is important to us and the tech community and will help us make sure we're delivering excellent quality content.

1
Getting Started with Machine Learning Using Amazon SageMaker

Machine learning (**ML**) is one of the most important topics in the world right now. Through the use of different algorithms and models, it can solve different practical problems and requirements, such as anomaly detection, forecasting, spam detection, image classification, and more. Performing a few experiments in your local machine will help get things started. However, once we need to deal with end-to-end experiments involving larger datasets, deep learning requirements, and production-grade model deployments, we will need a more dedicated set of solutions to help us effectively manage these experiments.

This is what Amazon SageMaker aims to accomplish. **Amazon SageMaker** is a fully managed ML service that brings together different solutions to speed up the process of preparing, building, training, and deploying ML models. As we go through each of the chapters in this book, we will see how it helps get things done much faster across the different phases of the ML process.

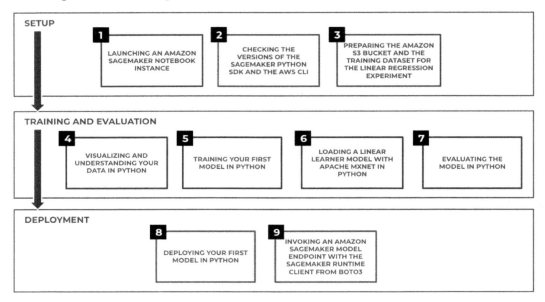

Figure 1.1 – Chapter 1 recipes

In this introductory chapter, we will perform a simplified end-to-end ML experiment using Amazon SageMaker. The goal of this chapter is to help get our feet wet and understand what it can do to help us train and deploy ML models quickly. As seen in *Figure 1.1*, we will see how Amazon SageMaker fits in the ML process. With a sample dataset, we will work our way toward building, analyzing, and deploying a model that predicts a professional's salary using the number of months of relevant managerial experience.

We will start by launching an Amazon SageMaker notebook instance where we will run our experiments. Once we have everything set up, we will visualize the data using the **pandas** data analysis and manipulation library and the **matplotlib** data visualization library. After that, we will use the **SageMaker Python SDK** when training our linear regression model. Once we have generated the model, we will then use the deep learning framework **Apache MXNet** to load the model file created and extract the parameters of the linear model. We will also use the metrics utilities from the **sklearn** library to evaluate the regression model we have prepared in this chapter. After we have performed the model evaluation step, we will use the **SageMaker Python SDK** to deploy our model to an inference endpoint. Finally, we will use the `SageMakerRuntime` client from `boto3` to invoke our deployed Amazon SageMaker inference endpoint.

We will cover the following recipes in this chapter:

- Launching an **Amazon SageMaker notebook instance**
- Checking the versions of the **SageMaker Python SDK** and the **AWS CLI**
- Preparing the **Amazon S3** bucket and the training dataset for the linear regression experiment
- Visualizing and understanding your data in Python
- Training your first model in Python
- Loading a **Linear Learner** model with **Apache MXNet** in Python
- Evaluating the model in Python
- Deploying your first model in Python
- Invoking an **Amazon SageMaker** model endpoint with the **SageMakerRuntime** client from `boto3`

Once we have completed the recipes in this chapter, we will have a good idea of how we can use **Amazon SageMaker** in the different phases of the ML process. With that in mind, let's get things started!

Technical requirements

As long as you have an existing **AWS** account, the next steps will be a piece of cake. If you still do not have an AWS account, feel free to check the **AWS Free Tier** page and click **Create a Free Account**: `https://aws.amazon.com/free/`. After clicking the **Create a Free Account** button, you will be redirected to the **Sign up for AWS** page, as shown in *Figure 1.2*:

Sign up for AWS

Explore Free Tier products with a new AWS account.

To learn more, visit aws.amazon.com/free.

Email address
You will use this email address to sign in to your new AWS account.

Password

Confirm password

AWS account name
Choose a name for your account. You can change this name in your account settings after you sign up.

Continue (step 1 of 5)

Sign in to an existing AWS account

Figure 1.2 – Sign up for AWS

All you need to do is complete the steps here to complete the sign-up process. The Jupyter notebooks, source code, and CSV files used for each chapter are available in this repository: `https://github.com/PacktPublishing/Machine-Learning-with-Amazon-SageMaker-Cookbook/tree/master/Chapter01`.

Check out the following link to see the relevant Code in Action video:

`https://bit.ly/38WpNDf`

Launching an Amazon SageMaker Notebook Instance

In this recipe, we will set up an **Amazon SageMaker notebook instance** where we can run our ML experiments using **Jupyter Notebooks**. The **SageMaker notebook instance** is a fully managed ML compute instance running a collection of tools and applications such as the **Jupyter Notebook app**. With several tools and libraries already installed and ready to use, we can go straight into working on our ML experiments without having to worry about the installation and maintenance work.

> **Important note**
>
> Take note that we can also perform our ML experiments in **Amazon SageMaker Studio**. We will take a closer look at **Amazon SageMaker Studio** in *Chapter 6*, *Automated Machine Learning in Amazon SageMaker*. From our end, it is critical to know how to use both of them as there will be features and capabilities such as **local mode**, which is supported in **notebook instances**, but not supported in **Amazon SageMaker Studio**.

Getting ready

Here are the prerequisites for this recipe — (1) an AWS account and (2) permissions to manage the **Amazon SageMaker** and **Amazon S3** resources if using an **AWS IAM** user with a custom URL. It is recommended to be signed in as an AWS IAM user instead of using the root account in most cases. For more information, feel free to take a look at `https://docs.aws.amazon.com/IAM/latest/UserGuide/best-practices.html`.

How to do it...

Before working on the next set of steps, make sure that the AWS region specified in the upper right-hand side of the screen is the region where you prefer your resources to be created. In this recipe, we will create our resources in the **N. Virginia** (us-east-1) region. Feel free to change this depending on your needs:

1. Click **Services** on the navigation bar. A list of services will be shown in the menu. Under **Machine Learning**, look for **Amazon SageMaker** and then click the link to navigate to the SageMaker console:

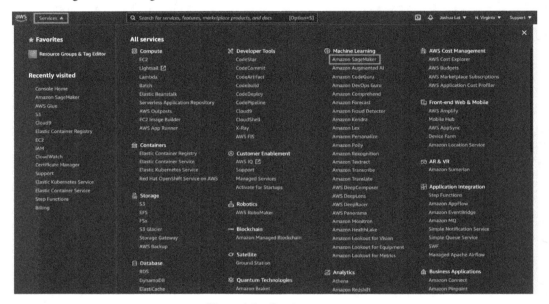

Figure 1.3 – Services menu

Figure 1.3 shows us what the menu looks like as of the time of writing. As AWS constantly improves the user experience and adds new services regularly, the UI may already be different by the time you read this!

2. In the navigation pane, click **notebook instances** under **Notebook**. This should open a page that shows all running notebooks (if any):

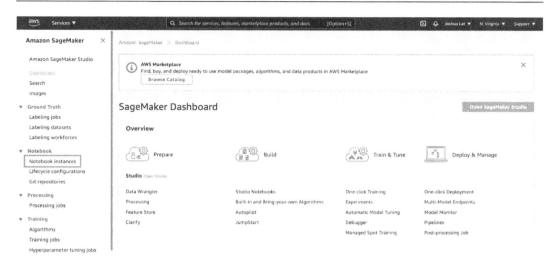

Figure 1.4 – Amazon SageMaker dashboard

We can see the **Amazon SageMaker dashboard** in *Figure 1.4*. The navigation pane is on the left side of the screen. Under **Notebook**, we have **notebook instances**, **Lifecycle configurations**, and **Git repositories**.

> **Note**
>
> We will discuss briefly how **Lifecycle configuration scripts** can help us automate notebook preparation and configuration steps in the *There's more...* section of the *Preparing the SageMaker Notebook instance for multiple deep learning local experiments* recipe in *Chapter 3, Using Machine Learning and Deep Learning Frameworks with Amazon SageMaker*.

3. On the **notebook instances** page, click the **Create notebook instance** button:

Figure 1.5 – Create notebook instance button

In *Figure 1.5*, we should see the **Create notebook instance** button on the upper right side of the image.

4.   Fill out the details in the **Create notebook instance** form. Specify the notebook instance name and choose **notebook-al1-v1** under **Platform identifier**:

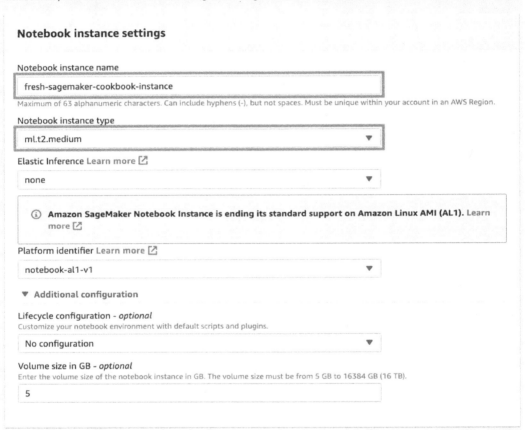

Figure 1.6 – Creating a notebook instance

In *Figure 1.6*, we have the form that allows us to manage the notebook instance settings before the notebook instance is created.

> **Note**
>
> Under **Elastic Inference** (**EI**), we have the option to specify an EI accelerator type (for example, `ml.eia2.medium`). EI accelerators help reduce the time it takes for an inference endpoint to perform a prediction. Attaching an EI accelerator to a notebook instance will allow us to test and evaluate inference performance while we are building the model locally. As we will not need this for now, we will leave this set to `none`. Note that EI accelerators can be attached to SageMaker-hosted endpoints as well to improve inference performance at a fraction of the cost. Why is this important? Of course, the faster it is for an inference endpoint to perform a prediction, the better. Feel free to refer to `https://docs.aws.amazon.com/sagemaker/latest/dg/ei.html` for more information on this topic.

5. Under **IAM role**, choose **Create a new role**:

Figure 1.7 – Create a new role

In *Figure 1.7*, we can see the different possible configuration options under **IAM role**. What's an **IAM role**? An **IAM role** is an IAM identity used to delegate access to entities and resources. This role can be assumed by a resource to gain the permissions needed to perform a specific task. In our case, we will create a role for the notebook instance to call other services and access specific resources.

6. Select **Any S3 bucket** under **S3 buckets you specify** and then click **Create role**:

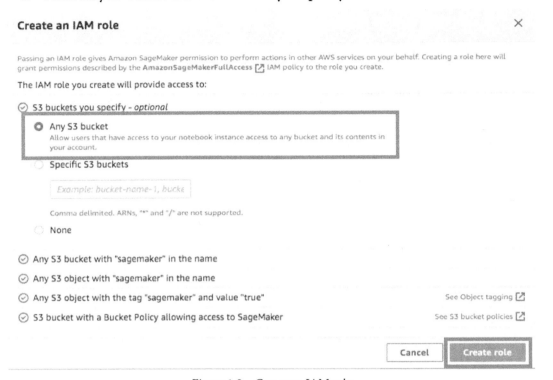

Figure 1.8 – Create an IAM role

In *Figure 1.8*, we have selected the **Any S3 bucket** option. We may opt to choose the **Specific S3 buckets** option for a more secure setup. For the recipes in this chapter, the **Any S3 bucket** option will do the trick.

> **Important note**
>
> Take note that the security configuration we specify for this example should only be used for development purposes. When dealing with production-level requirements, a more secure configuration needs to be implemented. In this case, we can choose **Specific S3 buckets** in this form. Another option is to prepare a stricter IAM role in the IAM console and use that when creating the notebook instance.

7. After successfully creating an IAM role, choose **Enable** for **Root access** and **No Custom Encryption** for **Encryption key**:

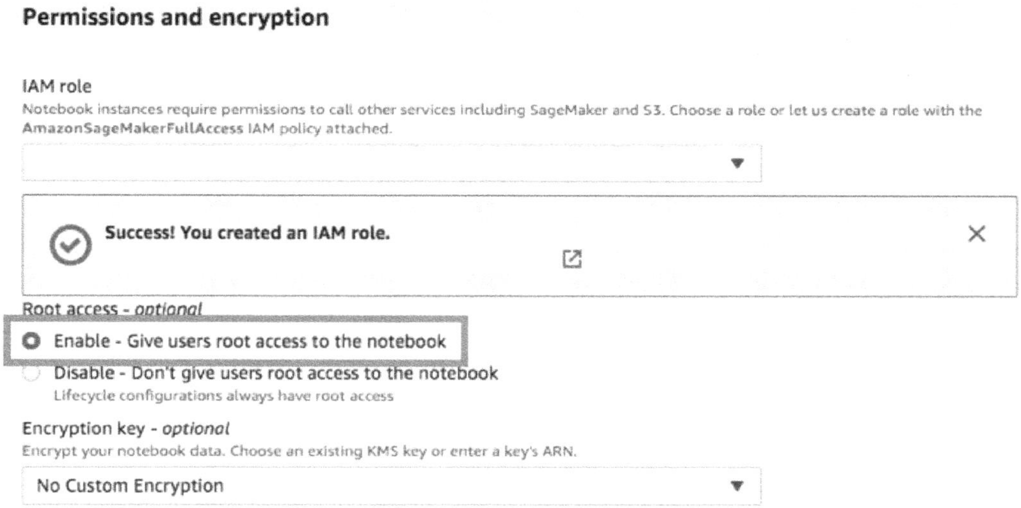

Permissions and encryption

IAM role

Notebook instances require permissions to call other services including SageMaker and S3. Choose a role or let us create a role with the AmazonSageMakerFullAccess IAM policy attached.

Figure 1.9 – Permissions and encryption

As we can see in *Figure 1.9*, we are allowing root access to the users of the notebook instance. Enabling root access means that users can install new software and even modify the system-critical files inside the instance.

8. Choose **No VPC** under **VPC - optional**:

▼ **Network** - *optional*

VPC - *optional*

Your notebook instance will be provided with SageMaker provided internet access because a VPC setting is not specified.

No VPC   ▼

Figure 1.10 – Network configuration

As can be seen in *Figure 1.10*, we have the network configuration set to **No VPC**. Given that we are just performing some test experiments using synthetic and dummy datasets, the **No VPC** option will do the trick. We can optionally select a VPC and specify whether we want to enable or disable direct internet access depending on the security needs and requirements. In this recipe, we will work with the **No VPC** configuration.

9. Choose **Clone a public Git repository to this notebook instance only** under **Repository** and set the field value under **Git repository URL** to `https://github.com/PacktPublishing/Machine-Learning-with-Amazon-SageMaker-Cookbook`:

▼ **Git repositories** - *optional*

▼ **Default repository**

Repository
Jupyter will start in this repository. Repositories are added to your home directory.

| Clone a public Git repository to this notebook instance only ▼ | C |

Git repository URL
Clone a repository to use for this notebook instance only.

| https://github.com/PacktPublishing/Machine-Learning-with-Amazon-SageMaker-Cookb(|

Add additional repository

Figure 1.11 – Cloning a public Git repository to the notebook instance

We have selected the option that allows us to clone a public Git repository to the notebook instance in *Figure 1.11*. The other option is the **Add a repository to Amazon SageMaker** option, which allows us to associate private Git repositories requiring credentials. It is possible for us to associate a public Git repository without credentials with the account as well. Given that we want to clone the notebook recipes of this book to the notebook instance, we will choose the **Clone a public Git repository to this notebook instance only** option.

10. Optionally, you may specify tags. Specify `Environment` under **Key** and `development` under **Value**:

▼ **Tags** - *optional*

Key	Value	
Environment	development	Remove

Add tag

Figure 1.12 – Adding tags

We can see in *Figure 1.12* that we have added an optional tag to the resource that we will create. If you have been using AWS for some time, you are probably aware that it is best to assign metadata to manage resources using tags. These tags are especially useful for cost allocation and security risk management. If you want to learn more about this, check out the following link: `https://d1.awsstatic.com/whitepapers/aws-tagging-best-practices.pdf`.

11. Click on the **Create notebook instance** button after the last section of the form to start the notebook creation process. You will be directed back to the **Notebook instances** page, as shown in the following screenshot:

Figure 1.13 – Notebook instance being created

In *Figure 1.13*, we can see that after completing the form, the notebook instance will be in the **Pending** state for a few minutes while it is being provisioned.

12. Once the status becomes **InService**, click **Open Jupyter**:

Figure 1.14 – Notebook instance with the InService status

As can be seen in *Figure 1.14*, once the status of the notebook instance changes to **InService**, the **Open Jupyter** and the **Open JupyterLab** links appear under **Actions**. Clicking **Open Jupyter** should open a new tab showing a page similar to what is shown in the following screenshot:

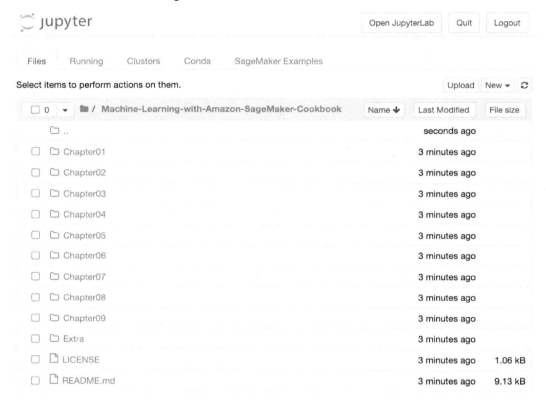

Figure 1.15 – Jupyter

In *Figure 1.15*, we can see that the Git repository we have specified during the notebook instance creation step has already been cloned to the Jupyter working directory.

Now that we have everything set up, we should be ready for the next recipes in this chapter!

How it works...

As mentioned at the start of this recipe, the **SageMaker notebook instance** is a fully managed compute instance running a collection of tools and applications to help data scientists and ML practitioners work on ML experiments right away. Here are some of the tools and applications already available inside the notebook instances. If we were to set this up ourselves, it may take a bit of time to get these tools working properly:

- **Jupyter Notebooks** and **JupyterLab**
- Jupyter kernels and Python packages such as the **SageMaker Python SDK**, **scikit-learn**, **TensorFlow**, and **Apache MXNet**
- The Jupyter R kernel along with the **reticulate** library pre-installed in Notebooks
- **AWS CLI**

When using **SageMaker Notebook instances**, it is recommended to use a smaller instance type (for example, `ml.t2.medium`) for the notebook instance whenever possible when performing experiments to reduce costs. Of course, when dealing with relatively large datasets, we can update and upgrade the notebook instance type as required. As seen in *Figure 1.16*, the notebook instance is generally running most of the time and the ML instance used for training is only running for a few minutes:

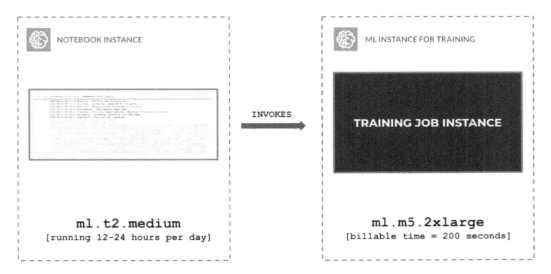

Figure 1.16 – Using a smaller instance type for the notebook instance and using
a larger instance type for the instances used for the training jobs

Note that using a larger instance type such as `ml.m5.2xlarge` is more expensive than when using the smaller instance types, including the `ml.t2.medium` and `ml.t3.medium` instances. Given that these training jobs only last for a few minutes, the cost impact of using the larger instance types is significantly reduced as these instances are automatically deleted once the training jobs have been completed.

> **Important note**
>
> To remove any confusion, it is important to note that the ML instances created during the training and processing jobs are different from the original notebook instance where we run the scripts. Before we run the training job using `estimator.fit()` with the **SageMaker Python SDK**, we only have one instance running — the notebook instance. Upon running the `estimator.fit()` function, we will have two or more instances running — the ML instance running for the notebook instance and one or more ML instances running for the training job. After the training jobs have been completed, the ML instances running for the training job are deleted and we are back to having one notebook instance. As we have not yet used the **SageMaker Python SDK** for model training and deployment, these details may not make complete sense yet. Do not worry as these notes will make more sense as we complete the recipes in this chapter.

In most cases, it is a better option to use larger instances during the data processing and training steps and smaller instances for the notebook instance where the scripts are executed and tested by the ML practitioner.

For more information, feel free to check the **Amazon SageMaker Pricing** page: `https://aws.amazon.com/sagemaker/pricing/`.

Checking the versions of the SageMaker Python SDK and the AWS CLI

In this recipe, we will check the **SageMaker Python SDK** version inside a notebook running the `conda_python3` kernel. The **SageMaker Python SDK** is a library that helps data scientists and ML practitioners to train and deploy ML models on **Amazon SageMaker**. Knowing the version of this is critical as there are several differences between Version 1.X and Version 2.X of the **SageMaker Python SDK**. In this book, we will use Version 2.X.

Getting ready

Make sure you have completed the *Launching an Amazon SageMaker notebook instance and preparing the prerequisites* recipe.

How to do it...

The first set of steps in this recipe focus on checking the **SageMaker Python SDK** version:

1. Click **New** and then choose **conda_python3** in the drop-down list:

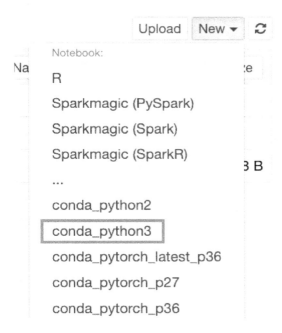

Figure 1.17 – Creating a new notebook using the conda_python3 kernel

We can see in *Figure 1.17* that there are several environments to choose from when creating a new notebook. At the top of the list, we have the **R** environment, which allows data scientists familiar and experienced with using the R language to perform ML experiments with R, `reticulate`, and the **SageMaker Python SDK**. You heard that right! We can use the Python libraries inside R and we get this to work using the `reticulate` package. We also have environments that allow us to use the different ML and deep learning libraries, frameworks, and tools right away. These include environments with **TensorFlow**, **PyTorch**, **Chainer**, and **MXNet** installed already.

2. A new tab will open. Click **File** > **Rename…**:

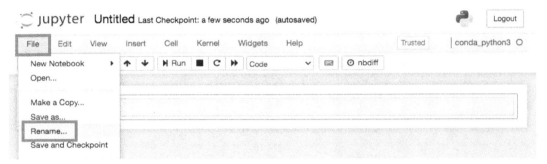

Figure 1.18 – New notebook

In *Figure 1.18*, we have a new notebook with its name initially set to `Untitled`. There are two ways to rename this notebook. The first one is by clicking the **File** menu and then clicking **Rename….** The other approach involves clicking the `Untitled` text between the Jupyter logo and the **Last Checkpoint** text.

3. A popup will open. Specify `First Notebook` as the new notebook name and then click **Rename**:

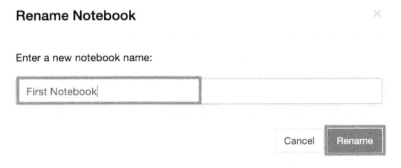

Figure 1.19 – Renaming the notebook

We specify the new notebook name as seen in *Figure 1.19*. Take note that even if this seems like a simple step, naming things accordingly and keeping things organized will help us get more things done in the long run.

4. In the code cell in the Jupyter notebook, run the following lines of code to see the version of the **SageMaker Python SDK** installed:

```
import sagemaker
sagemaker.__version__
```

We should get a 2.X.X value after running these lines of code:

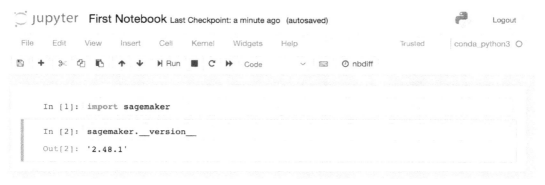

Figure 1.20 – Checking the SageMaker Python SDK version

As shown in *Figure 1.20*, we are using Version 2.X of the **SageMaker Python SDK**. In case you are using a lower version of the **SageMaker Python SDK**, install the latest version by running !pip install --upgrade sagemaker inside a notebook cell. The exclamation point before the command allows us to execute Terminal commands without having to open a separate Terminal.

> **Note**
>
> Refer to this page for more information on using Version 2.X of the **SageMaker Python SDK**: https://sagemaker.readthedocs.io/en/stable/v2.html.

The next set of steps in this recipe focus on checking the **AWS CLI** version and the operating system details. What's the AWS CLI? The **AWS Command-Line Interface** (AWS CLI) helps us to control multiple AWS services and manage resources from the command line.

5. Click on the Jupyter logo to go back to the root page. Next, create a new Terminal by clicking **New** and then choosing **Terminal** in the drop-down list:

Figure 1.21 – Creating a new Terminal

We can see in *Figure 1.21* that the **Terminal** option is at the bottom of the list. Choosing the **Terminal** option will open a new tab similar to what is shown in *Figure 1.22*:

Figure 1.22 – New Terminal

This Terminal allows us to execute bash commands and use command-line tools such as the **AWS CLI**. We can also use the `sudo yum install` command to install packages in the notebook instance.

6. Next, we check the version of the **AWS CLI** tool installed:

```
aws --version
```

You should get a value along the lines of aws-cli/1.19.22 Python/3.6.13 Linux/4.14.219-119.340.amzn1.x86_64 botocore/1.20.22.

7. Check the operating system details by reading the /etc/os-release file:

```
cat /etc/os-release
```

You should get operating system details similar to what is shown in the following screenshot:

```
sh-4.2$ cat /etc/os-release
NAME="Amazon Linux AMI"
VERSION="2018.03"
ID="amzn"
ID_LIKE="rhel fedora"
VERSION_ID="2018.03"
PRETTY_NAME="Amazon Linux AMI 2018.03"
ANSI_COLOR="0;33"
CPE_NAME="cpe:/o:amazon:linux:2018.03:ga"
HOME_URL="http://aws.amazon.com/amazon-linux-ami/"
```

Figure 1.23 – SageMaker notebook instance operating system details

In *Figure 1.23*, we can see that we are working with an instance using Amazon Linux AMI 2018.03. Now that we know the operating system details of the notebook instance we are using, it would be much easier to debug issues should shell commands or installation scripts not work right away.

Now that we have a better understanding of the tools and the versions we will use in this book, we can proceed with performing ML experiments in our next recipes!

How it works...

The **SageMaker Python SDK** helps data scientists and ML practitioners work on ML experiments using a Python library that abstracts the lower-level API operations, which distinguishes it from the **Boto3 AWS SDK for Python**. The **SageMaker Python SDK** makes use of abstraction layers and concepts such as models, estimators, and predictors, with fit() and deploy() functions similar to what libraries and frameworks such as **Keras** and **scikit-learn** have.

In this recipe, we have used `sagemaker.__version__` to check the version of the **SageMaker Python SDK**. This is important as we are trying to avoid the use of Version 1.X of the SDK before starting any of the experiments. If you need to migrate an existing notebook from Version 1.X to Version 2.X, refer to this link: `https://sagemaker.readthedocs.io/en/stable/v2.html`. The same goes for the **AWS CLI**. It is important to use the latest version of the command-line tool whenever possible. If you are using an older version of the **AWS CLI**, refer to this link and upgrade it to a newer version: `https://docs.aws.amazon.com/cli/latest/userguide/cli-chap-install.html`.

Toward the end of the recipe, we checked the operating system details by checking the `/etc/os-release` file. Given that we are working with an instance using the **Amazon Linux AMI**, we will use the `sudo yum install` command instead of `sudo apt install` when we need to install software packages. Feel free to check `https://aws.amazon.com/amazon-linux-ami/` for more information about this Linux image.

Preparing the Amazon S3 bucket and the training dataset for the linear regression experiment

In this recipe, we will create an **Amazon S3** bucket using the **AWS CLI** within the Terminal. This **S3** bucket will contain the input and output files when we are performing the different recipes in this chapter. If this is your first time hearing about **Amazon S3**, it is an object storage service that helps users store their files and their data. In the recipes in this book, we will store and download different files, datasets, and logs in **Amazon S3** while we are working on our ML experiments. The **AWS Command-Line Interface (CLI)**, on the other hand, is a command-line utility that helps to control and manage multiple AWS services and resources. In this recipe, we will use it to create an **Amazon S3** bucket with the `aws s3 mb` command in the Terminal.

> **Important note**
> Note that most of the recipes in this book will store and load files inside the S3 bucket we will create in this recipe. Inside this S3 bucket, we will create a folder for each chapter in this book to keep things organized.

We will also prepare the dataset we will use for the remaining recipes in this chapter. This dataset will only have three columns — `last_name`, `management_experience_months`, and `monthly_salary`. In addition to this, it will only have 20 records as this would be enough to prepare our first **Linear Learner** model. We are intentionally trying to keep things simple so that we can focus on getting the model trained, evaluated, and deployed without having to run through a lot of issues in our first attempt.

Getting ready

Make sure you have completed the *Launching an Amazon SageMaker notebook instance and preparing the prerequisites* recipe.

How to do it...

The first set of steps in this recipe focus on creating the S3 bucket using the **AWS CLI**:

1. Create a new Terminal by clicking **New** and then choosing **Terminal** in the drop-down list:

Figure 1.24 – Creating a new Terminal

In *Figure 1.24*, we can see that the Terminal option is at the bottom of the list. Choosing the **Terminal** option will open a new tab similar to what is shown in *Figure 1.25*:

Figure 1.25 – Terminal

In the next couple of steps, we will use this Terminal to create an S3 bucket using the **AWS CLI**.

2. Check the existing **Amazon S3** buckets by using the aws s3 ls command:

```
aws s3 ls
```

This will help us test whether the **AWS CLI** has been configured properly as well.

Important note

If you are getting the Unable to locate credentials error, this means that the AWS credentials are not properly configured in the **AWS CLI**. To resolve this, follow the steps on this page: https://aws.amazon.com/premiumsupport/knowledge-center/s3-locate-credentials-error/.

---

3. Create a new **Amazon S3** bucket by using the aws s3 mb command. Make sure to specify a unique BUCKET_NAME before running the following block of code:

```
BUCKET_NAME=my-custom-s3-bucket-abcdef
aws s3 mb s3://$BUCKET_NAME
```

If the bucket creation step is successful, you should see a line along the lines of make_bucket: <S3 bucket name> after using the aws s3 mb command. Again, do not forget to replace my-custom-s3-bucket-abcdef in the preceding block of code with a globally unique S3 bucket name of your choice.

Now that we are done creating the S3 bucket, the next set of steps focuses on creating the preferred directory structure along with an empty CSV file.

4. Now, let's print the working directory using the pwd command. Let's also check the files and directories in our current working directory using the ls command:

```
pwd
ls
```

Running these commands will generate output similar to what is shown in the following screenshot:

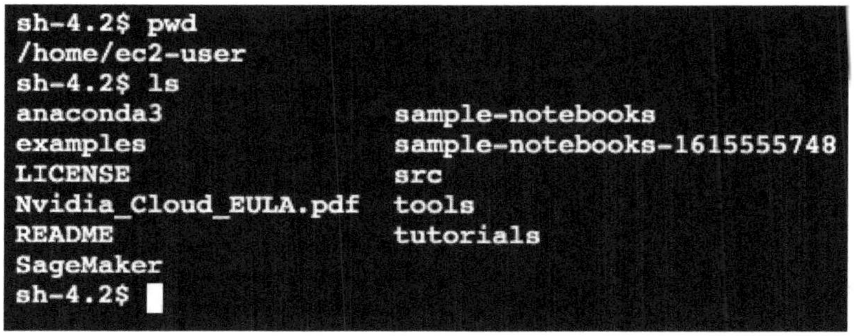

```
sh-4.2$ pwd
/home/ec2-user
sh-4.2$ ls
anaconda3                sample-notebooks
examples                 sample-notebooks-1615555748
LICENSE                  src
Nvidia_Cloud_EULA.pdf    tools
README                   tutorials
SageMaker
sh-4.2$ ▮
```

Figure 1.26 – Current working directory and the files inside the current working directory

In *Figure 1.26*, we can take a quick look at the initial directory structure after a **SageMaker notebook instance** has been created. We can see that we also have directories here for the SageMaker example notebooks. These example notebooks have been pulled automatically from this repository: https://github.com/aws/amazon-sagemaker-examples.

5. Create the required directory structure using the mkdir command. We use the -p option to automatically create the subdirectories as well:

```
cd SageMaker
mkdir -p my-experiments/chapter01/files
ls -1ahF
```

Running these commands will give us the directory structure similar to what is shown in *Figure 1.27*:

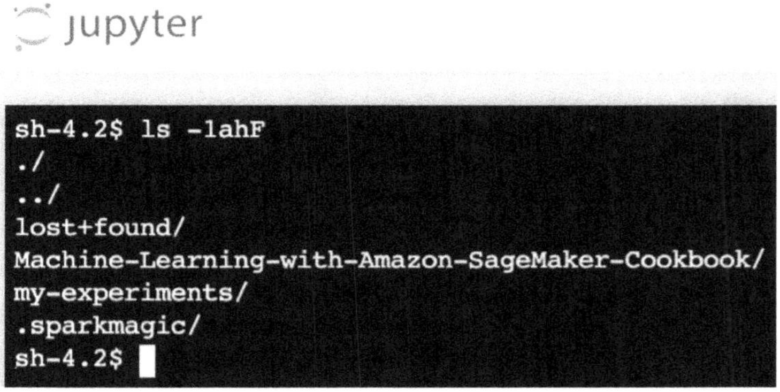

 Jupyter

```
sh-4.2$ ls -1ahF
./
../
lost+found/
Machine-Learning-with-Amazon-SageMaker-Cookbook/
my-experiments/
.sparkmagic/
sh-4.2$ ▮
```

Figure 1.27 – The my-experiments directory and its subdirectories have been created

Take note that in *Figure 1.27*, the command used is ls -1ahF and not ls –1ahF (the shorter *en dash* compared with the longer *em dash*). Make sure to use the number 1 instead of the small letter l as well.

6. Navigate to the `files` directory inside the `chapter01` directory using the following command:

```
cd my-experiments/chapter01/files
```

7. Using the `touch` command, create an empty file with the filename `management_experience_and_salary.csv`:

```
touch management_experience_and_salary.csv
```

8. Close the Terminal browser tab.

9. Navigate back to the Jupyter UI and click the **Jupyter** logo to go back to the home working directory:

Figure 1.28 – Clicking the Jupyter logo

You will be redirected back to the working directory showing both the `Machine-Learning-with-Amazon-SageMaker-Cookbook` directory and the `my-experiments` directory:

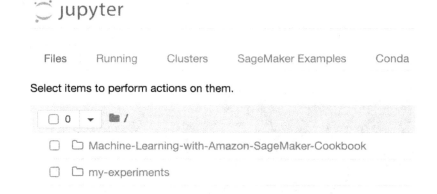

Figure 1.29 – Home directory showing both the Machine-Learning-with-Amazon-SageMaker-Cookbook and my-experiments directories

In *Figure 1.29*, we have our preferred directory structure. The `Machine-Learning-with-Amazon-SageMaker-Cookbook` directory contains our reference notebooks, and the `my-experiments` directory will contain the notebooks and files we will prepare from scratch while we are working on the recipes of this book. After we have completed this book, the contents of the `my-experiments` directory should be similar to what is inside the `Machine-Learning-with-Amazon-SageMaker-Cookbook` directory.

10. Navigate to the files directory by clicking `my-experiments`, `chapter01`, and then `files`:

Figure 1.30 – Navigating to the CSV file inside the files directory

As shown in *Figure 1.30*, we have the `management_experience_and_salary.csv` file inside the `files` directory. If we want to jump back to one of the parent directories, we can do this by clicking on the links in the breadcrumbs.

11. Click **management_experience_and_salary.csv**. This should open a new tab, as shown in the following screenshot:

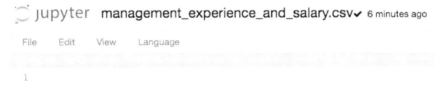

Figure 1.31 – Empty CSV file

In *Figure 1.31*, we can see an empty `management_experience_and_salary.csv` file.

12. Update the values inside the **comma-separated values (CSV)** file and make sure that it contains the following records:

last_name	management_experience_months	monthly_salary
Taylor	65	1630
Wang	61	1330
Brown	38	1290
Harris	71	1480
Jones	94	1590
Garcia	93	1750
Williams	15	1020
Lee	56	1290
White	59	1430
Tan	7	960
Chen	14	1090
Kim	67	1340
Davis	29	1170
James	49	1390
Perez	46	1240
Cruz	73	1390
Smith	19	960
Thompson	22	1040
Joseph	32	1090
Singh	37	1300

Figure 1.32 – The CSV data we will use in this chapter to prepare our linear regression model

We can see in *Figure 1.32* a table with 3 columns and 21 rows, including the header. The CSV file should have 3 columns: last_name, management_experience_months, and monthly_salary.

> **Tip**
>
> You may also optionally copy the contents of the `management_experience_and_salary.csv` file from the `Machine-Learning-with-Amazon-SageMaker-Cookbook/Chapter01` directory. You can also find the repository here and copy the necessary file(s) from the `Chapter01` directory: `https://github.com/PacktPublishing/Machine-Learning-with-Amazon-SageMaker-Cookbook`.

The CSV file should look like this before being saved:

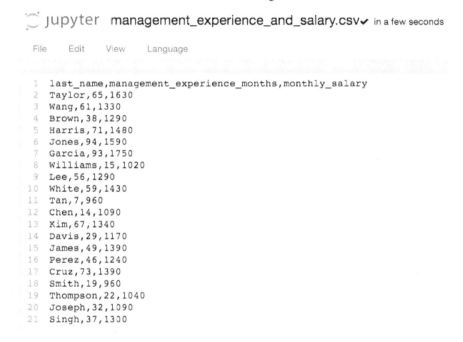

```
1   last_name,management_experience_months,monthly_salary
2   Taylor,65,1630
3   Wang,61,1330
4   Brown,38,1290
5   Harris,71,1480
6   Jones,94,1590
7   Garcia,93,1750
8   Williams,15,1020
9   Lee,56,1290
10  White,59,1430
11  Tan,7,960
12  Chen,14,1090
13  Kim,67,1340
14  Davis,29,1170
15  James,49,1390
16  Perez,46,1240
17  Cruz,73,1390
18  Smith,19,960
19  Thompson,22,1040
20  Joseph,32,1090
21  Singh,37,1300
```

Figure 1.33 – CSV file containing the training data

We have in *Figure 1.33* the CSV file with 20 rows of sample data. How do we interpret the contents of this CSV file? `Taylor` has `65` months of management experience and his monthly salary is USD `1,630`.

13. Finally, click **File** to open the **File** menu and then click **Save**. You may close the browser tab afterward or navigate back to the home directory by clicking the Jupyter logo.

Now that we have our training dataset ready, we can now proceed with the next steps in our ML process!

How it works...

The first part of this recipe involves creating an **Amazon S3** bucket using the **AWS CLI**. As you will see later in the succeeding recipes, we will use this S3 bucket to store the input and the output files of our ML experiment. Every time you see the `<insert bucket name here>` text in the blocks of code in the recipes, make sure to update and replace these with the bucket name used in this recipe.

The second part of this recipe involves creating the CSV file and filling it out with the dataset values. As you can see, we are intentionally working with a small and simplified dataset. When getting started with ML there is this misconception of having large datasets as a prerequisite when training models. Of course, the more data we can use, the better our results will be. However, while we are trying to make things work the first time around, we can use a smaller dataset to help us get the setup and experiments working quickly.

> **Note**
>
> In this chapter, we will make use of this relatively small and simple dataset to demonstrate linear regression in action using the **Linear Learner SageMaker** built-in algorithm. **Simple linear regression** involves predicting a target value using a single predictor variable. In this case, the target value is the monthly salary of a professional and the predictor variable is the management experience measured in months.

Before we proceed with training our **Linear Learner** model, let's first visualize and understand our data in the next recipe!

Visualizing and understanding your data in Python

In this recipe, we will load the sample dataset and generate a scatter plot to explore the relationship between the variables in the dataset. As you can see in the following screenshot, we have started with a DataFrame containing the `management_experience_months` and `monthly_salary` values and generated a visualization that allows us to observe the linear relationship between these two variables:

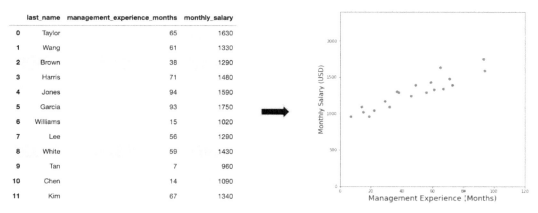

	last_name	management_experience_months	monthly_salary
0	Taylor	65	1630
1	Wang	61	1330
2	Brown	38	1290
3	Harris	71	1480
4	Jones	94	1590
5	Garcia	93	1750
6	Williams	15	1020
7	Lee	56	1290
8	White	59	1430
9	Tan	7	960
10	Chen	14	1090
11	Kim	67	1340

Figure 1.34 – Using matplotlib to generate a scatter plot chart from a DataFrame

The objective of this recipe is for us to understand the data first using plotting libraries (for example, `matplotlib`) before diving directly into the other steps of the ML process. We will start by loading a sample dataset from a CSV file to a pandas `DataFrame` and then use `matplotlib` to generate a scatter plot.

Getting ready

This recipe continues on from the *Preparing the Amazon S3 bucket and the training dataset for the linear regression experiment* recipe.

How to do it...

We will use the `pandas` data analysis library and the `matplotlib` plotting library in this recipe to load the data and generate a scatter plot to explore the relationship between `management_experience_months` (x) and `monthly_salary` (y). The following set of instructions show us how to load the x and y values of the dataset from a CSV file:

1. If you have not created a new notebook using the `conda_python3` kernel yet, create a new notebook inside the `my-experiments/chapter01` directory and rename it to the title of this recipe. We will run the code in the next set of steps inside this notebook.

2. Load the CSV file data into a `DataFrame` using **pandas**. The `read_csv()` function looks for and reads the file using the specified file path and returns the content as `pandas.DataFrame`.

```
import pandas as pd
filename = "files/management_experience_and_salary.csv"
df_all_data = pd.read_csv(filename)
```

Take note that you can configure a couple of arguments when using the `read_csv()` function. The most common ones include `sep` (separator), `header`, and `use_cols`.

3. Inspect the value of `df_all_data`:

```
df_all_data
```

The first few records of `df_all_data` will be displayed similar to what is shown in *Figure 1.35*:

	last_name	management_experience_months	monthly_salary
0	Taylor	65	1630
1	Wang	61	1330
2	Brown	38	1290
3	Harris	71	1480
4	Jones	94	1590
5	Garcia	93	1750
6	Williams	15	1020
7	Lee	56	1290
8	White	59	1430
9	Tan	7	960

Figure 1.35 – The df_all_data dataframe (showing 9 records out of 20 total records)

For larger datasets, you may use other available options in `pandas`, such as `head()` and `tail()`. You may also find the need to look for and handle *null* or *N/A* values depending on the dataset that you are working with, as shown in the following screenshot.

4. We use the `%store` magic from **IPython** to store `df_all_data` and make it available in the notebooks for our next recipes:

```
%store df_all_data
```

We will load and use `df_all_data` in the *Training your first model in Python* and *Evaluating the model in Python* recipes later in this chapter.

The next set of steps focus on generating the scatter plot showing the (linear) relationship of the *x* and *y* values using the **matplotlib** library. The **matplotlib** library is one of the most commonly used libraries for data visualization in Python.

5. Import **matplotlib**:

```
%matplotlib inline
import matplotlib.pyplot as plt
```

> **Tip**
>
> Take note that `%matplotlib inline` is an IPython magic function that allows the graphs generated by `matplotlib` to be included and stored within the Jupyter notebook.

6. Generate a scatter plot using **matplotlib**:

```
plt.rcParams["figure.figsize"] = (8,8)
plt.scatter(
    df_all_data.management_experience_months,
    df_all_data.monthly_salary)
plt.xlabel('Management Experience (Months)',
            fontsize=18)
plt.ylabel('Monthly Salary (USD)', fontsize=16)
plt.xlim(0, 120)
plt.ylim(0, 2400)
```

This should render a chart that helps us see the relationship between the two variables we are focusing on:

Figure 1.36 – Scatter plot using matplotlib where the x axis = management experience (months) and the y axis = monthly salary

In the scatter plot in *Figure 1.36*, we see that there is a linear trend. We also see that there is a positive correlation between the independent variable and the dependent variable. This correlation suggests a possible usage of a simple linear regression model to predict the value of the dependent variable (monthly salary) using the value of the independent variable (management experience in months).

While there are a few more assumptions and steps needed to justify the use of the linear regression model, we will skip these for now as we are focusing on the usage of the **Linear Learner SageMaker** built-in algorithm to solve this linear regression problem. Now, let's see how this works!

How it works...

In this recipe, we have used **pandas** to load the sample dataset from a CSV file. We will work only on a CSV file with a few records and a few columns. When dealing with experiments in real life, you will most likely deal with larger datasets with more columns. Generally, it takes hundreds to thousands of records to get good results in ML experiments. For more complex experiments dealing with text or images, you will definitely need more.

In the scatter plot generated, we can see that the *monthly salary* (generally) increases as the number of *months of management experience* increases. Depending on the type of problem we are solving, we may end up using other types of charts and, most likely, it can be visualized using **matplotlib**. In typical ML workloads and experiments, we usually perform data preprocessing and feature engineering steps before proceeding with the training step.

> **Important note**
>
> We have intentionally skipped the other data preprocessing and feature engineering steps so that we can focus on getting an inference endpoint running using the **Amazon SageMaker Python SDK**. In *Chapter 4, Preparing, Processing, and Analyzing the Data*, we will talk about a few different solutions, including **Amazon SageMaker Processing**, to process and transform the data.

Training your first model in Python

In the previous recipe, we generated a scatter plot diagram to explore the relationship between the two variables in the dataset. In this recipe, we will use the **SageMaker Linear Learner** built-in algorithm to build a linear regression model that predicts a professional's salary using the number of months of relevant managerial experience. This recipe aims to demonstrate how a **SageMaker** built-in algorithm is used in a ML experiment that involves the train-test split and running the training job:

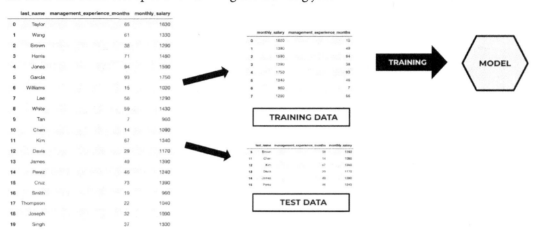

Figure 1.37 – Performing the train-test split and then running the training jobs to generate a model

Figure 1.37 shows us what we will do in this recipe. Using the `DataFrame` loaded from the *Visualizing and understanding your data in Python* recipe, we will perform the train-test split and use the training dataset to train and build the model.

Getting ready

This recipe continues on from *Visualizing and understanding your data in Python*. Make sure that you have completed the steps in that recipe as we will need the value of `df_all_data`.

How to do it...

The following set of instructions show us how to build and deploy a linear regression model using Python:

1. If you have not created a new notebook using the `conda_python3` kernel yet, create a new notebook inside the `my-experiments/chapter01` directory and rename it to the title of this recipe. We will run the code in the next set of steps inside this notebook.

2. Use the `%store` magic from **IPython** to read the stored variable, `df_all_data`, from the *Visualizing and understanding your data in Python* recipe:

```
%store -r df_all_data
```

3. Perform the train-test split:

```
from sklearn.model_selection import train_test_split
X = df_all_data['management_experience_months']
X = X.values
y = df_all_data['monthly_salary'].values

X_train, X_test, y_train, y_test = train_test_split(
    X, y, test_size=0.3, random_state=0)
```

The `train_test_split()` function accepts one or more arrays and splits them into mutually exclusive training and test `DataFrames`. With a test size of `0.3`, we allocate 30% of our dataset to be the test data. In our example, we passed two `DataFrames`, `X` and `y`. By way of output, we obtain `X_train` and `X_test` from `X`, and `y_train` and `y_test` from `y` in just a single line of code. Because we passed two `DataFrames`, `X` and `y`, to `train_test_split()`, we obtained splits for both DataFrames in just a single line of code. The `random_state` argument simply allows us to have the same set of results the next time we run the same line of code for reproducibility purposes.

Note

When working with a dataset, we usually split the dataset into training and test datasets. We use the training dataset to train the ML model. The test dataset is used to evaluate the model. In the forthcoming chapters in this book, we will see that in some recipes, we will split the dataset into training, validation, and test datasets. The validation dataset will be used to evaluate the model on the training data, while the model is being trained while tuning the model hyperparameters. The test dataset will then be used to evaluate the model after training has been completed using the train and validation datasets.

---

4. Ensure that the training dataset has the target column as the first column:

```
df_training_data = pd.DataFrame({
    'monthly_salary': y_train,
    'management_experience_months': X_train
})
```

This step is important as several algorithms, such as the **Linear Learner built-in algorithm**, expect the first column to contain the target variable data.

5. Inspect the training dataset:

```
df_training_data
```

Running the code will yield a quick view of the x (management_experience_months) and y (monthly_salary) values for the first few rows of the dataset:

	monthly_salary	management_experience_months
0	1020	15
1	1390	49
2	1590	94
3	1290	38
4	1750	93
5	1240	46
6	960	7
7	1290	56

Figure 1.38 – The df_training_data dataframe (showing 8 records out of 14 total records)

As you can see in *Figure 1.38*, the training dataset has a total of 14 records, which is 70% of the total number of records in our initial CSV file. This is due to having the test_size argument set to 0.3 when we used the train_test_split() function in an earlier step.

Once we have completed the steps for the train-test split, we will upload the training data to S3:

6. Create a temporary directory using the mkdir bash command. Be aware that this is different from the /tmp directory as this new tmp directory is created inside the my-experiments/chapter01 directory:

```
!mkdir -p tmp
```

7. Save the training data to CSV format (without the header and indices) using the to_csv() function. This saves the file in the tmp directory we just created. It is important to note that the header and index arguments are set to False as the training dataset to be used in the training job in a later step is expected to have no headers and indices:

```
df_training_data.to_csv(
    'tmp/training_data.csv',
    header=False, index=False)
```

8. Specify the bucket name and prefix. Make sure to replace the value of s3_bucket with the name of the S3 bucket you have created in the *Preparing the Amazon S3 bucket and the training dataset for the linear regression experiment* recipe:

```
s3_bucket = '<insert bucket name here>'
prefix = 'chapter01'
```

9. Upload the training data CSV to S3 using the **AWS CLI**. The following line of code runs the bash command that makes use of the AWS CLI tool to copy and upload the CSV file we have generated in a previous step to the S3 bucket. The first parameter after aws s3 cp is the source (tmp/training_data.csv). This is the file we generated and saved in the tmp directory after using the to_csv() function. The second parameter is the target destination (s3://<bucket + prefix>/training_data.csv):

```
!aws s3 cp tmp/training_data.csv s3://{s3_bucket}/
{prefix}/input/training_data.csv
```

It is important to note that the preceding block of code comprises just a single line (in case the following statement gets rendered as two lines because of the length of the command).

> **Tip**
>
> To specify a region to use when using the AWS CLI, you may use the --region option. For more information, you may visit https://docs.aws.amazon.com/cli/latest/userguide/cli-configure-options.html.

In the next set of steps, we will focus on using the **SageMaker Python SDK** to train our **Linear Learner** model:

10. Initialize and import the training prerequisites. The **SageMaker Python SDK**, along with the **Boto AWS Python SDK**, is already included when we use the *conda_python3* kernel in our Jupyter notebook. The **Boto AWS Python SDK** (boto3) is a service-level SDK that provides a way for us to access different AWS services (for example, **EC2**, **S3**, **IAM**, and more) programmatically. The **SageMaker Python SDK** focuses on operations we can do with SageMaker for our ML experiments:

```
import sagemaker
import boto3
from sagemaker import get_execution_role

role_arn = get_execution_role()
session = sagemaker.Session()
region_name = boto3.Session().region_name
```

The return values of get_execution_role() and sagemaker.Session() will be used in a later step. The get_execution_role() function from the **SageMaker Python SDK** returns the IAM role associated with the notebook instance. The return value of this function is used as an argument later when we initialize the Estimator object for the training job.

> **Note**
> What's an **IAM role**? **IAM** stands for **Identity and Access Management**. An IAM role is an identity with specific permissions used to delegate access to users and resources without having to use long-term credentials.

In addition to this, the value after we have called sagemaker.Session() is used as an argument later for a next step. It also provides the convenience functions and utilities used in SageMaker experiments, including the upload_data(), download_data(), default_bucket(), and account_id() functions.

> **Note**
>
> What's a SageMaker `Session` object? It is the object used to wcrk with the SageMaker API calls and requests along with other services (for example, **Amazon S3**) relevant to the ML experiment. For more information, feel free to check the available utility functions we can use with the `Session` object here: `https://sagemaker.readthedocs.io/en/stable/api/utility/session.html`.

11. Set the **S3** input location and the **S3** output location. It is important to note that the following block of code only comprises two lines (in case the following statement gets rendered as four or more lines because of the length of the statements):

```
training_s3_input_location = f"s3://{s3_bucket}/{prefix}/
input/training_data.csv"
training_s3_output_location = f"s3://{s3_bucket}/
{prefix}/output/"
```

12. Prepare the `S3 Input` parameter with `content_type="text/csv"`:

```
from sagemaker.inputs import TrainingInput
train = TrainingInput(training_s3_input_location,
content_type="text/csv")
```

This step is usually performed when we are using an input dataset format that is not the default format expected by the algorithm we are using. The **Linear Learner** algorithm supports the recordIO-wrapped protobuf format as well, which is more efficient to use than the CSV format. For the sake of making things simple and easier to absorb in this recipe, we will use the CSV format instead.

13. Prepare the image URI for **Linear Learner**. The `retrieve()` function returns the **Amazon ECR** URI of the **Linear Learner** built-in algorithm. Take note that the URI changes depending on the region and the experiments that you are running assume that all resources are in a single region. Otherwise, you will encounter issues during your training jobs. To solve these types of issues, simply specify the region name when using and configuring the different tools:

```
from sagemaker.image_uris import retrieve
container = retrieve("linear-learner", region_name, "1")
container
```

This should yield a value similar to `'382416733822.dkr.ecr.us-east-1.amazonaws.com/linear-learner:1'`.

14. Initialize the `Estimator` object. The `Estimator` class accepts a couple of arguments, including the container URI, SageMaker session object, and the role ARN we have obtained from the previous steps in this recipe. In the following code, we have also specified the arguments `instance_count`, `instance_type`, and `output_path`:

```
estimator = sagemaker.estimator.Estimator(
    container,
    role,
    instance_count=1,
    instance_type='ml.m5.xlarge',
    output_path=training_s3_output_location,
    sagemaker_session=session)
```

When running training jobs, **SageMaker** launches new instances outside of the Jupyter notebook instance you are using. These instances are dedicated to running the training jobs and are automatically destroyed after the training jobs have been completed. The number of training job instances used depends on the `instance_count` argument, and the size and type of the instances depend on the `instance_type` argument. That said, when the `fit()` function is called in a later step with this current configuration in the `Estimator`, SageMaker provisions a single `ml.m5.xlarge` instance to run the **Linear Learner** built-algorithm and store the results to `output_path`.

Important note

Note that this is one of the ways to initialize an `Estimator` object and configure training jobs. Another option would be to use the `LinearLearner` class, which abstracts the container image used when running the training job. We have decided to use the `Estimator` class in this chapter so that we will have a chance to show that a training job requires a (1) container image, (2) a training dataset, (3) hyperparameters, and (4) a few other configuration values to execute. We will see an example of how to use specific algorithm `Estimator` classes in the *Performing cluster analysis with the built-in KMeans algorithm*, *Performing dimensionality reduction with the built-in PCA algorithm*, and *Training a KNN model using the protobuf recordIO training input type* recipes from *Chapter 4, Preparing, Processing, and Analyzing the Data*.

---

15. Set the hyperparameters of the estimator using the set_hyperparameters() function:

```
estimator.set_hyperparameters(predictor_type='regressor',
mini_batch_size=4)
```

Hyperparameters are parameter-like values that can be tweaked and configured before the training jobs are executed. The configurable hyperparameter values depend on the algorithm used and, in this case, where we are using the Linear Learner built-in algorithm, we are setting the predictor_type and mini_batch_size hyperparameters.

> **Important note**
> Given that we are dealing with a relatively small training dataset (14 records), we have set the mini_batch_size value to 4. For regression problems, the predictor_type argument should be set to 'regressor', and for classification problems, the predictor_type argument is set to either 'binary_classifier' or 'multiclass_classifier' depending on the number of classes. Each built-in algorithm has its own set of valid hyperparameters and values, so be sure to check the official AWS documentation. As you get more familiar with the built-in algorithms and how they work internally, you will have a better feel for how to modify and tweak these hyperparameters.

16. Execute the training job using the fit() function. This runs the training job by provisioning the servers, running the algorithm on those servers, and then terminating the servers after. As this happens, log messages will be displayed in your notebook to update you on the status of the training job:

```
estimator.fit({'train': train})
```

The result of the training job is an ML model that we can use for the next steps. After the training job is completed, the path to the model.tar.gz file containing the output model artifacts can be accessed using the model_data attribute of the Estimator object.

> **Tip**
> When you have a larger dataset available, you may pass a validation dataset to the fit() function.

After about 8 to 15 minutes, we should see a set of logs similar to the logs of the training job shown in *Figure 1.39*:

```
In [20]: estimator.fit({'train': train})
```

```
2021-03-13 02:23:19 Starting - Starting the training job...
2021-03-13 02:23:43 Starting - Launching requested ML instancesProfilerReport-1615602198: InProgress
.........
2021-03-13 02:25:04 Starting - Preparing the instances for training......
2021-03-13 02:26:04 Downloading - Downloading input data...
2021-03-13 02:26:45 Training - Training image download completed. Training in progress.
2021-03-13 02:26:45 Uploading - Uploading generated training modelDocker entrypoint called with argument(s): train
Running default environment configuration script
[03/13/2021 02:26:42 INFO 140345420867392] Reading default configuration from /opt/amazon/lib/python2.7/site-package
s/algorithm/resources/default-input.json: {u'loss_insensitivity': u'0.01', u'epochs': u'15', u'feature_dim': u'auto',
u'init_bias': u'0.0', u'lr_scheduler_factor': u'auto', u'num_calibration_samples': u'10000000', u'accuracy_top_k':
u'3', u'_num_kv_servers': u'auto', u'use_bias': u'true', u'num_point_for_scaler': u'10000', u'_log_level': u'info',
u'quantile': u'0.5', u'bias_lr_mult': u'auto', u'lr_scheduler_step': u'auto', u'init_method': u'uniform', u'init_sigm
a': u'0.01', u'lr_scheduler_minimum_lr': u'auto', u'target_recall': u'0.8', u'num_models': u'auto', u'early_stopping_
patience': u'3', u'momentum': u'auto', u'unbias_label': u'auto', u'wd': u'auto', u'optimizer': u'auto', u'_tuning_obj
ective_metric': u'', u'early_stopping_tolerance': u'0.001', u'learning_rate': u'auto', u'_kvstore': u'auto', u'normal
ize_data': u'true', u'binary_classifier_model_selection_criteria': u'accuracy', u'use_lr_scheduler': u'true', u'targe
t_precision': u'0.8', u'unbias_data': u'auto', u'init_scale': u'0.07', u'bias_wd_mult': u'auto', u'f_beta': u'1.0',
u'mini_batch_size': u'1000', u'huber_delta': u'1.0', u'num_classes': u'1', u'beta_1': u'auto', u'loss': u'auto', u'be
```

Figure 1.39 – Results of the SageMaker training job execution in Python

In *Figure 1.39*, we have the logs generated by the `fit()` function. We can divide the log messages and steps into the following groups — (1) launching and preparing the ML instances for training, (2) downloading the input data and the training image, (3) running the *entrypoint* script inside the training Docker container, (4) saving the model, (5) specifying that the training job has been completed, and finally (6) logging the training and billable seconds. We will dive into the details on what is happening behind the scenes in the *There's more…* section of the *Setting up the Python and R experimentation environments* recipe from *Chapter 2, Building and Using Your Own Algorithm Container Image*.

In case you want to see the properties such as training job details and location of the model data of the `Estimator` object, feel free to run `estimator.__dict__` after the training job has been completed.

> **Important note**
>
> If you are using a fairly new AWS account, you may encounter the `ResourceLimitExceeded` error when launching ML instances using the `fit()` and `deploy()` functions. To resolve this, open the AWS Support Center and create a case. For more information, feel free to visit this link: `https://aws.amazon.com/premiumsupport/knowledge-center/resourcelimitexceeded-sagemaker/`.

In the final set of steps in this recipe, we will focus on using the %store magic to save the trained model S3 location, the image URI of the estimator, and the test dataset values.

17. Copy the value of estimator.model_data to a variable named model_data:

```
model_data = estimator.model_data
model_data
```

We should get a value similar to 's3://<S3 BUCKET NAME>/chapter01/output/linear-learner-2021-03-13-02-23-18-930/output/model.tar.gz'.

18. Use the %store magic to store the value of model_data so that we can use this in the *Loading a linear learner model with Apache MXNet in Python* and *Deploying your first model in Python* recipes:

```
%store model_data
```

19. Copy the value of estimator.image_uri to a variable named model_uri:

```
model_uri = estimator.image_uri
model_uri
```

We should get a value similar to '382416733822.dkr.ecr.us-east-1.amazonaws.com/linear-learner:1'.

20. Similar to model_data, we use the %store magic to store the value of model_uri, X_test, and y_test:

```
%store model_uri
%store X_test
%store y_test
```

With this, we have completed the steps required to train a **Linear Learner** model. Now let's see how this recipe works!

How it works...

In this recipe, we used the **SageMaker Python SDK** to build and deploy the linear regression model. This recipe is divided into the following major parts:

- **Train-test split**: The recipe started with the train-test split, which divided the dataset from the CSV file into the training and test datasets. Ideally, we should perform this step twice so that we will have the training and validation datasets for the model training phase. This way, we can evaluate the model using the test dataset, which was not used for training.

- **Saving and uploading the training dataset to S3**: The next step after splitting the dataset is to make sure that the data is formatted properly. Here, we ensured that the first column of the CSV file contains the target value for things to work correctly. The training data is uploaded to **Amazon S3** as it is the default source where the training data needs to be uploaded before running the **SageMaker** training jobs. We used the `text/csv` content type in this recipe, but we also have the option to use other content types, such as `application/x-recordio-protobuf` for the **Linear Learner** built-in algorithm. Using this allows us to use the **optimized protobuf recordIO** format during the training step and also take advantage of **Pipe mode**, which improves training job start times for larger datasets. We will have a closer look at the protobuf recordIO format in the *Converting CSV data into protobuf recordIO format* and *Training a KNN model using the protobuf recordIO training input type* recipes from *Chapter 4, Preparing, Processing, and Analyzing the Data*.

- **Training**: Once the training data has been uploaded to **S3**, the `Estimator` object is initialized and configured before starting the training job. The `Estimator` is simply a high-level interface that allows developers to initialize and configure training jobs with different parameters. Once the training job is complete, the `model.tar.gz` model file is uploaded to the target S3 bucket.

Take note that the Python code used to perform all these steps is similar across all **SageMaker built-in algorithms**. When dealing with different algorithms, however, we need to take note of the user input format along with the hyperparameters as these differ depending on the algorithm being used and the ML problem that we are solving.

There's more...

As you may have noticed, we are passing an **ECR** repository URI when initializing the `Estimator` object. Each of the **SageMaker Built-in algorithms** has corresponding container images that AWS has prepared and optimized for you already. You also have the option to use your own custom container images and algorithms, which we will talk about in detail in the next couple of chapters. The assumptions of **SageMaker** when dealing with these container images are pretty much the same regardless of whether we are using a built-in algorithm or a custom one we have built and pushed to our own ECR repository. We can create our own Docker image and let SageMaker use that image in our training jobs and model deployments. We would just need to follow a set of guidelines so that SageMaker can use that Docker image as it would when we are using a built-in algorithm instead. That said, here are some of the steps that **SageMaker** performs internally every time we perform the training step:

1. The hyperparameters, location of the training input and output files, and other arguments and configuration options specified using the **SageMaker Python SDK** are passed as parameters to the **SageMaker API** when using the `fit()` function.

2. ML instances are provisioned by **SageMaker** and the algorithm container image is pulled and started inside the ML instances. The containers have access to the files that SageMaker has downloaded, including the input files and other configuration options (for example, `hyperparameters.json`).

3. Once the training job has been completed, the output files stored in (`/opt/ml/model`) are automatically uploaded by **SageMaker** to the target S3 destination.

We will take a closer look at how this all works when we build our own custom container images in *Chapter 2, Building and Using Your Own Algorithm Container Image*.

See also

In case you are looking for practical applications of using the **Linear Learner** algorithm to build models that can be used to solve specific real-life problems using real datasets, feel free to check some of the notebooks in the **aws/amazon-sagemaker-examples** GitHub repository:

*   Breast Cancer Prediction — A **Linear Learner** model is trained to predict whether the properties of a breast mass image (for example, `radius_mean`, `smoothness_se`, and `concavity_worst`) indicate a benign or malignant tumor. Feel free to check this notebook using the following link: `https://github.com/aws/amazon-sagemaker-examples/tree/master/introduction_to_applying_machine_learning/breast_cancer_prediction`.

- Detecting Credit Card Fraud — A **Linear Learner** model is trained to predict high-risk or fraudulent transactions using the Credit Card Fraud Detection dataset from Kaggle, available here: `https://www.kaggle.com/mlg-ulb/creditcardfraud/data`. Feel free to check this notebook using the following link: `https://github.com/aws/amazon-sagemaker-examples/blob/master/scientific_details_of_algorithms/linear_learner_class_weights_loss_functions/linear_learner_class_weights_loss_functions.ipynb`.

With these notebooks, along with the recipes in this chapter, we need to be aware that we can use the **Linear Learner** algorithm to train models that can be used for (1) regression, (2) binary classification, and (3) multiclass classification. For more information, feel free to refer to the following link: `https://docs.aws.amazon.com/sagemaker/latest/dg/linear-learner.html`.

Loading a linear learner model with Apache MXNet in Python

In the previous recipe, we ran a training job using the **SageMaker Python SDK**. In this recipe, we will use **Apache MXNet** and **Gluon** to load the model, extract its parameters, and perform predictions locally. If you are wondering what **Gluon** is and how it differs from **Apache MXNet**, Gluon is a high-level API for deep learning, while **Apache MXNet** is the deep learning framework usually categorized with **TensorFlow** and **PyTorch**:

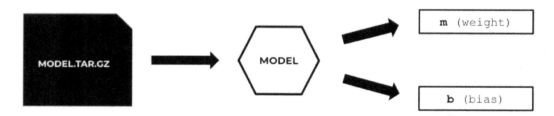

Figure 1.40 – Using Apache MXNet to load the model and extract the parameters of the model

That said, the objective of this recipe is to show that the model file uploaded to the Amazon S3 bucket after the training step can be loaded and analyzed using **Apache MXNet**, as shown in *Figure 1.40*:

Getting ready

Here are the prerequisites for this recipe:

- This recipe continues on from *Training your first model in Python*. Make sure that you have completed the steps in that recipe as we will need the value of `model_data`.

- A new notebook using the `conda_mxnet_p36` kernel.

How to do it...

For this recipe, we will need a new notebook and use the `conda_mxnet_p36` kernel to complete this recipe. The steps are as follows:

1. If you have not created a new notebook using the `conda_mxnet_p36` kernel yet, create a new notebook inside the `my-experiments/chapter01` directory and rename it to the title of this recipe. We will run the code in the next set of steps inside this notebook.

   > **Note**
   >
   > How is this different from the notebooks using the `conda_python3` kernel? The notebooks using the `conda_mxnet_p36` kernel can already make use of the **Apache MXNet** deep learning framework without having to install this separately.

2. Load `model_data` using the `%store` magic. If you can still recall, we used the `%store` magic to save the `model_data` value in the *Training your first model in Python* recipe:

   ```
   %store -r model_data
   model_data
   ```

   We should get a value similar to `'s3://<S3 BUCKET NAME>/chapter01/output/linear-learner-2021-03-13-02-23-18-930/output/model.tar.gz'`. What is inside the `model.tar.gz` file? It contains the model files generated after the SageMaker training job.

3. Prepare the SageMaker session using the following block of code. We will use the `session` object in the next step when we use the `S3Downloader.download()` function:

   ```
   import sagemaker
   session = sagemaker.Session()
   ```

4. Use the `S3Downloader.download()` function from the **SageMaker Python SDK** to download the `model.tar.gz` file from S3 to the `tmp` directory inside the `my-experiments/chapter01` directory:

```
from sagemaker.s3 import S3Downloader
S3Downloader.download(s3_uri=model_data,
                      local_path="tmp/",
                      sagemaker_session=session)
```

5. Quickly check what is inside the `tmp` directory with the help of the `ls` command:

```
!ls tmp
```

We should see the `model.tar.gz` file inside the `my-experiments/chapter01/tmp` directory.

The next instructions will allow us to extract the contents of the `model.tar.gz` file.

6. Extract the contents of the downloaded `model.tar.gz` file using the `tar` command:

```
!tar -xzvf tmp/model.tar.gz
```

We should get a value similar to `model_algo-1`.

7. Unzip the results of the previous step (`model_algo-1`):

```
!unzip model_algo-1
```

After running the `unzip` operation, you should see a similar set of extracted files from `model_algo-1` in the same directory:

```
Archive:  model_algo-1
  extracting: additional-params.json
  extracting: mx-mod-symbol.json
  extracting: mx-mod-0000.params
  extracting: manifest.json
```

Figure 1.41 – Extracted model output files after using the tar and unzip commands

In *Figure 1.41*, we can see that inside `model_algo-1`, we have several files — `additional-params.json`, `mx-mod-symbol.json`, `mx-mod-0000.params`, and `manifest.json`.

With the model files extracted already from the `model.tar.gz` file, we will now extract the weight and bias values of the linear regression model.

8. Import mxnet and the other required libraries and functions:

```
import mxnet

from mxnet import gluon
from json import load as json_load
from json import dumps as json_dumps
```

9. Load the model from the extracted contents of the zip file:

```
sym_json = json_load(open('mx-mod-symbol.json'))
sym_json_string = json_dumps(sym_json)

model = gluon.nn.SymbolBlock(
        outputs=mxnet.sym.load_json(
          sym_json_string
        ),
        inputs=mxnet.sym.var('data'))

model.load_parameters(
    'mx-mod-0000.params',
    allow_missing=True)
```

10. Initialize the model and prepare the local predict function. The mxnet_predict() function simply makes use of the model we've loaded in the previous steps to get the value of y for each specified value of x:

```
model.initialize()

def mxnet_predict(x, model=model):
    return model(mxnet.nd.array([x]))[0].asscalar()
```

11. Perform predictions using the mxnet_predict() function:

```
mxnet_predict(42)
```

We should get a value similar to 1226.6005. Later, when we run the *Deploying your first model in Python* recipe, we will see that it matches the result of what is returned by the inference endpoint.

12. Define the `extract_weight_and_bias()` function. This function accepts an MXNet linear model and returns the weight and bias:

```
def extract_weight_and_bias(model):
    params = model.collect_params()
    weight = params['fc0_weight'].data()[0].asscalar()
    bias = params['fc0_bias'].data()[0].asscalar()

    return {
        "weight": weight,
        "bias": bias
    }
```

13. Extract the *weight* and the *bias* of the linear regression model using the `extract_weight_and_bias()` function defined in the previous step:

```
weight_and_bias = extract_weight_and_bias(model)
weight_and_bias
```

We should get a value similar to `{'weight': 8.219234, 'bias': 881.3926}`.

14. Use the `%store` magic to save the value of `weight_and_bias`:

```
%store weight_and_bias
```

15. Finally, let's clean up a bit by deleting the files extracted from `model.tar.gz`:

```
%%bash
rm -f additional-params.json
rm -f manifest.json
rm -f model_algo-1
rm -f mx-mod-symbol.json
rm -f mx-mod-0000.params
```

With the steps in this recipe complete, we have demonstrated that using SageMaker to train our model does not prevent us from loading and analyzing the model artifacts produced by the training jobs.

Now let's see how this recipe works!

How it works...

In this recipe, we used **Apache MXNet** to load the model and perform local predictions. We also extracted the parameter values of the linear regression model (for example, weight and bias) and created our own prediction function using these parameter values.

> **Important note**
>
> The approach in terms of loading a trained model depends on the algorithm used along with the version of that algorithm. For example, if a model is trained with the **XGBoost** algorithm, the resulting model should be loaded by the corresponding XGBoost Python library.

The recipe might be a bit daunting at first, but we simply loaded the model and its parameters from the files created by the training job. This deserialized model allows us to perform predictions and additional analysis locally in the Jupyter notebook without having to rely on a deployed inference endpoint resource.

Evaluating the model in Python

In the previous recipes, we have trained the regression model using the **Linear Learner** algorithm and loaded the model using **MXNet** and **Gluon**. After the training step, the model needs to be evaluated, and the results and metric values need to be compared with other models. **Model evaluation** is a critical part of the ML process as this helps us find the best model, which will be used to perform predictions on future unseen values. This recipe aims to provide a simplified set of steps when evaluating regression models.

With the Python programming language, we will generate the visualization of the regression line over the original scatter plot chart and evaluate the ML model using the relevant metrics (for example, **Root Mean Squared Error(RMSE), Mean Squared Error (MSE), and Mean Absolute Error (MAE)**)

Getting ready

Here are the prerequisites for this recipe:

- This recipe continues on from *Loading a linear learner model with Apache MXNet in Python*. Make sure that you have completed the steps in that recipe as we will need the values of `weight_and_bias` along with the values from the previous recipes in Python.

- A new notebook using the `conda_python3` kernel.

How to do it...

The first set of steps focus on loading the saved data from the previous recipes and notebooks:

1. If you have not yet created a new notebook using the `conda_python3` kernel, create a new notebook inside the `my-experiments/chapter01` directory and rename it to the title of this recipe. We will run the code in the next set of steps inside this notebook.

2. Use the `%store` magic to load the `weight_and_bias` value from the *Loading a linear learner model with Apache MXNet in Python* recipe:

```
%store -r weight_and_bias
weight_and_bias
```

We should get a value similar to `{ 'weight': 8.219234, 'bias': 881.3926 }`.

3. Similarly, load the `df_all_data` value from the *Training your first model in Python* recipe:

```
%store -r df_all_data
df_all_data
```

This should show us a DataFrame of values similar to what is shown in *Figure 1.42*:

	last_name	management_experience_months	monthly_salary
0	Taylor	65	1630
1	Wang	61	1330
2	Brown	38	1290
3	Harris	71	1480
4	Jones	94	1590
5	Garcia	93	1750
6	Williams	15	1020
7	Lee	56	1290
8	White	59	1430
9	Tan	7	960
10	Chen	14	1090
11	Kim	67	1340
12	Davis	29	1170
13	James	49	1390
14	Perez	46	1240
15	Cruz	73	1390
16	Smith	19	960
17	Thompson	22	1040
18	Joseph	32	1090
19	Singh	37	1300

Figure 1.42 – df_all_data

We have the df_all_data DataFrame as seen in *Figure 1.42*. The good thing about %store magic is that it works on most simple serializable values, such as DataFrames, dictionaries, and strings. This allowed us to load the df_all_data values from a CSV file into the *Visualizing and understanding your data in Python* recipe and we can continue using df_all_data across different notebooks without having to load the data again from the CSV file.

4. Use the `%store` magic to load the `X_test` value from the *Training your first model in Python* recipe:

```
%store -r X_test
X_test
```

We should get a value for `X_test` similar to `array([32, 61, 37, 59, 14, 22])`.

5. We also load the value for `y_test`:

```
%store -r y_test
y_test
```

We should get a value for `y_test` similar to `array([1090, 1330, 1300, 1430, 1090, 1040])`.

6. Define the `manual_predict()` function. We will use this to perform predictions using the extracted weight and the bias. The `manual_predict()` function is similar to the `mxnet_predict()` function from the *Loading a linear learner model with Apache MXNet in Python* recipe, except that we are using the weight and bias directly to compute for the value of `y` instead of using the model directly for inference:

```
def manual_predict(x, weight_and_bias=weight_and_bias):
    params = weight_and_bias
    return params['weight'] * x + params['bias']
```

7. Use the `manual_predict()` function to perform a sample prediction:

```
manual_predict(42)
```

We should get a value similar to `1226.6004257202148`. Note that we got a similar result after using the `mxnet_predict()` function in the *Loading a linear learner model with Apache MXNet in Python* recipe.

8. Generate the regression line `DataFrame` and store it inside `regression_line_df`. Note that the code after the equals (=) symbol belongs to the same line and should not be treated as if it were a separate line of code (in case certain statements from the following code block get rendered as multiple lines because of the length of the command):

```
import pandas as pd
regression_line_df = pd.DataFrame(
    list(range(0, 121)),
```

```
        columns=['management_experience_months']
    )
```

```
regression_line_df['monthly_salary'] = manual_
predict(regression_line_df['management_experience_
months'])
```

In the preceding block of code, we generated a list of numbers from 1 to 121. Then, we used the manual_predict() function to compute the predicted monthly salary for each successive month of managerial experience.

9. Inspect the records of the generated regression line DataFrame:

```
regression_line_df
```

Running the preceding line of code would render the DataFrame showing the corresponding monthly_salary value for each of the values in the management_experience_months column:

	management_experience_months	monthly_salary
0	0	881.392578
1	1	889.611813
2	2	897.831047
3	3	906.050232
4	4	914.269516

Figure 1.43 – Regression line DataFrame (showing 5 records out of 121 total records)

In *Figure 1.43*, we can see the DataFrame containing the x and y values of the regression line.

In the next set of steps, we will focus on rendering the regression line on top of the scatter plot chart showing all the dataset points:

10. Import matplotlib:

```
%matplotlib inline
import matplotlib.pyplot as plt
```

11. Generate the visualization of the regression line with the original scatter plot chart:

```
plt.rcParams["figure.figsize"] = (8,8)
plt.scatter(
    df_all_data.management_experience_months,
```

```
        df_all_data.monthly_salary
)
r_line = regression_line_df
plt.plot(r_line['management_experience_months'],
          r_line['monthly_salary'],
          color='red',
          linewidth=3)
plt.xlabel('Management Experience (Months)', fontsize=18)
plt.ylabel('Monthly Salary (USD)', fontsize=16)
plt.xlim(0, 120)
plt.ylim(0, 2400)
```

Once you have executed the code blocks from the previous step, you will have a line chart representing the model on top of the scatter plot of the data points from the source CSV file:

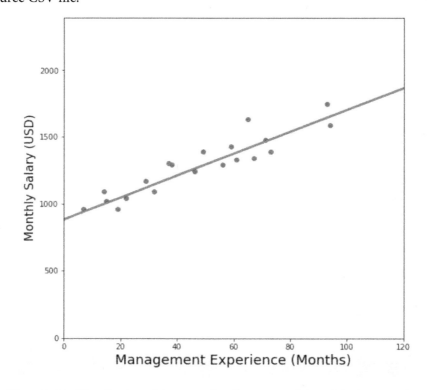

Figure 1.44 – Visualization of the regression line on top of the original scatter plot

In *Figure 1.44*, we can see the regression line on top of the scatter plot from the *Visualizing and understanding your data in Python* recipe.

In the final set of steps in this recipe, we will compute for the model evaluation metric values **RMSE**, **MSE**, and **MAE:**.

12. Import the `metrics` utilities from `scikit-learn`:

```
from sklearn import metrics
import numpy as np
```

13. View the records in the test dataset:

```
df_test_data = pd.DataFrame({
    'monthly_salary': y_test,
    'management_experience_months': X_test
})
df_test_data
```

Running the previous block of code would give us a `DataFrame` similar to what is shown in *Figure 1.45*:

	monthly_salary	management_experience_months
0	1090	32
1	1330	61
2	1300	37
3	1430	59
4	1090	14
5	1040	22

Figure 1.45 – df_test_data

In *Figure 1.45*, we have the df_test_data DataFrame. It should contain 6 records randomly selected from the dataset. If you can still remember what we did in the *Training your first model in Python* recipe, we performed the train-test split and allocated 14 records to the training dataset and 6 records to the test set.

14. Calculate the predicted monthly salary for each of the rows from the test dataset using the `manual_predict()` function. Store the results inside the `test_pred` variable. So far, we have only used the training data in each of the recipes of this chapter. Now, we will use the test data as the input to the trained model to compute each row's predicted monthly salary:

```
test_pred = manual_predict(
    df_test_data['management_experience_months']
)
```

15. Compute for **RMSE**, **MSE**, and **MAE** by using and comparing the predicted values from the actual values for the monthly salary data:

```
np.sqrt(metrics.mean_squared_error(df_test_data['monthly_
salary'], test_pred))

metrics.mean_squared_error(df_test_data['monthly_
salary'], test_pred)

metrics.mean_absolute_error(df_test_data['monthly_
salary'], test_pred)
```

We should get values similar to the following: `RMSE 73.2016`, `MSE 5358.4791`, `MAE 66.8494`.

Now let's see how this recipe works!

How it works...

Once we have the model, we must evaluate it on unseen data, such as the test set. The model evaluation step allows us to know the performance of a ML model and compare it with other models that solve the same type of problem.

To help us understand how our model performs and behaves, we generated a visualization of the regression line with the original scatter plot chart using `matplotlib`. With the testing data, we have the actual monthly salary value and the monthly salary value predicted by the model. From these two values, we generated evaluation metric values for **RMSE**, **MSE**, and **MAE**. These evaluation metrics measure how well our model predicts data that was not part of the training data. With these measures, we can compare the performance of other models we have built and trained so that we can assess which model is better.

> **Important note**
> What's the difference among the evaluation metrics **RMSE, MSE, and MAE?**
> RMSE and MSE penalize larger variations compared to MAE. When dealing
> with datasets with outliers, MAE may be a better choice to use.

There's more...

When dealing with significantly larger datasets, we can use **SageMaker Processing** so
that we can use dedicated ML instances to run the script that computes for the model
evaluation metric values. This is just one of the applications and possible use cases of
SageMaker Processing as it can also be used in data preparation and feature engineering.

Deploying your first model in Python

In the previous recipe, we performed the model evaluation step. In this recipe, we will
deploy the **Linear Learner** model to an inference endpoint using the **SageMaker Python
SDK**. What's an inference endpoint? An inference endpoint is a web application endpoint
that (1) accepts a set of values as input (for example, x value/s), (2) loads the trained
model, (3) uses the trained model to predict a value using the input, and finally, (4)
returns the predicted value in the preferred format.

After we have deployed the model, we will test the inference endpoint with a few test
predictions using sample `management_experience_months` values. We should
get the corresponding predicted `monthly_salary` values within a second or less!

Getting ready

This recipe continues on from the *Evaluating the model in Python* recipe. Make sure you
have completed the steps in that recipe along with the *Training your first model in Python*
recipe as we will need the values of `model_data` and `model_uri` in this recipe.

How to do it...

The next set of steps focus on deploying the model we have trained and evaluated in the previous recipes:

1. If you have not yet created a new notebook using the `conda_python3` kernel, create a new notebook inside the `my-experiments/chapter01` directory and rename it to the title of this recipe. We will run the code in the next set of steps inside this notebook.

2. Use the `%store` magic to read the value of `model_data`. Remember that we saved the value of `model_data` in the *Training your first model in Python* recipe:

    ```
    %store -r model_data
    model_data
    ```

    We should get a value similar to `'s3://<BUCKET NAME>/chapter01/output/linear-learner-2021-03-13-02-23-18-930/output/model.tar.gz'`.

3. Similarly, read the value of `model_uri`:

    ```
    %store -r model_uri
    model_uri
    ```

    We should get a value similar to `'382416733822.dkr.ecr.us-east-1.amazonaws.com/linear-learner:1'`.

4. Import and load a few prerequisites, including `role` and `session`:

    ```
    import sagemaker
    from sagemaker import get_execution_role


    role = get_execution_role()
    session = sagemaker.Session()
    ```

5. Initialize a `sagemaker.model.Model` object and use `model_uri`, `model_data`, `role`, and `session` as parameter values:

    ```
    from sagemaker.model import Model
    model = Model(image_uri=model_uri,
                  model_data=model_data,
                  role=role,
                  sagemaker_session=session)
    ```

6. Set `predictor_cls` of the model to `sagemaker.predictor.Predictor`:

```
from sagemaker.predictor import Predictor
model.predictor_cls = Predictor
```

If `predictor_cls` is set, calling the `deploy()` function in the next step would return a `Predictor` object. Otherwise, the `Predictor` object will not be returned after calling the `deploy()` function.

7. Call the `deploy()` function to deploy the **Linear Learner** model to an inference endpoint:

```
predictor = model.deploy(
    initial_instance_count=1,
    instance_type='ml.m5.xlarge',
    endpoint_name="linear-learner-python")
```

Feel free to specify the preferred `endpoint_name` value.

> **Important note**
>
> Running the `deploy()` function would launch an instance that would continue running until the delete resource operation is performed. While the instance is running, you will be charged for the amount of time it is running. Make sure to delete the inference endpoint after completing this chapter. Once you are done with this chapter, you may use `predictor. delete_endpoint()` to delete the inference endpoint. Note that in the last recipe in this chapter, *Invoking an Amazon SageMaker model endpoint with the SageMakerRuntime client from boto3*, we will need the inference endpoint active. That said, you may delete the inference endpoint after you have completed that recipe.

8. Update the `serializer` and `deserializer` configurations of `predictor`. Updating these in our predictor object changes the `serializer` and `deserializer` attributes for encoding and decoding data to and from the inference endpoint:

```
from sagemaker.serializers import CSVSerializer
from sagemaker.deserializers import JSONDeserializer
predictor.serializer = CSVSerializer()
predictor.deserializer = JSONDeserializer()
```

9. Perform a sample prediction using the `predict()` function:

```
predictor.predict("42")
```

We should get a return value with the structure and prediction value similar to `{'predictions': [{'score': 1226.6004638671875}]}`. This should match the result we got after using the `mxnet_predict()` function from the *Loading a linear learner model with Apache MXNet in Python* recipe.

10. Finally, let's try passing two `management_experience_months` values in our payload:

```
predictor.predict(["42", "81"])
```

We should get a return value with the structure and prediction values similar to `{'predictions': [{'score': 1226.6004638671875}, {'score': 1547.150634765625}]}`. Given that we have provided two `management_experience_months` values in our payload, we also got two predicted `monthly_salary` values in the response.

11. Delete the endpoint using the `delete_enpoint()` function:

```
predictor.delete_endpoint()
```

This will delete the deployed inference endpoint. Of course, once we run this line of code, we will no longer be able to use the `predict()` function. Feel free to experiment a bit before deleting the endpoint. In case you proceed with the *Invoking an Amazon SageMaker model endpoint with the SageMakerRuntime client from boto3* recipe in this chapter, the inference endpoint needs to be running, so do not delete the endpoint just yet. You may also just deploy the endpoint again by following the steps in this recipe and by calling the `deploy()` function if you have accidentally deleted it.

Be aware that there are different ways to perform predictions, but we will discuss the other options later in this book. Now, let's dive into the *How it works...* section!

How it works...

In this recipe, we used the `Model` and `Predictor` classes from the **SageMaker Python SDK** to deploy the model we trained in the *Training your first model in Python* recipe.

Here are the steps that SageMaker performs internally every time we use the `deploy()` function:

1. When we use the `deploy()` function, a new ML instance is provisioned by **SageMaker** to run the container. Inside this container, an inference web server starts running once the deployment step is complete. The inference web server simply serves a web endpoint that accepts the predictor values (for example, x – number of months of management experience) and returns the predicted values (for example, y – salary)

2. Behind the scenes, SageMaker downloads the model files from the S3 bucket. The model is then loaded by the web server that serves as the inference endpoint.

3. The container inside the ML instance exposes the API endpoint, which allows us to send a payload containing the predictor variables (for example, `management_experience_months`) and receive the response containing the predicted value (for example, `monthly_salary`)

You may probably be wondering why we did not go straight to using `estimator.deploy()` after using `estimator.fit()`. In the forthcoming chapters and recipes, we will use this approach instead. In this chapter, we must become aware of what is happening behind the scenes as we are trying to remove the perception that we are dealing with a black box using **Amazon SageMaker** and the **SageMaker Python SDK**.

Invoking an Amazon SageMaker model endpoint with the SageMakerRuntime client from boto3

With our model deployed in an inference endpoint using the SageMaker hosting services, we can now use the `SageMakerRuntime` client from `boto3` to invoke the endpoint. This will help us to invoke the SageMaker inference endpoint within any application code using `boto3` or a similar SDK. For example, we can use this in an **AWS Lambda** function with **Amazon API Gateway** to build a serverless API endpoint that accepts an HTTP request containing the number of months of management experience of a professional and returns a response with the predicted monthly salary of that individual.

In this recipe, we will use the `invoke_endpoint()` function from the `SageMakerRuntime` client from `boto3` to trigger an existing SageMaker inference endpoint. We can use the deployed endpoint from the *Deploying your first model in Python* recipe.

Getting ready

This recipe continues on from *Deploying your first model in Python*. We will need a running SageMaker inference endpoint for this recipe.

How to do it...

The next set of steps focus on invoking an existing SageMaker inference endpoint:

1. If you have not created a new notebook using the `conda_python3` kernel yet, create a new notebook inside the `my-experiments/chapter01` directory and rename it to the title of this recipe. We will run the code in the next set of steps inside this notebook.

2. Load the `SageMakerRuntime` client with **Boto3**:

   ```
   import boto3
   sagemaker_client = boto3.client('sagemaker-runtime')
   ```

3. Specify the endpoint. Feel free to modify the following line of code if your endpoint has a different name. In this recipe, we will use the endpoint we deployed in the *Deploying your first model in Python* recipe:

   ```
   endpoint = 'linear-learner-python'
   ```

4. Set the `42` string value as the payload:

   ```
   payload="42"
   ```

5. Trigger the SageMaker inference endpoint using the `invoke_endpoint()` function:

   ```
   response = sagemaker_client.invoke_endpoint(
       EndpointName=endpoint,
       ContentType='text/csv',
       Body=payload)
   ```

6. Inspect the structure and contents of `response`:

   ```
   response
   ```

We will get a response value similar to what is shown in *Figure 1.46*:

```
{'ResponseMetadata': {'RequestId': 'cbab66fd-30de-4781-a813-932ad1d662c7',
  'HTTPStatusCode': 200,
  'HTTPHeaders': {'x-amzn-requestid': 'cbab66fd-30de-4781-a813-932ad1d662c7',
   'x-amzn-invoked-production-variant': 'AllTraffic',
   'date': 'Sat, 13 Mar 2021 13:32:46 GMT',
   'content-type': 'application/json',
   'content-length': '48'},
  'RetryAttempts': 0},
 'ContentType': 'application/json',
 'InvokedProductionVariant': 'AllTraffic',
 'Body': <botocore.response.StreamingBody at 0x7fbdf4208ba8>}
```

Figure 1.46 – Response return value after using the invoke_endpcint() function

As we can see in *Figure 1.46*, the value for `Body` is not a string. It is wrapped with `StreamingBody` from `botocore`. `StreamingBody` is a wrapper class for an HTTP response body that provides a few convenience functions. We can also see here that the value for `InvokedProductionVariant` is `AllTraffic`. As we are dealing with a single model deployment, this is the expected value. When working with multiple deployed models using production variants, this value changes depending on the invoked model.

7. Finally, we use the following code block to convert the `StreamingBody` object to a dictionary:

```
import json
result = json.loads(
    response['Body'].read().decode('utf-8')
)
result
```

After running this block of code, we should get a structure and value similar to `{'predictions': [{'score': 1209.7744140625}]}`.

Now, let's see how this works!

How it works...

In this recipe, we demonstrated how to trigger an existing SageMaker inference endpoint using the `SageMakerRuntime` client from `boto3`. Here are some of the solutions that build on top of what we have in this recipe:

- `boto3` + **AWS Lambda** + **Amazon API Gateway** (Serverless REST API)
- `boto3` + **AWS Lambda** + **AWS AppSync** (Serverless GraphQL API)

- `boto3` + **Flask** web framework

- `boto3` + **Django** web framework

- `boto3` + **Celery** + **RabbitMQ**

Given that `boto3` and the `SageMakerRuntime` client provide a low-level abstraction layer for calling the SageMaker API, the same techniques and concepts used in this recipe also apply when using the AWS SDKs for other languages such as Java, JavaScript, PHP, Go, C++, .NET, and Ruby. For example, if we need to trigger the SageMaker inference endpoint from an application using Ruby, all we need to do is use the **AWS SDK for Ruby** and use the `invoke_endpoint()` function, which more or less behaves the same way as with the `invoke_endpoint()` function in `boto3`. You may check `https://docs.aws.amazon.com/sdk-for-ruby/v2/api/Aws/SageMakerRuntime/Client.html` for more information. Similar references should be available for the AWS SDKs for the other programming languages.

> **Important note**
>
> Now that we are done with this chapter, we can delete the existing inference endpoint deployed in the *Deploying your first model in Python* recipe using `predictor.delete_endpoint()`, or by deleting it through the user interface (**Inference > Endpoints**).

2
Building and Using Your Own Algorithm Container Image

In the previous chapter, we performed a simplified end-to-end machine learning experiment with the **Amazon SageMaker** built-in algorithm called **Linear Learner**. At the time of writing, there are 17 built-in algorithms to choose from! Depending on our requirements, we may simply choose one or more algorithms from these 17 built-in algorithms to solve our machine learning problem. In real life, we will be dealing with pre-trained models and other algorithms that are not in this list of built-in algorithms from SageMaker. One of the strengths of Amazon SageMaker is its flexibility and support for custom models and algorithms by using custom container images. Let's say that you want to use an algorithm that's not available in the list of built-in algorithms from SageMaker, such as **Support Vector Machines (SVM)**, to solve your machine learning problems. If that's the case, then this chapter is for you!

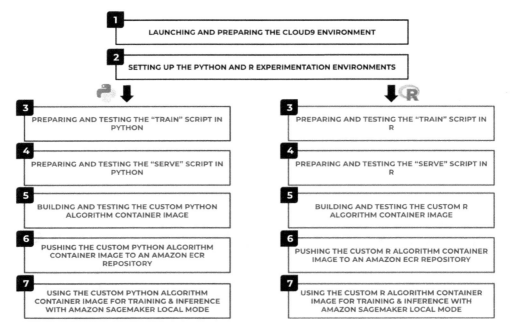

Figure 2.1 – Chapter 2 recipes

In this chapter, we will work on creating and using our own algorithm container images in **Amazon SageMaker**. With this approach, we can use any custom scripts, libraries, frameworks, or algorithms. This chapter will enlighten us on how we can make the most out of **Amazon SageMaker** through custom container images. As shown in the preceding diagram, we will start by setting up a cloud-based integrated development environment with **AWS Cloud9**, where we will prepare, configure, and test the scripts before building the container image. Once we have the environment ready, we will code the train and serve scripts inside this environment. The train script will be used during training, while the serve script will be used for the inference endpoint of the deployed model. We will then prepare a `Dockerfile` that makes use of the train and serve scripts that we generated in the earlier steps. Once this `Dockerfile` is ready, we will build the custom container image and use the container image for training and inference with the **SageMaker Python SDK**. We will work on these steps in both Python and R.

We will cover the following recipes in this chapter:

- Launching and preparing the **Cloud9** environment
- Setting up the Python and R experimentation environments
- Preparing and testing the train script in Python
- Preparing and testing the serve script in Python

- Building and testing the custom Python algorithm container image

- Pushing the custom Python algorithm container image to an **Amazon ECR** repository

- Using the custom Python algorithm container image for training and inference with **Amazon SageMaker Local Mode**

- Preparing and testing the train script in R

- Preparing and testing the serve script in R

- Building and testing the custom R algorithm container image

- Pushing the custom R algorithm container image to an **Amazon ECR** repository

- Using the custom R algorithm container image for training and inference with **Amazon SageMaker Local Mode**

After we have completed the recipes in this chapter, we will be ready to use our own algorithms and custom container images in **SageMaker**. This will significantly expand what we can do outside of the built-in algorithms and container images provided by **SageMaker**. At the same time, the techniques and concepts used in this chapter will give you the exposure and experience needed to handle similar requirements, as you will see in the upcoming chapters.

Technical requirements

You will need the following to complete the recipes in this chapter:

- A running **Amazon SageMaker** notebook instance (for example, `ml.t2.large`). Feel free to use the SageMaker notebook instance we launched in the *Launching an Amazon SageMaker Notebook instance* recipe of *Chapter 1, Getting Started with Machine Learning Using Amazon SageMaker.*

- Permission to manage the **Amazon SageMaker**, **Amazon S3**, and **AWS Cloud9** resources if you're using an **AWS IAM** user with a custom URL. It is recommended to be signed in as an AWS IAM user instead of using the root account in most cases. For more information, feel free to take a look at `https://docs.aws.amazon.com/IAM/latest/UserGuide/best-practices.html`.

The Jupyter Notebooks, source code, and CSV files used for each chapter are available in this book's GitHub repository: `https://github.com/PacktPublishing/Machine-Learning-with-Amazon-SageMaker-Cookbook/tree/master/Chapter02`.

Check out the following link to see the relevant Code in Action video:

`https://bit.ly/38Uvemc`

Launching and preparing the Cloud9 environment

In this recipe, we will launch and configure an **AWS Cloud9** instance running an **Ubuntu** server. This will serve as the experimentation and simulation environment for the other recipes in this chapter. After that, we will resize the volume attached to the instance so that we can build container images later. This will ensure that we don't have to worry about disk space issues while we are working with Docker containers and container images. In the succeeding recipes, we will be preparing the expected file and directory structure that our `train` and `serve` scripts will expect when they are inside the custom container.

> **Important note**
>
> Why go through all this effort of preparing an experimentation environment? Once we have finished preparing the experimentation environment, we will be able to prepare, test, and update the custom scripts quickly, without having to use the `fit()` and `deploy()` functions from the **SageMaker Python SDK** during the initial stages of writing the script. With this approach, the feedback loop is much faster, and we will detect the issues in our script and container image before we even attempt using these with the **SageMaker Python SDK** during training and deployment.

Getting ready

Make sure you have permission to manage the **AWS Cloud9** and **EC2** resources if you're using an **AWS IAM** user with a custom URL. It is recommended to be signed in as an AWS IAM user instead of using the root account in most cases.

How to do it...

The steps in this recipe can be divided into three parts:

- Launching a **Cloud9** environment
- Increasing the disk space of the environment
- Making sure that the volume configuration changes get reflected by rebooting the instance associated with the **Cloud9** environment

We'll begin by launching the Cloud9 environment with the help of the following steps:

1. Click **Services** on the navigation bar. A list of services will be shown in the menu. Under **Developer Tools**, look for **Cloud9** and then click the link to navigate to the Cloud9 console:

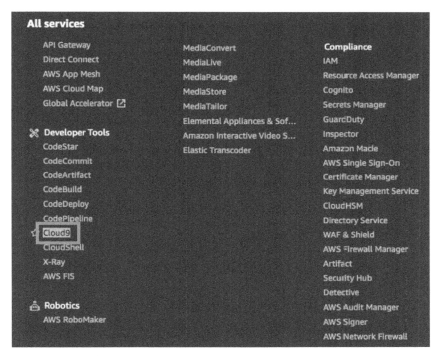

Figure 2.2 – Looking for the AWS Cloud9 service under Developer Tools

In the preceding screenshot, we can see the services after clicking the **Services** link on the navigation bar.

2. In the Cloud9 console, navigate to **Your environments** using the sidebar and click **Create environment**:

Figure 2.3 – Create environment button

Here, we can see that the **Create environment** button is located near the top-right corner of the page.

3. Specify the environment's name (for example, `Cookbook Experimentation Environment`) and, optionally, a description for your environment. Click **Next step** afterward:

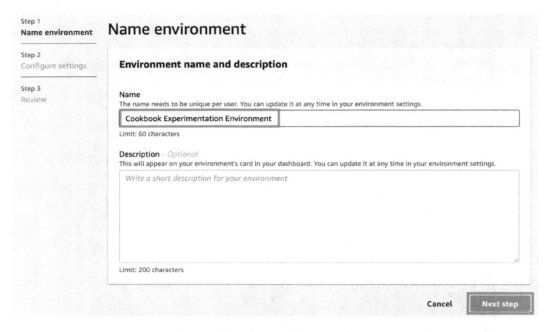

Figure 2.4 – Name environment form

Here, we have the **Name environment** form, where we can specify the name and description of our Cloud9 environment.

4. Select the **Create a new EC2 instance for environment (direct access)** option under **Environment type**, **t3.small** under **Instance type**, and **Ubuntu Server 18.04 LTS** under **Platform**:

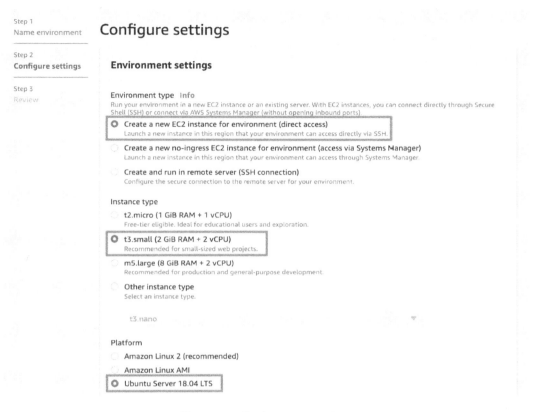

Step 1
Name environment

Step 2
Configure settings

Step 3
Review

Configure settings

Environment settings

Environment type   Info
Run your environment in a new EC2 instance or an existing server. With EC2 instances, you can connect directly through Secure Shell (SSH) or connect via AWS Systems Manager (without opening inbound ports).

○ **Create a new EC2 instance for environment (direct access)**
   Launch a new instance in this region that your environment can access directly via SSH.

○ **Create a new no-ingress EC2 instance for environment (access via Systems Manager)**
   Launch a new instance in this region that your environment can access through Systems Manager.

○ **Create and run in remote server (SSH connection)**
   Configure the secure connection to the remote server for your environment.

Instance type

○ **t2.micro (1 GiB RAM + 1 vCPU)**
   Free-tier eligible. Ideal for educational users and exploration.

○ **t3.small (2 GiB RAM + 2 vCPU)**
   Recommended for small-sized web projects.

○ **m5.large (8 GiB RAM + 2 vCPU)**
   Recommended for production and general-purpose development.

○ **Other instance type**
   Select an instance type.

   t3.nano   ▼

Platform

○ Amazon Linux 2 (recommended)

○ Amazon Linux AMI

○ Ubuntu Server 18.04 LTS

Figure 2.5 – Environment settings

We can see the different configuration settings here. Feel free to choose a different instance type as needed.

5. Under **Cost-saving setting**, select **After one hour**. Leave the other settings as-is and click **Next step**:

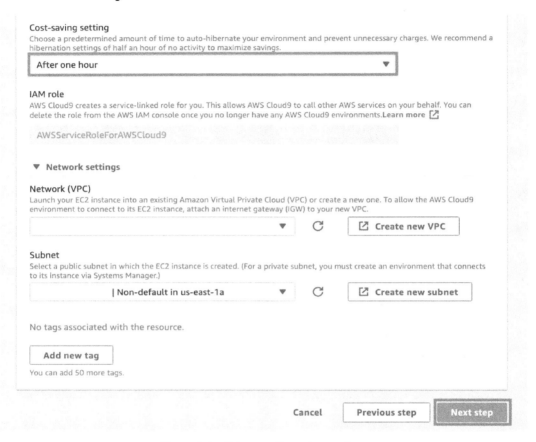

Figure 2.6 – Other configuration settings

Here, we can see that we have selected a **Cost-saving setting** of **After one hour**. This means that after an hour of inactivity, the EC2 instance linked to the Cloud9 environment will be automatically turned off to save costs.

6. Review the configuration you selected in the previous steps and then click
    Create environment:

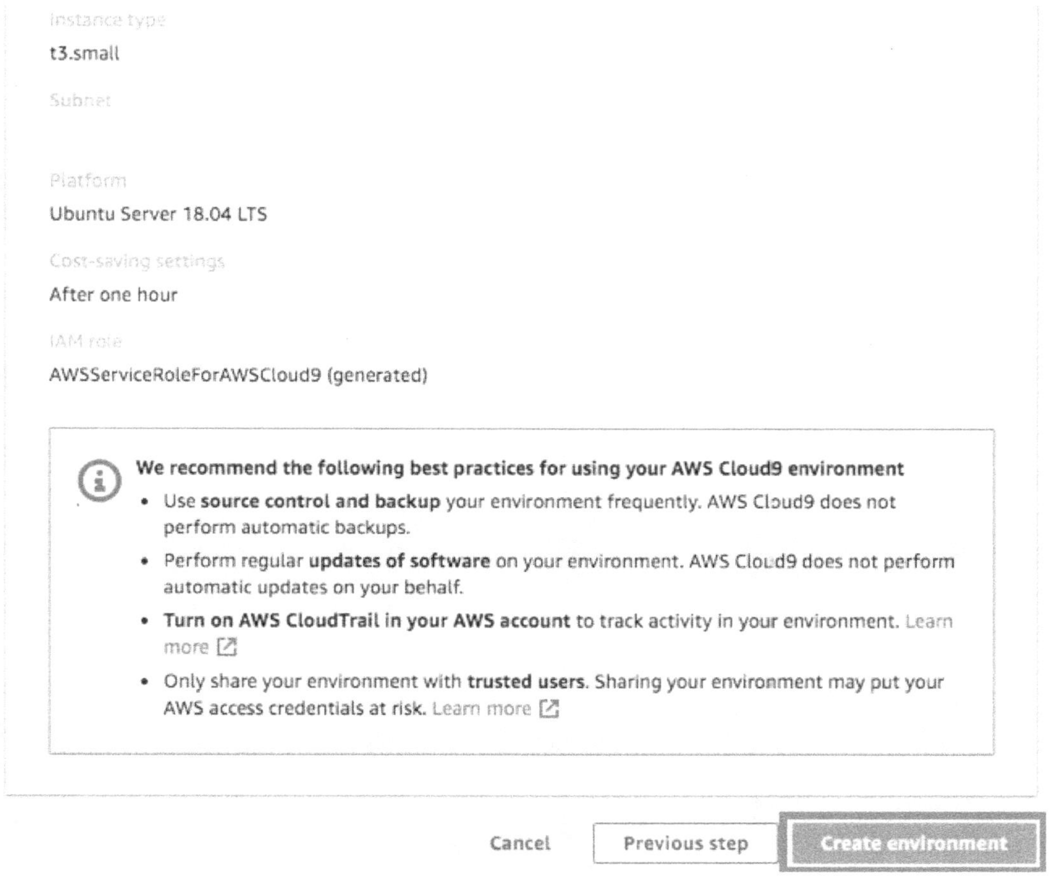

Instance type

t3.small

Subnet

Platform

Ubuntu Server 18.04 LTS

Cost-saving settings

After one hour

IAM role

AWSServiceRoleForAWSCloud9 (generated)

> (i) **We recommend the following best practices for using your AWS Cloud9 environment**
>   • Use **source control and backup** your environment frequently. AWS Cloud9 does not
>   perform automatic backups.
>   • Perform regular **updates of software** on your environment. AWS Cloud9 does not perform
>   automatic updates on your behalf.
>   • **Turn on AWS CloudTrail in your AWS account** to track activity in your environment. Learn
>   more [↗]
>   • Only share your environment with **trusted users**. Sharing your environment may put your
>   AWS access credentials at risk. Learn more [↗]

Cancel   Previous step   **Create environment**

Figure 2.7 – Create environment button

After clicking the **Create environment** button, it may take a minute or so for the environment to be ready. Once the environment is ready, check the different sections of the IDE:

Figure 2.8 – AWS Cloud9 development environment

As you can see, we have the **file tree** on the left-hand side. At the bottom part of the screen, we have the **Terminal**, where we can run our Bash commands. The largest portion, at the center of the screen, is the **Editor**, where we can edit the files.

Now, we need to increase the disk space.

7. Using the Terminal at the bottom section of the IDE, run the following command:

```
lsblk
```

With the `lsblk` command, we will get information about the available block devices, as shown in the following screenshot:

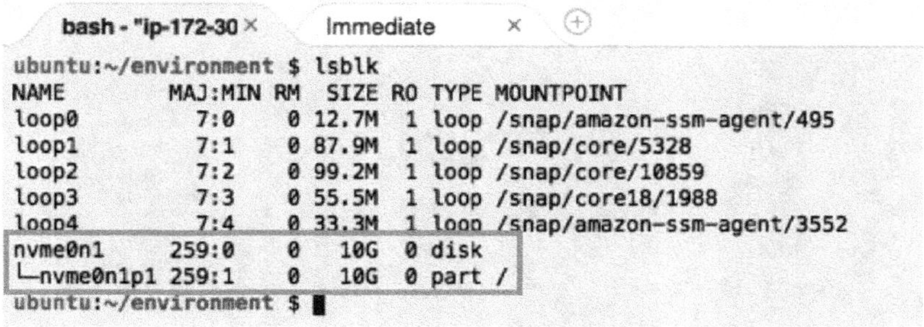

```
bash - "ip-172-30 ×        Immediate        ×    ⊕
ubuntu:~/environment $ lsblk
NAME          MAJ:MIN RM   SIZE RO TYPE MOUNTPOINT
loop0            7:0     0 12.7M  1 loop /snap/amazon-ssm-agent/495
loop1            7:1     0 87.9M  1 loop /snap/core/5328
loop2            7:2     0 99.2M  1 loop /snap/core/10859
loop3            7:3     0 55.5M  1 loop /snap/core18/1988
loop4            7:4     0 33.3M  1 loop /snap/amazon-ssm-agent/3552
nvme0n1        259:0     0   10G  0 disk
└─nvme0n1p1    259:1     0   10G  0 part /
ubuntu:~/environment $ ▮
```

Figure 2.9 – Result of the lsblk command

Here, we can see the results of the `lsblk` command. At this point, the root volume only has `10G` of disk space (minus what is already in the volume).

8. At the top left section of the screen, click **AWS Cloud9**. From the dropdown list, click **Go To Your Dashboard**:

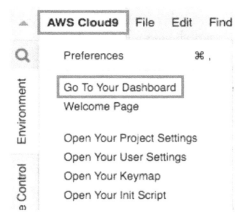

Figure 2.10 – How to go back to the AWS Cloud9 dashboard

This will open a new tab showing the Cloud9 dashboard.

9. Navigate to the EC2 console using the search bar. Type `ec2` in the search bar and click the **EC2** service from the list of results:

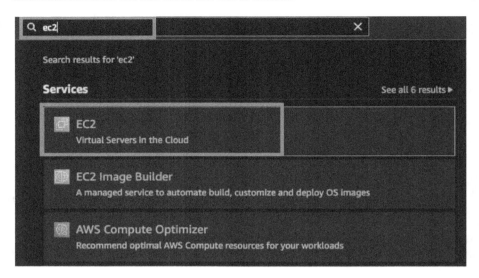

Figure 2.11 – Using the search bar to navigate to the EC2 console

Here, we can see that the search bar quickly gives us a list of search results after we have typed in `ec2`.

10. In the EC2 console, click **Instances (running)** under **Resources**:

Figure 2.12 – Instances (running) link under Resources

We should see the link we need to click under the **Resources** pane, as shown in the preceding screenshot.

11. Select the EC2 instance corresponding to the Cloud9 environment we launched in the previous set of steps. It should contain `aws-cloud9` and the name we specified while creating the environment. In the bottom pane showing the details, click the **Storage** tab to show **Root device details** and **Block devices**.

12. Inside the **Storage** tab, scroll down to the bottom of the page to locate the volumes under **Block devices**:

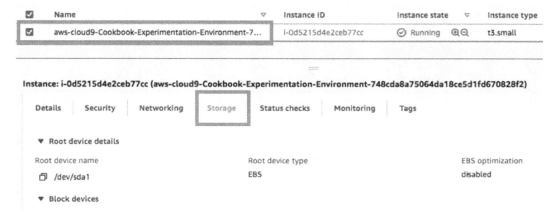

Figure 2.13 – Storage tab

Here, we can see the **Storage** tab showing **Root device details** and **Block devices**.

13. You should see an attached volume with 10 GiB for the volume size. Click the link under **Volume ID** (for example, vol-0130f00a6cf349ab37). Take note that this **Volume ID** will be different for your volume:

Figure 2.14 – Looking for the volume attached to the EC2 instance

You will be redirected to the **Elastic Block Store Volumes** page, which shows the details of the volume attached to your instance:

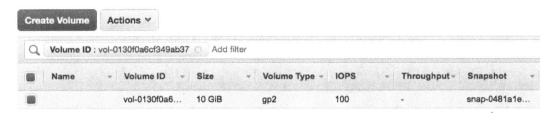

Figure 2.15 – Elastic Block Store Volumes page

Here, we can see that the size of the volume is currently set to **10 GiB**.

14. Click **Actions** and then **Modify Volume**:

Figure 2.16 – Modify Volume

This is where we can find the **Modify Volume** option.

15. Set **Size** to 100 and click **Modify**:

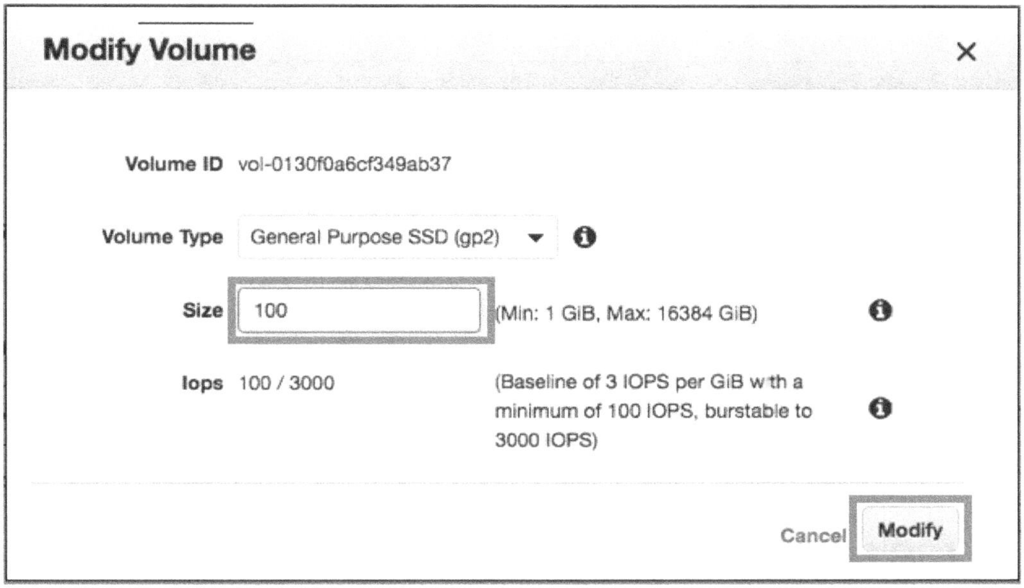

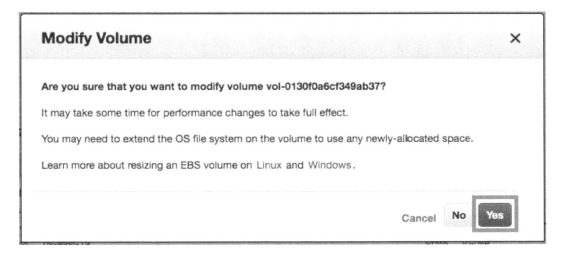

Figure 2.17 – Modifying the volume

As you can see, we specified a new volume size of 100 GiB. This should be more than enough to help us get through this chapter and build our custom algorithm container image.

16. Click **Yes** to confirm the volume modification action:

Figure 2.18 – Modify Volume confirmation dialog

We should see a confirmation screen here after clicking **Modify** in the previous step.

17. Click **Close** upon seeing the confirmation dialog:

Figure 2.19 – Modify Volume Request Succeeded message

Here, we can see a message stating **Modify Volume Request Succeeded**. At this point, the volume modification is still pending and we need to wait about 10-15 minutes for this to complete. Feel free to check out the *How it works...* section for this recipe while waiting.

18. Click the refresh button (the two rotating arrows) so that the volume state will change to the correct state accordingly:

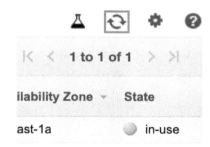

Figure 2.20 – Refresh button

Clicking the refresh button will update **State** from **in-use** (green) to **in-use – optimizing** (yellow):

Figure 2.21 – In-use state – optimizing (yellow)

Here, we can see that the volume modification step has not been completed yet.

19. After a few minutes, **State** of the volume will go back to **in-use** (green):

Figure 2.22 – In-use state (green)

When we see what is shown in the preceding screenshot, we should celebrate as this means that the volume modification step has been completed!

Now that the volume modification step has been completed, our next goal is to make sure that this change is reflected in our environment.

20. Navigate back to the browser tab of the AWS Cloud9 IDE. In the Terminal, run lsblk:

```
lsblk
```

Running `lsblk` should yield the following output:

```
ubuntu:~/environment $ lsblk
NAME           MAJ:MIN RM  SIZE RO TYPE MOUNTPOINT
loop0            7:0     0 12.7M  1 loop /snap/amazon-ssm-agent/495
loop1            7:1     0 87.9M  1 loop /snap/core/5328
loop2            7:2     0 99.2M  1 loop /snap/core/10859
loop3            7:3     0 55.5M  1 loop /snap/core18/1988
loop4            7:4     0 33.3M  1 loop /snap/amazon-ssm-agent/3552
nvme0n1        259:0     0  100G  0 disk
└─nvme0n1p1    259:1     0   10G  0 part /
ubuntu:~/environment $ █
```

Figure 2.23 – Partition not yet reflecting the size of the root volume

As you can see, while the size of the root volume, `/dev/nvme0n1`, reflects the new size, `100G`, the size of the `/dev/nvme0n1p1` partition reflects the original size, `10G`.

There are multiple ways to grow the partition, but we will proceed by simply rebooting the EC2 instance so that the size of the `/dev/nvme0n1p1` partition will reflect the size of the root volume, which is `100G`.

21. Navigate back to the **EC2 Volumes** page and select the EC2 volume attached to the Cloud9 instance. At the bottom portion of the screen showing the volume's details, locate the **Attachment information** value under the **Description** tab. Click the **Attachment information** link:

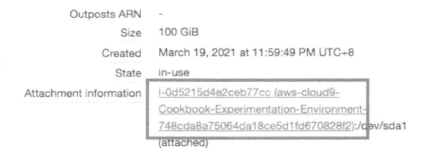

Figure 2.24 – Attachment information

Clicking this link will redirect us to the **EC2 Instances** page. It will automatically select the EC2 instance of our Cloud9 environment:

Figure 2.25 – EC2 instance of the Cloud9 environment

The preceding screenshot shows the EC2 instance linked to our Cloud9 environment.

22. Click **Instance state** at the top right of the screen and click **Reboot instance**:

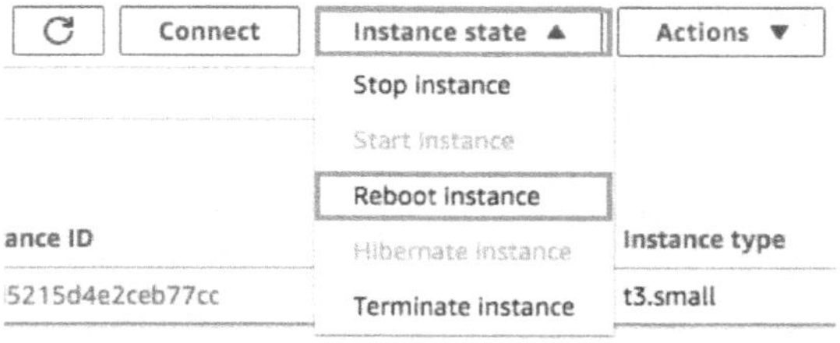

Figure 2.26 – Reboot instance

This is where we can find the **Reboot instance** option.

23. Navigate back to the browser tab showing the AWS Cloud9 environment IDE. It should take a minute or two to complete the reboot step:

Figure 2.27 – Instance is still rebooting

We should see a screen similar to the preceding one.

24. Once connected, run `lsblk` in the Terminal:

```
lsblk
```

We should get a set of results similar to what is shown in the following screenshot:

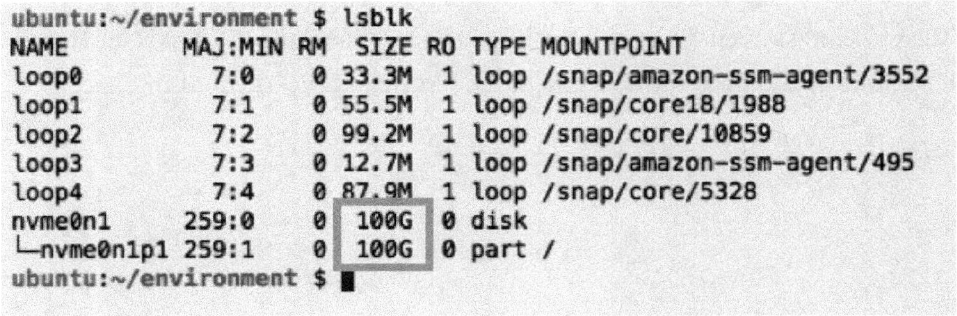

```
ubuntu:~/environment $ lsblk
NAME           MAJ:MIN RM   SIZE RO TYPE MOUNTPOINT
loop0            7:0     0  33.3M  1 loop /snap/amazon-ssm-agent/3552
loop1            7:1     0  55.5M  1 loop /snap/core18/1988
loop2            7:2     0  99.2M  1 loop /snap/core/10859
loop3            7:3     0  12.7M  1 loop /snap/amazon-ssm-agent/495
loop4            7:4     0  87.9M  1 loop /snap/core/5328
nvme0n1        259:0     0   100G  0 disk
└─nvme0n1p1    259:1     0   100G  0 part /
ubuntu:~/environment $ █
```

Figure 2.28 – Partition now reflecting the size of the root instance

As we can see, the `/dev/nvme0n1p1` partition now reflects the size of the root volume, which is `100G`.

That was a lot of setup work, but this will be definitely worth it, as you will see in the next few recipes in this chapter. Now, let's see how this works!

How it works...

In this recipe, we launched a **Cloud9** environment where we will prepare the custom container image. When building Docker container images, it is important to note that each container image consumes a bit of disk space. This is why we had to go through a couple of steps to increase the volume attached to the EC2 instance of our Cloud9 environment. This recipe was composed of three parts: launching a new Cloud9 environment, modifying the mounted volume, and rebooting the instance.

Launching a new Cloud9 environment involves using a **CloudFormation** template behind the scenes. This **CloudFormation** template is used as the blueprint when creating the EC2 instance:

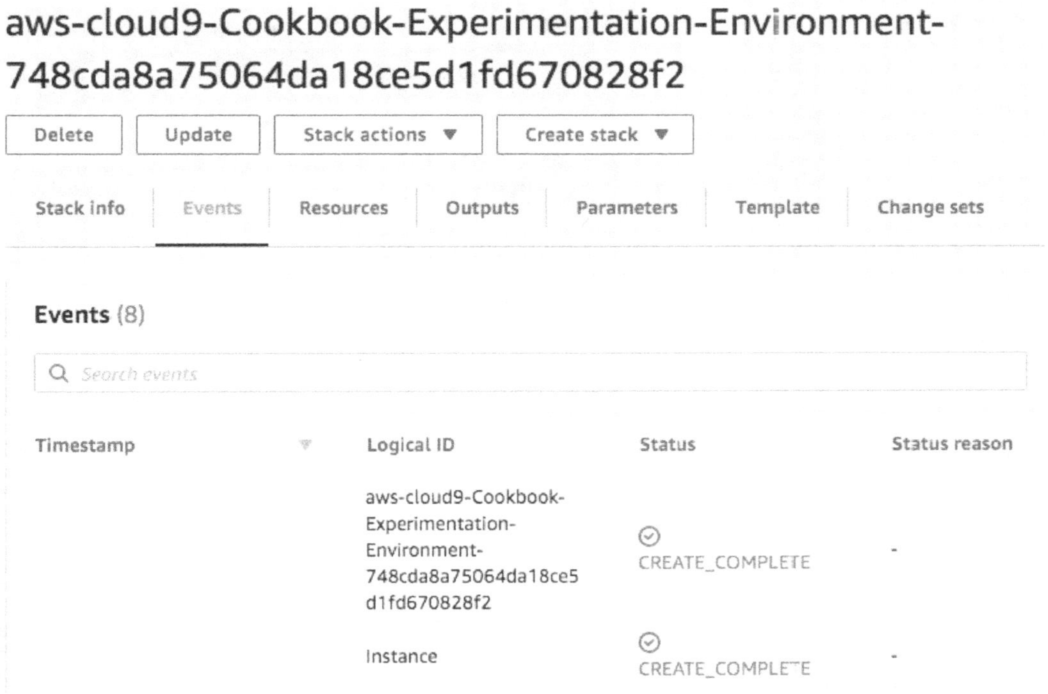

aws-cloud9-Cookbook-Experimentation-Environment-748cda8a75064da18ce5d1fd670828f2

| Delete | Update | Stack actions ▼ | Create stack ▼ |

| Stack info | Events | Resources | Outputs | Parameters | Template | Change sets |

Events (8)

🔍 *Search events*

Timestamp ▽	Logical ID	Status	Status reason
	aws-cloud9-Cookbook-Experimentation-Environment-748cda8a75064da18ce5d1fd670828f2	⊘ CREATE_COMPLETE	-
	Instance	⊘ CREATE_COMPLETE	-

Figure 2.29 – CloudFormation stack

Here, we have a **CloudFormation** stack that was successfully created. What's CloudFormation? **AWS CloudFormation** is a service that helps developers and DevOps professionals manage resources using templates written in JSON or YAML. These templates get converted into AWS resources using the **CloudFormation** service.

At this point, the EC2 instance should be running already and we can use the Cloud9 environment as well:

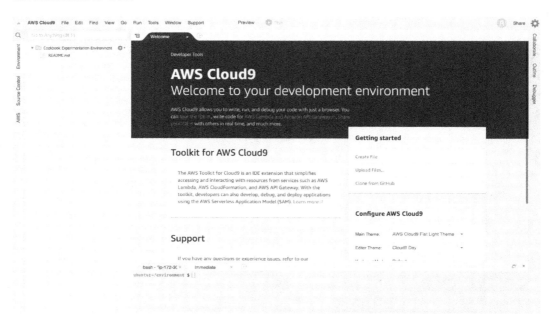

Figure 2.30 – AWS Cloud9 environment

We should be able to see the preceding output once the Cloud9 environment is ready. If we were to use the environment right away, we would run into disk space issues as we will be working with Docker images, which take up a bit of space. To prevent these issues from happening later on, we modified the volume in this recipe and restarted the EC2 instance so that this volume modification gets reflected right away.

> **Important note**
>
> In this recipe, we took a shortcut and simply restarted the EC2 instance. If we were running a production environment, we should avoid having to reboot and follow this guide instead: `https://docs.aws.amazon.com/ AWSEC2/latest/UserGuide/recognize-expanded-volume- linux.html`.

Note that we can also use a SageMaker Notebook instance that's been configured with root access enabled as a potential experimentation environment for our custom scripts and container images, before using them in SageMaker. The issue here is that when using a SageMaker Notebook instance, it reverts to how it was originally configured every time we turn off and reboot the instance. This makes us lose certain directories and installed packages, which is not ideal.

Setting up the Python and R experimentation environments

In the previous recipe, we launched a **Cloud9** environment. In this recipe, we will be preparing the expected file and directory structure inside this environment. This will help us prepare and test our `train` and `serve` scripts before running them inside containers and before using these with the **SageMaker Python SDK**:

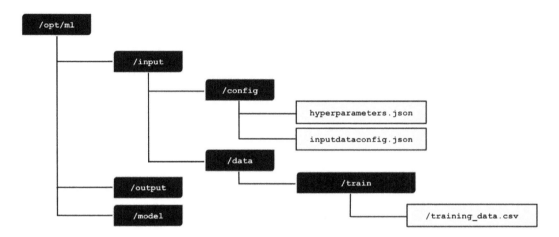

Figure 2.31 – Expected file and directory structure inside /opt/ml

We can see the expected directory structure in the preceding diagram. We will prepare the expected directory structure inside `/opt/ml`. After that, we will prepare the `hyperparameters.json`, `inputdataconfig.json`, and `training_data.csv` files. In the succeeding recipes, we will use these files when preparing and testing the `train` and `serve` scripts.

Getting ready

Here are the prerequisites for this recipe:

- This recipe continues from *Launching and preparing the Cloud9 environment*.

- We will need the S3 bucket from the *Preparing the Amazon S3 bucket and the training dataset for the linear regression experiment* recipe of *Chapter 1*. We will also need the `training_data.csv` file inside this S3 bucket. After performing the train-test split, we uploaded the CSV file to the S3 bucket in the *Training your first model in Python* recipe of *Chapter 1*. If you skipped this recipe, you can upload the `training_data.csv` file from this book's GitHub repository (`https://github.com/PacktPublishing/Machine-Learning-with-Amazon-SageMaker-Cookbook`) to the S3 bucket instead.

How to do it...

In the first set of steps in this recipe, we will use the Terminal to run the commands. We will continue where we left off in the previous *Launching and preparing the Cloud9 environment* recipe:

1. Use the `pwd` command to see the current working directory:

    ```
    pwd
    ```

2. Navigate to the `/opt` directory:

    ```
    cd /opt
    ```

3. Create the `/opt/ml` directory using the `mkdir` command. Make sure that you are inside the `/opt` directory before running the `sudo mkdir ml` command. Modify the ownership configuration of the `/opt/ml` directory using the `chown` command. This will allow us to manage the contents of this directory without using `sudo` over and over again in the succeeding steps:

    ```
    sudo mkdir -p ml
    sudo chown ubuntu:ubuntu ml
    ```

4. Navigate to the `ml` directory using the `cd` Bash command. Run the following commands to prepare the expected directory structure inside the `/opt/ml` directory. Make sure that you are inside the `ml` directory before running these commands. The `-p` flag will automatically create the required parent directories first, especially if some of the directories in the specified path do not exist yet. In this case, if the `input` directory does not exist, the `mkdir -p input/config` command will create it first before creating the `config` directory inside it:

    ```
    cd ml
    mkdir -p input/config
    mkdir -p input/data/train
    ```

```
mkdir -p output/failure
mkdir -p model
```

As we will see later, these directories will contain the files and configuration data that we'll pass as parameter values when we initialize the Estimator.

> **Important note**
>
> Again, if you are wondering why we are creating these directories, the answer is that we are preparing an environment where we can test and iteratively build our custom scripts first, before using the **SageMaker Python SDK** and **API**. It is hard to know if a script is working unless we run it inside an environment that has a similar set of directories and files. If we skip this step and use the custom training script directly with the **SageMaker Python SDK**, we will spend a lot of time debugging potential issues as we have to wait for the entire training process to complete (at least 5-10 minutes), before being able to fix a scripting bug and try again to see if the fix worked. With this simulation environment in place, we will be able to test our custom script and get results within a few seconds instead. As you can see, we can iterate rapidly if we have a simulation environment in place.

The following is the expected directory structure:

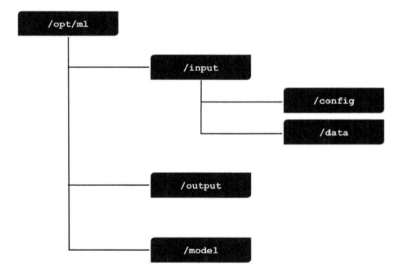

Figure 2.32 – Expected file and folder structure after running the mkdir commands

Here, we can see that there are /config and /data directories inside the /input directory. The /config directory will contain the hyperparameters.json file and the inputdataconfig.json file, as we will see later. We will not be using the /output directory in the recipes in this chapter, but this is where we can create a file called failure in case the training job fails. The failure file should describe why the training job failed to help us debug and adjust it in case the failure scenario happens.

5.   Install and use the tree command:

```
sudo apt install tree
tree
```

We should get a tree structure similar to the following:

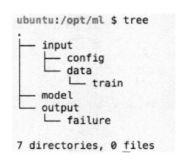

Figure 2.33 – Result of the tree command

Here, we can see the expected directory structure.

6.   Create the /home/ubuntu/environment/opt directory using mkdir and create two directories inside it called ml-python and ml-r:

```
mkdir -p /home/ubuntu/environment/opt
cd /home/ubuntu/environment/opt
mkdir -p ml-python ml-r
```

7. Create a soft symbolic link to make it easier to manage the files and directories using the **AWS Cloud9** interface:

```
sudo ln -s /opt/ml  /home/ubuntu/environment/opt/ml
```

Given that we are performing this step inside a **Cloud9** environment, we will be able to easily create and modify the files using the visual editor, instead of using `vim` or `nano` in the command line. What this means is that changes that are made inside the `/home/ubuntu/environment/opt/ml` directory will also be reflected inside the `/opt/ml` directory. This will allow us to use a visual editor to easily create and modify files:

Figure 2.34 – File tree showing the symlinked /opt/ml directory

We should see the directories inside the `/opt/ml` directory in the file tree, as shown in the preceding screenshot.

The next set of steps focus on adding the dummy files to the experimentation environment.

8. Using the file tree, navigate to the `/opt/ml/input/config` directory. Right-click on the **config** directory and select **New File**:

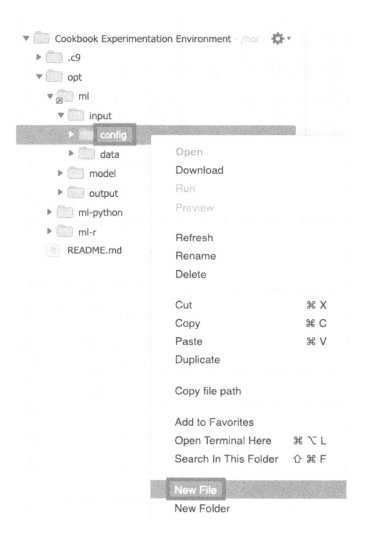

Figure 2.35 – Creating a new file inside the config directory

9. Name the new file `hyperparameters.json`. Double-click the new file to open it in the **Editor** pane:

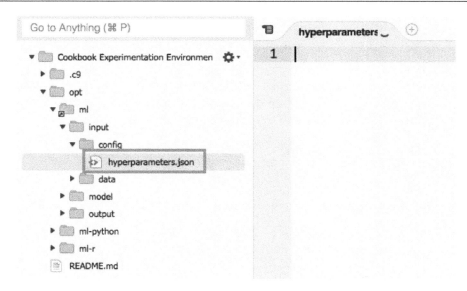

Figure 2.36 – Empty hyperparameters.json file

Here, we have an empty hyperparameters.json file inside the /opt/ml/input/config directory.

10. Set the content of the hyperparameters.json file to the following line of code:

```
{"a": 1, "b": 2}
```

Your Cloud9 environment IDE's file tree and **Editor** pane should look as follows:

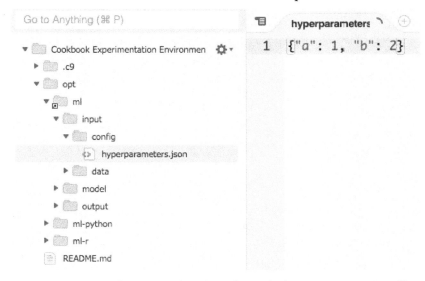

Figure 2.37 – Specifying a sample JSON value to the hyperparameters.json file

Make sure to save it by clicking the **File** menu and then clicking **Save**. You can also use *Cmd + S* or *Ctrl + S* to save the file, depending on the operating system you are using.

11. In a similar fashion, create a new file called `inputdataconfig.json` inside `/opt/ml/input/config`. Open the `inputdataconfig.json` file in the **Editor** pane and set its content to the following line of code:

```
{"train": {"ContentType": "text/csv",
"RecordWrapperType": "None", "S3DistributionType":
"FullyReplicated", "TrainingInputMode": "File"}}
```

Your Cloud9 environment IDE's file tree and **Editor** pane should look as follows:

Figure 2.38 – The inputdataconfig.json file

In the next set of steps, we will download the `training_data.csv` file from *Chapter 1, Getting Started with Machine Learning Using Amazon SageMaker*, to the experimentation environment. In the *Training your first model in Python* recipe from *Chapter 1, Getting Started with Machine Learning Using Amazon SageMaker*, we uploaded a `training_data.csv` file to an Amazon S3 bucket:

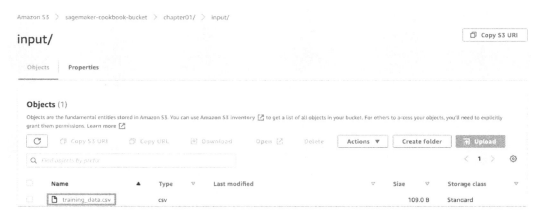

Figure 2.39 – The training_data.csv file inside the S3 bucket

In case you skipped these recipes in *Chapter 1*, make sure that you check out this book's GitHub repository (https://github.com/PacktPublishing/Machine-Learning-with-Amazon-SageMaker-Cookbook) and upload the training_data.csv file to the S3 bucket. Note that the recipes in this chapter assume that the training_data.csv file is inside s3://S3_BUCKET/PREFIX/input, where S3_BUCKET is the name of the S3 bucket and PREFIX is the folder's name. If you have not created an S3 bucket yet, follow the steps in the *Preparing the Amazon S3 bucket and the training dataset for the linear regression experiment* recipe of *Chapter 1* as we will need this S3 bucket for all the chapters in this book.

12. In the Terminal of the Cloud9 IDE, run the following commands to download the training_data.csv file from S3 to the /opt/ml/input/data/train directory:

```
cd /opt/ml/input/data/train
S3_BUCKET="<insert bucket name here>"
PREFIX="chapter01"
aws s3 cp s3://$S3_BUCKET/$PREFIX/input/training_data.csv training_data.csv
```

Make sure that you set the S3_BUCKET value to the name of the S3 bucket you created in the *Preparing the Amazon S3 bucket and the training dataset for the linear regression experiment* recipe of *Chapter 1*.

13. In the file tree, double-click the `training_data.csv` file inside the `/opt/ml/input/data/train` directory to open it in the **Editor** pane:

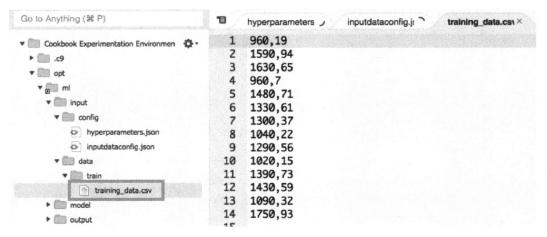

Figure 2.40 – The training_data.csv file inside the experimentation environment

As shown in the preceding screenshot, the `training_data.csv` file contains the y values in the first column and the x values in the second column.

In the next set couple of steps, we will install a few prerequisites in the Terminal.

14. In the Terminal, run the following scripts to make the R recipes work in the second half of this chapter:

```
sudo apt-get -y update
sudo apt-get install -y --no-install-recommends wget
sudo apt-get install -y --no-install-recommends r-base
sudo apt-get install -y --no-install-recommends r-base-dev
sudo apt-get install -y --no-install-recommends ca-certificates
```

15. Install the command-line JSON processor; that is, `jq`:

```
sudo apt install -y jq
```

In the last set of steps in this recipe, we will create the files inside the `ml-python` and `ml-r` directories. In the *Building and testing the custom Python algorithm container image* and *Building and testing the custom R algorithm container image* recipes, we will copy these files inside the container while building the container image with the `docker build` command.

16. Right-click on the `ml-python` directory and then click **New File** from the menu to create a new file, as shown here. Name the new file `train`:

Figure 2.41 – Creating a new file inside the ml-python directory

Perform this step two more times so that there are three files inside the `ml-python` directory called `train`, `serve`, and `Dockerfile`. Take note that these files are empty for now:

Figure 2.42 – Files inside the ml-python directory

The preceding screenshot shows these three empty files. We will work with these later in the Python recipes in this chapter.

17. Similarly, create four new files inside the `ml-r` directory called `train`, `serve`, `api.r`, and `Dockerfile`:

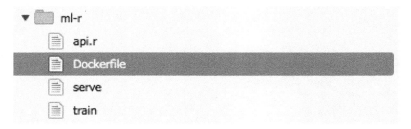

Figure 2.43 – Files inside the ml-r directory

The preceding screenshot shows these four empty files. We will be working with these later in the R recipes in this chapter.

Let's see how this works!

How it works...

In this recipe, we prepared the experimentation environment where we will iteratively build the `train` and `serve` scripts. Preparing the `train` and `serve` scripts is an iterative process. We will need an experimentation environment to ensure that the scripts work before using them inside a running container. Without the expected directory structure and the dummy files, it would be hard to test and develop the `train` and `serve` scripts in a way that seamlessly translates to using these with SageMaker.

Let's discuss and quickly describe how the `train` script should work. The `train` script may load one or more of the following:

- `hyperparameters.json`: Contains the hyperparameter configuration data set in `Estimator`
- `inputdataconfig.json`: Contains the information where the training dataset is stored
- `<directory>/<data file>`: Contains the training dataset's input (for example, `train/training.csv`)

We will have a closer look at preparing and testing train scripts in the *Preparing and testing the train script in Python* and *Preparing and testing the train script in R* recipes in this chapter.

Now, let's talk about the serve script. The serve script expects the model file(s) inside the /opt/ml/model directory. Take note that one or more of these files may not exist, and this depends on the configuration parameters and arguments we have set using the **SageMaker Python SDK**. This also depends on what we write our script to need. We will have a closer look at preparing and testing serve scripts in the *Preparing and testing the serve script in Python* and *Preparing and testing the serve script in R* recipes later in this chapter.

There's more...

As we are about to work on the recipes specific to Python and R, we need to have a high-level idea of how these all fit together. In the succeeding recipes, we will build a custom container image containing the train and serve scripts. This container image will be used during training and deployment using the **SageMaker Python SDK**. In this section, we will briefly discuss what happens under the hood when we run the fit() function while using a custom container image. I believe it would be instructive to reiterate here that we built those directories and dummy files to create the train and serve scripts that the fit() and deploy() commands will run.

If you are wondering what the train and serve script files are for, these script files are executed inside a container behind the scenes by **SageMaker** when the fit() and deploy() functions from the **SageMaker Python SDK** are used. We will write and test these scripts later in this chapter. When we use the fit() function, **SageMaker** starts the training job. Behind the scenes, **SageMaker** performs the following set of steps:

Preparation and configuration

1. One or more ML instances are launched. The number and types of ML instances for the training job depend on the instance_count and instance_type arguments specified when initializing the Estimator class:

```
container="<insert image uri of the custom container
image>"
estimator = sagemaker.estimator.Estimator(
    container,
    instance_count=1,
    instance_type='local',
    ...
)
estimator.fit({'train': train})
```

2. The hyperparameters specified using the `set_hyperparameters()` function are copied and stored as a JSON file called `hyperparameters.json` inside the `/opt/ml/input/config` directory. Take note that our custom container will not have this file at the start, and that **SageMaker** will create this file for us automatically when the training job starts.

Training

1. The input data we have specified in the `fit()` function will be loaded by **SageMaker** (for example, from the specified S3 bucket) and copied into `/opt/ml/input/data/`. For each of the input data channels, a directory containing the relevant files will be created inside the `/opt/ml/input/data` directory. For example, if we used the following line of code using the **SageMaker Python SDK**, then we would expect the `/opt/ml/input/data/apple` and `/opt/ml/data/banana` directories when the train script starts to run:

```
estimator.fit({'apple': TrainingInput(...), 'banana':
TrainingInput(...)})
```

2. Next, your custom `train` script runs. It loads the configuration files, `hyperparameters`, and the data files from the directories inside `/opt/ml`. It then trains a model using the training dataset and, optionally, a validation dataset. The model is then serialized and stored inside the `/opt/ml/model` directory.

> **Note**
> Do not worry if you have no idea how the train script looks like as we will discuss the train script in detail later, in the succeeding recipes.

3. **SageMaker** expects the model output file(s) inside the `/opt/ml/model` directory. After the training script has finished executing, **SageMaker** automatically copies the contents of the `/opt/ml/model` directory and stores it inside the target S3 bucket and path (inside `model.tar.gz`). Take note that we can specify the target S3 bucket and path by setting the `output_path` argument when initializing `Estimator` with the **SageMaker Python SDK**.

4. If there is an error running the script, **SageMaker** will look for a failure file inside the /opt/ml/output directory. If it exists, the text output stored in this file will be loaded when the DescribeTrainingJob API is used.

5. The created ML instances are deleted. The billable time is returned to the user.

Deployment

When we use the deploy() function, **SageMaker** starts the model deployment step. The assumption when running the deploy() function is that the model.tar.gz file is stored inside the target S3 bucket path.

1. One or more ML instances are launched. The number and types of ML instances for the deployment step depend on the instance_count and instance_type arguments specified when using the deploy() function:

```
predictor = estimator.deploy(
    initial_instance_count=1,
    instance_type='local',
    endpoint_name="custom-local-py-endpoint")
```

2. The model.tar.gz file is copied from the S3 bucket and the files are extracted inside the /opt/ml/model directory.

3. Next, your custom serve script runs. It uses the model files inside the /opt/ml/model directory to deserialize and load the model. The serve script then runs an API web server with the required /ping and /invocations endpoints.

Inference

1. After deployment, the predict() function calls the /invocations endpoint to use the loaded model for inference.

This should give us a better idea and understanding of the purpose of the files and directories we have prepared in this recipe. If you are a bit overwhelmed by the level of detail in this section, do not worry as things will become clearer as we work on the next few recipes in this chapter!

Preparing and testing the train script in Python

In this recipe, we will write a train script in Python that allows us to train a linear model with `scikit-learn`. Here, we can see that the train script inside a running custom container makes use of the hyperparameters, input data, and the configuration specified in the `Estimator` instance using the **SageMaker Python SDK**:

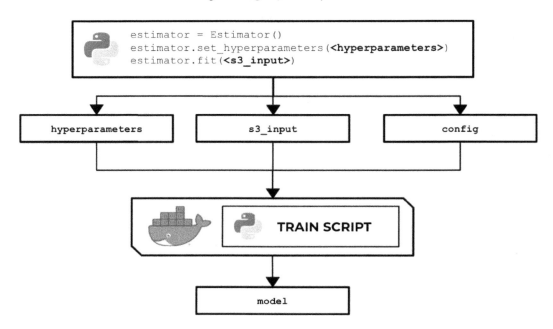

Figure 2.44 – How the train script is used to produce a model

There are several options when running a training job – use a built-in algorithm, use a custom train script and custom Docker container images, or use a custom train script and prebuilt Docker images. In this recipe, we will focus on the second option, where we will prepare and test a bare minimum training script in Python that builds a linear model for a specific regression problem.

Once we have finished working on this recipe, we will have a better understanding of how **SageMaker** works behind the scenes. We will see where and how to load and use the configuration and arguments we have specified in the **SageMaker Python SDK** `Estimator`.

Getting ready

Make sure you have completed the *Setting up the Python and R experimentation environments* recipe.

How to do it...

The first set of steps in this recipe focus on preparing the train script. Let's get started:

1. Inside the ml-python directory, double-click the train file to open the file inside the **Editor** pane:

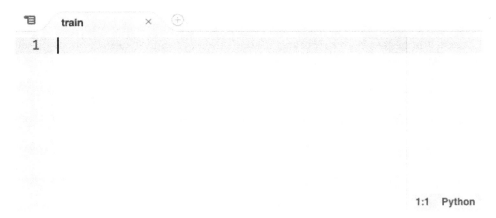

Figure 2.45 – Empty ml-python/train file

Here, we have an empty train file. In the lower right-hand corner of the **Editor** pane, you can change the syntax highlight settings to Python.

2. Add the following lines of code to start the train script to import the required packages and libraries:

```
#!/usr/bin/env python3
import json
import pprint
import pandas as pd
from sklearn.linear_model import LinearRegression
from joblib import dump, load
from os import listdir
```

In the preceding block of code, we imported the following:

- `json` for utility functions when working with JSON data

- `pprint` to help us "pretty-print" nested structures such as dictionaries

- `pandas` to help us read CSV files and work with DataFrames

- `LinearRegression` from the `sklearn` library for training a linear model when we run the train script

- `joblib` for saving and loading a model

- `listdir` from the `os` module to help us list the files inside a directory

3. Define the `PATHS` constant and the `get_path()` function. The `get_path()` function will be handy in helping us manage the paths and locations of the primary files and directories used in the script:

```
PATHS = {
    'hyperparameters': 'input/config/hyperparameters.
json',
    'input': 'input/config/inputdataconfig.json',
    'data': 'input/data/',
    'model': 'model/'
}


def get_path(key):
    return '/opt/ml/' + PATHS[key]
```

If we want to get the path of the `hyperparameters.json` file, we can use `get_path("hyperparameters")` instead of using the absolute path in our code.

> **Important note**
>
> In this chapter, we will intentionally use `get_path` for the function name. If you have been using Python for a while, you will probably notice that this is definitely not Pythonic code! Our goal is for us to easily find the similarities and differences between the Python and R scripts, so we made the function names the same for the most part.

4. Next, add the following lines just after the `get_path()` function definition from the previous step. These additional functions will help us later once we need to load and print the contents of the JSON files we'll be working with (for example, `hyperparameters.json`):

```python
def load_json(target_file):
    output = None

    with open(target_file) as json_data:
        output = json.load(json_data)

    return output

def print_json(target_json):
    pprint.pprint(target_json, indent=4)
```

5. Include the following functions as well in the train script (after the `print_json()` function definition):

```python
def inspect_hyperparameters():
    print('[inspect_hyperparameters]')
    hyperparameters_json_path = get_path(
        'hyperparameters'
    )
    print(hyperparameters_json_path)

    hyperparameters = load_json(
        hyperparameters_json_path
    )
    print_json(hyperparameters)

def list_dir_contents(target_path):
    print('[list_dir_contents]')
    output = listdir(target_path)
    print(output)

    return output
```

The `inspect_hyperparameters()` function allows us to inspect the contents of the `hyperparameters.json` file inside the `/opt/ml/input/config` directory. The `list_dir_contents()` function, on the other hand, allows us to display the contents of a target directory. We will use this later to check the contents of the training input directory.

6. After that, define the `inspect_input()` function. This allows us to inspect the contents of `inputdataconfig.json` inside the `/opt/ml/input/config` directory:

```
def inspect_input():
    print('[inspect_input]')
    input_config_json_path = get_path('input')
    print(input_config_json_path)
    input_config = load_json(input_config_json_path)
    print_json(input_config)
```

7. Define the `load_training_data()` function. This function accepts a string value pointing to the input data directory and returns the contents of a CSV file inside that directory:

```
def load_training_data(input_data_dir):
    print('[load_training_data]')
    files = list_dir_contents(input_data_dir)
    training_data_path = input_data_dir + files[0]
    print(training_data_path)

    df = pd.read_csv(
        training_data_path, header=None
    )
    print(df)

    y_train = df[0].values
    X_train = df[1].values
    return (X_train, y_train)
```

The flow inside the `load_training_data()` function can be divided into two parts – getting the specific path of the CSV file containing the training data, and then reading the contents of the CSV file using the `pd.read_csv()` function and returning the results inside a tuple of lists.

> **Note**
> Of course, the `load_training_data()` function we've implemented here assumes that there is only one CSV file inside that directory, so feel free to modify the following implementation when you are working with more than one CSV file inside the provided directory. At the same time, this function implementation only supports CSV files, so make sure to adjust the code block if you need to support multiple input file types.

8. Define the `get_input_data_dir()` function:

```python
def get_input_data_dir():
    print('[get_input_data_dir]')
    key = 'train'
    input_data_dir = get_path('data') + key + '/'
    return input_data_dir
```

9. Define the `train_model()` function:

```python
def train_model(X_train, y_train):
    print('[train_model]')
    model = LinearRegression()
    model.fit(X_train.reshape(-1, 1), y_train)
    return model
```

10. Define the `save_model()` function:

```python
def save_model(model):
    print('[save_model]')
    filename = get_path('model') + 'model'
    print(filename)
    dump(model, filename)
    print('Model Saved!')
```

11. Create the `main()` function, which executes the functions we created in the previous steps:

```python
def main():
    inspect_hyperparameters()
    inspect_input()
    input_data_dir = get_input_data_dir()
    X_train, y_train = load_training_data(
```

```
            input_data_dir
    )
    model = train_model(X_train, y_train)
    save_model(model)
```

This function simply inspects the hyperparameters and input configuration, trains a linear model using the data loaded from the input data directory, and saves the model using the save_model() function.

12. Finally, run the main() function:

```
if __name__ == "__main__":
    main()
```

The __name__ variable is set to "__main__" when the script is executed as the main program. This if condition simply tells the script to run if we're using it as the main program. If this script is being imported by another script, then the main() function will not run.

> **Tip**
>
> You can access a working copy of the train script file in the *Machine Learning with Amazon SageMaker Cookbook* GitHub repository: https://github.com/PacktPublishing/Machine-Learning-with-Amazon-SageMaker-Cookbook/blob/master/Chapter02/ml-python/train.

Now that we are done with the train script, we will use the Terminal to perform the last set of steps in this recipe.

The last set of steps focus on installing a few script prerequisites:

13. Open a new Terminal:

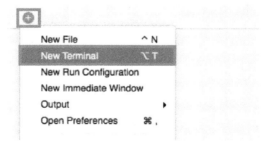

Figure 2.46 – New Terminal

Here, we can see how to create a new Terminal tab. We simply click the plus (+) button and then choose **New Terminal**.

14. In the Terminal at the bottom pane, run `python3 --version`:

```
python3 --version
```

Running this line of code should return a similar set of results to what is shown in the following screenshot:

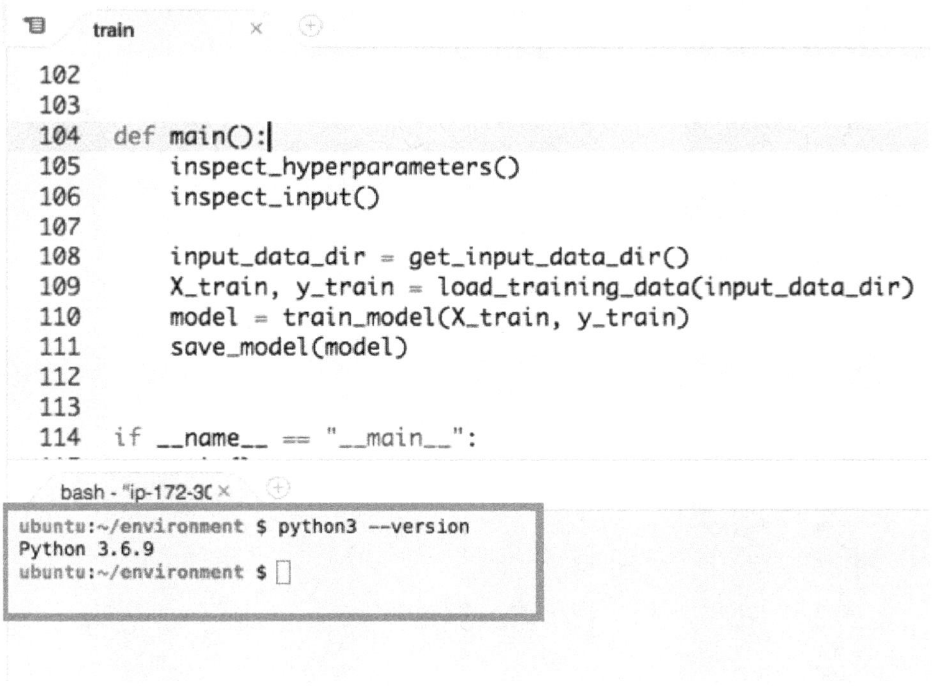

```
train                    ×    ⊕
102
103
104   def main():|
105         inspect_hyperparameters()
106         inspect_input()
107
108         input_data_dir = get_input_data_dir()
109         X_train, y_train = load_training_data(input_data_dir)
110         model = train_model(X_train, y_train)
111         save_model(model)
112
113
114   if __name__ == "__main__":
```

```
bash - "ip-172-3C ×    ⊕
ubuntu:~/environment $ python3 --version
Python 3.6.9
ubuntu:~/environment $ ▯
```

Figure 2.47 – Result of running python3 --version in the Terminal

Here, we can see that our environment is using Python version 3.6.9.

15. Install `pandas` using `pip`. The **pandas** library is used when working with DataFrames (tables):

```
pip3 install pandas
```

16. Install `sklearn` using `pip`. The **scikit-learn** library is a machine learning library that features several algorithms for classification, regression, and clustering problems:

```
pip3 install sklearn
```

17. Navigate to the `ml-python` directory:

```
cd /home/ubuntu/environment/opt/ml-python
```

18. To make the `train` script executable, run the following command in the Terminal:

```
chmod +x train
```

19. Test the `train` script in your **AWS Cloud9** environment by running the following command in the Terminal:

```
./train
```

Running the previous lines of code will yield results similar to the following:

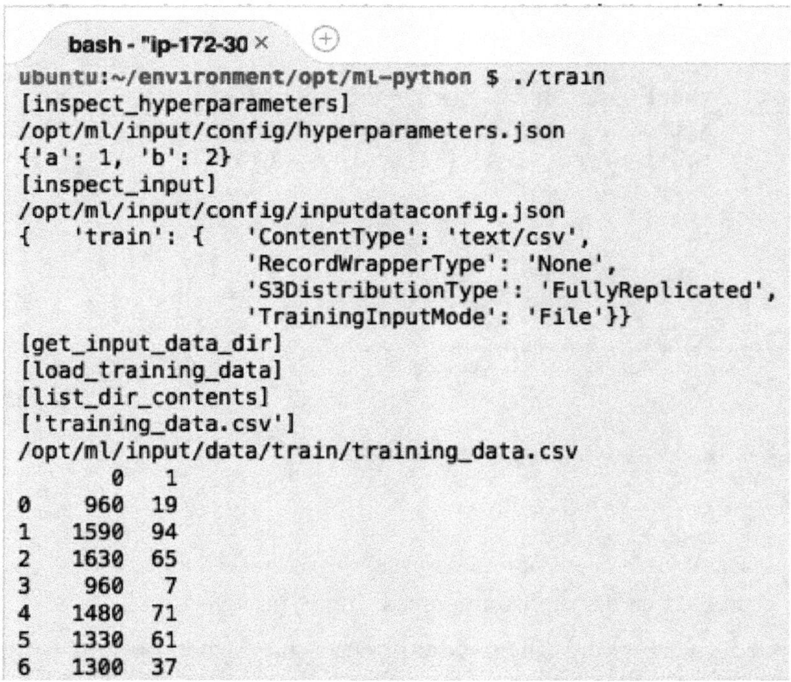

```
bash - "ip-172-30 ×    ⊕
ubuntu:~/environment/opt/ml-python $ ./train
[inspect_hyperparameters]
/opt/ml/input/config/hyperparameters.json
{'a': 1, 'b': 2}
[inspect_input]
/opt/ml/input/config/inputdataconfig.json
{    'train': {    'ContentType': 'text/csv',
                   'RecordWrapperType': 'None',
                   'S3DistributionType': 'FullyReplicated',
                   'TrainingInputMode': 'File'}}
[get_input_data_dir]
[load_training_data]
[list_dir_contents]
['training_data.csv']
/opt/ml/input/data/train/training_data.csv
        0    1
0    960   19
1   1590   94
2   1630   65
3    960    7
4   1480   71
5   1330   61
6   1300   37
```

Figure 2.48 – Result of running the train script

Here, we can see the logs that were produced by the `train` script. After the `train` script has been successfully executed, we expect the model files to be stored inside the `/opt/ml/model` directory.

Now, let's see how this works!

How it works...

In this recipe, we prepared a custom `train` script using Python. The script starts by identifying the input paths and loading the important files to help set the context of the execution. This `train` script demonstrates how the input and output values are passed around between the **SageMaker Python SDK** (or API) and the custom container. It also shows how to load the training data, train a model, and save a model.

When the `Estimator` object is initialized and configured, some of the specified values, including the hyperparameters, are converted from a Python dictionary into JSON format in an API call when invoking the `fit()` function. The API call on the SageMaker platform then proceeds to create and mount the JSON file inside the environment where the `train` script is running. It works the same way as it does with the other files loaded by the `train` script file, such as the `inputdataconfig.json` file.

If you are wondering what is inside the `inputdataconfig.json` file, refer to the following code block for an example of what it looks like:

```
{"<channel name>": {"ContentType": "text/csv",
       "RecordWrapperType": "None",
       "S3DistributionType": "FullyReplicated",
       "TrainingInputMode": "File"}}
```

For each of the input channels, a corresponding set of properties is specified in this file. The following are some of the common properties and values that are used in this file. Of course, the values here depend on the type of data and the algorithm being used in the experiment:

- `ContentType` – **Valid Values**: `text/csv`, `image/jpeg`, `application/x-recordio-protobuf`, and more.

- `RecordWrapperType` – **Valid Values**: `None` or `RecordIO`. The `RecordIO` value is set only when the `TrainingInputMode` value is set to `Pipe`. The training algorithm requires the `RecordIO` format for the input data, and the input data is not in `RecordIO` format yet.

- `S3DistributionType` – **Valid Values**: `FullyReplicated` or `ShardedByS3Key`. If the value is set to `FullyReplicated`, the entire dataset is copied on each ML instance that's launched during model training. On the other hand, when the value is set to `ShardedByS3Key`, each machine that's launched and used during model training makes use of a subset of the training data provided.

- `TrainingInputMode` – **Valid Values**: `File` or `Pipe`. When the `File` input mode is used, the entire dataset is downloaded first before the training job starts. On the other hand, the `Pipe` input mode is used to speed up training jobs, start faster, and requires less disk space. This is very useful when dealing with large datasets. If you are planning to support the `Pipe` input mode in your custom container, the directories inside the `/opt/ml/input/data` directory are a bit different and will be in the format of `<channel name>_<epoch number>`. If we used this example in our experimentation environment, we would have directories named `d_1`, `d_2`, ... instead inside the `/opt/ml/input/data` directory. Make sure that you handle scenarios dealing with data files that don't exist yet as you need to add some retry logic inside the `train` script.

In addition to files stored inside a few specific directories, take note that there are a couple of environment variables that can be loaded and used by the `train` script as well. These include `TRAINING_JOB_NAME` and `TRAINING_JOB_ARN`.

The values for these environment variables can be loaded by using the following lines of Python code:

```
import os
training_job_name = os.environ['TRAINING_JOB_NAME']
```

We can test our script by running the following code in the Terminal:

```
TRAINING_JOB_NAME=abcdef ./train
```

Feel free to check out the following reference on how SageMaker provides training information: `https://docs.aws.amazon.com/sagemaker/latest/dg/your-algorithms-training-algo-running-container.html`.

There's more...

If you are dealing with distributed training where datasets are automatically split across different instances to achieve data parallelism and model parallelism, another configuration file that can be loaded by the `train` script is the `resourceconfig.json` file. This file can be found inside the `/opt/ml/input/config` directory. This file contains details regarding all running containers when the training job is running and provides information about `current_host`, `hosts`, and `network_interface_name`.

> **Important note**
>
> Take note that the `resourceconfig.json` file only exists when
> distributed training is used, so check the existence of this file (as well as other
> files) before performing the load operation.

If you want to update your train script with the proper support for distributed
training, simply use the experiment environment from the *Setting up the Python
and R experimentation environments* recipe and create a dummy file named
`resourceconfig.json` inside the `/opt/ml/input/config` directory:

```
{
    "current_host": "host-1",
    "hosts": ["host-1","host-2"],
    "network_interface_name":"eth1"
}
```

The preceding code will help you create that dummy file.

Preparing and testing the serve script in Python

In this recipe, we will create a sample `serve` script using Python that loads the model
and sets up a Flask server for returning predictions. This will provide us with a template to
work with and test the end-to-end training and deployment process before adding more
complexity to the `serve` script. The following diagram shows the expected behavior of the
Python `serve` script that we will prepare in this recipe. The Python `serve` script loads the
model file from the `/opt/ml/model` directory and runs a **Flask** web server on port `8080`:

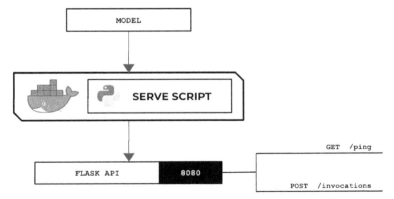

Figure 2.49 – The Python serve script loads and deserializes the model and runs
a Flask API server that acts as the inference endpoint

The web server is expected to have the /ping and /invocations endpoints. This standalone Python script will run inside a custom container that allows the Python train and serve scripts to run.

Getting ready

Make sure you have completed the *Preparing and testing the train script in Python* recipe.

How to do it...

We will start by preparing the serve script:

1. Inside the ml-python directory, double-click the serve file to open it inside the **Editor** pane:

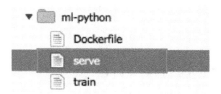

Figure 2.50 – Locating the empty serve script inside the ml-python directory

Here, we can see three files under the ml-python directory. Remember that in the *Setting up the Python and R experimentation environments* recipe, we prepared an empty serve script:

Figure 2.51 – Empty serve file

In the next couple of steps, we will add the lines of code for the serve script.

2. Add the following code to the serve script to import and initialize the prerequisites:

```python
#!/usr/bin/env python3
import numpy as np
from flask import Flask
from flask import Response
from flask import request

from joblib import dump, load
```

3. Initialize the Flask app. After that, define the get_path() function:

```python
app = Flask(__name__)


PATHS = {
    'hyperparameters': 'input/config/hyperparameters.
json',
    'input': 'input/config/inputdataconfig.json',
    'data': 'input/data/',
    'model': 'model/'
}

def get_path(key):
    return '/opt/ml/' + PATHS[key]
```

4. Define the load_model() function by adding the following lines of code to the serve script:

```python
def load_model():
    model = None

    filename = get_path('model') + 'model'
    print(filename)

    model = load(filename)
    return model
```

Note that the filename of the model here is `model` as we specified this model artifact filename when we saved the model using the `dump()` function in the *Preparing and testing the train script in Python* recipe.

> **Important note**
>
> Note that it is important to choose the right approach when saving and loading machine learning models. In some cases, machine learning models from untrusted sources may contain malicious instructions that cause security issues such as **arbitrary code execution**! For more information on this topic, feel free to check out `https://joblib.readthedocs.io/en/latest/persistence.html`.

5. Define a function that accepts the POST requests for the `/invocations` route:

```python
@app.route("/invocations", methods=["POST"])
def predict():
    model = load_model()
    post_body = request.get_data().decode("utf-8")
    payload_value = float(post_body)

    X_test = np.array(
        [payload_value]
    ).reshape(-1, 1)
    y_test = model.predict(X_test)

    return Response(
        response=str(y_test[0]),
        status=200
    )
```

This function has five parts: loading the trained model using the `load_model()` function, reading the POST request data using the `request.get_data()` function and storing it inside the `post_body` variable, transforming the prediction payload into the appropriate structure and type using the `float()`, `np.array()`, and `reshape()` functions, making a prediction using the `predict()` function, and returning the prediction value inside a `Response` object.

> **Important note**
>
> Note that the implementation of the `predict()` function in the preceding code block can only handle predictions involving single payload values. At the same time, it can't handle different types of input similar to how built-in algorithms handle CSV, JSON, and other types of request formats. If you need to provide support for this, additional lines of code need to be added to the implementation of the `predict()` function.

6. Prepare the `/ping` route and handler by adding the following lines of code to the serve script:

```
@app.route("/ping")
def ping():
    return Response(response="OK", status=200)
```

7. Finally, use the `app.run()` method and bind the web server to port `8080`:

```
app.run(host="0.0.0.0", port=8080)
```

> **Tip**
>
> You can access a working copy of the `serve` script file in this book's GitHub repository: `https://github.com/PacktPublishing/Machine-Learning-with-Amazon-SageMaker-Cookbook/blob/master/Chapter02/ml-python/serve`.

8. Create a new Terminal in the bottom pane, below the **Editor** pane:

Figure 2.52 – New Terminal

Here, we can see a Terminal tab already open. If you need to create a new one, simply click the plus (+) sign and then click **New Terminal**. We will run the next few commands in this Terminal tab.

9. Install the Flask framework using pip. We will use Flask for our inference API endpoint:

```
pip3 install flask
```

10. Navigate to the ml-python directory:

```
cd /home/ubuntu/environment/opt/ml-python
```

11. Make the serve script executable using chmod:

```
chmod +x serve
```

12. Test the serve script using the following command:

```
./serve
```

This should start the Flask app, as shown here:

```
serve                  ×    ⊕
1   #!/usr/bin/env python3
2
3   import numpy as np
4
5   from flask import Flask
6   from flask import Response
7   from flask import request
8
9   from joblib import dump, load
10
11  app = Flask(__name__)
12
13
```

```
python3 - "ip-172 ×   ⊕
ubuntu:~/environment/opt/ml-python $ ./serve
 * Serving Flask app "serve" (lazy loading)
 * Environment: production
   WARNING: This is a development server. Do not use it in a production deployment.
   Use a production WSGI server instead.
 * Debug mode: off
 * Running on http://0.0.0.0:8080/ (Press CTRL+C to quit)
```

Figure 2.53 – Running the serve script

Here, we can see that our `serve` script has successfully run a `flask` API web server on port `8080`.

Finally, we will trigger this running web server.

13. Open a new Terminal window:

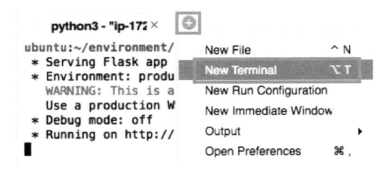

Figure 2.54 – New Terminal

As we can see, we are creating a new Terminal tab as the first tab is already running the `serve` script.

14. In a separate Terminal window, test the `ping` endpoint URL using the `curl` command:

```
SERVE_IP=localhost
curl http://$SERVE_IP:8080/ping
```

Running the previous line of code should yield an `OK` message from the `/ping` endpoint.

15. Test the invocations endpoint URL using the `curl` command:

```
curl -d "1" -X POST http://$SERVE_IP:8080/invocations
```

We should get a value similar or close to `881.3428400857507` after invoking the `invocations` endpoint.

Now, let's see how this works!

How it works...

In this recipe, we prepared the `serve` script in Python. The `serve` script makes use of the **Flask** framework to generate an API that allows `GET` requests for the `/ping` route and `POST` requests for the `/invocations` route.

The `serve` script is expected to load the model file(s) from the `/opt/ml/model` directory and run a backend API server inside the custom container. It should provide a `/ping` route and an `/invocations` route. With these in mind, our bare minimum Flask application template may look like this:

```python
from flask import Flask
app = Flask(__name__)


@app.route("/ping")
def ping():
    return <RETURN VALUE>


@app.route("/invocations", methods=["POST"])
def predict():
    return <RETURN VALUE>
```

The `app.route()` decorator maps a specified URL to a function. In this template code, whenever the `/ping` URL is accessed, the `ping()` function is executed. Similarly, whenever the `/invocations` URL is accessed with a POST request, the `predict()` function is executed.

> **Note**
>
> Take note that we are free to use any other web framework (for example, the **Pyramid Web Framework**) for this recipe. So long as the custom container image has the required libraries for the script that's been installed, then we can import and use these libraries in our script files.

Building and testing the custom Python algorithm container image

In this recipe, we will prepare a `Dockerfile` for the custom Python container image. We will make use of the `train` and `serve` scripts that we prepared in the previous recipes. After that, we will run the `docker build` command to prepare the image before pushing it to an **Amazon ECR** repository.

> **Tip**
>
> Wait! What's a `Dockerfile`? It's a text document containing the directives (commands) used to prepare and build a container image. This container image then serves as the blueprint when running containers. Feel free to check out `https://docs.docker.com/engine/reference/builder/` for more information on Dockerfiles.

Getting ready

Make sure you have completed the *Preparing and testing the serve script in Python* recipe.

How to do it...

The initial steps in this recipe focus on preparing a `Dockerfile`. Let's get started:

1. Double-click the `Dockerfile` file in the file tree to open it in the **Editor** pane. Make sure that this is the same `Dockerfile` that's inside the `ml-python` directory:

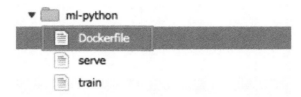

Figure 2.55 – Opening the Dockerfile inside the ml-python directory

Here, we can see a `Dockerfile` inside the `ml-python` directory. Remember that we created an empty `Dockerfile` in the *Setting up the Python and R experimentation environments* recipe. Clicking it in the file tree should open an empty file in the **Editor** pane:

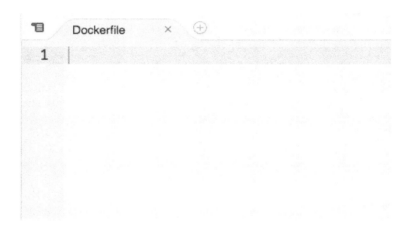

Figure 2.56 – Empty Dockerfile in the Editor pane

Here, we have an empty `Dockerfile`. In the next step, we will update this by adding three lines of code.

2. Update `Dockerfile` with the following block of configuration code:

```
FROM arvslat/amazon-sagemaker-cookbook-python-base:1
COPY train /usr/local/bin/train
COPY serve /usr/local/bin/serve
```

Here, we are planning to build on top of an existing image called `amazon-sagemaker-cookbook-python-base`. This image already has a few prerequisites installed. These include the `Flask`, `pandas`, and `Scikit-learn` libraries so that you won't have to worry about getting the installation steps working properly in this recipe. For more details on this image, check out `https://hub.docker.com/r/arvslat/amazon-sagemaker-cookbook-python-base`:

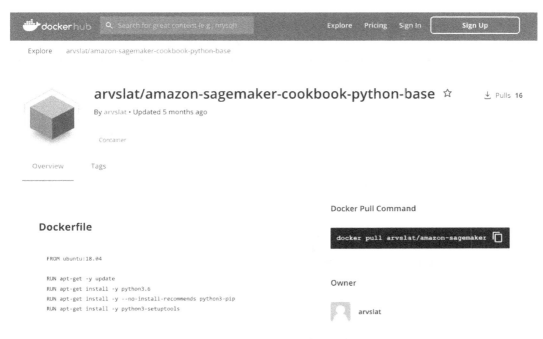

Figure 2.57 – Docker Hub page for the base image

Here, we can see the **Docker Hub** page for the **amazon-sagemaker-cookbook-python-base** image.

> Tip
>
> You can access a working copy of this `Dockerfile` in the *Machine Learning with Amazon SageMaker Cookbook* GitHub repository: `https://github.com/PacktPublishing/Machine-Learning-with-Amazon-SageMaker-Cookbook/blob/master/Chapter02/ml-python/serve`.

With the `Dockerfile` ready, we will proceed with using the Terminal until the end of this recipe:

3. You can use a new Terminal tab or an existing one to run the next set of commands:

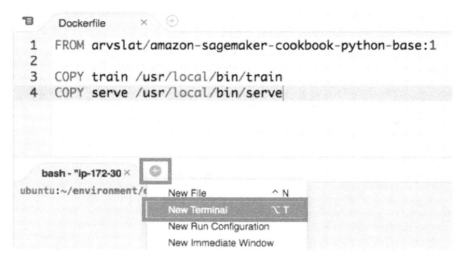

Figure 2.58 – New Terminal

Here, we can see how to create a new Terminal. Note that the **Terminal** pane is under the **Editor** pane in the AWS Cloud9 IDE.

4. Navigate to the `ml-python` directory containing our `Dockerfile`:

```
cd /home/ubuntu/environment/opt/ml-python
```

5. Specify the image name and the tag number:

```
IMAGE_NAME=chap02_python
TAG=1
```

6. Build the **Docker** container using the `docker build` command:

```
docker build --no-cache -t $IMAGE_NAME:$TAG .
```

The `docker build` command makes use of what is written inside our `Dockerfile`. We start with the image specified in the `FROM` directive and then we proceed by copying the file into the container image.

7. Use the `docker run` command to test if the `train` script works:

```
docker run --name pytrain --rm -v /opt/ml:/opt/ml $IMAGE_
NAME:$TAG train
```

Let's quickly discuss some of the different options that were used in this command. The `--rm` flag makes Docker clean up the container after the container exits. The `-v` flag allows us to mount the `/opt/ml` directory from the host system to the `/opt/ml` directory of the container:

```
ubuntu:~/environment/opt/ml-python $ docker run --name pytrain --rm -v
[inspect_hyperparameters]
/opt/ml/input/config/hyperparameters.json
{'a': 1, 'b': 2}
[inspect_input]
/opt/ml/input/config/inputdataconfig.json
{    'train': {    'ContentType': 'text/csv',
                   'RecordWrapperType': 'None',
                   'S3DistributionType': 'FullyReplicated',
                   'TrainingInputMode': 'File'}}
[get_input_data_dir]
[load_training_data]
[list_dir_contents]
['training_data.csv']
/opt/ml/input/data/train/training_data.csv
        0    1
0     960   19
1    1590   94
2    1630   65
3     960    7
4    1480   71
5    1330   61
6    1300   37
7    1040   22
8    1290   56
9    1020   15
```

Figure 2.59 – Result of the docker run command (train)

Here, we can see the results after running the `docker run` command. It should show logs similar to what we had in the *Preparing and testing the train script in Python* recipe.

8. Use the `docker run` command to test if the `serve` script works:

```
docker run --name pyserve --rm -v /opt/ml:/opt/ml $IMAGE_
NAME:$TAG serve
```

After running this command, the Flask API server starts successfully. We should see logs similar to what we had in the *Preparing and testing the serve script in Python* recipe:

```
docker - "ip-172-: ×      ⊕

ubuntu:~/environment/opt/ml-python $ docker run --name pyserve --rm -v /opt/ml:/opt/ml
 * Serving Flask app "serve" (lazy loading)
 * Environment: production
   WARNING: This is a development server. Do not use it in a production deployment.
   Use a production WSGI server instead.
 * Debug mode: off
 * Running on http://0.0.0.0:8080/ (Press CTRL+C to quit)
```

Figure 2.60 – Result of the docker run command (serve)

Here, we can see that the API is running on port 8080. In the base image we used, we added EXPOSE 8080 to allow us to access this port in the running container.

9. Open a new Terminal tab:

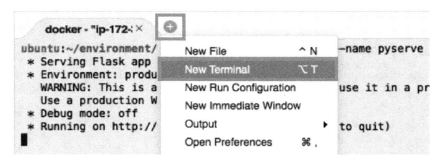

Figure 2.61 – New Terminal

As the API is running already in the first Terminal, we have created a new one.

10. In the new Terminal tab, run the following command to get the IP address of the running Flask app:

```
SERVE_IP=$(docker network inspect bridge | jq -r ".[0].
Containers[].IPv4Address" | awk -F/ '{print $1}')
echo $SERVE_IP
```

We should get an IP address that's equal or similar to 172.17.0.2. Of course, we may get a different IP address value.

11. Next, test the ping endpoint URL using the curl command:

```
curl http://$SERVE_IP:8080/ping
```

We should get an `OK` after running this command.

12. Finally, test the `invocations` endpoint URL using the `curl` command:

```
curl -d "1" -X POST http://$SERVE_IP:8080/invocations
```

We should get a value similar or close to `881.3428400857507` after invoking the `invocations` endpoint.

At this point, it is safe to say that the custom container image we have prepared in this recipe is ready. Now, let's see how this works!

How it works...

In this recipe, we built a custom container image using the `Dockerfile` configuration we specified. When you have a `Dockerfile`, the standard set of steps would be to use the `docker build` command to build the Docker image, authenticate with ECR to gain the necessary permissions, use the `docker tag` command to tag the image appropriately, and use the `docker push` command to push the Docker image to the ECR repository.

Let's discuss what we have inside our `Dockerfile`. If this is your first time hearing about Dockerfiles, they are simply text files containing commands to build the image. In our `Dockerfile`, we did the following:

- We used `arvslat/amazon-sagemaker-cookbook-python-base` as the base image. Check out `https://hub.docker.com/repository/docker/arvslat/amazon-sagemaker-cookbook-python-base` for more details about this image.

- We copied the `train` and `serve` scripts to the `/usr/local/bin` directory inside the container image. These scripts are executed when we use `docker run`.

Using the `arvslat/amazon-sagemaker-cookbook-python-base` image as the base image allowed us to write a shorter `Dockerfile` that focuses only on copying the `train` and `serve` files to the directory inside the container image. Behind the scenes, we have already pre-installed the `flask`, `pandas`, `scikit-learn`, and `joblib` packages, along with their prerequisites, inside this container image so that we will not run into issues when building the custom container image. Here is a quick look at the `Dockerfile` file we used as the base image that we are using in this recipe:

```
FROM ubuntu:18.04

RUN apt-get -y update
```

```
RUN apt-get install -y python3.6
RUN apt-get install -y --no-install-recommends python3-pip
RUN apt-get install -y python3-setuptools

RUN ln -s /usr/bin/python3 /usr/bin/python & \
    ln -s /usr/bin/pip3 /usr/bin/pip

RUN pip install flask
RUN pip install pandas
RUN pip install scikit-learn
RUN pip install joblib

WORKDIR /usr/local/bin
EXPOSE 8080
```

In this Dockerfile, we can see that we are using Ubuntu:18.04 as the base image. Note that we can use other base images as well, depending on the libraries and frameworks we want to be installed in the container image.

Once we have the container image built, the next step will be to test if the train and serve scripts will work inside the container once we use docker run. Getting the IP address of the running container may be the trickiest part, as shown in the following block of code:

```
SERVE_IP=$(docker network inspect bridge | jq -r ".[0].
Containers[].IPv4Address" | awk -F/ '{print $1}')
```

We can divide this into the following parts:

- docker network inspect bridge: This provides detailed information about the bridge network in JSON format. It should return an output with a structure similar to the following JSON value:

```
[
    {
        ...
        "Containers": {
            "1b6cf4a4b8fc5ea5...": {
                "Name": "pyserve",
                "EndpointID": "ecc78fb63c1ad32f0...",
```

```
                            "MacAddress":  "02:42:ac:11:00:02",
                            "IPv4Address":  "172.17.0.2/16",
                            "IPv6Address":  ""
                        }
                    },
                    ...
                }
            ]
```

- `jq -r ".[0].Containers[].IPv4Address"`: This parses through the JSON response value from `docker network inspect bridge`. Piping this after the first command would yield an output similar to `172.17.0.2/16`.

- `awk -F/ '{print $1}'`: This splits the result from the `jq` command using the `/` separator and returns the value before `/`. After getting the `AA.BB.CC.DD/16` value from the previous command, we get `AA.BB.CC.DD` after using the `awk` command.

Once we have the IP address of the running container, we can ping the `/ping` and `/invocations` endpoints, similar to how we did in the *Preparing and testing the serve script in Python* recipe.

In the next recipes in this chapter, we will use this custom container image when we do training and deployment with the **SageMaker Python SDK**.

Pushing the custom Python algorithm container image to an Amazon ECR repository

In the previous recipe, we have prepared and built the custom container image using the `docker build` command. In this recipe, we will push the custom container image to an **Amazon ECR** repository. If this is your first time hearing about **Amazon ECR**, it is simply a fully managed container registry that helps us manage our container images.

After pushing the container image to an Amazon ECR repository, we can use this image for training and deployment in the *Using the custom Python algorithm container image for training and inference with Amazon SageMaker Local Mode* recipe.

Getting ready

Here are the prerequisites for this recipe:

- This recipe continues from the *Building and testing the custom Python algorithm container image* recipe.

- You will need the necessary permissions to manage the **Amazon ECR** resources if you're using an **AWS IAM** user with a custom URL.

How to do it...

The initial steps in this recipe focus on creating the ECR repository. Let's get started:

1. Use the search bar in the **AWS Console** to navigate to the **Elastic Container Registry** console. Click **Elastic Container Registry** when it appears in the search results:

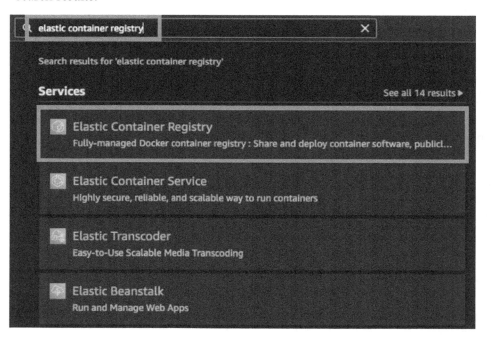

Figure 2.62 – Navigating to the ECR console

As you can see, we can use the search bar to quickly navigate to the **Elastic Container Registry** service. If we type in ecr, the **Elastic Container Registry** service in the search results may come up in third or fourth place.

2. Click the **Create repository** button:

Figure 2.63 – Create repository button

Here, the **Create repository** button is at the top right of the screen.

3. In the **Create repository** form, specify a **Repository name**. Use the value of
    $IMAGE_NAME from the *Building and testing the custom Python algorithm container
    image* recipe. In this case, we will use chap02_python:

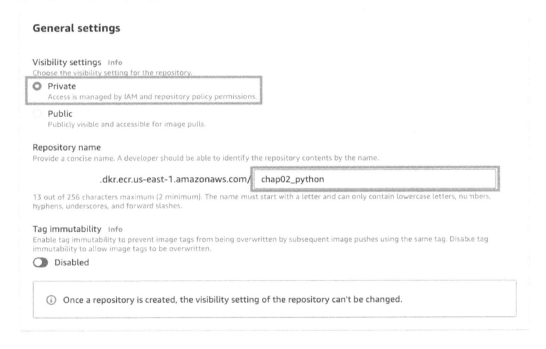

Figure 2.64 – Create repository form

Here, we have the **Create repository** form. For **Visibility settings**, we will choose
Private and set the **Tag immutability** configuration to **Disabled**.

4. Scroll down until you see the **Create repository** button. Leave the other configuration settings as-is and click **Create repository**:

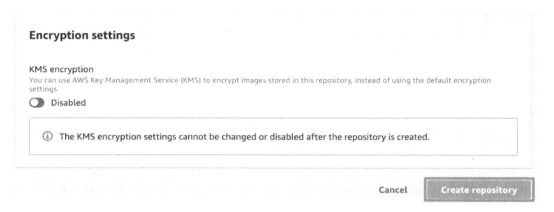

Figure 2.65 – Create repository button

As we can see, the **Create repository** button is at the bottom of the page.

5. Click **chap02_python**:

Figure 2.66 – Link to the ECR repository page

Here, we have a link under the **Repository name** column. Clicking this link should redirect us to the repository's details page.

6. Click **View push commands**:

Figure 2.67 – View push commands button (upper right)

As we can see, the **View push commands** button is at the top right of the page, beside the **Edit** button.

7. You may optionally copy the first command, `aws ecr get-login-password` ..., from the dialog box.

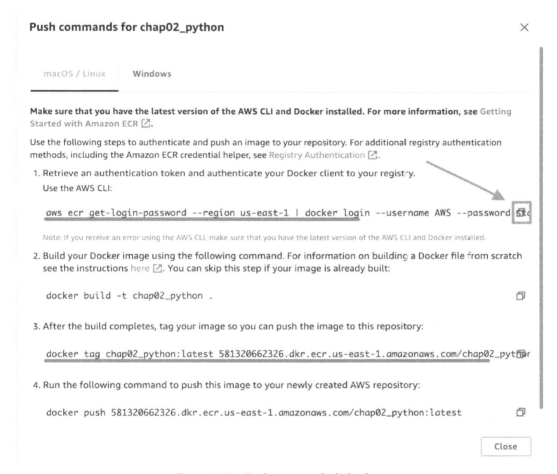

Figure 2.68 – Push commands dialog box

Here, we can see multiple commands that we can use. We will only need the first one (`aws ecr get-login-password` ...). Click the icon with two overlapping boxes on the right-hand side of the code box to copy the entire line to the clipboard.

8. Navigate back to the AWS Cloud9 environment IDE and create a new Terminal. You may also reuse an existing one:

Figure 2.69 – New Terminal

The preceding screenshot shows us how to create a new Terminal. Click the green plus button and then select **New Terminal** from the list of options. Note that the green plus button is directly under the **Editor** pane.

9. Navigate to the `ml-python` directory:

```
cd /home/ubuntu/environment/opt/ml-python
```

10. Get the account ID using the following commands:

```
ACCOUNT_ID=$(aws sts get-caller-identity | jq -r
".Account")
echo $ACCOUNT_ID
```

11. Specify the `IMAGE_URI` value and use the ECR repository name we specified while creating the repository in this recipe. In this case, we will run `IMAGE_URI="chap02_python"`:

```
IMAGE_URI="<insert ECR Repository URI>"
TAG="1"
```

12. Authenticate with **Amazon ECR** so that we can push our **Docker** container image to an **Amazon ECR** repository in our account later:

```
aws ecr get-login-password --region us-east-1 | docker
login --username AWS --password-stdin $ACCOUNT_ID.dkr.
ecr.us-east-1.amazonaws.com
```

> **Important note**
> Note that we have assumed that our repository is in the `us-east-1` region. Feel free to modify the region in the command if needed. This applies to all the commands in this chapter.

13. Use the `docker tag` command:

```
docker tag $IMAGE_URI:$TAG $ACCOUNT_ID.dkr.ecr.us-east-1.
amazonaws.com/$IMAGE_URI:$TAG
```

14. Push the image to the **Amazon ECR** repository using the `docker push` command:

```
docker push $ACCOUNT_ID.dkr.ecr.us-east-1.amazonaws.
com/$IMAGE_URI:$TAG
```

At this point, our custom container image should now be successfully pushed into the ECR repository.

Now that we have completed this recipe, we can proceed with using this custom container image for training and inference with SageMaker in the next recipe. But before that, let's see how this works!

How it works...

In the previous recipe, we used the `docker build` command to prepare the custom container image. In this recipe, we created an **Amazon ECR** repository and pushed our custom container image to the repository. With Amazon ECR, we can store, manage, share, and run custom container images anywhere. This includes using these container images in SageMaker during training and deployment.

When pushing the custom container image to the Amazon ECR repository, we need the account ID, region, repository name, and tag. Once we have these, the `docker push` command will look something like this:

```
docker push <ACCOUNT_ID>.dkr.ecr.<REGION>.amazonaws.
com/<REPOSITORY NAME>:<TAG>
```

When working with container image versions, make sure to change the version number every time you modify this `Dockerfile` and push a new version to the **ECR** repository. This will be helpful when you need to use a previous version of a container image.

Using the custom Python algorithm container image for training and inference with Amazon SageMaker Local Mode

In this recipe, we will perform the training and deployment steps in **Amazon Sagemaker** using the custom container image we pushed to the ECR repository in the *Pushing the custom Python algorithm container image to an Amazon ECR repository* recipe. In *Chapter 1, Getting Started with Machine Learning Using Amazon SageMaker*, we used the image URI of the container image of the built-in **Linear Learner**. In this chapter, we will use the image URI of the custom container image instead.

The following diagram shows how **SageMaker** passes data, files, and configuration to and from each custom container when we use the `fit()` and `predict()` functions with the **SageMaker Python SDK**:

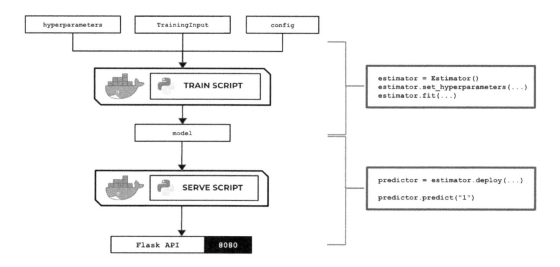

Figure 2.70 – The train and serve scripts inside the custom container make use of the hyperparameters, input data, and config specified using the SageMaker Python SDK

We will also take a look at how to use **local mode** in this recipe. This capability of SageMaker allows us to test and emulate the CPU and GPU training jobs inside our local environment. Using **local mode** is useful while we are developing, enhancing, and testing our custom algorithm container images and scripts. We can easily switch to using ML instances that support the training and deployment steps once we are ready to roll out the stable version of our container image.

Once we have completed this recipe, we will be able to run training jobs and deploy inference endpoints using Python with custom `train` and `serve` scripts inside custom containers.

Getting ready

Here are the prerequisites for this recipe:

- This recipe continues from the *Pushing the custom Python algorithm container image to an Amazon ECR repository* recipe.

- We will use the SageMaker Notebook instance from the *Launching an Amazon SageMaker Notebook instance and preparing the prerequisites* recipe of *Chapter 1, Getting Started with Machine Learning Using Amazon SageMaker*.

How to do it...

The first couple of steps in this recipe focus on preparing the Jupyter Notebook using the `conda_python3` kernel:

1. Inside your SageMaker Notebook instance, create a new directory called `chapter02` inside the `my-experiments` directory. As shown in the following screenshot, we can perform this step by clicking the **New** button and then choosing **Folder** (under **Other**):

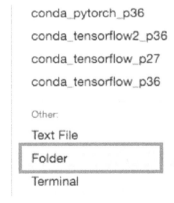

Figure 2.71 – New > Folder

This will create a directory named `Untitled Folder`.

2. Click the checkbox and then click **Rename**. Change the name to `chapter02`:

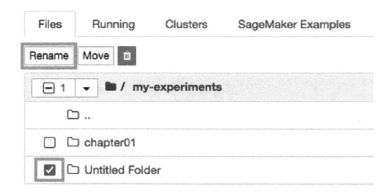

Figure 2.72 – Renaming "Untitled Folder" to "chapter02"

After that, we should get the desired directory structure, as shown in the preceding screenshot. Now, let's look at the following directory structure:

Figure 2.73 – Directory structure

This screenshot shows how we want to organize our files and notebooks. As we go through each chapter, we will add more directories using the same naming convention to keep things organized.

3. Click the **chapter02** directory to navigate to **/my-experiments/chapter02**.

4. Create a new notebook by clicking **New** and then clicking **conda_python3**:

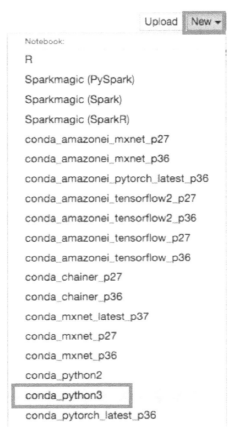

Figure 2.74 – Creating a new notebook using the conda_python3 kernel

Now that we have a fresh Jupyter Notebook using the `conda_python3` kernel, we will proceed with preparing the prerequisites for the training and deployment steps.

5. In the first cell of the Jupyter Notebook, use `pip install` to upgrade `sagemaker[local]`:

```
!pip install 'sagemaker[local]' --upgrade
```

This will allow us to use **local mode**. We can use local mode when working with framework images such as **TensorFlow**, **PyTorch**, **scikit-learn**, and **MXNet**, and custom images we have built ourselves.

> **Important note**
>
> Note that we can NOT use **local mode** in **SageMaker Studio**. We also can NOT use **local mode** with built-in algorithms.

6. Specify the bucket name where the `training_data.csv` file is stored. Use the bucket name we created in the *Preparing the Amazon S3 bucket and the training dataset for the linear regression experiment* recipe of *Chapter 1, Getting Started with Machine Learning Using Amazon SageMaker*:

```
s3_bucket = "<insert bucket name here>"
prefix = "chapter01"
```

Note that our `training_data.csv` file should exist already inside the S3 bucket and should have the following path:

```
s3://<S3 BUCKET NAME>/<PREFIX>/input/training_data.csv
```

7. Set the variable values for `training_s3_input_location` and `training_s3_output_location`:

```
training_s3_input_location = \ f"s3://{s3_bucket}/
{prefix}/input/training_data.csv"
training_s3_output_location = \ f"s3://{s3_bucket}/
{prefix}/output/custom/"
```

8. Import the **SageMaker Python SDK** and check its version:

```
import sagemaker
sagemaker.__version__
```

We should get a value equal to or near `2.31.0` after running the previous block of code.

9. Set the value of the container image. Use the value from the *Pushing the custom Python algorithm container image to an Amazon ECR repository* recipe. The `container` variable should be set to a value similar to `<ACCOUNT_ID>.dkr.ecr.us-east-1.amazonaws.com/chap02_python:1`. Make sure to replace `<ACCOUNT_ID>` with your AWS account ID:

```
container="<insert image uri and tag here>"
```

To get the value of `<ACCOUNT_ID>`, run `ACCOUNT_ID=$(aws sts get-caller-identity | jq -r ".Account")` and then `echo $ACCOUNT_ID` inside a Terminal. Remember that we performed this step in the *Pushing the custom Python algorithm container image to an Amazon ECR repository* recipe, so you should get the same value for `ACCOUNT_ID`.

10. Import a few prerequisites such as `role` and `session`. You will probably notice one of the major differences between this recipe and the recipes in *Chapter 1, Getting Started with Machine Learning Using Amazon SageMaker* – the usage of `LocalSession`. The `LocalSession` class allows us to use **local mode** in the training and deployment steps:

```
import boto3
from sagemaker import get_execution_role
role = get_execution_role()
from sagemaker.local import LocalSession
session = LocalSession()
session.config = {'local': {'local_code': True}}
```

11. Initialize the `TrainingInput` object for the train data channel:

```
from sagemaker.inputs import TrainingInput
train = TrainingInput(training_s3_input_location,
content_type="text/csv")
```

Now that we have the prerequisites, we will proceed with initializing `Estimator` and using the `fit()` and `predict()` functions.

12. Initialize `Estimator` and use `container`, `role`, `session`, and `training_s3_output_location` as the parameter values when initializing the `Estimator` object:

```
estimator = sagemaker.estimator.Estimator(
    container,
    role,
    instance_count=1,
    instance_type='local',
    output_path=training_s3_output_location,
    sagemaker_session=session)
```

Here, we set the `instance_type` value to `local` and the `sagemaker_session` value to `session` (which is a `LocalSession` object). This means that when we run the `fit()` function later, the training job is performed locally and no ML instances will be provisioned for the training job.

> **Important note**
>
> If we want to perform the training job in a dedicated ML instance, simply replace the instance_type value with ml.m5.xlarge (or an alternative ML instance type) and the sagemaker_session value with a Session object. To make sure that we do not encounter training job name validation issues (as we used an underscore in the ECR repository name), specify the base_job_name parameter value with the appropriate value when initializing Estimator.

13. Set a few dummy hyperparameters by using the set_hyperparameters() function. Behind the scenes, these values will passed to the hyperparameters. json file inside the /opt/ml/input/config directory, which will be loaded and used by the train script when we run the fit() function later:

```
estimator.set_hyperparameters(a=1, b=2, c=3)
```

14. Start the training job using fit():

```
estimator.fit({'train': train})
```

This should generate a set of logs similar to the following:

```
Creating 7fb7uq7z4r-algo-1-nlh5m ...
Creating 7fb7uq7z4r-algo-1-nlh5m ... done
Attaching to 7fb7uq7z4r-algo-1-nlh5m
7fb7uq7z4r-algo-1-nlh5m | [inspect_hyperparameters]
7fb7uq7z4r-algo-1-nlh5m | /opt/ml/input/config/hyperparameters.json
7fb7uq7z4r-algo-1-nlh5m | {'a': '1', 'b': '2', 'c': '3'}
7fb7uq7z4r-algo-1-nlh5m | [inspect_input]
7fb7uq7z4r-algo-1-nlh5m | /opt/ml/input/config/inputdataconfig.json
7fb7uq7z4r-algo-1-nlh5m | {'train': {'ContentType': 'text/csv', 'TrainingInputMode': 'File'}}
7fb7uq7z4r-algo-1-nlh5m | [get_input_data_dir]
7fb7uq7z4r-algo-1-nlh5m | [load_training_data]
7fb7uq7z4r-algo-1-nlh5m | [list_dir_contents]
7fb7uq7z4r-algo-1-nlh5m | ['training_data.csv']
7fb7uq7z4r-algo-1-nlh5m | /opt/ml/input/data/train/training_data.csv
7fb7uq7z4r-algo-1-nlh5m |          0    1
7fb7uq7z4r-algo-1-nlh5m | 0      960   19
7fb7uq7z4r-algo-1-nlh5m | 1     1590   94
7fb7uq7z4r-algo-1-nlh5m | 2     1630   65
7fb7uq7z4r-algo-1-nlh5m | 3      960    7
7fb7uq7z4r-algo-1-nlh5m | 4     1480   71
7fb7uq7z4r-algo-1-nlh5m | 5     1330   61
```

Figure 2.75 – Using fit() with local mode

Here, we can see a similar set of logs are generated when we run the train script inside our experimentation environment. As with the *Training your first model in Python* recipe of *Chapter 1, Getting Started with Machine Learning Using Amazon SageMaker*, the fit() command will prepare an instance for the duration of the training job to train the model. In this recipe, we are using **local mode**, so no instances are created.

> **Important note**
>
> To compare this to what we did in *Chapter 1, Getting Started with Machine
> Learning Using Amazon SageMaker*, we used the `fit()` function with the
> `Model` class from *Chapter 1*, while we used the `fit()` function with the
> `Estimator` class in this chapter. We can technically use either of these but
> in this recipe, we went straight ahead and used the `fit()` function after the
> `Estimator` object was initialized, without initializing a separate `Model`
> object.

15. Use the `deploy()` function to deploy the inference endpoint:

```
predictor = estimator.deploy(
    initial_instance_count=1,
    instance_type='local',
    endpoint_name="custom-local-py-endpoint")
```

As we are using **local mode**, no instances are created and the container is run inside
the SageMaker Notebook instance:

```
Attaching to hy2t7ds2h2-algo-1-ai4yn
hy2t7ds2h2-algo-1-ai4yn  |  * Serving Flask app "serve" (lazy loading)
hy2t7ds2h2-algo-1-ai4yn  |  * Environment: production
hy2t7ds2h2-algo-1-ai4yn  |    WARNING: This is a development server. Do not use it in a production deployment.
hy2t7ds2h2-algo-1-ai4yn  |    Use a production WSGI server instead.
hy2t7ds2h2-algo-1-ai4yn  |  * Debug mode: off
hy2t7ds2h2-algo-1-ai4yn  |  * Running on http://0.0.0.0:8080/ (Press CTRL+C to quit)
hy2t7ds2h2-algo-1-ai4yn  |  172.18.0.1 - - [16/Mar/2021 18:14:06] "GET /ping HTTP/1.1" 200 -
!
```

Figure 2.76 – Using deploy() with local mode

As we can see, we are getting log messages in a similar way to how we got them in
the *Building and testing the custom Python algorithm container image* recipe. This
means that if we couldn't get the container running in that recipe, then we will not
get the container running in this recipe either.

16. Once the endpoint is ready, we can use the `predict()` function to test if the
inference endpoint is working as expected. This will trigger the `/invocations`
endpoint behind the scenes and pass a value of `"1"` in the POST body:

```
predictor.predict("1")
```

This should yield a set of logs similar to the following:

```
hy2t7ds2h2-algo-1-ai4yn | /opt/ml/model/model
hy2t7ds2h2-algo-1-ai4yn | 172.18.0.1 - - [16/Mar/2021 18:14:33] "POST /invocations HTTP/1.1" 200 -
```

Figure 2.77 – Using predict() with local mode

Here, we can see the logs from the API web server that was launched by the serve script inside the container. We should get a value similar or close to '881.3428400857507'. We will get the return value of the sample invocations endpoint we have coded in a previous recipe:

```
@app.route("/invocations", methods=["POST"])
def predict():
    model = load_model()
    ...
    return Response(..., status=200)
```

If we go back and check the *Preparing and testing the serve script in Python* recipe in this chapter, we will see that we have full control of how the invocations endpoint works by modifying the code inside the predict() function in the serve script. We have copied a certain portion of the function in the preceding code block for your convenience.

17. Use delete_endpoint() to delete the local prediction endpoint:

```
predictor.delete_endpoint()
```

We should get a message similar to the following:

```
Gracefully stopping... (press Ctrl+C again to force)
```

Figure 2.78 – Using delete_endpoint() with local mode

As we can see, using delete_endpoint() will result in the **Gracefully stopping...** message. Given that we are using **local mode** in this recipe, the delete_endpoint() function will stop the running API server in the SageMaker Notebook instance. If **local mode** is not used, the SageMaker inference endpoint and the ML compute instance(s) that support it will be deleted.

Now, let's check how this works!

How it works...

In this recipe, we used the custom container image we prepared in the previous sections while training and deploying Python, instead of the built-in algorithms of **SageMaker**. All the steps are similar to the ones we followed for the built-in algorithms; the only changes you will need to take note of are the container image, the input parameters, and the hyperparameters.

Take note that we have full control of the hyperparameters we can specify in `Estimator` as this depends on the hyperparameters that are expected by our custom script. If you need a more realistic example of these hyperparameters, here are the hyperparameters from *Chapter 1, Getting Started with Machine Learning Using Amazon SageMaker*:

```
estimator.set_hyperparameters(
    predictor_type='regressor',
    mini_batch_size=4)
```

In this example, the `hyperparameters.json` file, which contains the following content, is created when the `fit()` function is called:

```
{"predictor_type": "regressor", "mini_batch_size": 4}
```

The arguments we can use and configure in this recipe are more or less the same as the ones we used for the built-in algorithms of SageMaker. The only major difference is that we are using the container image URI of our ECR repository instead of the container image URI for the built-in algorithms.

When we're using our custom container images, we have the option to use **local mode** when performing training and deployment. With local mode, no additional instances outside of the SageMaker Notebook instance are created. This allows us to test if the custom container image is working or not, without having to wait for a couple of minutes compared to using real instances (for example, `ml.m5.xlarge`). Once things are working as expected using local mode, we can easily switch to using the real instances by replacing `session` and `instance_type` in `Estimator`.

Preparing and testing the train script in R

In this recipe, we will write a custom train script in R that allows us to inspect the input and configuration parameters set by **Amazon SageMaker** during the training process. The following diagram shows the `train` script inside the custom container, which makes use of the hyperparameters, input data, and configuration specified in the `Estimator` instance using the **SageMaker Python SDK** and the `reticulate` package:

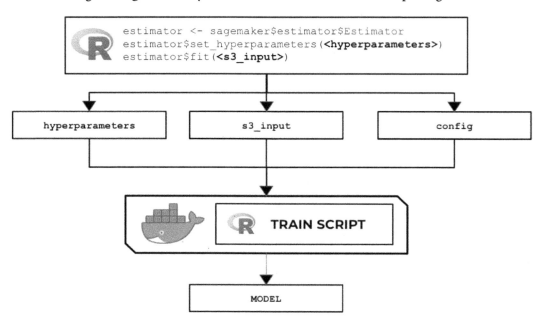

Figure 2.79 – The R train script inside the custom container makes use of the input parameters, configuration, and data to train and output a model

There are several options when running a training job – use a built-in algorithm, use a custom train script and custom Docker images, or use a custom train script and prebuilt Docker images. In this recipe, we will focus on the second option, where we will prepare and test a bare minimum training script in R that builds a linear model for a specific regression problem.

Once we have finished working on this recipe, we will have a better understanding of how **SageMaker** works behind the scenes. We will see where and how to load and use the configuration and arguments we have specified in the **SageMaker Python SDK** `Estimator` instance.

> **Important note**
> Later on, you will notice a few similarities between the Python and R recipes in this chapter. What is critical here is noticing and identifying both major and subtle differences in certain parts of the Python and R recipes. For example, when working with the serve script in this chapter, we will be dealing with two files in R (`api.r` and `serve`) instead of one in Python (`serve`). As we will see in the other recipes of this book, working on the R recipes will help us have a better understanding of the internals of SageMaker's capabilities, as there is a big chance that we will have to prepare custom solutions to solve certain requirements. As we get exposed to more machine learning requirements, we will find that there are packages in R for machine learning without direct counterparts in Python. That said, we must be familiar with how to get custom R algorithm code working in SageMaker. Stay tuned for more!

Getting ready

Make sure you have completed the *Setting up the Python and R experimentation environments* recipe.

How to do it...

The first set of steps in this recipe focus on preparing the `train` script. Let's get started:

1. Inside the `ml-r` directory, double-click the `train` file to open it inside the **Editor** pane:

Figure 2.80 – Empty ml-r/train file

Here, we have an empty train file. In the lower-right-hand corner of the **Editor** pane, you can change the syntax highlight settings to R.

2. Add the following lines of code to start the `train` script in order to import the required packages and libraries:

```
#!/usr/bin/Rscript
library("rjson")
```

3. Define the `prepare_paths()` function, which we will use to initialize the `PATHS` variable. This will help us manage the paths of the primary files and directories used in the script:

```
prepare_paths <- function() {
    keys <- c('hyperparameters',
              'input',
              'data',
              'model')

    values <- c('input/config/hyperparameters.json',
                'input/config/inputdataconfig.json',
                'input/data/',
                'model/')

    paths <- as.list(values)
    names(paths) <- keys

    return(paths);
}

PATHS <- prepare_paths()
```

This function allows us to initialize the `PATHS` variable with a dictionary-like data structure where we can get the absolute paths of the required file.

4. Next, define the `get_path()` function, which makes use of the `PATHS` variable from the previous step:

```
get_path <- function(key) {
    output <- paste('/opt/ml/', PATHS[[key]],
                    sep="")
```

```
    return(output);
}
```

When referring to the location of a specific file, such as hyperparameters.json, we will use get_path('hyperparameters') instead of the absolute path.

5. Next, add the following lines of code just after the get_path() function definition from the previous step. These functions will be used to load and print the contents of the JSON files we will work with later:

```
load_json <- function(target_file) {
    result <- fromJSON(file = target_file)
}
print_json <- function(target_json) {
    print(target_json)
}
```

6. After that, define the inspect_hyperparameters() and list_dir_contents() functions after the print_json() function definition:

```
inspect_hyperparameters <- function() {
    hyperparameters_json_path <- get_path(
        'hyperparameters'
    )
    print(hyperparameters_json_path)
    hyperparameters <- load_json(
        hyperparameters_json_path
    )
    print(hyperparameters)
}
list_dir_contents <- function(target_path) {
    print(list.files(target_path))
}
```

The inspect_hyperparameters() function inspects the contents of the hyperparameters.json file inside the /opt/ml/input/config directory. The list_dir_contents() function, on the other hand, displays the contents of a target directory.

7. Define the `inspect_input()` function. It will help us inspect the contents of `inputdataconfig.json` inside the `/opt/ml/input/config` directory:

```r
inspect_input <- function() {
    input_config_json_path <- get_path('input')
    print(input_config_json_path)
    input_config <- load_json(
        input_config_json_path
    )
    print_json(input_config)

    for (key in names(input_config)) {
        print(key)

        input_data_dir <- paste(get_path('data'),
                                key, '/', sep="")
        print(input_data_dir)
        list_dir_contents(input_data_dir)
    }
}
```

This will be used to list the contents of the training input directory inside the `main()` function later.

8. Define the `load_training_data()` function:

```r
load_training_data <- function(input_data_dir) {
    print('[load_training_data]')
    files <- list_dir_contents(input_data_dir)
    training_data_path <- paste0(
        input_data_dir, files[[1]])
    print(training_data_path)

    df <- read.csv(training_data_path, header=FALSE)
    colnames(df) <- c("y","X")
    print(df)
    return(df)
}
```

This function can be divided into two parts – preparing the specific path pointing to the CSV file containing the training data and reading the contents of the CSV file using the `read.csv()` function. The return value of this function is an R DataFrame (a two-dimensional table-like structure).

9. Next, define the `get_input_data_dir()` function:

```
get_input_data_dir <- function() {
    print('[get_input_data_dir]')
    key <- 'train'
    input_data_dir <- paste0(
        get_path('data'), key, '/')

    return(input_data_dir)
}
```

10. After that, define the `train_model()` function:

```
train_model <- function(data) {
    model <- lm(y ~ X, data=data)
    print(summary(model))
    return(model)
}
```

This function makes use of the `lm()` function to fit and prepare linear models, which can then be used for regression tasks. It accepts a formula such as `y ~ X` as the first parameter value and the training dataset as the second parameter value.

Note

Formulas in R involve a tilde symbol (~) and one or more independent variables at the right of the tilde (~), such as `X1 + X2 + X3`. In the example in this recipe, we only have one variable on the right-hand side of the tilde (~), meaning that we will only have a single predictor variable for this model. On the left-hand side of the tilde (~) is the dependent variable that we are trying to predict using the predictor variable(s). That said, the `y ~ X` formula simply expresses a relationship between the predictor variable, X, and the y variable we are trying to predict. Since we are dealing with the same dataset as we did for the recipes in *Chapter 1, Getting Started with Machine Learning Using Amazon SageMaker*, the y variable here maps to `monthly_salary`, while X maps to `management_experience_months`.

---

11. Define the `save_model()` function:

```
save_model <- function(model) {
    print('[save_model]')
    filename <- paste0(get_path('model'), 'model')
    print(filename)
    saveRDS(model, file=filename)
    print('Model Saved!')
}
```

Here, we make use of the `saveRDS()` function, which accepts an R object and writes it to a file. In this case, we will accept a trained model object and save it inside the `/opt/ml/model` directory.

12. Define the `main()` function, as shown here. This function triggers the functions defined in the previous steps:

```
main <- function() {
    inspect_hyperparameters()
    inspect_input()
    input_data_dir = get_input_data_dir()
    print(input_data_dir)
    data <- load_training_data(input_data_dir)
    model <- train_model(data)
    save_model(model)
}
```

This `main()` function can be divided into four parts – inspecting the hyperparameters and the input, loading the training data, training the model using the `train_model()` function, and saving the model using the `save_model()` function.

13. Finally, call the `main()` function at the end of the script:

```
main()
```

> **Tip**
>
> You can access a working copy of the train file in the *Machine Learning with Amazon SageMaker Cookbook* GitHub repository: https://github.com/PacktPublishing/Machine-Learning-with-Amazon-SageMaker-Cookbook/blob/master/Chapter02/ml-r/train.

Now that we are done with the `train` script, we will use the Terminal to perform the last set of steps in this recipe. The last set of steps focus on installing a few script prerequisites.

14. Open a new Terminal:

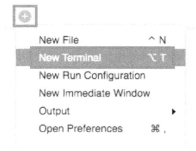

Figure 2.81 – New Terminal

Here, we can see how to create a new Terminal tab. We simply click the plus (+) button and then choose **New Terminal**.

15. Check the version of R in the Terminal:

```
R --version
```

Running this line of code should return a similar set of results to what is shown here:

```
121        save_model(model)
122    }
123
124
125    main()
126    |
```

```
bash - "ip-172-3C ×        ⊕

ubuntu:~/environment $ R --version
R version 3.4.4 (2018-03-15) -- "Someone to Lean On"
Copyright (C) 2018 The R Foundation for Statistical Computing
Platform: x86_64-pc-linux-gnu (64-bit)

R is free software and comes with ABSOLUTELY NO WARRANTY.
You are welcome to redistribute it under the terms of the
GNU General Public License versions 2 or 3.
For more information about these matters see
http://www.gnu.org/licenses/.

ubuntu:~/environment $ []
```

Figure 2.82 – Result of the R --version command in the Terminal

Here, we can see that our environment is using R version 3.4.4.

16. Install the rjson package:

```
sudo R -e "install.packages('rjson',repos='https://
cloud.r-project.org')"
```

The rjson package provides the utilities for handling JSON data in R.

17. Use the following commands to make the train script executable and then run the train script:

```
cd /home/ubuntu/environment/opt/ml-r
chmod +x train
./train
```

Running the previous lines of code will yield results similar to what is shown here:

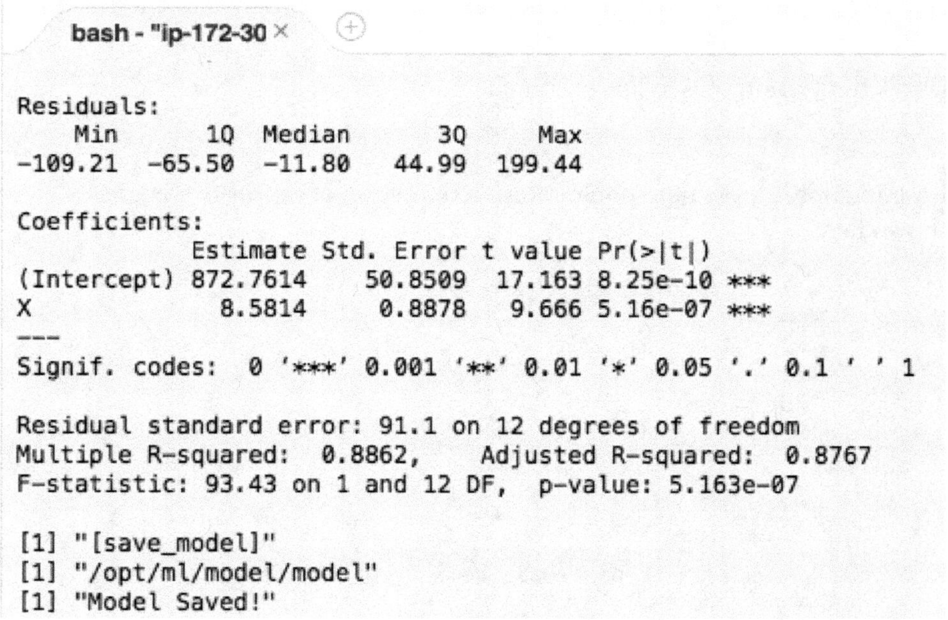

```
bash - "ip-172-30 ×     ⊕

Residuals:
    Min      1Q  Median      3Q     Max
-109.21  -65.50  -11.80   44.99  199.44

Coefficients:
            Estimate Std. Error t value Pr(>|t|)
(Intercept) 872.7614    50.8509  17.163 8.25e-10 ***
X             8.5814     0.8878   9.666 5.16e-07 ***
---
Signif. codes:  0 '***' 0.001 '**' 0.01 '*' 0.05 '.' 0.1 ' ' 1

Residual standard error: 91.1 on 12 degrees of freedom
Multiple R-squared:  0.8862,    Adjusted R-squared:  0.8767
F-statistic: 93.43 on 1 and 12 DF,  p-value: 5.163e-07

[1] "[save_model]"
[1] "/opt/ml/model/model"
[1] "Model Saved!"
```

Figure 2.83 – R train script output

Here, we can see the logs that were produced by the train script. Once the train script has been successfully executed, we expect the model files to be stored inside the /opt/ml/model directory.

At this point, we have finished preparing and testing the train script. Now, let's see how this works!

How it works...

The train script in this recipe demonstrates how the input and output values are passed around between the SageMaker API and the custom container. It also performs a fairly straightforward set of steps to train a linear model using the training data provided.

When you are required to work on a more realistic example, the train script will do the following:

- Load and use a few environment variables using the Sys.getenv() function in R. We can load environment variables set by **SageMaker** automatically, such as TRAINING_JOB_NAME and TRAINING_JOB_ARN.

- Load the contents of the hyperparameters.json file using the fromJSON() function.

- Load the contents of the inputdataconfig.json file using the fromJSON() function. This file contains the properties of each of the input data channels, such as the file type and usage of the file or pipe mode.

- Load the data file(s) inside the /opt/ml/input/data directory. Take note that there's a parent directory named after the input data channel in the path before the actual files themselves. An example of this would be /opt/ml/input/data/<channel name>/<filename>.

- Perform model training using the hyperparameters and training data that was loaded in the previous steps.

- Save the model inside the /opt/ml/model directory:

      saveRDS(model, file="/opt/ml/model/model.RDS")

- We can optionally evaluate the model using the validation data and log the results.

Now that have finished preparing the train script in R, let's quickly discuss some possible solutions we can prepare using what we learned in this recipe.

There's more...

It is important to note that we are free to use any algorithm in the train script to train our model. This level of flexibility gives us an edge once we need to work on more complex examples. Here is a quick example of what the train function may look like if the neuralnet R package is used in the train script:

```
train <- function(df.training_data, hidden_layers=4) {
    model <- neuralnet(
```

```
        label ~ .,
        df.training_data,
        hidden=c(hidden_layers,1),
        linear.output = FALSE,
        threshold=0.02,
        stepmax=1000000,
        act.fct = "logistic")
    return(model)
}
```

In this example, we allow the number of hidden layers to be set while we are configuring the Estimator object using the set_hyperparameters() function. The following example shows how to implement a train function to prepare a time series forecasting model in R:

```
train <- function(data) {
    model <- snaive(data)
    print(summary(model))
    return(model)
}
```

Here, we simply used the snaive() function from the forecast package to prepare the model. Of course, we are free to use other functions as well, such as ets() and auto.arima() from the forecast package.

Preparing and testing the serve script in R

In this recipe, we will create a serve script using R that runs an inference API using the plumber package. This API loads the model during initialization and uses the model to perform predictions during endpoint invocation.

The following diagram shows the expected behavior of the R serve script that we will prepare in this recipe. The R serve script loads the model file from the /opt/ml/ model directory and runs a plumber web server on port 8080:

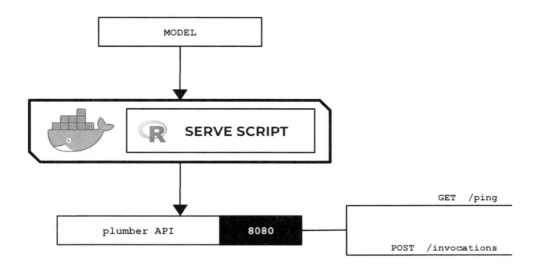

Figure 2.84 – The R serve script loads and deserializes the model and
runs a plumber API server that acts as the inference endpoint

The web server is expected to have the /ping and /invocations endpoints. This
standalone R backend API server will run inside a custom container later.

Getting ready

Make sure you have completed the *Preparing and testing the train script in R* recipe.

How to do it...

We will start by preparing the api.r file:

1. Double-click the api.r file inside the ml-r directory in the file tree:

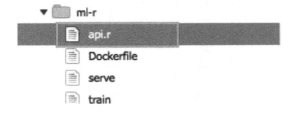

Figure 2.85 – An empty api.r file inside the ml-r directory

Here, we can see four files under the ml-r directory. Remember that we created an empty api.r file in the *Setting up the Python and R experimentation environments* recipe:

Figure 2.86 – Empty api.r file

In the next couple of steps, we will add a few lines of code inside this api.r file. Later, we will learn how to use the **plumber** package to generate an API from this api.r file.

2. Define the prepare_paths() function, which we will use to initialize the PATHS variable. This will help us manage the paths of the primary files and directories used in the script. This function allows us to initialize the PATHS variable with a dictionary-like data structure, which we can use to get the absolute paths of the required files:

```
prepare_paths <- function() {
    keys <- c('hyperparameters',
              'input',
              'data',
              'model')
    values <- c('input/config/hyperparameters.json',
                'input/config/inputdataconfig.json',
                'input/data/',
                'model/')
```

```
    paths <- as.list(values)
    names(paths) <- keys
    return(paths);
}

PATHS <- prepare_paths()
```

3. Next, define the get_path() function, which makes use of the PATHS variable from the previous step:

```
get_path <- function(key) {
    output <- paste(
        '/opt/ml/', PATHS[[key]], sep="")
    return(output);
}
```

4. Create the following function (including the comments), which responds with "OK" when triggered from the /ping endpoint:

```
#* @get /ping
function(res) {
  res$body <- "OK"
  return(res)
}
```

The line containing #* @get /ping tells plumber that we will use this function to handle the GET requests with the /ping route.

5. Define the load_model() function:

```
load_model <- function() {
  model <- NULL
  filename <- paste0(get_path('model'), 'model')
  print(filename)
  model <- readRDS(filename)
  return(model)
}
```

6. Define the following /invocations function, which loads the model and uses it to perform a prediction on the input value from the request body:

```
#* @post /invocations
function(req, res) {
  print(req$postBody)
  model <- load_model()
  payload_value <- as.double(req$postBody)
  X_test <- data.frame(payload_value)
  colnames(X_test) <- "X"

  print(summary(model))
  y_test <- predict(model, X_test)
  output <- y_test[[1]]
  print(output)

  res$body <- toString(output)
  return(res)
}
```

Here, we loaded the model using the load_model() function, transformed and prepared the input payload before passing it to the predict() function, used the predict() function to perform the actual prediction when given an X input value, and returned the predicted value in the request body.

> **Tip**
>
> You can access a working copy of the api.r file in the *Machine Learning with Amazon SageMaker Cookbook* GitHub repository: https://github.com/PacktPublishing/Machine-Learning-with-Amazon-SageMaker-Cookbook/blob/master/Chapter02/ml-r/api.r.

Now that the api.r file is ready, let's prepare the serve script:

7. Double-click the serve file inside the ml-r directory in the file tree:

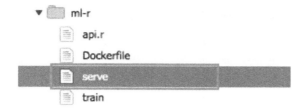

Figure 2.87 – The serve file inside the ml-r directory

It should open an empty serve file, similar to what is shown in the following screenshot:

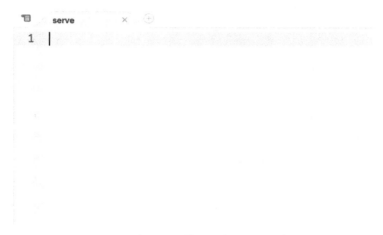

Figure 2.88 – The serve file inside the ml-r directory

We will add the necessary code to this empty serve file in the next set of steps.

8. Start the serve script with the following lines of code. Here, we are loading the plumber and here packages:

```
#!/usr/bin/Rscript
suppressWarnings(library(plumber))
library('here')
```

The here package provides utility functions to help us easily build paths to files (for example, api.r).

9. Add the following lines of code to start the `plumber` API server:

```
path <- paste0(here(), "/api.r")
pr <- plumb(path)
pr$run(host="0.0.0.0", port=8080)
```

Here, we used the `plumb()` and `run()` functions to launch the web server. It is important to note that the web server endpoint needs to run on port `8080` for this to work correctly.

> **Tip**
> You can access a working copy of the serve script in the *Machine Learning with Amazon SageMaker Cookbook* GitHub repository: `https://github.com/PacktPublishing/Machine-Learning-with-Amazon-SageMaker-Cookbook/blob/master/Chapter02/ml-r/serve`.

10. Open a new Terminal tab:

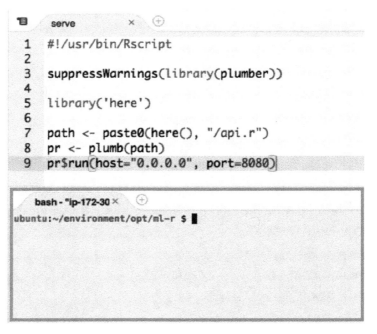

Figure 2.89 – Locating the Terminal

Here, we see that a Terminal tab is already open. If you need to create a new one, simply click the plus (+) sign and then click **New Terminal**.

11. Install `libcurl4-openssl-dev` and `libsodium-dev` using `apt-get install`. These are some of the prerequisites for installing the `plumber` package:

    ```
    sudo apt-get install -y --no-install-recommends libcurl4-
    openssl-dev
    sudo apt-get install -y --no-install-recommends
    libsodium-dev
    ```

12. Install the `here` package:

    ```
    sudo R -e "install.packages('here',repos='https://
    cloud.r-project.org')"
    ```

    The `here` package helps us get the string path values we need to locate specific files (for example, `api.r`). Feel free to check out `https://cran.r-project.org/web/packages/here/index.html` for more information.

13. Install the `plumber` package:

    ```
    sudo R -e "install.packages('plumber',repos='https://
    cloud.r-project.org')"
    ```

    The **plumber** package allows us to generate an HTTP API in R. For more information, feel free to check out `https://cran.r-project.org/web/packages/plumber/index.html`.

14. Navigate to the `ml-r` directory:

    ```
    cd /home/ubuntu/environment/opt/ml-r
    ```

15. Make the `serve` script executable using `chmod`:

    ```
    chmod +x serve
    ```

16. Run the `serve` script:

```
./serve
```

This should yield log messages similar to the following ones:

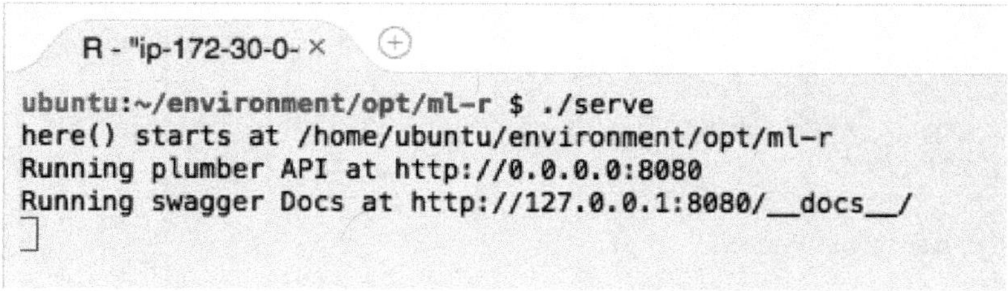

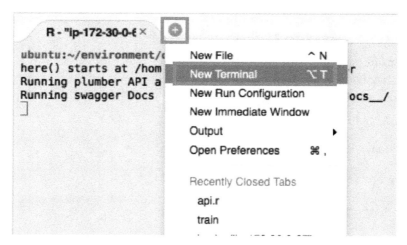

Figure 2.90 – The serve script running

Here, we can see that our serve script has successfully run a `plumber` API web server on port `8080`.

Finally, we must trigger this running web server.

17. Open a new Terminal tab:

Figure 2.91 – New Terminal

Here, we are creating a new Terminal tab as the first tab is already running the `serve` script.

18. Set the value of the `SERVE_IP` variable to `localhost`:

```
SERVE_IP=localhost
```

19. Check if the `ping` endpoint is available with `curl`:

```
curl http://$SERVE_IP:8080/ping
```

Running the previous line of code should yield an `OK` from the `/ping` endpoint.

20. Test the `invocations` endpoint with `curl`:

```
curl -d "1" -X POST http://$SERVE_IP:8080/invocations
```

We should get a value close to `881.342840085751`.

Now, let's see how this works!

How it works...

In this recipe, we prepared the `serve` script in R. The `serve` script makes use of the `plumber` package to serve an API that allows GET requests for the `/ping` route and POST requests for the `/invocations` route. The `serve` script is expected to load the model file(s) from the specified model directory and run a backend API server inside the custom container. This should provide a `/ping` route and an `/invocations` route.

Compared to its Python recipe counterpart, we are dealing with two files instead of one as that's how we used **plumber** in this recipe:

- The `api.r` file defines what the API looks like and how it behaves.
- The `serve` script uses the `api.r` file to initialize and launch a web server using the `plumb()` function from the `plumber` package. Note that with **Flask**, there is no need to create a separate file to define the API routes.

When working with the `plumber` package, we start with an R file describing how the API will behave (for example, `api.r`). This R file follows the following format:

```
#* @get /ping
function(res) {
  res$body <- "OK"
  return(<RETURN VALUE>)
}
```

```
#* @post /invocations
function(req, res) {
    return(<RETURN VALUE>)
}
```

Once this R file is ready, we simply create an R script that makes use of the `plumb()` function from the `plumber` package. This will launch a web server using the configuration and behavior coded in the `api.r` file:

```
pr <- plumb(<PATH TO API.R>)
pr$run(host="0.0.0.0", port=8080)
```

With this, whenever the `/ping` URL is accessed, the mapped function defined in the `api.r` file is executed. Similarly, whenever the `/invocations` URL is accessed with a POST request, the corresponding mapped function is executed. For more information on the `plumber` package, feel free to check out `https://www.rplumber.io/`.

Building and testing the custom R algorithm container image

In the previous two recipes, we prepared and tested the `train`, `serve`, and `api.r` files. With these ready, we can now proceed with crafting the Dockerfile and building the custom algorithm container image.

> **Tip**
>
> Wait! What's a `Dockerfile`? It is a text document containing the directives (commands) used to prepare and build a container image. This container image then serves as the blueprint when running containers. Feel free to check out `https://docs.docker.com/engine/reference/builder/` for more information.

In this recipe, we will prepare a `Dockerfile` for the custom R container image. We will make use of the `api.r` file, as well as the `train` and `serve` scripts we prepared in the *Preparing and testing the train script in R* and *Preparing and testing the serve script in R* recipes. After that, we will use the `docker build` command to prepare the image before pushing it to an **Amazon ECR** repository.

Getting ready

Make sure you have completed the *Preparing and testing the serve script in R* recipe.

How to do it...

The initial steps in this recipe focus on preparing the `Dockerfile`. Let's get started:

1. Double-click the `Dockerfile` file in the file tree to open it in the **Editor** pane. Make sure that this is the same Dockerfile that's inside the `ml-r` directory:

Figure 2.92 – Opening the Dockerfile inside the ml-r directory

Here, we can see that there's a `Dockerfile` inside the `ml-r` directory. Remember that we created an empty `Dockerfile` in the *Setting up the Python and R experimentation environments* recipe. Clicking on it in the file tree should open an empty file in the **Editor** pane:

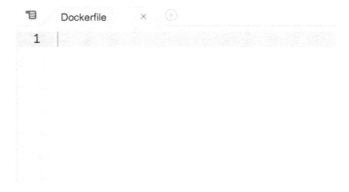

Figure 2.93 – Empty Dockerfile

Here, we have an empty `Dockerfile`. In the next step, we will update this by adding four lines of code.

2. Update the `Dockerfile` with the following block of configuration code:

```
FROM arvslat/amazon-sagemaker-cookbook-r-base:1
COPY train /usr/local/bin/train
```

```
COPY serve /usr/local/bin/serve
COPY api.r /usr/local/bin/api.r
```

Here, we are planning to build on top of an existing image called `amazon-sagemaker-cookbook-r-base`. This image already has a few prerequisites installed. These include the `rjson`, `here`, and `plumber` packages so that you don't have to worry about getting the installation steps working properly in this recipe. For more details on this image, check out `https://hub.docker.com/r/arvslat/amazon-sagemaker-cookbook-r-base`:

Figure 2.94 – Docker Hub page for the amazon-sagemaker-cookbook-r-base image

Here, we can see the **Docker Hub** page for the `amazon-sagemaker-cookbook-r-base` image.

> **Tip**
>
> You can access a working copy of this `Dockerfile` in the *Amazon SageMaker Cookbook* GitHub repository: `https://github.com/PacktPublishing/Machine-Learining-with-Amazon-SageMaker-Cookbook/blob/master/Chapter02/ml-r/Dockerfile`.

With our `Dockerfile` ready, we will proceed by using the Terminal until the end of this recipe.

3. You may use a new Terminal tab or an existing one to run the next set of commands:

Figure 2.95 – New Terminal

The preceding screenshot shows how to create a new Terminal. Note that the **Terminal** pane is right under the **Editor** pane in the AWS Cloud9 IDE.

4. Navigate to the `ml-python` directory containing our `Dockerfile`:

```
cd /home/ubuntu/environment/opt/ml-r
```

5. Specify the image name and the tag number:

```
IMAGE_NAME=chap02_r
TAG=1
```

6. Build the Docker container using the `docker build` command:

```
docker build --no-cache -t $IMAGE_NAME:$TAG .
```

The `docker build` command makes use of what is written inside our `Dockerfile`. We start with the image specified in the `FROM` directive and then proceed by copying the file files into the container image.

7. Use the `docker run` command to test if the `train` script works:

```
docker run --name rtrain --rm -v /opt/ml:/opt/ml $IMAGE_
NAME:$TAG train
```

Let's quickly discuss some of the different options that are used in this command. The - -rm flag makes Docker clean up the container after the container exits, while the -v flag allows us to mount the /opt/ml directory from the host system to the /opt/ml directory of the container:

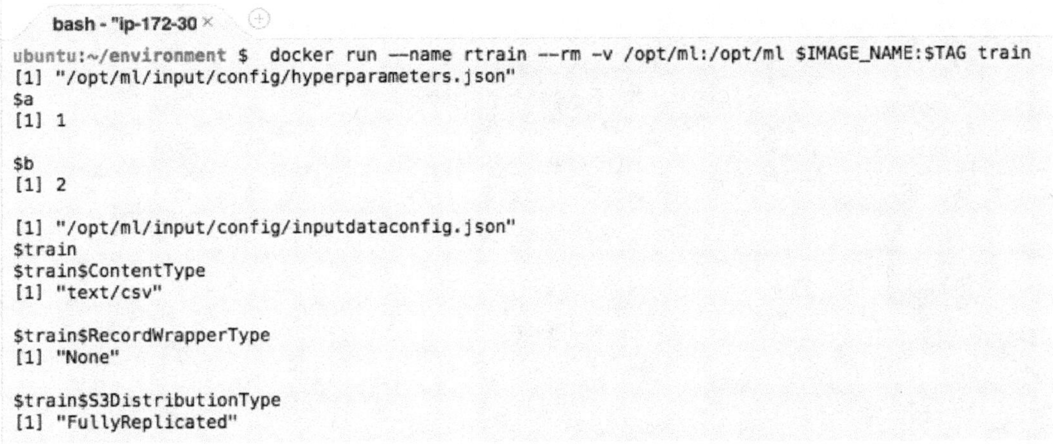

```
bash - "ip-172-30 ×        ⊕
ubuntu:~/environment $ docker run —name rtrain --rm -v /opt/ml:/opt/ml $IMAGE_NAME:$TAG train
[1] "/opt/ml/input/config/hyperparameters.json"
$a
[1] 1

$b
[1] 2

[1] "/opt/ml/input/config/inputdataconfig.json"
$train
$train$ContentType
[1] "text/csv"

$train$RecordWrapperType
[1] "None"

$train$S3DistributionType
[1] "FullyReplicated"
```

Figure 2.96 – Result of the docker run command (train)

Here, we can see the logs and results after running the docker run command.

8. Use the docker run command to test if the serve script works:

```
docker run --name rserve --rm -v /opt/ml:/opt/ml $IMAGE_
NAME:$TAG serve
```

After running this command, the plumber API server will start successfully, as shown in the following screenshot:

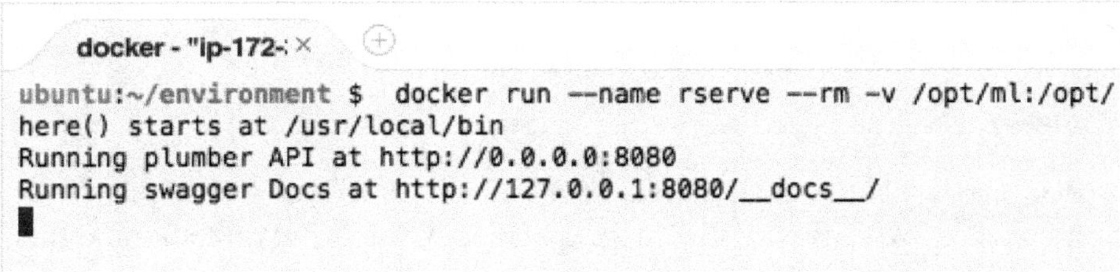

```
docker - "ip-172-; ×        ⊕
ubuntu:~/environment $  docker run —name rserve --rm -v /opt/ml:/opt/
here() starts at /usr/local/bin
Running plumber API at http://0.0.0.0:8080
Running swagger Docs at http://127.0.0.1:8080/__docs__/
```

Figure 2.97 – Result of the docker run command (serve)

Here, we can see that the API is running on port 8080. In the base image we used, we added EXPOSE 8080 to allow us to access this port in the running container.

9. Open a new Terminal tab:

Figure 2.98 – New Terminal

As the API is running already in the first Terminal, we have created a new Terminal here.

10. In the new Terminal tab, run the following command to get the IP address of the running Plumber API:

```
SERVE_IP=$(docker network inspect bridge | jq -r ".[0].
Containers[].IPv4Address" | awk -F/ '{print $1}')
echo $SERVE_IP
```

What happened here? Check out the *How it works…* section of this recipe for a detailed explanation of what happened in the previous block of code! In the meantime, let's think of this line as using multiple commands to get the IP address of the running API server. We should get an IP address equal or similar to 172.17.0.2. Of course, we may get a different IP address value altogether.

11. Next, test the ping endpoint URL using the curl command:

```
curl http://$SERVE_IP:8080/ping
```

We should get an OK after running this command.

12. Finally, test the `invocations` endpoint URL using the `curl` command:

```
curl -d "1" -X POST http://$SERVE_IP:8080/invocations
```

We should get a value similar or close to `881.342840085751` after invoking the `invocations` endpoint.

Now, let's see how this works!

How it works...

In this recipe, we built a custom container image with our `Dockerfile`. In our `Dockerfile`, we did the following:

- We used the `arvslat/amazon-sagemaker-cookbook-r-base` image as the base image. Check out `https://hub.docker.com/repository/docker/arvslat/amazon-sagemaker-cookbook-r-base` for more details on this image.

- We copied the `train`, `serve`, and `api.r` files to the `/usr/local/bin` directory inside the container image. These scripts are executed when we use `docker run`.

Using the `arvslat/amazon-sagemaker-cookbook-r-base` image as the base image allowed us to write a shorter `Dockerfile` that focuses only on copying the `train`, `serve`, and `api.r` files to the directory inside the container image. Behind the scenes, we have already pre-installed the `rjson`, `plumber`, and `here` packages, along with their prerequisites, inside this container image so that we will not run into issues when building the custom container image. Here is a quick look at the `Dockerfile` file that was used for the base image that we are using in this recipe:

```
FROM r-base:4.0.2
RUN apt-get -y update
RUN apt-get install -y --no-install-recommends wget
RUN apt-get install -y --no-install-recommends libcurl4-openssl-dev
RUN apt-get install -y --no-install-recommends libsodium-dev

RUN R -e "install.packages('rjson',repos='https://cloud.r-project.org')"
RUN R -e "install.packages('plumber',repos='https://cloud.r-project.org')"
```

```
RUN R -e "install.packages('here',repos='https://cloud.r-
project.org')"

ENV PATH "/opt/ml:$PATH"
WORKDIR /usr/local/bin
EXPOSE 8080
```

In this `Dockerfile`, we can see that we are using `r-base:4.0.2` as the base image. If we were to use a higher version, there's a chance that the `plumber` package will not install properly, which is why we had to stick with a lower version of this base image.

With these potential blockers out of the way, we were able to build a custom container image in a short amount of time. In the *Using the custom R algorithm container image for training and inference with Amazon SageMaker Local Mode* recipe of this chapter, we will use this custom container image when we do training and deployment with `reticulate` so that we can use the **SageMaker Python SDK** with our R code.

Pushing the custom R algorithm container image to an Amazon ECR repository

In the previous recipe, we prepared and built the custom container image using the `docker build` command. In this recipe, we will push the custom container image to an **Amazon ECR** repository. If this is your first time hearing about **Amazon ECR**, it is simply a fully managed container registry that helps us manage our container images. After pushing the container image to an Amazon ECR repository, we will use this image for training and deployment in the *Using the custom R algorithm container image for training and inference with Amazon SageMaker Local Mode* recipe.

Getting ready

Here are the prerequisites for this recipe:

- This recipe continues from the *Building and testing the custom R algorithm container image* recipe.

- Permission to manage the **Amazon ECR** resources if you're using an **AWS IAM** user with a custom URL.

How to do it...

The initial steps in this recipe focus on creating the ECR repository. Let's get started:

1. Use the search bar in the **AWS Console** to navigate to the **Elastic Container Registry** console. Click **Elastic Container Registry** when you see it in the search results:

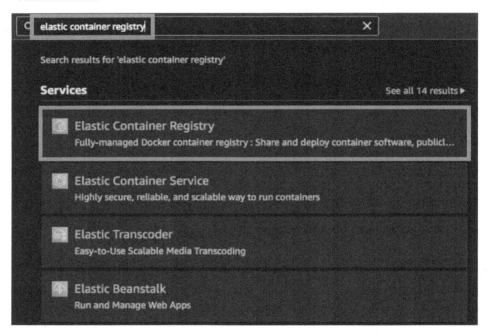

Figure 2.99 – Navigating to the ECR console

As we can see, we can use the search bar to quickly navigate to the **Elastic Container Registry** service.

2. Click the **Create repository** button:

Figure 2.100 – Create repository button

Here, the **Create repository** button is at the top right of the screen.

3. In the **Create repository** form, specify a **Repository name**. Use the value of
    $IMAGE_NAME from the *Building and testing the custom R algorithm container
    image* recipe. In this case, we will use chap02_r:

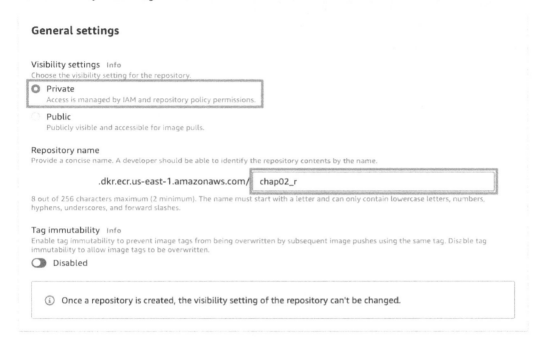

Figure 2.101 – Create repository form

Here, we have the **Create repository** form. For **Visibility settings**, we chose **Private**
and we set the **Tag immutability** configuration to **Disabled**.

4. Scroll down until you see the **Create repository** button. Leave the other configuration settings as-is and click **Create repository**:

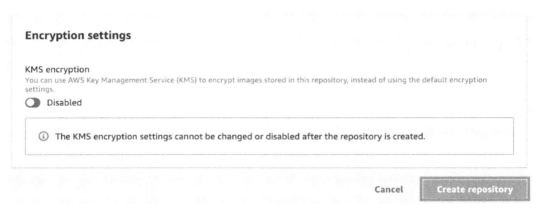

Figure 2.102 – Create repository button

Finally, to complete the repository creation process, click the **Create repository** button at the bottom of the page.

5. Click **chap02_r**:

Figure 2.103 – Link to the ECR repository page

Here, we have a link under the **Repository name** column. Clicking this link should redirect us to a page containing details about the repository.

6. Click **View push commands**:

Figure 2.104 – View push commands button (upper right)

The **View push commands** button can be found at the top right of the page.

7. You can optionally copy the first command, `aws ecr get-login-password` ..., from the dialog box:

Figure 2.105 – Push commands dialog box

Here, we can see multiple commands that we can use. We will only need the first one (`aws ecr get-login-password` ...). Click the icon with two overlapping boxes at the right-hand side of the code box to copy the entire line to the clipboard.

8. Navigate back to the AWS Cloud9 environment IDE and create a new Terminal. You can also reuse an existing one:

Figure 2.106 – New Terminal

The preceding screenshot shows us how to create a new Terminal. We click the green plus button and then select **New Terminal** from the list of options. Note that the green plus button is right under the **Editor** pane.

9. Navigate to the ml-r directory:

```
cd /home/ubuntu/environment/opt/ml-r
```

10. Get the account ID using the following commands:

```
ACCOUNT_ID=$(aws sts get-caller-identity | jq -r
".Account")
echo $ACCOUNT_ID
```

11. Specify the IMAGE_URI value and use the ECR repository name we specified while creating the repository in this recipe. In this case, we will run IMAGE_URI="chap02_r":

```
IMAGE_URI="<insert ECR Repository URI>"
TAG="1"
```

12. Authenticate with **Amazon ECR** so that we can push our **Docker** container image to an **Amazon ECR** repository in our account later:

```
aws ecr get-login-password --region us-east-1 | docker
login --username AWS --password-stdin $ACCOUNT_ID.dkr.
ecr.us-east-1.amazonaws.com
```

> **Important note**
>
> Note that we have assumed that our repository is in the us-east-1 region. Feel free to modify this region in the command if needed. This applies to all the commands in this chapter.

13. Use the `docker tag` command:

```
docker tag $IMAGE_URI:$TAG $ACCOUNT_ID.dkr.ecr.us-east-1.
amazonaws.com/$IMAGE_URI:$TAG
```

14. Push the image to the **Amazon ECR** repository using the `docker push` command:

```
docker push $ACCOUNT_ID.dkr.ecr.us-east-1.amazonaws.
com/$IMAGE_URI:$TAG
```

Now that we have completed this recipe, we can proceed with using this custom algorithm container image with SageMaker in the next recipe. But before that, let's see how this works!

How it works...

In the *Building and testing the custom R algorithm container image* recipe, we used `docker build` to prepare the custom container image. In this recipe, we created an **Amazon ECR** repository and pushed our custom container image to it. We also used the `docker push` command to push the custom container image we built to the ECR repository.

> **Important note**
>
> Don't forget to include the `api.r` file inside the container when writing this `Dockerfile` and running the build step. The Python counterpart recipe copies the `train` and `serve` scripts to the `/opt/ml` directory inside the container, while the R recipe copies the `train`, `serve`, and `api.r` files to the `/opt/ml` directory. If the `api.r` file is not included, the following line in the `serve` script file will trigger an error and cause the script to fail: `pr <- plumb("/opt/ml/api.r")`.

Using the custom R algorithm container image for training and inference with Amazon SageMaker Local Mode

In the previous recipe, we pushed the custom R container image to an Amazon ECR repository. In this recipe, we will perform the training and deployment steps in **Amazon SageMaker** using this custom container image. In the first chapter, we used the image URI of the container image of the built-in **Linear Learner** algorithm. In this chapter, we will use the image URI of the custom container image instead:

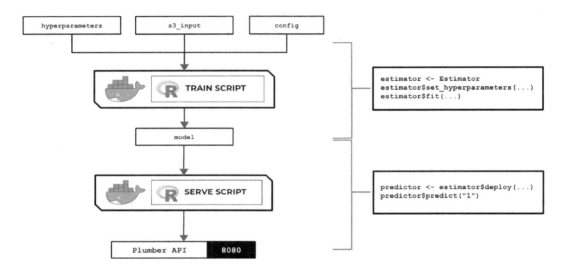

Figure 2.107 – The train and serve scripts inside the custom container make use of the hyperparameters, input data, and config specified using the SageMaker Python SDK

The preceding diagram shows how SageMaker passes data, files, and configuration to and from each custom container when we use the fit() and predict() functions in our R code, which we do with the **reticulate** package and the **SageMaker Python SDK**.

We will also look at how to use **local mode** in this recipe. This capability of SageMaker allows us to test and emulate the CPU and GPU training jobs inside our local environment. Using **local mode** is useful while we are developing, enhancing, and testing our custom algorithm container images and scripts. We can easily switch to using ML instances that support the training and deployment steps once we are ready to roll out the stable version of our container image.

Once we have completed this recipe, we will be able to run training jobs and deploy inference endpoints using R with custom `train` and `serve` scripts inside custom containers.

Getting ready

Here are the prerequisites for this recipe:

- This recipe continues from the *Pushing the custom R algorithm container image to an Amazon ECR repository* recipe.

- We will use the SageMaker Notebook instance from the *Launching an Amazon SageMaker Notebook instance and preparing the prerequisites* recipe of *Chapter 1, Getting Started with Machine Learning Using Amazon SageMaker*.

How to do it...

The first couple of steps in this recipe focus on preparing the Jupyter Notebook using the R kernel. Let's get started:

1. Inside your SageMaker Notebook instance, create a new directory called `chapter02` inside the `my-experiments` directory if it does not exist yet:

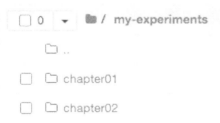

Figure 2.108 – Preferred directory structure

The preceding screenshot shows how we want to organize our files and notebooks. As we go through each chapter, we will add more directories using the same naming convention to keep things organized.

2. Click the `chapter02` directory to navigate to `/my-experiments/chapter02`.

3. Create a new notebook by clicking **New** and then clicking **R**:

Figure 2.109 – Creating a new notebook using the R kernel

The preceding screenshot shows how to create a new Jupyter Notebook using the **R** kernel.

Now that we have a fresh Jupyter Notebook using the **R** kernel, we will proceed with preparing the prerequisites for the training and deployment steps.

4. Prepare the `cmd` function, which will help us run the Bash commands in the subsequent steps:

```
cmd <- function(bash_command) {
    output <- system(bash_command, intern=TRUE)
    last_line = ""
    for (line in output) {
        cat(line)
        cat("\n")
        last_line = line
    }
    return(last_line)
}
```

Given that we are using the R kernel, we will not be able to use the `!` prefix to run Bash commands. Instead, we have created a `cmd()` function that helps us perform a similar operation. This `cmd()` function makes use of the `system()` function to invoke system commands.

5. Next, let's use the cmd function to run the pip install command to install and upgrade sagemaker[local]:

```
cmd("pip install 'sagemaker[local]' --upgrade")
```

This will allow us to use **local mode**. We can use local mode when working with framework images such as **TensorFlow**, **PyTorch**, and **MXNet** and custom container images we build ourselves.

> **Important note**
> At the time of writing, we can't use **local mode** in **SageMaker Studio**. We also can't use **local mode** with built-in algorithms.

6. Specify the values for s3.bucket and s3.prefix. Make sure that you set the s3.bucket value to the name of the S3 bucket we created in the *Preparing the Amazon S3 bucket and the training dataset for the linear regression experiment* recipe of *Chapter 1, Getting Started with Machine Learning Using Amazon SageMaker*:

```
s3.bucket <- "<insert S3 bucket name here>"
s3.prefix <- "chapter01"
```

Remember that our training_data.csv file should already exist inside the S3 bucket and that it should have the following path:

```
s3://<S3 BUCKET NAME>/<PREFIX>/input/training_data.csv
```

7. Now, let's specify the input and output locations in training.s3_input_location and training.s3_output_location, respectively:

```
training.s3_input_location <- paste0('s3://', s3.bucket,
'/', s3.prefix, '/input/training_data.csv')
training.s3_output_location <- paste0('s3://', s3.bucket,
'/', s3.prefix, '/output/custom/')
```

8. Load the reticulate package using the library() function. The reticulate package allows us to use the **SageMaker Python SDK** and other libraries in Python inside R. This gives us a more powerful arsenal of libraries in R. We can use these with other R packages such as ggplot2, dplyr, and caret:

```
library('reticulate')
sagemaker <- import('sagemaker')
```

> **Tip**
>
> For more information on the `reticulate` package, feel free to check out the *How it works...* section at the end of this recipe.

9. Check the **SageMaker Python SDK** version:

    ```
    sagemaker[['__version__']]
    ```

    We should get a value equal to or greater than `2.31.0`.

10. Set the value of the container image. Use the value from the *Pushing the custom R algorithm container image to an Amazon ECR repository* recipe. The `container` variable should be set to a value similar to `<ACCOUNT_ID>.dkr.ecr.us-east-1.amazonaws.com/chap02_r:1`. Make sure that you replace `<ACCOUNT_ID>` with your AWS account ID:

    ```
    container <- "<insert container image URI here>"
    ```

> **Tip**
>
> To get the value of `<ACCOUNT_ID>`, run `ACCOUNT_ID=$(aws sts get-caller-identity | jq -r ".Account")` and then `echo $ACCOUNT_ID` inside a Terminal. Remember that we performed this step in the *Pushing the custom R algorithm container image to an Amazon ECR repository* recipe, so you should get the same value for `ACCOUNT_ID`.

11. Import a few prerequisites, such as the role and the session. For `session`, we will use the `LocalSession` class, which will allow us to use **local mode** in the training and deployment steps:

    ```
    role <- sagemaker$get_execution_role()
    LocalSession <- sagemaker$local$LocalSession
    session <- LocalSession()
    session$config <- list(local=list(local_code=TRUE))
    ```

12. Prepare the train input so that it points to the **Amazon S3** path with `content_type="text/csv"`:

    ```
    TrainingInput <- sagemaker$inputs$TrainingInput
    sagemaker.train_input <- TrainingInput(training.s3_input_
    location, content_type="text/csv")
    ```

    Now that we have the prerequisites ready, we will proceed with initializing `Estimator` and using the `fit()` and `predict()` functions.

13. Initialize `Estimator` with the relevant arguments, as shown in the following code block. Take note that the `container` variable contains the **Amazon ECR** image URI of the custom R container image:

```
Estimator <- sagemaker$estimator$Estimator
estimator <- Estimator(
    container,
    role,
    instance_count=1L,
    instance_type="local",
    output_path=training.s3_output_location,
    sagemaker_session=session)
```

14. Set a few dummy hyperparameters using the `set_hyperparameters()` function:

```
estimator$set_hyperparameters(a=1L, b=2L, c=3L)
```

Behind the scenes, these values will be passed to the `hyperparameters.json` file inside the `/opt/ml/input/config` directory, which will be loaded and used by the `train` script when we run the `fit()` function later.

15. Perform the training step by calling the `fit()` function with the `train` argument set to the `sagemaker.train_input` variable value from the previous step:

```
estimator$fit(list(train = sagemaker.train_input))
```

Compared to our experiment in *Chapter 1, Getting Started with Machine Learning Using Amazon SageMaker*, the `fit()` function in this recipe will run the training job inside the SageMaker Notebook instance because of **local mode**. Given that we are not using local mode, we are launching ML instances that support the training jobs. As we discussed in *Chapter 1*, these ML instances are normally deleted automatically after the training jobs have been completed.

> **Important note**
>
> Even if we are using **local mode**, the model files generated by the `train` script are NOT stored inside the SageMaker notebook instance. The `model.tar.gz` file that contains the model files is still uploaded to the specified `output_path` in Amazon S3. You can check the value of `estimator$model_data` to verify this!

16. Deploy the model using the `deploy()` function. We set `instance_type` to `'local'` and `initial_instance_count` to 1L. Note that the L makes the number an explicit integer:

```
predictor <- estimator$deploy(
    initial_instance_count=1L,
    instance_type="local",
    endpoint_name="custom-local-r-endpoint"
)
```

Given that we are using **local mode**, the `deploy()` function will run the container and the `serve` script inside the SageMaker Notebook instance.

> **Important note**
>
> Note that if we change `instance_type` to a value such as "ml.m5.xlarge" (in addition to not using the `LocalSession` object), we will be launching a dedicated ML instance outside the SageMaker Notebook instance for the inference endpoint. Of course, the best practice would be to get things working first using **local mode**. Once we have ironed out the details and fixed the bugs, we can deploy the model to an inference endpoint supported by a dedicated ML instance.

17. Finally, test the `predict()` function. This triggers the `invocations` API endpoint you prepared in the previous step and passes `"1"` as the parameter value:

```
predictor$predict("1")
```

We should get a value similar or close to `881.342840085751` after invoking the `invocations` endpoint using the **predict()** function. Expect the predicted value here to be similar to what we have in the *Building and testing the custom R algorithm container image* recipe.

Now that we have a model and an inference endpoint, we can perform some post-processing, visualization, and evaluation steps using R and packages such as `ggplot2`, `dplyr`, and `Metrics`.

18. Delete the endpoint:

```
predictor$delete_endpoint()
```

Given that we are using **local mode** in this recipe, the `delete_endpoint()` function will stop the running API server in the SageMaker Notebook instance. If **local mode** is not being used, the SageMaker inference endpoint and the ML compute instance(s) that support it will be deleted.

Now, let's check out how this works!

How it works...

In this recipe, we used the `reticulate` R package to use the **SageMaker Python SDK** inside our R code. This will help us train and deploy our machine learning model. Instead of using the built-in algorithms of **Amazon SageMaker**, we used the custom container image we prepared in the previous recipes.

> **Note**
>
> Feel free to check out the *How it works...* section of the *Using the custom Python algorithm container image for training and inference with Amazon SageMaker Local Mode* recipe if you need a quick explanation on how training jobs using custom container images work.

To help us understand this recipe better, here are a few common conversions from Python to R you need to be familiar with when using `reticulate`:

- Dot (`.`) to dollar sign (`$`): `estimator.fit()` to `estimator$fit()`
- Python dictionary to R lists: `{'train': train}` to `list(train=train)`
- Integer values: `1` to `1L`
- Built-in constants: `None` to `NULL`, `True` to `TRUE`, and `False` to `FALSE`

Why spend the effort trying to perform machine learning experiments in R when you can use Python instead? There are a couple of possible reasons for this:

- Research papers and examples written by data scientists using R may use certain packages that do not have proper counterpart libraries in Python.

- Professionals and teams already familiar with the R language and using it for years should be able to get an entire ML experiment to work from end to end, without having to learn another language, especially when under time constraints. This happens a lot in real life, where teams are not easily able to shift from using one language to another due to time constraints and language familiarity.

- Migrating existing code from R to Python may not be practical or possible due to time constraints, as well as differences in the implementation of existing libraries in R and Python.

Other data scientists and ML engineers simply prefer to use R over Python. That said, it is important to be ready with solutions that allow us to use R when performing end-to-end machine learning and machine learning engineering tasks. Refer to the following diagram for a quick comparison of the tools and libraries that are used when performing end-to-end experiments in Python and R:

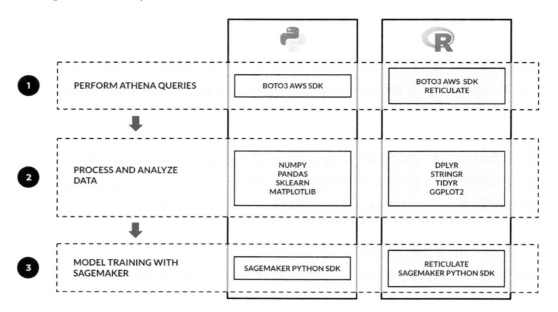

Figure 2.110 – Sample guide for tech stack selection when using Python and R
in machine learning experiments

As we can see, we can use the `reticulate` package to use the **Boto3 AWS SDK** and the **SageMaker Python SDK** inside our R code. "Note that the diagram does NOT imply a one-to-one mapping of the presented sample libraries and packages in Python and R." between "As we can see, we can use the reticulate package to use the Boto3 AWS SDK and the SageMaker Python SDK inside our R code." We used Amazon Athena in this example as it is one of the services we can use to help us prepare and query our data before the training phase. With the `reticulate` package, we can seamlessly use `boto3` to execute Athena queries in our R code.

> **Note**
>
> We will take a look at how we can use Amazon Athena with deployed models for data preparation and processing in the *Invoking machine learning models with Amazon Athena using SQL queries* recipe of *Chapter 4, Preparing, Processing, and Analyzing the Data*.

When using R, packages such as `dplyr`, `tidyr`, and `ggplot2` can easily be used with `reticulate` and the **AWS SDKs** to solve machine learning requirements from start to finish. That said, machine learning practitioners and teams already using R in their workplace may no longer need to learn another language (for example, Python) and migrate existing code from R to Python.

3
Using Machine Learning and Deep Learning Frameworks with Amazon SageMaker

Neural networks and **deep learning** are some of the hottest topics in the tech industry right now. If this is your first time hearing about an **artificial neural network**, it is simply a network of interconnected units called neurons used to solve specific machine learning problems. These neural network models have been used in solving different practical real-life problems including image classification, time-series forecasting, and even language translation. One of the properties of neural networks is the number of node layers these networks have. Generally, having more layers helps improve a model's performance up to a certain point. When a neural network has around three or more layers, we consider that artificial neural network a deep neural network. When dealing with larger datasets, deep learning models achieve better performance as these models scale effectively with data. These past couple of years, there has been widespread adoption of deep learning frameworks as more professionals understand the power of deep neural networks.

In this chapter, we will work on defining, training, and deploying our own models using several machine learning and deep learning frameworks with the **SageMaker Python SDK**. This will allow us to use any custom models we have prepared using libraries and frameworks such as **TensorFlow, Keras, scikit-learn**, and **PyTorch** and port these to **SageMaker**. This will enable us to use the **SageMaker** features and infrastructure abstraction capabilities with the custom algorithms we will prepare using the frameworks mentioned above. We will port our custom deep learning network code to **SageMaker** and use the **SageMaker Python SDK** for training, deployment, and inference. We will generate and use a synthetic training dataset during the training of custom neural networks. After the training step has been completed, the resulting model will be deployed in **local mode**.

For each of the machine learning and deep learning libraries and frameworks we have just mentioned, we will prepare the `entrypoint` script, which will be used during the training step. The `entrypoint` script is then used as an argument to the corresponding framework estimator from the **SageMaker Python SDK**. After that, the training, deployment, and inference steps proceed as usual using the `fit()`, `deploy()`, and `predict()` functions from the SDK. Note that there are a couple of subtle but important differences when using these frameworks with the **SageMaker Python SDK** in the recipes in this chapter. We will cover them in this chapter. Finally, we will work with two recipes that will help us debug and fix some of the most common errors after calling the `fit()` and `deploy()` functions.

We will cover the following recipes in this chapter:

- Preparing the **SageMaker notebook instance** for multiple deep learning local experiments
- Generating a synthetic dataset for deep learning experiments
- Preparing the `entrypoint` **TensorFlow** and Keras training script
- Training and deploying a **TensorFlow** and **Keras** model with **Amazon SageMaker local mode**
- Preparing the `entrypoint` **PyTorch** training script
- Preparing the `entrypoint` **PyTorch** inference script
- Training and deploying a **PyTorch** model with **Amazon SageMaker local mode**
- Preparing the `entrypoint` **scikit-learn** training script
- Training and deploying a **scikit-learn** model with **Amazon SageMaker local mode**
- Debugging disk space issues when using **local mode**
- Debugging container execution issues when using **local mode**

Once we have completed the recipes in this chapter, we will be able to easily work on more specific and more complex examples involving **image classification**, **time-series forecasting**, and **natural language processing** (**NLP**) requirements later on.

Let's get started!

Technical requirements

To execute the recipes in the chapter, make sure that you have the following ready:

- A running **Amazon SageMaker** notebook instance (for example, ml.t2.large)
- An Amazon S3 bucket

If you do not have these prerequisites ready yet, feel free to check the *Launching an Amazon SageMaker notebook instance* and *Preparing the Amazon S3 bucket and the training dataset for the linear regression experiment* recipes from *Chapter 1, Getting Started with Machine Learning Using Amazon SageMaker.*

As the recipes in this chapter involve a bit of code, we have made these scripts and notebooks available in this repository: https://github.com/PacktPublishing/Machine-Learning-with-Amazon-SageMaker-Cookbook/tree/master/Chapter03.

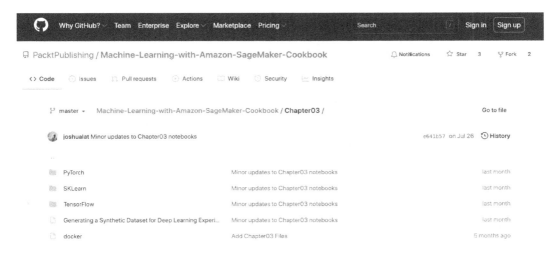

Figure 3.1 – Machine-Learning-with-Amazon-SageMaker-Cookbook GitHub repository

As seen in *Figure 3.1*, we have the source code for the recipes in this chapter organized inside the `Chapter03` directory of the *Machine-Learning-with-Amazon-SageMaker-Cookbook* GitHub repository.

Check out the following link to see the relevant Code in Action video:

`https://bit.ly/38QZnmn`

Preparing the SageMaker notebook instance for multiple deep learning local experiments

When working with deep learning experiments in **Amazon SageMaker**, it is important to note that the custom scripts developed and used to train and deploy our models can be tested inside a running deep learning container using **local mode**. This allows us to fix any issues in the custom scripts right away without having to use dedicated ML training instances. However, working with deep learning containers involves pulling container images, which may cause disk space issues. That said, it is critical that we prepare the **SageMaker notebook instance** first and configure it to prevent any disk space issues later on.

In this recipe, we will (1) modify the volume size of the notebook instance, (2) create the directories where we will store the notebooks and scripts in this chapter, and (3) configure the Docker service to help us prevent potential disk space issues when we are pulling container images and running the deep learning containers.

> **Important note**
>
> If you are wondering why we are not using **SageMaker Studio** in this chapter, it is important to note that SageMaker **local mode** is not supported in **SageMaker Studio**. Without **local mode**, it would be harder and it would take more time to debug and fix our custom deep learning scripts.

Getting ready

All we need to work on this recipe is an existing **SageMaker notebook instance**.

How to do it

The first set of steps focus on updating the settings of the SageMaker notebook instance:

1. Navigate to the **Notebook Instances** page using the sidebar.

2. Select the notebook instance we wish to modify. Under the **Actions** dropdown, click **Update settings**. If the selected notebook instance is still running, click **Stop** first and wait for about a minute or two before clicking **Update settings**. Make sure to save your progress first (if any) before stopping the notebook instance.

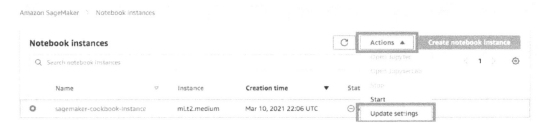

Figure 3.2 – Update settings option in the Actions dropdown

In *Figure 3.2*, we can see the **Update settings** option in the **Actions** drop-down list in the upper right-hand corner of the screen.

3.   Change the **Volume size in GB - optional** field to 200.

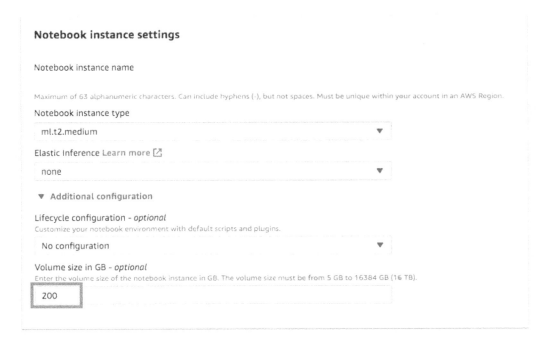

Figure 3.3 – Modifying the Notebook instance volume size

We have in *Figure 3.3* the **Notebook instance settings** form where we can modify the volume size of the notebook instance. Note that modifying the volume size configuration of the **SageMaker notebook instance** is limited to at most once every 6 hours as of writing.

4. Scroll down towards the end of the page and click **Update notebook instance**.

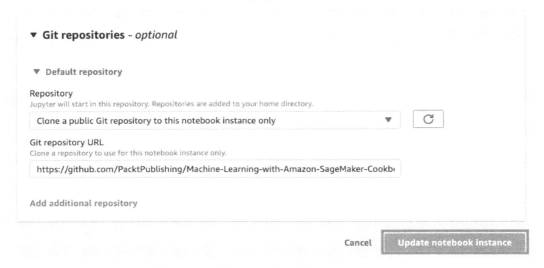

Figure 3.4 – Update notebook instance button

As we scroll down towards the end of the page, we will see the **Update notebook instance** button in the lower right-hand corner of the page.

5. Wait for about a minute or two for the notebook instance to update and then click **Start**. Once the notebook instance is in the **InService** state, click **Open Jupyter**.

6. A new tab similar to what is shown in *Figure 3.5* will open. After that, navigate to the / directory by clicking the folder icon.

Figure 3.5 – Root working directory

We can see two directories in *Figure 3.5* — the `Machine-Learning-with-Amazon-SageMaker-Cookbook` directory containing the reference scripts and notebooks of the cloned GitHub repository and the `my-experiments` directory where we will put our scripts and notebooks as we work on the recipes of this chapter.

The next set of steps focus on preparing the preferred directory structure inside the `my-experiments` directory. This will make it much easier to organize the scripts and notebooks that will be introduced in this chapter.

7. Navigate to the `/my-experiments` directory. Click **New** and then **Folder**.

Figure 3.6 – Creating a new folder inside the my-experiments directory

In *Figure 3.6*, we can see that the **Folder** option is located near the end of the drop-down list. Note that some of the other kernel options have been removed from the image to keep the image size smaller.

8. Select the **Untitled Folder** directory. After that, click the **Rename** button.

Figure 3.7 – Renaming the Untitled Folder

Renaming a file or directory should be straightforward, as seen in *Figure 3.7*.

9. A popup will appear. Rename the directory to chapter03. After that, click the **Rename** button.

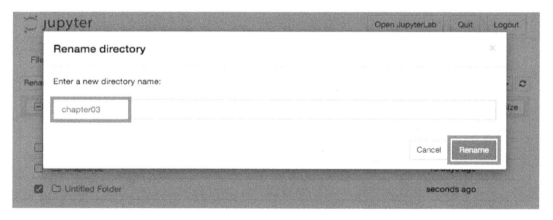

Figure 3.8 – Rename directory popup

In *Figure 3.8*, we can see the popup where we can change the name of the directory to chapter03.

10. Navigate to the /my-experiments/chapter03 directory. Click the **New** button, which opens a drop-down list of options, and then select **Folder** from the list.

Figure 3.9 – Creating a new directory inside the chapter03 directory

We can see in *Figure 3.9* that there are still no files and directories inside the chapter03 directory. Once we are done with the recipes in this book, this directory will contain (more or less) what is inside the Chapter03 directory of the *Machine-Learning-with-Amazon-SageMaker-Cookbook* GitHub repository. Feel free to check the source code inside our official GitHub repository here: https://github.com/PacktPublishing/Machine-Learning-with-Amazon-SageMaker-Cookbook/tree/master/Chapter03.

11. Using a similar set of steps, create three directories inside the /my-experiments/ chapter03 directory — PyTorch, SKLearn, and TensorFlow.

Figure 3.10 – Preferred directory structure

We have in *Figure 3.10* the preferred directory structure inside the my-experiments/chapter03 directory. In each of these directories, we will have the entrypoint script file(s) (for example, tensorflow_script.py) and the Jupyter notebook that makes use of the **SageMaker Python SDK** to train and deploy the custom neural network models. We will introduce these files later in each of the succeeding recipes in this chapter.

> **Note**
>
> What are entrypoint files? An entrypoint or entry_point file is a script that contains custom functions that get used by the deep learning container during training and deployment. When working with the deep learning framework estimator classes from the **SageMaker Python SDK**, the path to the entrypoint script file is specified as the value to the entry_point parameter when initializing the Estimator object.

12. Click the **New** button to open the drop-down options. Select **Terminal** from the list of drop-down options, as shown in *Figure 3.11*.

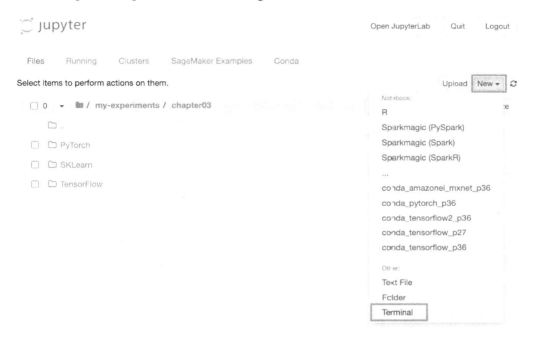

Figure 3.11 – Creating a new Terminal

A new tab where we can run Terminal commands will open.

Figure 3.12 – New Terminal tab

The succeeding steps in this recipe will be performed inside this Terminal.

The next set of steps focus on changing the data directory for the **Docker** service. Given that the current **Docker** data directory is stored inside a partition with a relatively limited storage capacity, our goal is to change this data directory to a new directory on a partition that can store more files.

13. Run the following command to create a new directory:

```
sudo mkdir -p /home/ec2-user/SageMaker/docker-stuff
```

14. Copy the contents of the /var/lib/docker directory to the new directory created:

```
sudo cp -r /var/lib/docker/. /home/ec2-user/SageMaker/
docker-stuff/ --verbose
```

We should get a similar set of logs as shown in *Figure 3.13*.

Figure 3.13 – Verbose log messages while copying the contents of the
/var/lib/docker directory to the docker-stuff directory

What's happening here? As we have used the --verbose option in the cp command, we will see a log entry for each file being copied.

15. Create a backup copy of the `docker` configuration file:

```
sudo cp /etc/sysconfig/docker /etc/sysconfig/docker.bak
```

This will allow us to revert things back to the original configuration using the backup copy if we make a mistake updating the configuration file.

16. Run the following command to open the `docker` configuration file using Vim. Vim is a text editor that can be used to edit files inside a Terminal:

```
sudo vim /etc/sysconfig/docker
```

We should see the `docker` configuration file opened inside Vim similar to what is shown in *Figure 3.14*.

⌣ Jupyter

```
# The max number of open files for the daemon itself, and all
# running containers.  The default value of 1048576 mirrors the value
# used by the systemd service unit.
DAEMON_MAXFILES=1048576

# Additional startup options for the Docker daemon, for example:
# OPTIONS="--ip-forward=true --iptables=true"
# By default we limit the number of open files per container
OPTIONS="--default-ulimit nofile=1024:4096"

# How many seconds the sysvinit script waits for the pidfile to appear
# when starting the daemon.
DAEMON_PIDFILE_TIMEOUT=10
~
~
~
```

Figure 3.14 – The docker configuration file in Vim

If this is your first time using Vim, it may be a bit intimidating at first. If you make mistakes and accidentally save an incorrectly configured `docker` configuration file, feel free to run `sudo cp /etc/sysconfig/docker.bak /etc/sysconfig/docker` to return the configuration file back to its original configuration. Restarting the notebook instance also resets the configuration settings.

17. Type `:`.

18. Type `set nu`.

19. Press the *Enter* key.

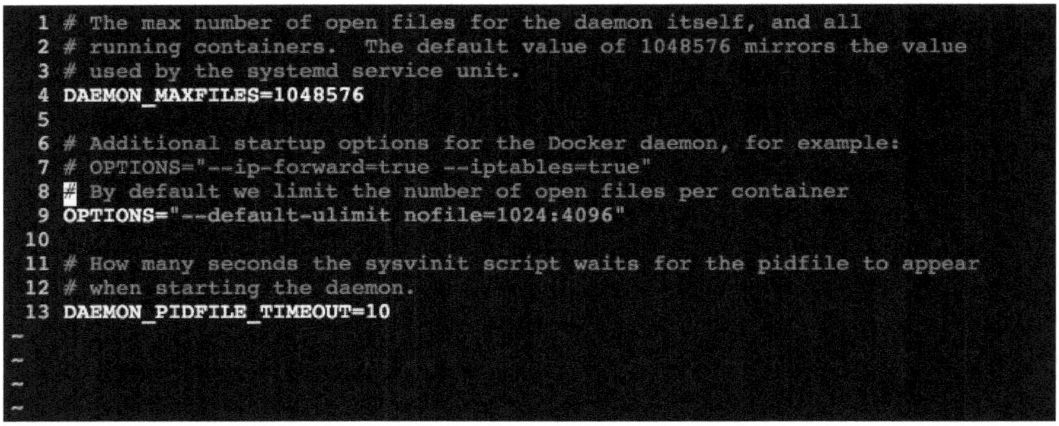

```
1 # The max number of open files for the daemon itself, and all
2 # running containers.  The default value of 1048576 mirrors the value
3 # used by the systemd service unit.
4 DAEMON_MAXFILES=1048576
5
6 # Additional startup options for the Docker daemon, for example:
7 # OPTIONS="--ip-forward=true --iptables=true"
8 # By default we limit the number of open files per container
9 OPTIONS="--default-ulimit nofile=1024:4096"
10
11 # How many seconds the sysvinit script waits for the pidfile to appear
12 # when starting the daemon.
13 DAEMON_PIDFILE_TIMEOUT=10
~
~
~
```

Figure 3.15 – Using line numbers in Vim

As seen in *Figure 3.15*, the line numbers appear on the left side of the Terminal. Typing :set nu simply sets the line numbers in command mode.

20. Use the arrow keys to put the cursor at the end of line 9, right before the double quotes. After that, press *i*.

21. Add `-g /home/ec2-user/SageMaker/docker-stuff` right after `--default-ulimit nofile=1024:4096`. After performing this step, line 9 should look like the following line of code:

```
OPTIONS="--default-ulimit nofile=1024:4096 -g /home/
ec2-user/SageMaker/docker-stuff"
```

22. Press *Esc*. Type :. Then, type wq!. Press *Enter* afterward. This will save and exit Vim in the Terminal.

> **Tip**
>
> You may also skip the previous six steps involving the use of Vim commands as long as you run: `sudo wget https://raw.githubusercontent.com/PacktPublishing/Machine-Learning-with-Amazon-SageMaker-Cookbook/master/Chapter03/docker -O /etc/sysconfig/docker`. Note that the previous block of code involves a single line of command with the `syntax sudo wget <source> <destination>` without the dot (`.`) after `docker`. What this command does is download and replace the existing `docker` configuration file with the desired configuration already available in the GitHub repository. Try opening this file to see what we mean by this: `https://raw.githubusercontent.com/PacktPublishing/Machine-Learning-with-Amazon-SageMaker-Cookbook/master/Chapter03/docker`. Note that using Vim is still the preferred approach as the `docker` configuration file in the GitHub repository may not work if there are breaking changes for later versions of **Docker**.

23. Run the following command to check whether we have updated the file successfully:

```
cat /etc/sysconfig/docker
```

We should see the `docker` configuration file similar to what is shown in *Figure 3.16*.

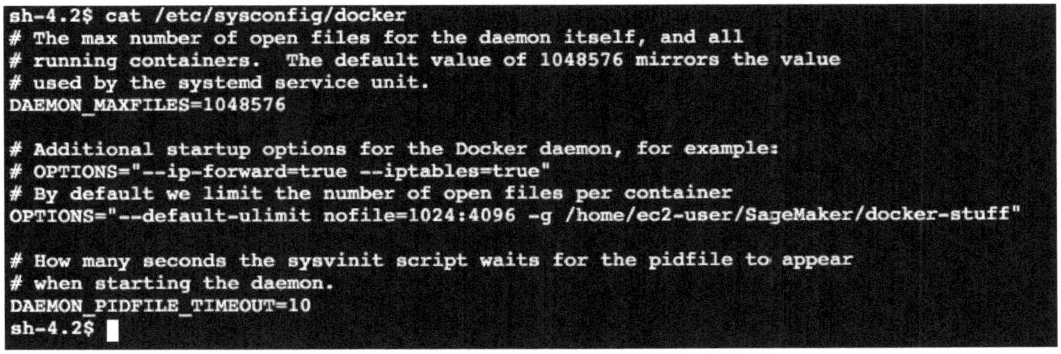

```
sh-4.2$ cat /etc/sysconfig/docker
# The max number of open files for the daemon itself, and all
# running containers.  The default value of 1048576 mirrors the value
# used by the systemd service unit.
DAEMON_MAXFILES=1048576

# Additional startup options for the Docker daemon, for example:
# OPTIONS="--ip-forward=true --iptables=true"
# By default we limit the number of open files per container
OPTIONS="--default-ulimit nofile=1024:4096 -g /home/ec2-user/SageMaker/docker-stuff"

# How many seconds the sysvinit script waits for the pidfile to appear
# when starting the daemon.
DAEMON_PIDFILE_TIMEOUT=10
sh-4.2$
```

Figure 3.16 – The updated docker configuration file

In *Figure 3.16*, we can see the desired configuration in `/etc/sysconfig/docker`.

24. Run the following command to restart the `docker` service:

```
sudo service docker restart
```

Running the preceding command would restart the Docker service, as shown in *Figure 3.17*.

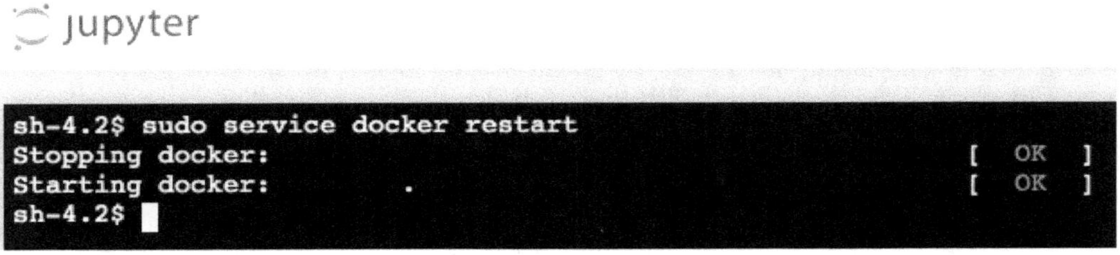

```
sh-4.2$ sudo service docker restart
Stopping docker:                                      [   OK   ]
Starting docker:                .                     [   OK   ]
sh-4.2$ 
```

Figure 3.17 – Restarting the Docker service

Now that we have restarted the service, the modifications we have made in the `/etc/sysconfig/docker` configuration file should now take effect.

> **Important note**
>
> Note that when the **SageMaker notebook instance** is restarted, the `/etc/sysconfig/docker` file reverts to how it was originally configured by AWS. Feel free to repeat the last set of steps with the `docker` configuration file if the SageMaker notebook instance has been rebooted. Another option would be to use **lifecycle configuration scripts** to automate this process, which we will discuss in the *There's more…* section in this recipe.

How it works...

This recipe is composed of three parts — volume size modification, Jupyter Notebook directory setup for the upcoming recipes in this chapter, and Docker configuration update.

The first part of this recipe involves increasing the volume size of the notebook instance. Given that we will be downloading several Docker container images in the succeeding recipes in this chapter, we will definitely need more disk space than what we originally configured in *Chapter 1, Getting Started with Machine Learning Using Amazon SageMaker*. A container image would most likely be around 1 GB or more in size. The `fit()` function downloads and makes use of the container image for training. In the same way, the `deploy()` function downloads and makes use of a container image for inference.

The second part of this recipe focuses more on the directory setup for this chapter. As we will be running different experiments involving different deep learning frameworks, it is important that the source code and files used in these experiments will not conflict with each other. This will also make things easier to manage as we will be working with different notebooks and entrypoint script files.

The last part of this recipe focuses more on updating the **Docker** configuration file. This is critical as modifying the volume size of the SageMaker notebook instance is not enough to prevent disk space issues from happening. When we use the docker pull command, the container images are pulled in the /var/lib/docker data directory. This directory is inside a volume with limited available storage space. The volume size increase modification actually affects a different volume mounted on /home/ec2-user/ SageMaker. If we were to run df -h /var/lib/docker, we would get results similar to what is shown in *Figure 3.18*.

```
sh-4.2$ df -h /var/lib/docker
Filesystem        Size   Used  Avail  Use%  Mounted on
/dev/xvda1        94G    84G   9.7G   90%   /
```

Figure 3.18 – Results of running df -h /var/lib/docker

As we have in *Figure 3.18*, we can see that we have approximately 9.7 GB left of storage in /. This means that if we perform a few experiments and use **local mode** for training and deployment, we may encounter disk space issues sooner than later as we will pull a certain number of container images that are approximately 1 GB along the way. If we run df -h /home/ec2-user/SageMaker/docker-stuff, we will get results similar to what is shown in *Figure 3.19*.

```
sh-4.2$ df -h /home/ec2-user/SageMaker/docker-stuff
Filesystem        Size   Used  Avail  Use%  Mounted on
/dev/xvdf         197G   20G   169G   11%   /home/ec2-user/SageMaker/docker-stuff
```

Figure 3.19 – Results of running df -h /home/ec2-user/SageMaker/docker-stuff

As we have in *Figure 3.19*, we can see that we have only used approximately 11% of the total size of what we can store in /home/ec2-user/SageMaker. Once we have set the /home/ec2-user/SageMaker/docker-stuff directory as the new data directory, the container images will be pulled here automatically instead. This means that we can perform multiple experiments with **SageMaker local mode** without having to worry too much about disk space issues.

There's more...

Given that the configuration changes introduced in this recipe are reset when the SageMaker notebook instance is restarted, performing these steps manually every time a SageMaker notebook instance is started is time-consuming and error-prone.

Instead of doing this manually, we can automate some of the steps in this recipe through the use of **lifecycle configuration scripts**. With a lifecycle configuration script, the commands in the specified script are executed when a **SageMaker notebook instance** is created or started.

Here is a sample script we can use to automate the steps discussed in the *How to do it...* section of this recipe:

```
sudo mkdir -p /home/ec2-user/SageMaker/docker-stuff
sudo cp -r /var/lib/docker/. /home/ec2-user/SageMaker/docker-stuff/ --verbose
sudo cp /etc/sysconfig/docker /etc/sysconfig/docker.bak
sudo wget https://raw.githubusercontent.com/PacktPublishing/Machine-Learning-with-Amazon-SageMaker-Cookbook/master/Chapter03/docker -O /etc/sysconfig/docker
sudo service docker restart
```

Where do we put these commands? We can create a lifecycle configuration by navigating to the **Lifecycle configurations** page using the sidebar (**Notebook** > **Lifecycle configurations**) and clicking the **Create configuration** button.

Amazon SageMaker > Lifecycle configurations > Create lifecycle configuration

Create lifecycle configuration

Configuration setting

Name

change-docker-configuration

Alphanumeric characters and "-", no spaces. Maximum 63 characters.

Scripts

Start notebook   **Create notebook**

This script will be run each time an associated notebook instance is started, including during initial creation If the associated notebook instance is already started, it will be run the next time it is stopped and started. See **a curated list of sample scripts**

```
1    sudo mkdir -p /home/ec2-user/SageMaker/docker-stuff
2    sudo cp -r /var/lib/docker/. /home/ec2-user/SageMaker/docker-stuff/ --verbose
3    sudo cp /etc/sysconfig/docker /etc/sysconfig/docker.bak
4    sudo wget https://raw.githubusercontent.com/PacktPublishing/Amazon-SageMaker-Cookbook/master/Chapte
5    sudo service docker restart
```

Figure 3.20 – Create lifecycle configuration form

We have in *Figure 3.20* the form where we can specify the name of the lifecycle configuration along with the script to be executed when the notebook instance is started. After a lifecycle configuration has been created, we can simply stop and modify the **SageMaker notebook instance** we are using and link the lifecycle configuration we just created to that notebook instance.

Figure 3.21 – Customizing the notebook environment with a lifecycle configuration

When the SageMaker notebook instance is started, we can check the information and error logs produced by the lifecycle configuration script in **CloudWatch Logs**.

Figure 3.22 – CloudWatch log events generated by the lifecycle configuration script

In *Figure 3.22*, we can see that the **lifecycle configuration script** has generated a set of logs similar to what we had when we executed the statements in the script manually. Feel free to play around with lifecycle configuration scripts as these will help you automate a lot of tasks and come in handy in scenarios like this. There is a lot more practical automation work we can perform through the use of lifecycle configuration scripts. For more information, refer to `https://docs.aws.amazon.com/sagemaker/latest/dg/notebook-lifecycle-config.html`.

Generating a synthetic dataset for deep learning experiments

Synthetic data generation is the process of programmatically generating artificial data with the purpose of helping data scientists and machine learning engineers test different algorithms and perform machine learning experiments without using real collected data. As we will work with neural networks and deep learning frameworks, we will need an acceptably large dataset. The dataset we have in *Chapter 1, Getting Started with Machine Learning Using Amazon SageMaker,* has only 20 records and will definitely not be a good fit for the recipes in this chapter. In this recipe, we will generate training, validation, and test dummy data using a custom synthetic data generator and store these datasets in **Amazon S3**.

> **Important note**
>
> Why generate and use synthetic datasets? Working with synthetic datasets will allow us to focus more on the tasks that we are working on as we can simply generate a bare-minimum synthetic dataset to help us demonstrate a concept or technique. Sometimes, using real datasets when working on **proof of concept** (**PoC**) code may complicate things a bit as using these may require some prerequisite knowledge on the dataset being used, especially the columns and records in that dataset. Using synthetic datasets will help us avoid any additional data cleaning and data processing steps as well so that we can go straight into using the different solutions in this book. Do not worry as we will share some notes and examples on how certain recipes and solutions in this book are used on real-life datasets along the way.

After completing this recipe, we will be able to proceed with performing several machine learning experiments involving different machine learning and deep learning frameworks using the synthetic datasets we have prepared.

Getting ready

This recipe continues from the recipe *Preparing the SageMaker notebook instance for multiple deep learning local experiments.*

How to do it

We will be running the following set of steps inside a Jupyter notebook running in an **Amazon SageMaker notebook instance**. We will start by defining the `formula()` function, which will serve as our sample synthetic data generator:

1. Navigate to the `/my-experiments/chapter03` directory in the SageMaker notebook instance. Create a new notebook using the `conda_python3` kernel by clicking on **New** and then choosing **conda_python3** from the drop-down options.

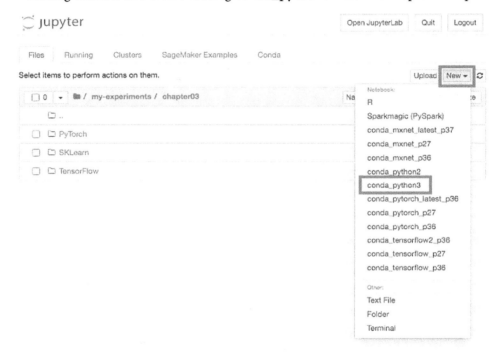

Figure 3.23 – Creating a new Jupyter notebook using the conda_python3 kernel

2. Import `numpy`:

```
import numpy as np
```

NumPy is a library with a collection of functions and utilities that help machine learning practitioners work with numerical data and arrays. In this recipe, we will make use of the `random.randint()`, `vectorize()`, and `random.normal()` functions from NumPy to generate the synthetic dataset.

3. Define the `formula()` function as shown in the following code block:

```
def formula(x):
    if x >= -2000:
        return x
    else:
        return -x - 4000
```

4. Test whether the function `formula()` works by running the following line of code:

```
formula(100)
```

The next couple of steps focus on generating values of y from x using the formula function prepared in the previous steps.

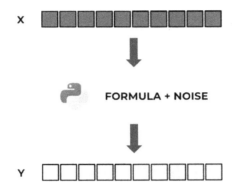

Figure 3.24 – Generating a synthetic dataset using a formula and noise functions

Figure 3.24 shows us that the y values are obtained by passing the x values as an input to the `formula()` function and adding noise to it.

5. Prepare the data generation function called `generate_synthetic_data()`. Internally, it uses the `formula()` function previously defined and adds a bit of noise using `np.random.normal()`:

```
def generate_synthetic_data(n_samples=1000,
                            start=-5000,
                            end=5000):
    np.random.seed(42)
    x = np.random.randint(
        low=start,
```

```
                high=end,
                size=(n_samples,)).astype(int)
    y = np.vectorize(formula)(x) + \
        np.random.normal(150, 150, n_samples)
    return (x, y)
```

The np.vectorize() function accepts a function and makes it accept and return numpy arrays. That said, our formula() function that accepts a single value and returns a single value would be vectorized and would then be able to operate on more values.

> **Tip**
>
> For more information about np.vectorize(), feel free to check this link: https://numpy.org/doc/stable/reference/generated/numpy.vectorize.html

6. Next, we use the generate_synthethic_data() function prepared previously and assign the array results to X and y respectively:

```
X, y = generate_synthetic_data()
```

7. Render the scatterplot using matplotlib:

```
from matplotlib import pyplot
pyplot.rcParams["figure.figsize"] = (10,8)
pyplot.scatter(X,y,s=1)
pyplot.show()
```

In *Figure 3.25*, we have a scatterplot of the synthetic dataset. Our goal over the next couple of recipes is to define and train our custom neural network and see how it fits the data we have generated in this recipe:

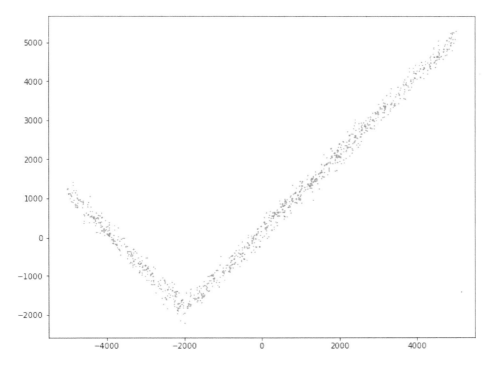

Figure 3.25 – Scatterplot of the synthetic dataset

The next two steps focus on performing the `train-test split` on the generated dataset to prepare the training, validation, and test sets.

8. Use the `train_test_split()` function from `sklearn` to split the generated dataset into training, validation, and test sets:

```
from sklearn.model_selection import train_test_split

X_train, X_test, y_train, y_test = train_test_split(
    X, y, test_size=0.2, random_state=0)

X_train, X_validation, y_train, y_validation = train_
test_split(
    X_train, y_train, test_size=0.2, random_state=0)

print(X_train.shape)
print(X_validation.shape)
print(X_test.shape)
```

This should give us the output (640,) (160,) (200,).

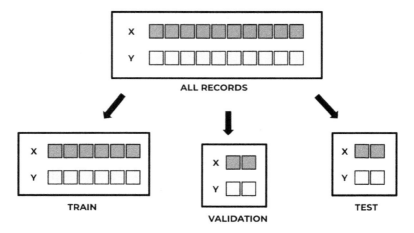

Figure 3.26 – Dividing the generated synthetic dataset into train, validation, and test datasets

Figure 3.26 shows us a quick visual of how the dataset containing all the records generated by the generate_synthethic_data() function are split into training, validation, and test sets.

9. Create the tmp directory using the mkdir command if it does not exist yet:

```
!mkdir -p tmp
```

10. Import the pandas library:

```
import pandas as pd
```

The **pandas** (panel data) library is a library used for data analysis in Python. It provides the DataFrame and Series classes, which help machine learning practitioners analyze and manipulate data.

11. Use the to_csv() function to generate a CSV file with the first column containing y values and the second column containing X values:

```
df_all_data = pd.DataFrame({ 'y': y, 'x': X})
df_all_data.to_csv('all_data.csv', header=False,
index=False)
```

Note that the `header` and `index` parameters are set to `False`, which means that the generated CSV file will only contain the y and x values separated by a comma.

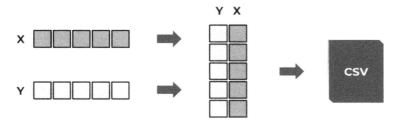

Figure 3.27 – The NumPy arrays containing the x and y values are first combined into a single DataFrame object before being saved into a CSV file with the y values in the first column.

Figure 3.27 shows us how the x and y **NumPy** arrays are combined in a single `DataFrame` object before being exported to a CSV file. Take note that the y values are in the first column. Similar to one of the earlier recipes in this book, the headers and indexes are removed and excluded from the CSV file.

12. We perform the same set of steps on the training, validation, and test datasets:

```
df_training_data = pd.DataFrame({ 'y': y_train, 'x': X_
train})
df_training_data.to_csv('training_data.csv',
    header=False, index=False)
df_validation_data = pd.DataFrame({ 'y': y_validation,
'x': X_validation})
df_validation_data.to_csv('validation_data.csv',
    header=False, index=False)
df_test_data = pd.DataFrame({ 'y': y_test, 'x': X_test})
df_test_data.to_csv('test_data.csv',
    header=False, index=False)
```

The next couple of steps focus on getting the generated CSV files uploaded to the S3 bucket. Storing these files in the S3 bucket will allow us to use them in the succeeding recipes when training and evaluating the models.

13. Set the **S3** bucket value to a bucket that exists in your AWS account. Use the bucket name created in the recipe *Preparing the Amazon S3 bucket and the training dataset for the linear regression experiment* from *Chapter 1, Getting Started with Machine Learning Using Amazon SageMaker*:

```
s3_bucket = "<insert s3 bucket name here>"
prefix = "chapter03"
```

14. We use the **AWS CLI** to save the generated data CSV file containing all the generated records to S3:

```
!aws s3 cp tmp/all_data.csv \
s3://{s3_bucket}/{prefix}/synthetic/all_data.csv
```

15. Next, we use the same approach to upload the training, validation, and test datasets to **S3**:

```
!aws s3 cp tmp/training_data.csv \
s3://{s3_bucket}/{prefix}/synthetic/training_data.csv
!aws s3 cp tmp/validation_data.csv \
s3://{s3_bucket}/{prefix}/synthetic/validation_data.csv
!aws s3 cp tmp/test_data.csv \
s3://{s3_bucket}/{prefix}/synthetic/test_data.csv
```

At this point, our CSV files should be uploaded in the S3 bucket.

Now let's see how this recipe works!

How it works...

In this recipe, we have generated sample values of x and y using a custom synthetic generator function. We have added a bit of noise using np.random.normal(). Without the noise, the synthetic data generation function will generate the same value of y for each value of x. To ensure the robustness of the model we will be training and deploying later, we have added some noise. We then performed the train and test split twice so that we will have three datasets: train, validation, and test sets. We saved the x and y values inside their corresponding CSV files and uploaded them to S3. The files uploaded to S3 will be used in the succeeding recipes.

The `formula()` function we have defined in this recipe is basically a piecewise function, which is defined by several sub-functions applied at different intervals. This piecewise function has the following parts:

- `f(x) = x IF x >= -2000`
- `f(x) = - x - 4000 OTHERWISE`

As seen in the scatterplot in the *How to do it…* section of this recipe, this dataset may not easily be solvable using a simple linear regression model. In this chapter, we will use custom neural networks using different machine learning and deep learning frameworks with **Amazon SageMaker** to train and deploy a model that generalizes and represents the data generated.

Preparing the entrypoint TensorFlow and Keras training script

TensorFlow is a popular open source software library for machine learning. **Keras**, on the other hand, is a user-friendly high-level neural network library that helps build and train models faster.

In this recipe, we will define a custom **TensorFlow** and **Keras** neural network model and prepare the `entrypoint` training script. In the next recipe, we will use the `TensorFlow` estimator class from the **SageMaker Python SDK** with this script as the `entrypoint` argument for training and deployment. If you are planning to migrate your custom **TensorFlow** and **Keras** neural network code from your local machine and perform training and deployment with the **SageMaker** platform, then this recipe (and the next) is for you!

Getting ready

This recipe continues from *Generating a synthetic dataset for deep learning experiments*.

How to do it

The instructions in this recipe focus on preparing the `entrypoint` script. Let's start by creating an empty file named `tensorflow_script.py` inside the Jupyter notebook instance and proceed with the next set of steps:

1. Navigate to the `my-experiments/chapter03/TensorFlow` directory.

Figure 3.28 – Navigating to the my-experiments/chapter03/TensorFlow directory

We can see in *Figure 3.28* the **TensorFlow** directory. Remember that we created this directory in the recipe *Preparing the SageMaker notebook instance for multiple deep learning local experiments*.

2. Create a new file – `tensorflow_script.py`. Open the file using the editor provided in Jupyter.

Figure 3.29 – Creating a new file

Let's quickly check how the `entryscript` file will look before we start with writing the code in `tensorflow_script.py`. If you are wondering why it is called the `entryscript` file, we will pass the `tensorflow_script.py` filename as the value to the parameter `entry_script` when we initialize the TensorFlow estimator object.

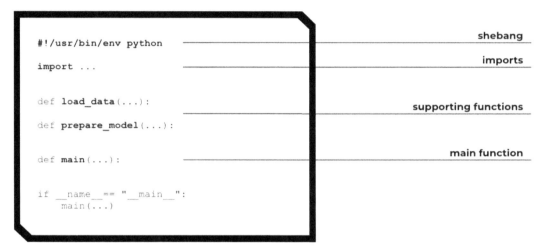

Figure 3.30 – TensorFlow entrypoint script code structure

Figure 3.30 shows us how our training script with our custom **TensorFlow** model will look after we have completed the instructions in this recipe.

3.   Start the Python script with the shebang line:

```
#!/usr/bin/env python
```

4.   Add the following lines of code to import `tensorflow`, `numpy`, and the `os` Python module:

```
import os
import tensorflow as tf
import numpy as np
```

5.   Next, add the following lines of code to import `Sequential` and `Dense`:

```
from tensorflow.keras import datasets, layers, models
from tensorflow.keras.models import Sequential
from tensorflow.keras.layers import Dense,
BatchNormalization
from numpy.random import seed
```

6. Define the `set_seed()` function, which will help us generate the same set of results every time we run the experiment:

```
def set_seed():
    seed(42)
    tf.random.set_seed(42)
```

7. Define the `load_data()` function that loads a target CSV file, extracts the x and y columns, and returns these values as the function output:

```
def load_data(training_data_location):
    result = np.loadtxt(
        open(training_data_location, "rb"),
        delimiter=",")

    y = result[:, 0]
    x = result[:, 1]
    return (x, y)
```

8. Define the `prepare_model()` function that returns the model's architecture. In this example, we have prepared a model with an arbitrary network architecture using the **Keras** `Sequential` class:

```
def prepare_model():
    model = Sequential([
        Dense(100, activation=tf.nn.leaky_relu,
                   input_shape=[1]),
        Dense(100, activation=tf.nn.leaky_relu),
        Dense(100, activation=tf.nn.leaky_relu),
        Dense(100, activation=tf.nn.leaky_relu),
        Dense(100, activation=tf.nn.leaky_relu),
        Dense(100, activation=tf.nn.leaky_relu),
        Dense(100, activation=tf.nn.leaky_relu),
        Dense(1)
    ])

    model.compile(loss='mean_squared_error',
                  optimizer='adam')
    return model
```

The `Sequential` class helps prepare a model using a linear stack of layers. The `Dense` class helps prepare a **dense layer** (a layer that accepts input from all the neurons from the previous layer in a neural network). For more information, feel free to check the **Keras** developer guide: `https://keras.io/guides/sequential_model/`.

Note that we are just using a dummy neural network and we can easily replace this with other **TensorFlow** and **Keras** models. We can even load pre-trained **TensorFlow** models and use **transfer learning** to build a new model.

> **Important note**
>
> Take note that the `prepare_model()` function returns the initial version of the model, which has not yet undergone the training step.

9. Prepare the `main()` function. This `main()` function will be called at the end of the script:

```
def main(model_dir, train_path, val_path,
         batch_size=200, epochs=2000):
    set_seed()

    model = prepare_model()
    model.summary()

    x, y = load_data(train_path)
    print("x.shape:", x.shape)
    print("y.shape:", y.shape)

    x_val, y_val = load_data(val_path)
    print("x_val.shape:", x_val.shape)
    print("y_val.shape:", y_val.shape)

    model.fit(x=x,
              y=y,
              batch_size=batch_size,
              epochs=epochs,
              validation_data=(x_val, y_val))
```

```
tf.saved_model.save(
    model,
    os.path.join(model_dir, '000000001'))
```

Figure 3.31 shows the groups of actions performed inside the main() function:

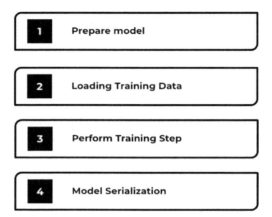

Figure 3.31 – High-level set of actions performed inside the main() function

10. The model files saved using the tf.saved_mode.save() function will be loaded inside the container using **TensorFlow Serving**. Compared to the implementation using other frameworks in this chapter, there is no need to define a model_fn() function to load the architecture and weights of the serialized model. Execute the main() function if the script is run directly and not imported:

```
if __name__ == "__main__":
    data_path = "/opt/ml/input/data"
    model_dir = "/opt/ml/model"
    train_csv = "train/training_data.csv"
    train_path = f"{data_path}/{train_csv}"
    val_csv = "validation/validation_data.csv"
    val_path = f"{data_path}/{val_csv}"

    main(model_dir=model_dir,
        train_path=train_path,
        val_path=val_path,
        batch_size=200,
        epochs=1000)
```

> **Tip**
>
> You can access a working copy of the `tensorflow_script.py` file in the official Machine Learning with Amazon SageMaker Cookbook: `https://github.com/PacktPublishing/Machine-Learning-with-Amazon-SageMaker-Cookbook/blob/master/Chapter03/TensorFlow/tensorflow_script.py`.

Now let's see how this recipe works!

How it works...

The path to the script we have prepared in this recipe will be the value to the `entry_point` parameter when initializing the `TensorFlow` estimator object from the **SageMaker Python SDK** in the next recipe. The `entrypoint` script is expected to perform the following key steps:

- Define the model (architecture)

- Load the training data from `/opt/ml/input/data` directories

- Perform model training and use the *hyperparameters* for the configuration of the training job

- Save the model (for example, parameters and weights) inside the `/opt/ml/model` directory

In this recipe, we have used an arbitrary neural network architecture using several `Dense` layers for the sake of demonstration. Feel free to modify or replace the model used in this script.

The `entrypoint` training script follows the same assumptions as the training scripts from *Chapter 2, Building and Using Your Own Algorithm Container Image,* with several environment variables, folder structure, and files (for example, training input files) already in place before the script is executed. Take note that this `entrypoint` script will be executed inside a container once we use the **SageMaker Python SDK** and will not see the same files and directories you have prepared inside the SageMaker notebook instance.

There are several approaches we can use if we need to use other packages with the training script.

The first one involves installing the additional dependencies inside the training script. Refer to the following code for the utility function, which helps us install and import the additional dependencies:

```
def install_and_load(target):
    sequence = [executable, "-m",
                "pip", "install", target]
    subprocess.call(sequence)
    return importlib.import_module(target)
```

This function makes use of the `subprocess.call()` function to execute Bash commands and install packages using the `pip install` command. We will see this in action in the next chapter.

The second approach involves the usage of a `requirements.txt` file, placing it in the same directory as the `entrypoint` training script, and specifying the `source_dir` argument when initializing the `TensorFlow` estimator from the **SageMaker Python SDK**. Take note that this works only for **TensorFlow** version *1.15.2+* using *Python 3.7.*

There's more...

We have excluded loading the environment variables from the *How to do it...* section, but here are some of the environment variables that **SageMaker** will automatically set inside the script's container training environment:

- `SM_MODEL_DIR` – Target path where the model artifacts should be saved. Set to `/opt/ml/model`.

- `SM_NUM_GPUS` – Reflects the number of GPUs available. If no GPUs are available, then this is set to `0`.

- `SM_INPUT_ DIR` – Target path to load input files and configuration data. Set to `/opt/ml/input`.

- `SM_INPUT_DATA_CONFIG` – Contains the input data configuration JSON value from `inputdataconfig.json`. This is where we can check whether the mode of an input data channel is `Pipe` or `File`.

- `SM_INPUT_CONFIG_DIR` – Target path to load configuration data. Set to `/opt/ml/input/config`.

- `SM_OUTPUT_DATA_DIR` – Target path to save output files excluding model artifacts.

- SM_CHANNEL_<channel name> – Target path to the directory containing the input data as specified in the fit() function. If we used fit({"a": s3_input(…), "b": s3_input(…)}, then we have two channels, A and B. This means that we will have the SM_CHANNEL_A and SM_CHANNEL_B environment variables containing /opt/ml/input/data/a and /opt/ml/input/data/b values.

If you want to load the values stored in these environment variables, you may use the following block of code:

```
import os
model_directory = os.environ["SM_MODEL_DIR"]
print(model_directory)
```

Now let's check the next recipe on training and deploying the **TensorFlow** entrypoint script we have prepared in this recipe.

Training and deploying a TensorFlow and Keras model with the SageMaker Python SDK

Performing the training and deployment of a custom **TensorFlow** and **Keras** model with **SageMaker** is fairly straightforward. Step 1 involves creating the entrypoint script where our custom neural network and training logic are defined and coded. Step 2 involves using this script as an argument to the TensorFlow estimator from the **SageMaker Python SDK** to proceed with the training and deployment steps.

In this recipe, we will focus on *step 2* and proceed with the training and deployment of our custom **TensorFlow** and **Keras** neural network model in **SageMaker**. If you are looking for *step 1*, feel free to check the previous recipe, *Preparing the entrypoint TensorFlow and Keras training script*.

Getting ready

This recipe continues from *Preparing the entrypoint TensorFlow and Keras training script*.

How to do it

The instructions in this recipe focus on using the custom `entrypoint` training script from the previous recipe when initializing the `TensorFlow` estimator. Once we have initialized the `TensorFlow` estimator, we will use the `fit()` and `deploy()` functions before using the `predict()` function to use our custom neural network for inference:

1. Navigate to the `/my-experiments/chapter03/TensorFlow` directory.

Figure 3.32 – Navigating to the TensorFlow directory

As we can see in *Figure 3.32*, there is a file already in this directory called `tensorflow_script.py`. We will use the path to `tensorflow_script.py` as the value to the `entry_point` parameter when initializing the `TensorFlow` estimator object in a later step.

2. Click the **New** button to open a dropdown of notebook kernel options. Select `conda_tensorflow_p36`.

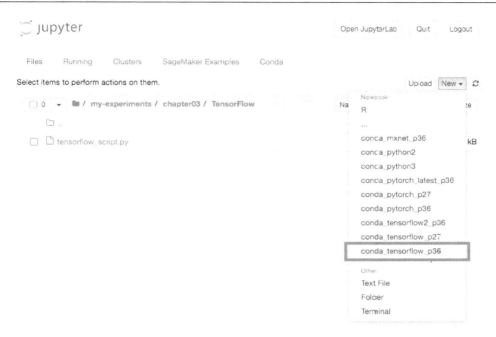

Figure 3.33 – Creating a new Jupyter notebook using the conda_tensorflow_p36 kernel

As seen in *Figure 3.33*, the new Jupyter notebook must be in the same directory as the `tensorflow_script.py` file.

3. Update the installation of **SageMaker local mode**:

```
!pip install 'sagemaker[local]' --upgrade
```

4. Restart the Docker service to make sure that we will not encounter any issues due to previously run containers:

```
!sudo service docker restart
```

5. Delete all Docker container images to free up a bit of space:

```
!docker rmi -f $(docker images -a -q)
```

With some of the preparation and installation work complete, we will proceed with the steps focused on preparing the prerequisites and arguments of the **SageMaker** training job:

6. Set the `s3_bucket` variable containing the S3 bucket path. Use the bucket created in the recipe *Preparing the Amazon S3 bucket and the training dataset for the linear regression experiment* from *Chapter 1, Getting Started with Machine Learning Using Amazon SageMaker*:

```
s3_bucket = '<insert bucket name here>'
prefix = 'chapter03'
```

Note that our `training_data.csv` file should exist already inside the S3 bucket and it should have this path:

```
s3://<S3 BUCKET NAME>/<PREFIX>/synthetic/training_data.
csv
```

The same goes for the `validation_data.csv` file.

7. Store the S3 paths of the training and validation datasets to the `train_s3` and `val_s3` variables respectively:

```
train_s3 = \
f"s3://{s3_bucket}/{prefix}/synthetic/training_data.csv"
val_s3 = \
f"s3://{s3_bucket}/{prefix}/synthetic/validation_data.
csv"
s3_output_location = \
f"s3://{s3_bucket}/{prefix}/output/tensorflow/"
```

8. Use the `TrainingInput` class to prepare the input parameters with `content_type` set to `"text/csv"`:

```
from sagemaker.inputs import TrainingInput
train_input = TrainingInput(train_s3, content_type="text/
csv")
val_input = TrainingInput(val_s3, content_type="text/
csv")
```

9. Import a few prerequisites to run the training job. In addition to this, we will use `LocalSession` to initialize the `sagemaker_session` object. The `LocalSession` class allows us to use **local mode** in the training and deployment steps. As we will see in a later step, we will specify `'local'` as the parameter value for `instance_type` when initializing the estimator to make us perform the training job locally in the SageMaker notebook instance:

```
import sagemaker
from sagemaker import get_execution_role
from sagemaker.local import LocalSession
sagemaker_session = LocalSession()
sagemaker_session.config = {'local': {'local_code':
True}}
role = get_execution_role()
```

The next couple of steps follow the standard SageMaker training and deployment steps.

10. Initialize the `TensorFlow` estimator from the **SageMaker Python SDK**. Note that we set the `instance_type` parameter value to `'local'`:

```
from sagemaker.tensorflow.estimator import TensorFlow
estimator = TensorFlow(
    entry_point='tensorflow_script.py',
    output_path=s3_output_location,
    role=role,
    sesion=sagemaker_session,
    instance_count=1,
    instance_type='local',
    framework_version='2.1.0',
    py_version='py3')
```

As you can see, we are not specifying the container image where our script file will run. The `TensorFlow` estimator makes use of one of the prebuilt Docker deep learning container images in **SageMaker**. This means that we will not have to worry about preparing the container image as **SageMaker** has already prepared this for us. If this is not sufficient, we also have the option to extend an existing **prebuilt SageMaker Docker image**.

> **Note**
>
> Another option would be for us to build and use our custom container images from scratch. Feel free to check the recipes in *Chapter 2, Building and Using Your Own Algorithm Container Image.*

11. Start the training job by running the `fit()` method:

```
estimator.fit({'train': train_input, 'validation': val_
input})
```

Figure 3.34 shows us how the training logs look a few minutes after using the `fit()` function from the `TensorFlow` estimator.

```
Creating p63wxxlqyg-algo-1-q3n6u ...
Creating p63wxxlqyg-algo-1-q3n6u ... done
Attaching to p63wxxlqyg-algo-1-q3n6u
p63wxxlqyg-algo-1-q3n6u | 2021-08-28 21:09:51,053 sagemaker-containers INFO     Imported framework sagemaker_tensorfl
ow_container.training
p63wxxlqyg-algo-1-q3n6u | 2021-08-28 21:09:51,060 sagemaker-containers INFO     No GPUs detected (normal if no gpus i
nstalled)
p63wxxlqyg-algo-1-q3n6u | 2021-08-28 21:09:52,090 sagemaker-containers INFO     No GPUs detected (normal if no gpus i
nstalled)
p63wxxlqyg-algo-1-q3n6u | 2021-08-28 21:09:52,108 sagemaker-containers INFO     No GPUs detected (normal if no gpus i
nstalled)
p63wxxlqyg-algo-1-q3n6u | 2021-08-28 21:09:52,126 sagemaker-containers INFO     No GPUs detected (normal if no gpus i
nstalled)
p63wxxlqyg-algo-1-q3n6u | 2021-08-28 21:09:52,139 sagemaker-containers INFO     Invoking user script
p63wxxlqyg-algo-1-q3n6u |
p63wxxlqyg-algo-1-q3n6u | Training Env:
p63wxxlqyg-algo-1-q3n6u |
p63wxxlqyg-algo-1-q3n6u | {
p63wxxlqyg-algo-1-q3n6u |     "additional_framework_parameters": {},
```

Figure 3.34 – Training logs after running the fit() function after initializing the TensorFlow estimator

If you encounter issues when running the `fit()` function, you may check the recipes *Debugging disk space issues when using local mode* and *Debugging container execution issues when using local mode* at the end of this chapter.

> **Note**
>
> Note that there are a few key differences when using local mode and when using dedicated ML training instances. One notable difference is that the training job name is validated (for example, pattern = `^[a-zA-Z0-9]` `(-*[a-zA-Z0-9]){0,62})` when not using **local mode**. Given that the prefix of training job names may sometimes be derived from the ECR container image repository name, a validation error may be encountered when using dedicated ML instances if the ECR repository name contains an underscore (_). This means that training experiments working in local mode may not necessarily work right away when using ML training instances due to this potential validation blocker. To resolve this, simply specify a valid `base_` `job_name` value to replace the default training job name prefix.

12. After the training job completes, deploy the model using the deploy() function:

```
predictor = estimator.deploy(
    initial_instance_count=1,
    instance_type='local'
)
```

Note that we specify 'local' as the parameter value for instance_type when using the deploy() function. When the deploy() function is called, a container optimized for **TensorFlow Serving** (a flexible and high-performance ML model serving system for production environments) is started. After that, a **TensorFlow Serving** process is configured and started to run the model, and an HTTP server is started inside the container. This HTTP server bridges the **TensorFlow Server** process with the SageMaker API, especially when the InvokeEndpoint API is called. For more information about **TensorFlow Serving**, feel free to check https://www.tensorflow.org/tfx/guide/serving.

> **Tip**
>
> If you are wondering which container image is used when running the local prediction endpoint, simply run !docker container ls to show the running containers. We should get a value equal or similar to 763104351884.dkr.ecr.us-east-1.amazonaws.com/tensorflow-inference:2.1.0-cpu. For more information, check the available images at https://github.com/aws/deep-learning-containers/blob/master/available_images.md.

13. Perform a few test predictions using the predict() function:

```
input = {
    'instances': [[100], [200]]
}
result = predictor.predict(input)
result
```

We should get a set of logs and output values similar to the one in *Figure 3.35*:

```
{'predictions': [[262.650543], [370.939178]]}
v1696a2cqz-algo-1-oxnt/ | 172.18.0.1 - - [28/Aug/2021:21:11:13 +0000] "POST /invocations HTTP/1.1" 200 56 "-" "python
-urllib3/1.26.6"
```

Figure 3.35 – Logs and output after using the predict() function in local mode

As seen in *Figure 3.35*, when calling the `predict()` function in local mode, the `/invocations` endpoint inside a running container is triggered with a `POST` request. After the request has been processed, the response is returned back to the user by the `predict()` function.

14. Make a new directory called `tmp`:

```
!mkdir -p tmp
```

15. Download the data CSV file to the `tmp` directory:

```
all_s3 = f"s3://{s3_bucket}/{prefix}/synthetic/all_data.
csv"
```

```
!aws s3 cp {all_s3} tmp/all_data.csv
```

The next couple of steps focus on checking the predicted values of **Y** for a set of **X** values that would represent the "prediction line."

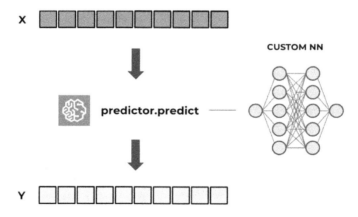

Figure 3.36 – The predict() function using the custom neural network deployed in the inference endpoint to predict the y values from the x input values

In *Figure 3.36*, we have the `predict()` function generating predicted **Y** values from **X** values using the custom **TensorFlow** and **Keras** neural network we trained and deployed to an inference endpoint.

16. Load the content of the CSV file using the `read_csv()` function from pandas to the x and y variables:

```
import pandas as pd
all_data = pd.read_csv("tmp/all_data.csv", header=None)
x = all_data[[1]].values
y = all_data[[0]].values
```

17. Prepare the `line_x` array variable that contains the x values between -5000 and 5000:

```
from numpy import arange
line_x = arange(-5000, 5000, 10)
```

The `arange()` function from **NumPy** generates an array of evenly spaced values within the specified interval (for example, between -5000 and 5000).

18. Perform predictions using the deployed model using the `predict()` function with the x array values (reshaped into a 1D array) from the previous step:

```
input = { 'instances': line_x.reshape(-1, 1) }
result = predictor.predict(input)
result
```

We should get a set of values similar to the one in *Figure 3.37*:

```
                            172.18.0.1 - - [28/Aug/2021:21:11:15 +0000] "POST /invocations HTTP/1.1" 200 14228 "-" "pyt
hon-urllib3/1.26.6"
{'predictions': [[1074.38831],
 [1068.75928],
 [1063.10559],
 [1057.45178],
 [1051.79858],
 [1046.14441],
 [1040.49072],
 [1034.83716],
 [1029.18347],
 [1023.5296],
 [1017.87592],
 [1012.12665],
 [1005.83295],
 [998.338867],
 [989.38031],
 [979.480286],
```

Figure 3.37 – Results after using the predict() function on a range of x values

In *Figure 3.37*, we can see the predicted y values inside an array. This array is nested inside a dictionary with predictions as the key and the array as the value.

19. Store the predicted values inside line_y:

```
import numpy as np
line_y = np.array(result['predictions']).flatten()
line_y
```

We should get a set of logs and output values similar to the ones in *Figure 3.38*:

```
Out[18]: array([ 1.07438831e+03,  1.06875928e+03,  1.06310559e+03,  1.05745178e+03,
                 1.05179858e+03,  1.04614441e+03,  1.04049072e+03,  1.03483716e+03,
                 1.02918347e+03,  1.02352960e+03,  1.01787592e+03,  1.01212665e+03,
                 1.00583295e+03,  9.98338867e+02,  9.89380310e+02,  9.79480286e+02,
                 9.69380066e+02,  9.59249000e+02,  9.49117126e+02,  9.38985291e+02,
                 9.28855530e+02,  9.18724487e+02,  9.08592834e+02,  8.98461731e+02,
                 8.88330383e+02,  8.78199036e+02,  8.68029541e+02,  8.57807129e+02,
                 8.47585266e+02,  8.37363953e+02,  8.27108215e+02,  8.16812744e+02,
                 8.06516907e+02,  7.96221191e+02,  7.85925110e+02,  7.75628540e+02,
                 7.65332581e+02,  7.55037537e+02,  7.44740112e+02,  7.34444885e+02,
                 7.24148621e+02,  7.13852234e+02,  7.03556152e+02,  6.93260193e+02,
                 6.82963806e+02,  6.72667358e+02,  6.62373108e+02,  6.52076782e+02,
                 6.41779968e+02,  6.31482666e+02,  6.21179999e+02,  6.10876221e+02,
                 6.00571289e+02,  5.90267151e+02,  5.79963440e+02,  5.69659363e+02,
                 5.59356079e+02,  5.49052612e+02,  5.38747864e+02,  5.28437317e+02,
                 5.18127808e+02,  5.07819946e+02,  4.97510132e+02,  4.87201935e+02,
                 4.76893066e+02,  4.66584534e+02,  4.56276520e+02,  4.45967041e+02,
                 4.35658356e+02,  4.25349579e+02,  4.15041107e+02,  4.04733246e+02,
                 3.94423828e+02,  3.84115000e+02,  3.73806244e+02,  3.63497467e+02,
```

Figure 3.38 – Array of values inside line_y after using the flatten() function

In *Figure 3.38*, we have a flattened array of values, which is stored in the line_y variable.

20. Finally, we render the line on top of the scatterplot of the original dataset using matplotlib:

```
from matplotlib import pyplot
pyplot.plot(line_x, line_y, 'r')
pyplot.scatter(x, y, s=1)
pyplot.show()
```

In *Figure 3.39*, we can see the prediction line along with the scatterplot of the dataset to have a quick view of how the predicted values match up with the actual values.

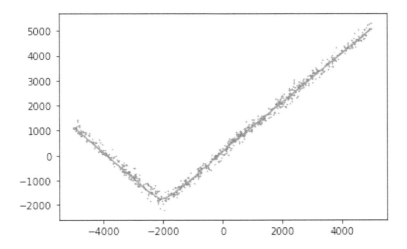

Figure 3.39 – Prediction line using the TensorFlow and Keras custom neural network over the scatterplot showing the actual values from the generated synthetic dataset.

As we can see in *Figure 3.39*, our custom `TensorFlow` model seems to do just fine in generalizing over the dataset. We can perform additional model evaluation steps to measure the model's performance properly. We will skip this to prevent this recipe from getting too long.

21. Delete the endpoint after running the experiment:

```
predictor.delete_endpoint()
```

This should stop the running container with the local inference endpoint.

Let's see how this works!

How it works...

Using the **TensorFlow** support of the **SageMaker Python SDK**, we are able to use the **TensorFlow** `entrypoint` script directly without having to build the container images and containers used for the experiment as the SDK abstracts this for us.

Let's discuss the different options and arguments when using the `TensorFlow` estimator. In our example in this recipe, we have specified the following arguments— `entry_point`, `role`, `instance_count`, `instance_type`, `framework_version`, and `py_version`. Most of these options map directly to the options available for the `Estimator` class. Using the same `TensorFlow` estimator class, we can specify the `distribution`, `input_mode`, and other options, which we will discuss shortly.

> **Tip**
>
> What if we want to use other Python packages in the `entrypoint` script? For TensorFlow version 1.15.2 with Python 3.7 or higher, and TensorFlow version 2.2 or higher, we simply create and include a `requirements.txt` file inside the same directory where we stored the `entrypoint` script. After that, we specify the path to the directory as the value to the `source_dir` parameter when initializing the `TensorFlow` estimator. For more information, feel free to check `https://sagemaker.readthedocs.io/en/stable/frameworks/tensorflow/using_tf.html#use-third-party-libraries`.

Using extended prebuilt container images: If we have extended a prebuilt framework container image and we want to use that container image instead of the prebuilt container image that the **SageMaker Python SDK** sets for us automatically, we can specify the `image_name` argument and specify the **Amazon ECR URI** similar to how we initialize an `Estimator` object. Do not forget to set `script_mode` to `True` when using this approach.

Local mode: During the training and deployment step, we have the option to perform these steps "locally" by setting the `instance_type` to `local`. By default, training and deployment involve spinning up new ML instances outside of the Jupyter notebook instance, which takes a few minutes to complete for each experiment. When using the **local mode** option, the entire process can be completed in a much shorter period of time as no instances and hardware need to be launched and provisioned behind the scenes. This allows us to test, debug, and modify the **TensorFlow** `entrypoint` script at a much faster rate.

> **Important note**
>
> Using **local mode** has its own set of limitations. This includes working only on relatively light training and deployment workloads since only workloads that can be supported by the running SageMaker notebook instance can be executed in local mode. For example, it would be hard to run a training job that involves fine-tuning a **DistilBERT** model inside an `ml.t2.medium` SageMaker notebook instance as we will need a larger notebook instance to support this training job locally.

Deploying existing models: In this recipe, we have performed the deployment step right after the training step. In cases where we have existing serialized model files inside our S3 bucket, we can perform the deployment step directly. We can use `TensorFlowModel` from the **SageMaker Python SDK** as shown in the following block of code:

```
from sagemaker.tensorflow import TensorFlowModel
model = TensorFlowModel(model_data='s3://<insert S3 bucket name + prefix>/model.tar.gz', ...)
predictor = model.deploy(...)
```

`TensrFlowModel` allows us to make use of an optional custom `entrypoint` inference script. With a custom `entrypoint` inference script, we will be able to implement and modify the default preprocessing and post-processing handlers in order to change how the input and output data is processed.

Take note that even if we performed the training step in **local mode**, the model files are still uploaded to the Amazon S3 target output location. Similar to how we did it in *Chapter 1, Getting Started with Machine Learning Using Amazon SageMaker,* we can locate and download the `model.tar.gz` file using the S3 path value in `estimator.model_data` after running the `fit()` function. Once we have downloaded the `model.tar.gz` file and extracted its contents, we can then use the **TensorFlow** and **Keras** Python APIs to load and analyze the model. For more information, check out this link, which contains more information on this topic: `https://www.tensorflow.org/tutorials/keras/save_and_load`.

There's more...

There are several additional options we can set using the `TensorFlow` estimator from the **SageMaker Python SDK**.

Distributed training: If we are planning to make use of **distributed training** and the `instance_count` is greater than 1, we can either train with parameter servers or train with a distributed training framework called **Horovod**.

When training with parameter servers, we add the following code when initializing the `TensorFlow` estimator:

```
distribution={
    "parameter_server": {"enabled": True}
}
```

Otherwise, we add the following code to the `TensorFlow` estimator initialization if we want to use **Horovod** instead:

```
distribution={
    "mpi": {
        "custom_mpi_options": "--NCCL_DEBUG INFO",
        "processes_per_host": <Insert Number Here>,
        "enabled": True
    }
}
```

Pipe mode: For larger datasets, we can make use of **Pipe mode** to speed things up further. When initializing our `TensorFlow` estimator, we can set `input_mode="Pipe"` and update the `entrypoint` training script to use `PipeModeDataset` from the `sagemaker_tensorflow` package. We will not discuss the different steps to use Pipe mode in this chapter so please check the **SageMaker Python SDK** documentation pages for more information here: `https://sagemaker.readthedocs.io/en/stable/frameworks/tensorflow/using_tf.html#training-with-pipe-mode-using-pipemodedataset`.

See also

If you are looking for examples of training and deploying **TensorFlow** models in SageMaker using real datasets, feel free to check some of the notebooks in the **aws/amazon-sagemaker-examples** GitHub repository:

- Training a classification model on the MNIST dataset: `https://github.com/aws/amazon-sagemaker-examples/blob/master/sagemaker-python-sdk/tensorflow_script_mode_training_and_serving/tensorflow_script_mode_training_and_serving.ipynb`

- Horovod distributed training: `https://github.com/aws/amazon-sagemaker-examples/blob/master/sagemaker-python-sdk/tensorflow_script_mode_horovod/tensorflow_script_mode_horovod.ipynb`

For more information on this topic, feel free to check `https://sagemaker.readthedocs.io/en/stable/frameworks/tensorflow/using_tf.html`.

Preparing the entrypoint PyTorch training script

PyTorch is a popular open source software library for machine learning. In this recipe, we will define a custom **PyTorch** neural network model and prepare the `entrypoint` training script. In one of the succeeding recipes, we will use the `PyTorch` estimator class from the **SageMaker Python SDK** with this script as the `entrypoint` argument when initializing the `estimator` object. Take note that we will have a separate `entrypoint` inference script to deploy a **PyTorch** model with **SageMaker** as we will see in the succeeding recipes.

If you are planning to migrate your custom **PyTorch** neural network code and perform training and deployment with the **SageMaker** platform, then this recipe and the next ones are for you!

Getting ready

This recipe continues from *Generating a synthetic dataset for deep learning experiments*.

How to do it

The instructions in this recipe focus on preparing the training `entrypoint` script. Let's start by creating an empty file named `pytorch_training.py` inside the Jupyter notebook instance and proceed with the next set of steps:

1. Navigate to the `/ml-experiments/chapter03/PyTorch` directory.

Figure 3.40 – Navigating to the my-experiments/chapter03/PyTorch directory

We can see in *Figure 3.40* the **PyTorch** directory. Remember that we created this directory in the recipe *Preparing the SageMaker notebook instance for multiple deep learning local experiments.*

2. Create a new file and rename it to pytorch_training.py. After that, open the empty file as we will be adding a few lines of code in the next set of steps.

    Let's quickly check how the entryscript file will look before we start with writing the code in pytorch_training.py. If you are wondering why it is called the entryscript file, we will pass the pytorch_training.py filename as the value to the parameter entry_script when we initialize the PyTorch estimator.

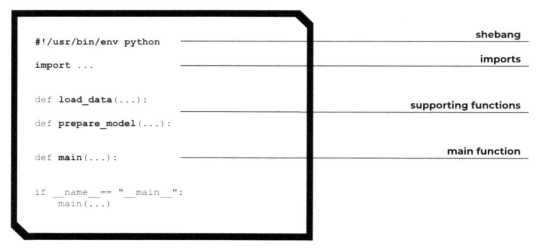

Figure 3.41 – PyTorch training entrypoint script code structure

The next couple of steps focus on preparing the code inside the entrypoint script. *Figure 3.41* shows us how our training script with our custom **PyTorch** model will look after we've completed the instructions in this recipe.

3. Start the script with the shebang line:

```
#!/usr/bin/env python
```

4. Import the prerequisites for the training script:

```
import os
import numpy as np
import torch
import torch.utils.data as Data
import random
```

5. Define the set_seed() function:

```
def set_seed():
    torch.manual_seed(0)
    random.seed(0)
    np.random.seed(0)
```

6. Prepare a function called load_data() that loads a CSV file and returns tensors for the x and y values:

```
def load_data(training_data_location):
    result = np.loadtxt(
        open(training_data_location, "rb"),
        delimiter=","
    )
    x = result[:, 1]
    xt = torch.Tensor(x.reshape(-1, 1))
    y = result[:, 0]
    yt = torch.Tensor(y.reshape(-1, 1))
    return (xt, yt)
```

7. Define a function that prepares the model:

```
def prepare_model():
    model = torch.nn.Sequential(
        torch.nn.Linear(1, 50),
        torch.nn.ReLU(),
        torch.nn.Linear(50, 50),
        torch.nn.Dropout(0.01),
        torch.nn.ReLU(),
        torch.nn.Linear(50, 50),
        torch.nn.Dropout(0.01),
        torch.nn.ReLU(),
        torch.nn.Linear(50, 50),
        torch.nn.Dropout(0.01),
        torch.nn.ReLU(),
        torch.nn.Linear(50, 50),
        torch.nn.Dropout(0.01),
        torch.nn.ReLU(),
```

```
        torch.nn.Linear(50, 50),
        torch.nn.Dropout(0.01),
        torch.nn.ReLU(),
        torch.nn.Linear(50, 50),
        torch.nn.Dropout(0.01),
        torch.nn.ReLU(),
        torch.nn.Linear(50, 50),
        torch.nn.Dropout(0.01),
        torch.nn.ReLU(),
        torch.nn.Linear(50, 1),
    )

    return model
```

Note that we are just using a dummy neural network and we can easily replace this with other **PyTorch** models. We can even load pre-trained **PyTorch** models and use **transfer learning** to build a new model.

> **Important note**
>
> Take note that the `prepare_model()` function returns the initial version of the model, which has not yet undergone the training step.

8. Define a function that accepts the x and y values and returns a data loader, which will be used in a later step. We will name this function `prepare_data_loader()`:

```
def prepare_data_loader(x, y, batch_size):
    dataset = Data.TensorDataset(x, y)
    data_loader = Data.DataLoader(
        dataset=dataset,
        batch_size=batch_size,
        shuffle=False, num_workers=2)
    return data_loader
```

9. Prepare the `train()` model function:

```
def train(model, x, y, epochs=200,
          learning_rate = 0.001, batch_size=100):
```

```python
    data_loader = prepare_data_loader(
        x=x, y=y, batch_size=batch_size
    )
    loss_fn = torch.nn.MSELoss(reduction='sum')
    optimizer = torch.optim.Adam(
        model.parameters(),
        lr=learning_rate
    )
    for e in range(epochs):
        for step, (batch_x, batch_y) in \
            enumerate(data_loader):
            prediction = model(batch_x)

            loss = loss_fn(prediction, batch_y)

            optimizer.zero_grad()
            loss.backward()
            optimizer.step()

        if (e % 10 == 0):
            print("Iteration:", e,
                    "\t| Loss:", loss.item())

    return model
```

In the inner `for` loop, the training dataset (stored in `batch_x`) is passed as the payload for the predict step with the model and then, given the prediction error calculated from the previous step, backpropagation is performed to adjust the parameters of the said model.

> **Tip**
>
> For more information on how to use **PyTorch**, feel free to check the following link: `https://pytorch.org/tutorials/beginner/blitz/neural_networks_tutorial.html`.

10. Define the `main()` function that makes use of the functions we have defined in the previous steps to prepare, train, and save the model:

```
def main(model_dir, train_path, epochs=200,
         learning_rate=0.001, batch_size=100):

    set_seed()
    model = prepare_model()

    x, y = load_data(train_path)
    print("x.shape:", x.shape)
    print("y.shape:", y.shape)

    model = train(model=model,
                  x=x,
                  y=y,
                  epochs=epochs,
                  learning_rate=learning_rate,
                  batch_size=batch_size)

    print(model)

    torch.save(model.state_dict(),
               os.path.join(model_dir, "model.pth"))
```

In the `main()` function, we perform the following set of actions: (1) prepare the model using the `prepare_model()` function, (2) load the data using the `load_data()` function, (3) train the model using the `train()` function, and finally, (4) save the model using the `torch.save()` function. Take note that the `torch.save()` function in the previous code block saves only the model parameters in the `model.pth` file.

> **Note**
>
> How does **SageMaker** know the model architecture when the model is loaded back again? During the deployment and inference steps, **SageMaker** looks for the `model_fn()` function in the inference `entrypoint` script file as we will see in the next recipe. This `model_fn()` function makes use of the same `prepare_model()` function that defines our custom model architecture before loading the parameters and weights from the `model.pth` file.

11. Finally, add the lines of code to trigger the `main()` function:

```
if __name__ == "__main__":
    model_dir = "/opt/ml/model"
    train_csv = "train/training_data.csv"
    train_path = f"/opt/ml/input/data/{train_csv}"

    main(model_dir=model_dir,
         train_path=train_path,
         epochs=1000,
         learning_rate=0.001,
         batch_size=100)
```

> **Tip**
>
> You can access a working copy of the `pytorch_training.py` file here in the *Machine Learning with Amazon SageMaker Cookbook* GitHub repository: `https://github.com/PacktPublishing/Machine-Learning-with-Amazon-SageMaker-Cookbook/blob/master/Chapter03/PyTorch/pytorch_training.py`.

Now let's see how this recipe works.

How it works...

The path to the script we have prepared in this recipe will be the input argument to the `entry_point` parameter to initialize the **PyTorch** class from the **SageMaker Python SDK** in the next recipe. When saving **PyTorch** models, you must be aware that we are just using one of the approaches in this recipe. If we are using **Elastic Inference** (**EI**), then we must do the following:

1. Convert the trained model to the **TorchScript** format.
2. Use the `torch.jit.save()` function instead of `torch.save()` when saving the model inside the `entrypoint` script.

To attach an EI accelerator to the endpoint during the deployment step, we specify the `accelerator_type` argument when using the `deploy()` function. For more information on using **Amazon Elastic Inference**, feel free to check this link from the AWS documentation site: `https://docs.aws.amazon.com/sagemaker/latest/dg/ei.html`.

> **Note**
>
> Feel free to take a look at the *How it works...* and *There's more...* section of the recipe *Preparing the entrypoint TensorFlow and Keras training script* for additional information on `entrypoint` scripts. We will not discuss it here in this section to avoid repetitive content.

Similar to the **TensorFlow** `entrypoint` training script, the **PyTorch** `entrypoint` training script can make use of the environment variables set by SageMaker. These include `SM_MODEL_DIR`, `SM_INPUT_DATA_CONFIG`, and `SM_OUTPUT_DATA_DIR`. For more information on these environment variables, refer to the *There's more...* section of the *Preparing the entrypoint TensorFlow and Keras training script* recipe.

Preparing the entrypoint PyTorch inference script

In the previous recipe, we prepared the `entrypoint` training script to train a **PyTorch** model. In this recipe, we will prepare the `entrypoint` inference script to deploy a **PyTorch** model with **SageMaker** as we will see in the next recipe.

If you are planning to migrate your custom **PyTorch** neural network code and perform training and deployment with the **SageMaker** platform, then this recipe, the previous one, and the next one are for you!

Getting ready

This recipe continues from *Preparing the entrypoint PyTorch training script.*

How to do it

The instructions in this recipe focus on preparing the inference `entrypoint` script. Let's start by creating an empty file named `pytorch_inference.py` inside the Jupyter notebook instance and proceed with the next set of steps:

1. Navigate to the `/ml-experients/chapter03/PyTorch` directory.

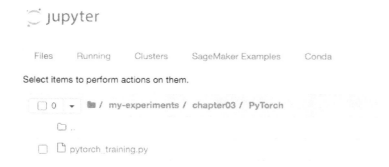

Figure 3.42 – Navigating to the my-experiments/chapter03/SKLearn directory

We can see in *Figure 3.42* the PyTorch directory. It already contains the pytorch_training.py, which we created in the recipe *Preparing the entrypoint PyTorch training script*.

2. Create a new file and name it `pytorch_inference.py`. After that, open the empty file as we will be adding a few lines of code in the next set of steps.

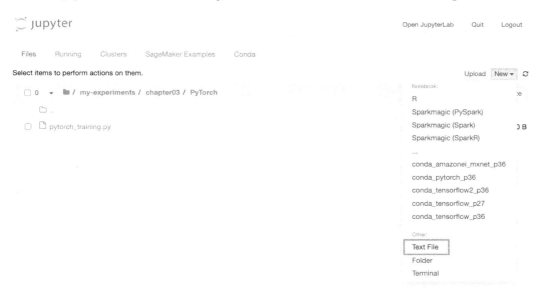

Figure 3.43 – Creating a new file

In *Figure 3.43*, we select **Text File** in the list of options under the **New** dropdown. After this step, we should have two files inside the `PyTorch` directory—`pytorch_training.py` and `pytorch_inference.py`. Make sure to open the `pytorch_inference.py` file before proceeding with the next set of steps.

3. Start the script with the shebang line:

```
#!/usr/bin/env python
```

4. Import the prerequisites:

```
import os
import torch
import numpy as np
```

5. Define the function that prepares the model. We'll call this function prepare_ model():

```
def prepare_model():
    model = torch.nn.Sequential(
        torch.nn.Linear(1, 50),
        torch.nn.ReLU(),
        torch.nn.Linear(50, 50),
        torch.nn.Dropout(0.01),
        torch.nn.ReLU(),
        torch.nn.Linear(50, 50),
        torch.nn.Dropout(0.01),
        torch.nn.ReLU(),
        torch.nn.Linear(50, 50),
        torch.nn.Dropout(0.01),
        torch.nn.ReLU(),
        torch.nn.Linear(50, 50),
        torch.nn.Dropout(0.01),
        torch.nn.ReLU(),
        torch.nn.Linear(50, 50),
        torch.nn.Dropout(0.01),
        torch.nn.ReLU(),
        torch.nn.Linear(50, 50),
        torch.nn.Dropout(0.01),
        torch.nn.ReLU(),
        torch.nn.Linear(50, 50),
        torch.nn.Dropout(0.01),
        torch.nn.ReLU(),
        torch.nn.Linear(50, 1),
```

```
    )

    return model
```

Take note that this needs to have the same model architecture as the `prepare_model()` function defined in the `entrypoint` training script in the recipe *Preparing the entrypoint PyTorch training script*.

6. Prepare the function `model_fn()` that loads and returns the model:

```
def model_fn(model_dir):
    model = prepare_model()
    path = os.path.join(model_dir, 'model.pth')
    model.load_state_dict(torch.load(path))
    model.eval()
    return model
```

Make sure to save the file and use `pytorch_inference.py` as the filename. The next and final step for this recipe is to run the script prepared.

> **Tip**
>
> You can access a working copy of the `pytorch_inference.py` file here in the *Machine Learning with Amazon SageMaker Cookbook* Github Repository: `https://github.com/PacktPublishing/Machine-Learning-with-Amazon-SageMaker-Cookbook/blob/master/Chapter03/PyTorch/pytorch_inference.py`.

Now let's see how this recipe works.

How it works...

The script we have prepared will be the input argument to the `entrypoint` parameter to initialize the `PyTorchModel` object from the **SageMaker Python SDK**, which we will see in the next recipe.

In the inference `entrypoint` script, you are expected to do the following:

-   Replace the default `model_fn()` implementation (required).
-   Replace the `input_fn()` implementation (optional) – this function accepts the request data and performs the deserialization step, which converts the data used before the prediction step.

- Replace the `output_fn()` implementation (optional) – this function accepts the prediction results and performs the serialization step, which converts the results depending on the response content type.

- Replace the `predict_fn()` implementation (optional) – this function accepts the deserialized request object after the `input_fn()` function has completed the deserialization step. The loaded model is used to perform predictions on the deserialized request object and the prediction results are returned.

Take note that the model architecture used in the training script (`pytorch_script.py`) needs to be the same as with the inference script (`pytorch_inference.py`). Since we are only saving the state in the model file(s), we need the model architecture defined via code before loading the state data using the `load_state_dict()` function. That said, this model architecture needs to be the same upon saving the model in `pytorch_script.py` and when loading the model in `pytorch_inference.py`.

Training and deploying a PyTorch model with the SageMaker Python SDK

Performing the training and deployment of a custom **PyTorch** model with **SageMaker** is fairly straightforward. Step 1 involves creating the `entrypoint` script where our custom neural network and training logic are defined and coded. Step 2 involves creating the inference `entrypoint` script, which helps us load the trained model. Step 3 involves using these scripts as arguments when initializing the `PyTorch` and `PyTorchModel` objects respectively.

In this recipe, we will focus on *step 3* and proceed with the training and deployment of our custom **PyTorch** neural network model in **SageMaker**. If you are looking for *step 1*, feel free to check the recipe *Preparing the entrypoint PyTorch training script*. If you are looking for *step 2* instead, please check the recipe *Preparing the entrypoint PyTorch inference script*.

Getting ready

This recipe continues from *Preparing the entrypoint PyTorch inference script*.

How to do it

The instructions in this recipe focus on using the custom `entrypoint` training script from the *Preparing the entrypoint PyTorch training script* recipe when initializing the `PyTorch` estimator. Once we have initialized the `PyTorch` estimator, we will use the `fit()` function and then use the `PyTorchModel` for deployment before using the `predict()` function to use our custom neural network for inference:

1. Navigate to the `/my-experiments/chapter03/PyTorch` directory. If you can still remember, we created this directory in the recipe *Preparing the SageMaker notebook instance for multiple deep learning local experiments*.

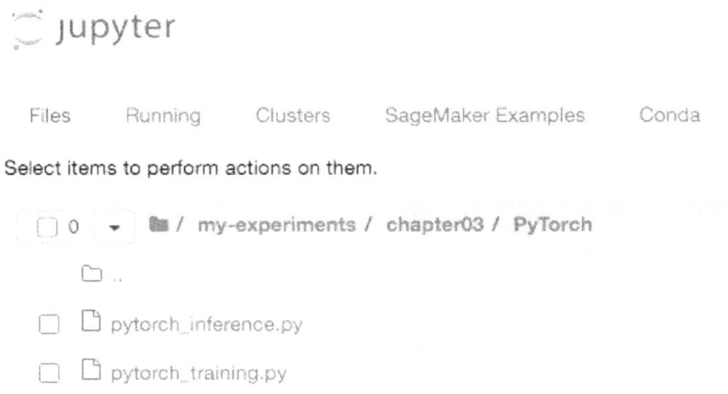

Figure 3.44 – Navigating to the PyTorch directory

As we can see in *Figure 3.44*, there are files already in this directory called `pytorch_inference.py` and `pytorch_training.py`. We will use these files as the values to the `entry_point` parameters when initializing the `PyTorch` estimator and `PyTorchModel` in a later step.

2. Click the **New** button to open a dropdown of notebook kernel options. Select `conda_pytorch_p36`.

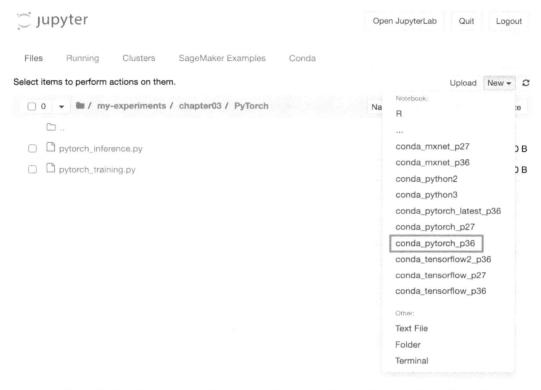

Figure 3.45 – Creating a new Jupyter notebook using the conda_pytorch_p36 kernel

As seen in *Figure 3.45*, the new Jupyter notebook must be in the same directory as the `pytorch_inference.py` and `pytorch_training.py` files.

3. Update the installation of **SageMaker local mode**:

```
!pip install 'sagemaker[local]' --upgrade
```

4. Restart the Docker service to make sure that we will not encounter any issues due to previously run containers:

```
!sudo service docker restart
```

5. Delete all Docker container images to free up a bit of space:

```
!docker rmi -f $(docker images -a -q)
```

With some of the preparation and installation work complete, we will proceed with the steps focused on preparing the prerequisites and arguments of the **SageMaker** training job:

6. Set the `s3_bucket` variable containing the S3 bucket path. Use the bucket created in the recipe *Preparing the Amazon S3 bucket and the training dataset for the linear regression experiment* from *Chapter 1, Getting Started with Machine Learning Using Amazon SageMaker*:

    ```
    s3_bucket = '<insert bucket name here>'
    prefix = 'chapter03'
    ```

    Note that our `training_data.csv` file should exist already inside the S3 bucket and it should have the path `s3://<S3 BUCKET NAME>/<PREFIX>/synthetic/training_data.csv`.

7. Set the value of the variable `train_s3` that contains the S3 path of the training data CSV file we uploaded in a previous recipe:

    ```
    train_s3 = \
    f"s3://{s3_bucket}/{prefix}/synthetic/training_data.csv"
    ```

8. Use the `TrainingInput` class to prepare the training input parameter with `content_type` set to `"text/csv"`:

    ```
    from sagemaker.inputs import TrainingInput
    train_input = TrainingInput(train_s3, content_type="text/
    csv")
    ```

9. Import a few prerequisites to run the training job. In addition to this, we will use `LocalSession` to initialize the `sagemaker_session` object. The `LocalSession` class allows us to use **local mode** in the training and deployment steps. As we will see in a later step, we will specify `'local'` as the parameter value for `instance_type` when initializing the estimator to make us perform the training job locally in the SageMaker notebook instance:

    ```
    import sagemaker
    from sagemaker import get_execution_role
    from sagemaker.local import LocalSession
    sagemaker_session = LocalSession()
    sagemaker_session.config = {'local': {'local_code':
    True}}
    role = get_execution_role()
    ```

10. Initialize the `PyTorch` estimator from the **SageMaker Python SDK**:

```
from sagemaker.pytorch import PyTorch
estimator = PyTorch(
    entry_point='pytorch_training.py',
    session=sagemaker_session,
    role=role,
    instance_count=1,
    instance_type='local',
    framework_version='1.5.0',
    py_version='py3')
```

As you can see, we are not specifying the container where our script file will run. The `PyTorch` estimator makes use of one of the prebuilt Docker container images in **SageMaker**. This means that we will not have to worry about preparing the container as **SageMaker** has already prepared this for us. If this is not sufficient, we also have the option to extend an existing **prebuilt SageMaker Docker image**.

> **Note**
>
> Another option would be for us to build and use our custom container images from scratch. Feel free to check the recipes in *Chapter 2, Building and Using Your Own Algorithm Container Image*.

11. Execute the training job using the `fit()` function:

```
estimator.fit({'train': train_input})
```

Figure 3.46 shows us how the training logs look a few minutes after using the `fit()` function from the `PyTorch` estimator:

```
Creating gx1bku3uf0-algo-1-1vsd7 ...
Creating gx1bku3uf0-algo-1-1vsd7 ... done
Attaching to gx1bku3uf0-algo-1-1vsd7
gx1bku3uf0-algo-1-1vsd7  | 2021-08-28 21:21:29,017 sagemaker-training-toolkit INFO     Imported framework sagemaker_py
torch_container.training
gx1bku3uf0-algo-1-1vsd7  | 2021-08-28 21:21:29,031 sagemaker-training-toolkit INFO     No GPUs detected (normal if no
gpus installed)
gx1bku3uf0-algo-1-1vsd7  | 2021-08-28 21:21:29,043 sagemaker_pytorch_container.training INFO     Block until all host
DNS lookups succeed.
gx1bku3uf0-algo-1-1vsd7  | 2021-08-28 21:21:29,175 sagemaker_pytorch_container.training INFO     Invoking user trainin
g script.
gx1bku3uf0-algo-1-1vsd7  | 2021-08-28 21:21:30,505 sagemaker-training-toolkit INFO     No GPUs detected (normal if no
gpus installed)
gx1bku3uf0-algo-1-1vsd7  | 2021-08-28 21:21:30,520 sagemaker-training-toolkit INFO     No GPUs detected (normal if no
gpus installed)
gx1bku3uf0-algo-1-1vsd7  | 2021-08-28 21:21:30,534 sagemaker-training-toolkit INFO     No GPUs detected (normal if no
gpus installed)
gx1bku3uf0-algo-1-1vsd7  | 2021-08-28 21:21:30,547 sagemaker-training-toolkit INFO     Invoking user script
```

Figure 3.46 – Training logs after running the fit() function after initializing the PyTorch estimator

If you encounter issues when running the `fit()` function, you may check the recipes *Debugging disk space issues when using local mode* and *Debugging container execution issues when using local mode* at the end of this chapter.

12. Initialize a `PyTorchModel` object the value of `estimator.model_data`:

```
from sagemaker.pytorch.model import PyTorchModel
pytorch_model = PyTorchModel(
    model_data=estimator.model_data,
    role=role,
    entry_point='pytorch_inference.py',
    framework_version='1.5.0',
    py_version="py3")
```

This is where **PyTorch** differs a bit from the usage of other framework estimators. As you can see in the preceding code block, we initialized a `PyTorchModel` instance using the `model_data` from the `PyTorch` estimator instance. We have also used a separate inference `entrypoint` script for the deployment and inference steps. Note that we prepared the inference `entrypoint` script in the recipe *Preparing the entrypoint PyTorch inference script*.

13. Deploy the model using the `deploy()` function:

```
predictor = pytorch_model.deploy(instance_type='local',
                                 initial_instance_
count=1)
```

We should get a set of logs and output values similar to the one in *Figure 3.47*:

```
Attaching to zdli64cbsd-algo-1-shq8w
zdli64cbsd-algo-1-shq8w  | 2021-08-28 21:24:54,182 [INFO ] main com.amazonaws.ml.mms.ModelServer -
zdli64cbsd-algo-1-shq8w  | MMS Home: /opt/conda/lib/python3.6/site-packages
zdli64cbsd-algo-1-shq8w  | Current directory: /
zdli64cbsd-algo-1-shq8w  | Temp directory: /home/model-server/tmp
zdli64cbsd-algo-1-shq8w  | Number of GPUs: 0
zdli64cbsd-algo-1-shq8w  | Number of CPUs: 2
zdli64cbsd-algo-1-shq8w  | Max heap size: 878 M
zdli64cbsd-algo-1-shq8w  | Python executable: /opt/conda/bin/python3.6
zdli64cbsd-algo-1-shq8w  | Config file: /etc/sagemaker-mms.properties
zdli64cbsd-algo-1-shq8w  | Inference address: http://0.0.0.0:8080
zdli64cbsd-algo-1-shq8w  | Management address: http://0.0.0.0:8080
zdli64cbsd-algo-1-shq8w  | Model Store: /.sagemaker/mms/models
zdli64cbsd-algo-1-shq8w  | Initial Models: ALL
zdli64cbsd-algo-1-shq8w  | Log dir: /logs
zdli64cbsd-algo-1-shq8w  | Metrics dir: /logs
```

Figure 3.47 – Logs and output after calling the deploy() function

This will run a container with the inference endpoint locally using the **PyTorch** inference container image.

> **Note**
> If you are wondering which container image is used when running the local prediction endpoint, simply run `!docker container ls` to show the running containers. We should get a value equal or similar to `763104351884.dkr.ecr.us-east-1.amazonaws.com/pytorch-inference:1.5.0-cpu-py3`. For more information, check the images available at `https://github.com/aws/deep-learning-containers/blob/master/available_images.md`.

14. Import **NumPy**. Perform a few test predictions using the `predict()` function:

```
import numpy as np
predictor.predict(np.array([[100], [200]], dtype=np.float32))
```

We should get a set of logs and output values similar to the one in *Figure 3.48*:

```
zd1i64cbsd-algo-1-shq8w | 2021-08-28 21:24:56,614 [WARN ] W-9000-model com.amazonaws.ml.mms.wlm.WorkerLifeCycle - att
achIOStreams() threadName=W-model-2
zd1i64cbsd-algo-1-shq8w | 2021-08-28 21:24:58,263 [INFO ] W-9000-model com.amazonaws.ml.mms.wlm.WorkerThread - Backen
d response time: 1647
Out[12]:  array([[227.69871521],
         [386.82015991]])
zd1i64cbsd-algo-1-shq8w | 2021-08-28 21:24:58,263 [INFO ] W-9000-model ACCESS_LOG - /172.18.0.1:56092 "POST /invocati
ons HTTP/1.1" 200 2420
```

Figure 3.48 – Logs and output after using the predict() function in local mode

As seen in *Figure 3.48*, when calling the `predict()` function in local mode, the `/invocations` endpoint inside a running container is triggered with a `POST` request. After the request has been processed, the response is returned to the user by the `predict()` function.

15. Create a `tmp` directory where we will download and store all the data for a quick visualization in a later step. Take note that this is different from the `/tmp` directory:

```
!mkdir -p tmp
```

16. Prepare a variable that contains the path where the CSV File with all the records are stored:

```
all_s3 = \
f"s3://{s3_bucket}/{prefix}/synthetic/all_data.csv"
!aws s3 cp {all_s3} tmp/all_data.csv
```

The next couple of steps focus on checking the predicted values of y for a set of x values that would represent the "prediction line."

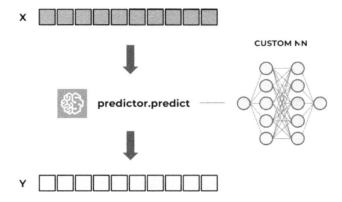

Figure 3.49 – The predict() function using the custom neural network deployed in the inference endpoint to predict the y values from the x input values

In *Figure 3.49*, we have the `predict()` function generating predicted y values from x values using the custom **PyTorch** neural network we have trained and deployed to an inference endpoint.

17. Load the CSV file containing all records and store the x and y values as arrays inside the x and y variables respectively:

```
import pandas as pd
all_data = pd.read_csv("tmp/all_data.csv", header=None)
x = all_data[[1]].values
y = all_data[[0]].values
```

18. Using the `arange()` function, generate an array with values between `-5000` and `5000` and store these values inside `line_x`:

```
from numpy import arange
line_x = arange(-5000, 5000, 10)
```

19. Execute the `predict()` function using the deployed model with the input values from the previous step:

```
input_data = np.array(line_x.reshape(-1, 1), dtype=np.
float32)
result = predictor.predict(input_data)
result
```

We should get a set of logs and output values similar to the one in *Figure 3.50*:

```
zd:i64cbsd-algo-i-shq8w | 2021-08-28 21:25:00,056 [INFO ] W-9000-model com.amazonaws.ml.mms.wlm.WorkerThread - Backen
d response time: 5
zd:i64cbsd-algo-i-shq8w | 2021-08-28 21:25:00,056 [INFO ] W-9000-model ACCESS_LOG - /172.18.0.1:56092 "POST /invocati
ons HTTP/1.1" 200 8
```

```
array([[ 1.02461316e+03],
       [ 1.02212274e+03],
       [ 1.01962878e+03],
       [ 1.01677557e+03],
       [ 1.01345319e+03],
       [ 1.00987567e+03],
       [ 1.00624750e+03],
       [ 1.00225098e+03],
       [ 9.97802979e+02],
       [ 9.93195801e+02],
       [ 9.87457764e+02],
       [ 9.81670532e+02],
       [ 9.75475891e+02],
       [ 9.68850159e+02],
```

Figure 3.50 – Results after using the predict() function on an array of values

In *Figure 3.50*, we can see the predicted y values after passing an array of x values.

20. Store the results in a variable named `line_y`:

```
line_y = result
```

21. Render the line on top of the scatterplot of the original dataset using **Matplotlib**:

```
from matplotlib import pyplot
pyplot.plot(line_x, line_y, 'r')
pyplot.scatter(x,y,s=1)
pyplot.show()
```

This should render a chart similar to what is shown in *Figure 3.51*.

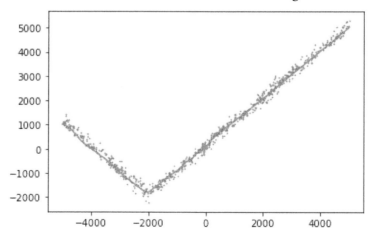

Figure 3.51 – Prediction line over the scatterplot of points

As we can see in *Figure 3.51*, our custom **PyTorch** model seems to do just fine in generalizing over the dataset. Of course, we would need a more comprehensive set of steps for model evaluation, but we will skip this in this recipe.

22. Stop the running container with the inference endpoint:

```
predictor.delete_endpoint()
```

This will stop the running container with the local inference endpoint.

Let's see how this works!

How it works...

Using the **PyTorch** support of the **SageMaker Python SDK**, we are able to use the **PyTorch** entrypoint script directly without having to build and manage the containers used for the experiment as the SDK abstracts this for us. In this recipe, we initialized and used PyTorchModel from the **SageMaker Python SDK** to get this recipe working. This is somewhat different from other frameworks where we can directly perform the deployment using the deploy() function of the estimator. Here are some of the arguments we can specify when using PyTorchModel:

- model_data
- role
- entry_point (expects the inference entrypoint script)
- framework_version
- py_version
- image_uri
- model_server_workers

After the PyTorchModel object has been initialized, the PyTorchModel object has the deploy() function that accepts the same set of arguments as the deploy() functions from other estimators.

> **Note**
>
> For more information, feel free to check the *How it works...* section of the recipe *Training and deploying a TensorFlow and Keras model with Amazon SageMaker local mode*. We will not discuss some of the details in this section to avoid repetitive content.

See also

If you are looking for examples of training and deploying **PyTorch** models in SageMaker using real datasets, feel free to check some of the notebooks in the **aws/amazon-sagemaker-examples** GitHub repository:

- Deploying pre-trained **PyTorch** models: `https://github.com/aws/amazon-sagemaker-examples/blob/master/sagemaker_neo_compilation_jobs/pytorch_torchvision/pytorch_torchvision_neo.ipynb`

- Using **SageMaker Debugger** when training **PyTorch** models: `https://github.com/aws/amazon-sagemaker-examples/blob/master/sagemaker-debugger/pytorch_custom_container/pytorch_byoc_smdebug.ipynb`

Note that we will deal with **SageMaker Debugger** in the recipes *Identifying issues with SageMaker Debugger* and *Inspecting SageMaker Debugger logs and results* from *Chapter 5, Effectively Managing Machine Learning Experiments.*

Preparing the entrypoint scikit-learn training script

Scikit-learn is a popular open source software library for machine learning. In this recipe, we will define a custom **scikit-learn** neural network model and prepare the `entrypoint` training script. In the next recipe, we will use the `SKLearn` estimator from the **SageMaker Python SDK** with this script as the `entrypoint` argument for training and deployment. If you are planning to migrate your custom **scikit-learn** neural network code and perform training and deployment with the **SageMaker** platform, then this recipe (and the next) is for you!

Getting ready

This recipe continues from *Generating a synthetic dataset for deep learning experiments.*

How to do it

The instructions in this recipe focus on preparing the `entrypoint` script. Let's start by creating an empty file named `sklearn_script.py` inside the Jupyter notebook instance and then proceed with the next set of steps:

1. Navigate to the `/ml-experiments/chapter03/SKLearn` directory.

Figure 3.52 – Navigating to the my-experiments/chapter03/SKLearn directory

We can see in *Figure 3.52* the SKLearn directory. Remember that we created this directory in the recipe *Preparing the SageMaker notebook instance for multiple deep learning local experiments*.

2. Create a new file and name it sklearn_script.py. After that, open the empty file as we will be adding a few lines of code in the next set of steps.

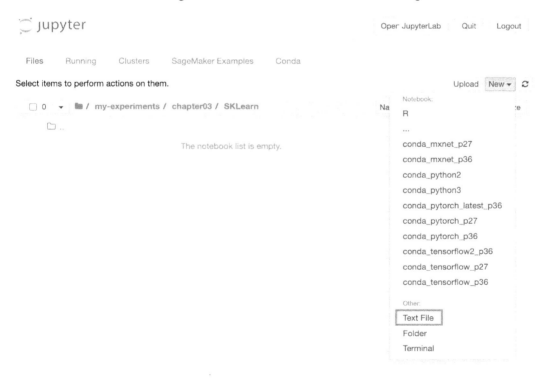

Figure 3.53 – Creating a new file

Let's quickly check how the `entryscript` file will look before we start with writing the code in `sklearn_script.py`. If you are wondering why it's called the `entryscript` file, we will pass the `sklearn_script.py` filename as the value to the parameter `entry_script` when we initialize the `SKLearn` estimator.

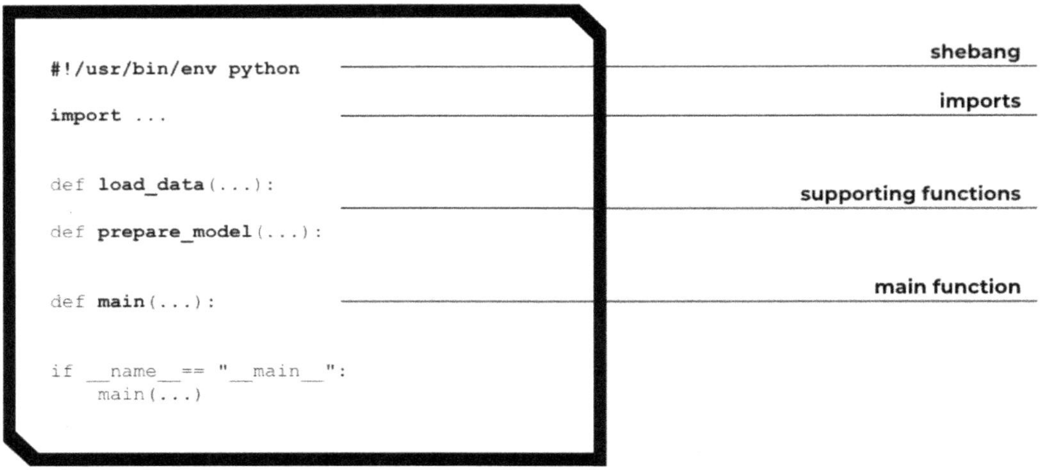

Figure 3.54 – SKLearn entrypoint script code structure

Figure 3.54 shows us how our training script with our custom **SKLearn** model will look after we have completed the instructions in this recipe.

3. Start by importing a few essential libraries and utilities such as `numpy`, `json`, and the `os` libraries and packages:

```
#!/usr/bin/env python

import os
import numpy as np
from sklearn.neural_network import MLPRegressor
from sklearn.externals import joblib
```

For this recipe, we will be using the **scikit-learn multi-layer perceptron regressor**. We do this by using `MLPRegressor` from the **scikit-learn** package.

Define the `set_seed()` function, which will help us generate the same set of results every time we run the experiment:

```
def set_seed():
    np.random.seed(0)
```

4. Prepare the `model_fn` function that loads an existing model given a path:

```
def model_fn(model_dir):
    path = os.path.join(model_dir, "model.joblib")
    model = joblib.load(path)
    return model
```

This `model_fn()` function simply loads and returns the model from the `model.joblib` file stored inside `/opt/ml/model`.

> **Important note**
>
> In the `model_fn()` function in the previous block of code, we used `joblib.load()` to load a trained model. Be careful when using the `joblib.load()` function as machine learning models from untrusted sources may contain malicious instructions that cause security issues (such as arbitrary code execution)! For more information on this topic, feel free to check `https://joblib.readthedocs.io/en/latest/persistence.html`.

5. Prepare the `load_data()` function that loads the contents of a CSV file and returns the x and y values:

```
def load_data(training_data_location):
    result = np.loadtxt(
        open(training_data_location, "rb"),
        delimiter=","
    )
    y = result[:, 0]
    x = result[:, 1]
    return (x, y)
```

6. Define the model inside the `prepare_model()` function:

```
def prepare_model(epochs=1000):
    model = MLPRegressor(
        hidden_layer_sizes=(10,10,10,10,10),
        activation='relu',
        solver='adam',
        max_iter=2000,
        verbose=True,
```

```
        batch_size=100,
        learning_rate='adaptive',
        n_iter_no_change=2000,
        early_stopping=True,
        tol=0.01,
        random_state=0
    )

    return model
```

Note that we are just using a dummy neural network and we may replace this with other **SKLearn** models. For more information on `MLPRegressor`, feel free to check `https://scikit-learn.org/stable/modules/neural_networks_supervised.html`.

7. Define the `train()` function:

```
def train(model, x, y):
    model.fit(x.reshape(-1, 1),y.reshape(-1, 1))
    return model
```

8. Define the `main()` function that prepares, trains, and saves the model when executed:

```
def main(model_dir, train_path, epochs=2000):
    set_seed()
    model = prepare_model(epochs=epochs)
    x, y = load_data(train_path)
    print(x.shape)
    print(y.shape)

    model = train(model, x, y)
    print(model)

    path = os.path.join(model_dir, "model.joblib")
    joblib.dump(model, path)
```

Take note that the `joblib.dump()` function serializes and saves the trained model in the `model.joblib` file. During deployment and inference, **SageMaker** looks for the `model_fn()` function in the same script file and uses the loaded model for inference. The `model_fn()` function we've defined in the same file makes use of the `joblib.load()` function that deserializes and loads the model from the `model.joblib` file.

9. Finally, add the following lines of code to trigger the `main()` function when the script is used directly:

```
if __name__ == "__main__":
    model_dir = "/opt/ml/model"
    train_csv = "train/training_data.csv"
    train_path = f"/opt/ml/input/data/{train_csv}"

    main(model_dir=model_dir,
         train_path=train_path,
         epochs=1000)
```

With that, the `sklearn_script.py` is complete and we are ready to proceed with the next section.

> **Tip**
>
> You can access a working copy of the `sklearn_script.py` file here in the *Machine Learning with Amazon SageMaker Cookbook* GitHub repository: `https://github.com/PacktPublishing/Machine-Learning-with-Amazon-SageMaker-Cookbook/blob/master/Chapter03/SKLearn/sklearn_script.py`.

Now let's see how this recipe works!

How it works...

In this recipe, we have used an arbitrary neural network architecture using the `MLPRegressor` for the sake of demonstration. Feel free to modify or replace the model used in this script.

Compared to the **TensorFlow** `entrypoint` script, the **scikit-learn** `entrypoint` script needs to have the `model_fn()` function defined. When the `deploy()` function is used in a later recipe, the **SageMaker scikit-learn model server** loads the model by invoking the `model_fn()` function defined in the `entrypoint` script.

In addition to the `model_fn()` function, there are a few other optional functions we can define in the `entrypoint` script:

- `input_fn()` – This function accepts the request data and performs the deserialization step, which converts the data used before the prediction step.

- `output_fn()` – This function accepts the prediction results and performs the serialization step, which converts the results depending on the response content type.

- `predict_fn()` – This function accepts the deserialized request object after the `input_fn()` function has completed the deserialization step. The loaded model is used to perform predictions on the deserialized request object and the prediction results are returned.

Feel free to check the *How it works...* section of the recipe *Preparing the entrypoint TensorFlow and Keras training script* for additional details on `entrypoint` scripts. We will not discuss some of the details in this section to avoid repetitive content.

Training and deploying a scikit-learn model with the SageMaker Python SDK

Performing the training and deployment of a custom **scikit-learn** model with **SageMaker** is fairly straightforward. *Step 1* involves creating the `entrypoint` script where our custom neural network and training logic are defined and coded. *Step 2* involves using this script as an argument to the `SKLearn` estimator from the **SageMaker Python SDK** to proceed with the training and deployment steps.

In this recipe, we will focus on *step 2* and proceed with the training and deployment of our custom **scikit-learn** neural network model in **SageMaker**. If you are looking for *step 1*, feel free to check the previous recipe, *Preparing the entrypoint scikit-learn training script*.

Getting ready

This recipe continues from *Preparing the entrypoint scikit-learn training script*.

How to do it

The instructions in this recipe focus on using the custom `entrypoint` training script from the previous recipe when initializing the `SKLearn` estimator. Once we have initialized the `SKLearn` estimator, we will use the `fit()` and `deploy()` functions before using the `predict()` function to use our custom neural network for inference:

1. Navigate to the `/my-experiments/chapter03/SKLearn` directory.

Figure 3.55 – Navigating to the SKLearn directory

As we can see in *Figure 3.55*, there is a file already in this directory called `sklearn_script.py`. We will use `sklearn_script.py` as the value to the `entry_point` parameter when initializing the `SKLearn` estimator in a later step.

2. Click the **New** button to open a dropdown of notebook kernel options. Select `conda_python3`.

Figure 3.56 – Creating a new Jupyter notebook using the conda_python3 kernel

As we have in *Figure 3.56*, the new Jupyter notebook must be in the same directory as the `sklearn_script.py` file.

3. Update the installation of **SageMaker local mode**:

```
!pip install 'sagemaker[local]' --upgrade
```

4. Restart the Docker service to make sure that we will not encounter any issues due to previously run containers:

```
!sudo service docker restart
```

5. Delete all Docker container images to free up a bit of space:

```
!docker rmi -f $(docker images -a -q)
```

With some of the preparation and installation work complete, we will proceed with the steps focused on preparing the prerequisites and arguments of the **SageMaker** training job:

6. Set the `s3_bucket` variable containing the S3 bucket path. Use the bucket created in the recipe *Preparing the Amazon S3 bucket and the training dataset for the linear regression experiment* from *Chapter 1, Getting Started with Machine Learning Using Amazon SageMaker*:

```
s3_bucket = '<insert bucket name here>'
prefix = 'chapter03'
```

Note that our `training_data.csv` file should (1) exist already inside the S3 bucket and (2) it should have the path `s3://<S3 BUCKET NAME>/<PREFIX>/synthetic/training_data.csv`.

7. Set the `train_s3` variable value:

```
train_s3 = \
f"s3://{s3_bucket}/{prefix}/synthetic/training_data.csv"
```

8. Use the `TrainingInput` class to prepare the training input parameter with `content_type` set to `"text/csv"`:

```
from sagemaker.inputs import TrainingInput
train_input = TrainingInput(train_s3, content_type="text/
csv")
```

9. Import a few prerequisites to run the training job. In addition to this, we will use LocalSession to initialize the sagemaker_session object. The LocalSession class allow us to use **local mode** in the training and deployment steps. As we will see in a later step, we will specify 'local' as the parameter value for instance_type when initializing the estimator to make us perform the training job locally in the SageMaker notebook instance:

```
import sagemaker
from sagemaker import get_execution_role
from sagemaker.local import LocalSession

sagemaker_session = LocalSession()
sagemaker_session.config = {'local': {'local_code':
True}}

role = get_execution_role()
```

10. Initialize the SKLearn estimator from the **SageMaker Python SDK**:

```
from sagemaker.sklearn.estimator import SKLearn

estimator = SKLearn(entry_point='sklearn_script.py',
                    session=sagemaker_session,
                    role=role,
                    instance_type='local',
                    instance_count=1,
                    py_version='py3',
                    framework_version='0.20.0')
```

As you can see, we are not specifying the container where our script file will run. The SKLearn estimator makes use of one of the prebuilt Docker container images in **SageMaker**. This means that we will not have to worry about preparing the container as **SageMaker** has already prepared this for us. If this is not sufficient, we also have the option to extend an existing **prebuilt SageMaker Docker image**.

11. Perform the training step using the `fit()` function:

```
estimator.fit({'train': train_input})
```

This should yield a set of logs similar to what is shown in *Figure 3.57*.

```
Creating puf5tta0o7-algo-1-jadq8 ...
Creating puf5tta0o7-algo-1-jadq8 ... done
Attaching to puf5tta0o7-algo-1-jadq8
puf5tta0o7-algo-1-jadq8 | 2021-08-28 21:39:28,695 sagemaker-containers INFO      Imported framework sagemaker_sklearn_
container.training
puf5tta0o7-algo-1-jadq8 | 2021-08-28 21:39:28,697 sagemaker-training-toolkit INFO      No GPUs detected (normal if no
gpus installed)
puf5tta0o7-algo-1-jadq8 | 2021-08-28 21:39:28,708 sagemaker_sklearn_container.training INFO      Invoking user trainin
g script.
puf5tta0o7-algo-1-jadq8 | 2021-08-28 21:39:29,526 sagemaker-training-toolkit INFO      No GPUs detected (normal if no
gpus installed)
puf5tta0o7-algo-1-jadq8 | 2021-08-28 21:39:29,541 sagemaker-training-toolkit INFO      No GPUs detected (normal if no
gpus installed)
puf5tta0o7-algo-1-jadq8 | 2021-08-28 21:39:29,555 sagemaker-training-toolkit INFO      No GPUs detected (normal if no
gpus installed)
puf5tta0o7-algo-1-jadq8 | 2021-08-28 21:39:29,567 sagemaker-training-toolkit INFO      Invoking user script
puf5tta0o7-algo-1-jadq8 |
puf5tta0o7-algo-1-jadq8 | Training Env:
```

Figure 3.57 – Training logs after running the fit() function after initializing the SKlearn estimator

If you encounter issues when running the `fit()` function, you may check the recipes *Debugging disk space issues when using local mode* and *Debugging container execution issues when using local mode* at the end of this chapter.

12. Once the training step has been completed, deploy the model using the `deploy()` function:

```
predictor = estimator.deploy(initial_instance_count=1,
instance_type='local')
```

We should get a set of logs and output values similar to the one in *Figure 3.58*:

```
Attaching to n7ytm8rtm6-algo-1-9yagj
n7ytm8rtm6-algo-1-9yagj | Processing /opt/ml/code
n7ytm8rtm6-algo-1-9yagj | 2021/08/28 21:39:57 [crit] 21#21: *1 connect() to unix:/tmp/gunicorn.sock failed (2: No suc
h file or directory) while connecting to upstream, client: 172.18.0.1, server: , request: "GET /ping HTTP/1.1", upstr
eam: "http://unix:/tmp/gunicorn.sock:/ping", host: "localhost:8080"
n7ytm8rtm6-algo-1-9yagj | 172.18.0.1 - - [28/Aug/2021:21:39:57 +0000] "GET /ping HTTP/1.1" 502 182 "-" "python-urllib
3/1.26.6"
n7ytm8rtm6-algo-1-9yagj | Building wheels for collected packages: sklearn-script
n7ytm8rtm6-algo-1-9yagj |   Building wheel for sklearn-script (setup.py) ... done
n7ytm8rtm6-algo-1-9yagj |   Created wheel for sklearn-script: filename=sklearn_script-1 0.0-py2.py3-none-any.whl size
=4173 sha256=d4e0b8bcd5edeff0a2837e70565e3e0a59c2500f6bab79438251e5c8f4247e8e
n7ytm8rtm6-algo-1-9yagj |   Stored in directory: /tmp/pip-ephem-wheel-cache-nh8d066d/wheels/3e/0f/51/2f1df833dd0412c1
bc2f5ee56baac195b5be563353d11ldca6
n7ytm8rtm6-algo-1-9yagj | Successfully built sklearn-script
n7ytm8rtm6-algo-1-9yagj | Installing collected packages: sklearn-script
n7ytm8rtm6-algo-1-9yagj | Successfully installed sklearn-script-1.0.0
n7ytm8rtm6-algo-1-9yagj | /miniconda3/lib/python3.7/site-packages/sklearn/externals/joblib/externals/cloudpickle/clou
dpickle.py:47: DeprecationWarning: the imp module is deprecated in favour of importlib; see the module's documentatio
n for alternative uses
n7ytm8rtm6-algo-1-9yagj |   import imp
n7ytm8rtm6-algo-1-9yagj | [2021-08-28 21:40:01 +0000] [30] [INFO] Starting gunicorn 20.0.4
n7ytm8rtm6-algo-1-9yagj | [2021-08-28 21:40:01 +0000] [30] [INFO] Listening at: unix:/tmp/gunicorn.sock (30)
n7ytm8rtm6-algo-1-9yagj | [2021-08-28 21:40:01 +0000] [30] [INFO] Using worker: gevent
n7ytm8rtm6-algo-1-9yagj | [2021-08-28 21:40:01 +0000] [33] [INFO] Booting worker with pid: 33
n7ytm8rtm6-algo-1-9yagj | [2021-08-28 21:40:01 +0000] [34] [INFO] Booting worker with pid: 34
n7ytm8rtm6-algo-1-9yagj | 2021-08-28 21:40:02,700 INFO - sagemaker-containers - No GPUs detected (normal if no gpus i
nstalled)
n7ytm8rtm6-algo-1-9yagj | /miniconda3/lib/python3.7/site-packages/sklearn/externals/joblib/externals/cloudpickle/clou
dpickle.py:47: DeprecationWarning: the imp module is deprecated in favour of importlib; see the module's documentatio
n for alternative uses
n7ytm8rtm6-algo-1-9yagj |   import imp
n7ytm8rtm6-algo-1-9yagj | 172.18.0.1 - - [28/Aug/2021:21:40:03 +0000] "GET /ping HTTP/1.1" 200 0 "-" "python-urllib
3/1.26.6"
```

Figure 3.58 – Logs and output after using the predict() function in local mode

As seen in *Figure 3.58*, when calling the `predict()` function in **local mode**, the `/invocations` endpoint inside a running container is triggered with a `POST` request. After the request has been processed, the response is returned to the user by the `predict()` function.

> **Note**
>
> If you are wondering which container image is used when running the local prediction endpoint, simply run `!docker container ls` to show the running containers. We should get a value equal or similar to `683313688378.dkr.ecr.us-east-1.amazonaws.com/sagemaker-scikit-learn:0.20.0-cpu-py3`. For more information, check the images available at `https://github.com/aws/deep-learning-containers/blob/master/available_images.md`.

13. Perform a test prediction using the `predict()` function:

```
import numpy as np


input_data = np.array([[100], [200]], dtype=np.float32)
result = predictor.predict(input_data)


result
```

We should get a set of logs similar to what is shown in *Figure 3.59*:

```
n7ytm8rtm6-algo-1-9yag} | 2021-08-28 21:40:03,055 INFO - sagemaker-containers - No GPUs detected (normal if no gpus i
nstalled)
n7ytm8rtm6-algo-1-9yag} | /miniconda3/lib/python3.7/site-packages/sklearn/externals/joblib/externals/cloudpickle/clou
dpickle.py:47: DeprecationWarning: the imp module is deprecated in favour of importlib; see the module's documentatio
n for alternative uses
n7ytm8rtm6-algo-1-9yag} |     import imp
```

Out[16]: `array([205.57487957, 307.21642029])`

```
n7ytm8rtm6-algo-1-9yag} | 172.18.0.1 - - [28/Aug/2021:21:40:03 +0000] "POST /invocations HTTP/1.1" 200 144 "-" "pytho
n-urllib3/1.26.6"
```

Figure 3.59 – Logs and output after using the predict() function in local mode

As seen in *Figure 3.59*, when calling the `predict()` function in local mode, the `/invocations` endpoint inside a running container is triggered with a POST request. After the request has been processed, the response is returned to the user by the `predict()` function.

14. Create an empty directory called `tmp`:

```
!mkdir -p tmp
```

15. Download the `all_data.csv` file from the S3 bucket to the `tmp` directory:

```
all_s3 = \
f"s3://{s3_bucket}/{prefix}/synthetic/all_data.csv"
!aws s3 cp {all_s3} tmp/all_data.csv
```

The next couple of steps focus on checking the predicted values of y for a set of x values that would represent the "prediction line."

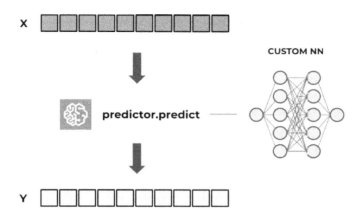

Figure 3.60 – The predict() function using the custom neural network deployed in the inference endpoint to predict the y values from the x input values

In *Figure 3.60*, we have the predict() function generating predicted y values from x values using the custom **SKLearn** MLPRegressor neural network we have trained and deployed to an inference endpoint.

16. Load the all_data.csv file:

```
import pandas as pd
all_data = pd.read_csv("tmp/all_data.csv", header=None)
x = all_data[[1]].values
y = all_data[[0]].values
```

17. Prepare the line_x variable containing values for x between -5000 and 5000:

```
from numpy import arange
line_x = arange(-5000, 5000, 10)
line_x
```

18. Use the deployed model to predict the values of y for each x in line_x:

```
input_data = np.array(line_x.reshape(-1, 1), dtype=np.float32)
result = predictor.predict(input_data)
result
```

19. Store the results inside the line_y variable:

```
line_y = result
```

20. Render the line on top of the scatterplot of the original dataset:

```
from matplotlib import pyplot
pyplot.plot(line_x, line_y, 'r')
pyplot.scatter(x,y,s=1)
pyplot.show()
```

In *Figure 3.61*, we can see the prediction line along with the scatterplot of the dataset to have a quick view of how the predicted values match up with the actual values:

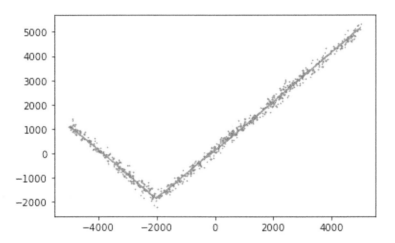

Figure 3.61 – Prediction line using the scikit-learn custom neural network over the scatterplot showing the actual values from the generated synthetic dataset

As we can see in *Figure 3.61*, our custom scikit-learn model seems to do just fine in generalizing over the dataset. We can perform additional model evaluation steps to measure the model's performance properly. We will skip this to prevent this recipe from getting too long.

21. Finally, stop the running container with the inference endpoint using the `delete_endpoint()` function:

```
predictor.delete_endpoint()
```

This should stop the container from running our local prediction endpoint.

Let's see how this works!

How it works...

Using the **scikit-learn** support of the **SageMaker Python SDK**, we are able to use the **scikit-learn** `entrypoint` script from the recipe *Preparing the entrypoint scikit-learn training script* directly without having to build and manage the containers used for the experiment as the SDK abstracts this for us.

> **Note**
>
> For more information, feel free to check the *How it works...* section of the recipe *Training and deploying a TensorFlow and Keras model with Amazon SageMaker local mode*. We will not discuss some of the details in this section to avoid repetitive content.

See also

If you are looking for examples of training and deploying **scikit-learn** models in SageMaker using real datasets, feel free to check some of the notebooks in the **aws/amazon-sagemaker-examples** GitHub repository:

- Training and deploying a classifier using **scikit-learn**: `https://github.com/aws/amazon-sagemaker-examples/blob/master/sagemaker-python-sdk/scikit_learn_iris/scikit_learn_estimator_example_with_batch_transform.ipynb`

- Building your own algorithm container: `https://github.com/aws/amazon-sagemaker-examples/blob/master/advanced_functionality/scikit_bring_your_own/scikit_bring_your_own.ipynb`

In addition to **TensorFlow, PyTorch, and scikit-learn,** there are other frameworks and libraries we can easily use with the **SageMaker Python SDK** using the framework-specific estimator classes — MXNet, Chainer, Hugging Face, and XGBoost. In this book, we will not deep dive into training and deploying models with **Apache MXNet**, **Chainer**, and **Hugging Face**.

Debugging disk space issues when using local mode

Sometimes, when running the `estimator.fit()` function in an experiment using **local mode**, we may encounter issues similar to what is shown in *Figure 3.62*.

```
~/anaconda3/envs/tensorflow_p36/lib/python3.6/site-packages/sagemaker/local/image.py in _pull_image(image)
   1089        logger.info("docker command: %s", pull_image_command)
   1090
-> 1091        subprocess.check_output(pull_image_command.split())
   1092        logger.info("image pulled: %s", image)

~/anaconda3/envs/tensorflow_p36/lib/python3.6/subprocess.py in check_output(timeout, *popenargs, **kwargs)
    354
    355        return run(*popenargs, stdout=PIPE, timeout=timeout, check=True,
--> 356                **kwargs).stdout
    357
    358

~/anaconda3/envs/tensorflow_p36/lib/python3.6/subprocess.py in run(input, timeout, check, *popenargs, **kwargs)
    436            if check and retcode:
    437                raise CalledProcessError(retcode, process.args,
--> 438                            output=stdout, stderr=stderr)
    439        return CompletedProcess(process.args, retcode, stdout, stderr)
    440

CalledProcessError: Command '['docker', 'pull', '763104351884.dkr.ecr.us-east-1.amazonaws.com/tensorflow-training:2.
1.0-cpu-py3']' returned non-zero exit status 1.
```

Figure 3.62 – CalledProcessError potentially due to a disk space issue

Note that this may or may not be caused by disk space issues but there is a big chance that the root cause is that there is no space left. The error message may include the following error message:

```
CalledProcessError: Command '['docker', 'pull', '763104351884.
dkr.ecr.us-east-1.amazon.com/<image-uri>:<tag>'] ' returned
non-zero exit status 1.
```

In this recipe, we will take a look at how to debug this issue.

> **Tip**
>
> If everything went smoothly when running the recipes in this chapter, feel free to check the *There's more...* section for instructions on how to replicate this issue. Once this issue has been replicated, you may proceed with following the instructions in the *How to do it* section.

Getting ready

The prerequisite of this recipe is a `CalledProcessError` triggered when using the `fit()` function. You may try to trigger this error yourself by checking the *There's more...* section before proceeding with the steps in the *How to do it* section.

How to do it

We need to run the following statements inside the Jupyter notebook where the issue occurred.

1. Create a new cell after the `Estimator` object has been initialized. Run the following statement so that we can check which Docker image was being downloaded when the error occurred:

   ```
   estimator.training_image_uri()
   ```

   We expect the output of the preceding statement to be equal or similar to `'763104351884.dkr.ecr.us-east-1.amazonaws.com/tensorflow-training:2.1.0-cpu-py3'`.

2. Run the following command to see the actual error message when pulling the container image:

   ```
   !docker pull {estimator.training_image_uri()}
   ```

   We should see a set of logs similar to the one in *Figure 3.63*.

```
2.1.0-cpu-py3: Pulling from tensorflow-training

21c94847: Pulling fs layer
eae58dfc: Pulling fs layer
0eb9f140: Pulling fs layer
7ea0da2e: Pulling fs layer
72a8de99: Pulling fs layer
e1f109aa: Pulling fs layer
763be64c: Pulling fs layer
0ba9ccf8: Pulling fs layer
bf7b3e53: Pulling fs layer
e5edfb40: Pulling fs layer
03d744d3: Pulling fs layer
b55a1844: Pulling fs layer
df39c07b: Pulling fs layer
23e6b10f: Pulling fs layer
29ae34f6: Pulling fs layer
45b30c31: Pulling fs layer
85083f0d: Pulling fs layer
write /var/lib/docker/tmp/GetImageBlob880968931: no space left on device18A
```

Figure 3.63 – The "no space left on deviceA" issue

As seen in *Figure 3.63*, we encountered the disk space issue after running the `docker pull` command.

3.   Next, use the `df -h` command to check the available disk space:

```
!df -h
```

This should give us disk space statistics similar to what is shown in *Figure 3.64*.

```
Filesystem      Size  Used Avail Use% Mounted on
devtmpfs        2.0G   64K  2.0G   1% /dev
tmpfs           2.0G     0  2.0G   0% /dev/shm
/dev/xvda1      104G  104G   20M 100% /
/dev/xvdf       197G   23G  167G  12% /home/ec2-user/SageMaker
```

Figure 3.64 – /dev/xvda1 has 1.9 GB of available disk space left

In *Figure 3.64*, we can see that we have 20 MB of available disk space left. When we performed the `docker pull` command, this pulled a container large enough to trigger the disk space issue.

4.   You may initially try running the following command to clear up a bit of space as a quick fix:

```
!docker rmi -f $(docker images -a -q)
```

Running this command will remove all Docker container images in the system. We should get an output similar to what is shown in *Figure 3.65*.

```
Untagged: 763104351884.dkr.ecr.us-east-1.amazonaws.com/mxnet-training:1.8.0-gpu-py37-cu110-ubuntu16.04
Untagged: 763104351884.dkr.ecr.us-east-1.amazonaws.com/mxnet-training@sha256:66bbbf4b3f271403ce6d32f3e3756d057bcac6b4
c99c548dbc3208d571f91d54
Deleted: sha256:0596b9ed7682e8cf64999b2f3d3057d90e1417a43c4668db77234e6b79f8c7a2
Deleted: sha256:3612ef94279453542837649dc68c6c832fa9b93d233ced2d55ed5da3f9de26c1
Deleted: sha256:bd5cecf38e7bc3be76f1ab7262cab89b48b0f84756013f34bd20b88bba9dcea6
Deleted: sha256:eda02b464a727b9c359484f325301d332fa6385b353de506ed321a9190b8ee32
Deleted: sha256:e62df8b18d979dff5311eff3ee9eba34710d01684e864aa0915bc39e2be517b9
Deleted: sha256:a612529d7ddf3029880472ae8707423a674b69a40ffb3255a4c50f50b7a64ea1
Deleted: sha256:30315694fb15cf7a2a006761d9aa088766bfa8aa4b4fdca748c113e58ccbeb16
Deleted: sha256:23612f2927e0a6f448d94251eb70bcf9f7ff1e939e01aea0e9b5aff500cdc730
Deleted: sha256:8729e9c188d0dff628bc85230264d970132d21d7c47d934fa4eb1b3b322922b3
Deleted: sha256:6806549e19e7133e26aac27d90ce06fadf19a877052c428434995b1dd44c3150
Deleted: sha256:97bd62bb9ccf4c1945ec76734fcf3615a087f3ee56944e881d45c7b4fddc2d56
Deleted: sha256:0209b59e96636b4880d1e569c4e8f787985cf7d5bf33f63b38717ab7b6c1603f
```

Figure 3.65 – Docker container images deleted

We can see that we have cleaned up a bit of space using `docker rmi`.

> **Tip**
> To see the image space used by the downloaded deep learning containers, you may use the command `!docker system df -v` in the Jupyter notebook cell.

5. Run the following command to see how much space we have freed up:

```
!df -h
```

This should give us disk space statistics similar to what is shown in *Figure 3.66*.

```
Filesystem      Size  Used Avail Use% Mounted on
devtmpfs        2.0G   64K  2.0G   1% /dev
tmpfs           2.0G     0  2.0G   0% /dev/shm
/dev/xvda1      104G   91G   14G  88% /
/dev/xvdf       197G   23G  167G  12% /home/ec2-user/SageMaker
```

Figure 3.66 – /dev/xvda1 has 9.6 GB of available disk space left

We can see in *Figure 3.66* that we have recovered 14 GB of disk space after running the docker rmi command. Earlier we only had 20 MB of disk space. Now we have 14 GB!

6. Run the cell again with the fit() or predict() function call. If that succeeds, then that's great! This means that we freed up just enough space to pull the container image. Otherwise, follow the steps in the recipe *Preparing the SageMaker notebook instance for multiple deep learning local experiments* at the start of this chapter to give us the extra disk space we need to run our experiments.

> **Tip**
> If all else fails, restart the notebook instance to see if this fixes your problem. Restarting the notebook instance should reset the configuration of the running services (for example, docker).

Now, let's see how this works!

How it works...

In this recipe, we performed a couple of steps to debug and fix a potential disk space issue when calling the fit() or predict() function. When the fit() function is called while in **local mode**, the training container image is pulled from the ECR repository and then run with the appropriate parameters. When there is a disk space issue, we may get an error message that looks like this:

```
CalledProcessError: Command '['docker', 'pull', '763104351884.
dkr.ecr.us-east-1.amazon.com/<image-uri>:<tag>'] ' returned
non-zero exit status 1.
```

It is also possible to get an error that looks like this:

```
RuntimeError: Failed to run: ['docker-compose', '-f', '/
tmp/abcdefghij12345/docker-compose.yaml', 'up', '--build',
'--abort-on-container-exit'], Process exited with code: 1
```

The error message depends on when the disk space issue happened. In the first scenario, the disk space issue prevented the `docker pull` command from succeeding due to the lack of space available. In the second scenario, the container has successfully been downloaded and the disk space issue caused the script within the container to fail. In the next recipe, we will see the other possible causes of the `fit()` function failing.

Finally, we may also get error messages that may seem like a different issue altogether but are a side effect of a disk space issue. As seen in the *How to do it* section in this recipe, a good first step is to check first if we have a disk space issue. If it is a disk space issue, we may solve the problem with one or both of the following:

- A short-term solution, for example, freeing up just enough space for the experiment
- A long-term solution, for example, increasing the volume size of the SageMaker notebook instance so that we will not need to delete large files every time we encounter a disk space issue blocker

For the long-term solution, check the recipe *Preparing the SageMaker notebook instance for multiple deep learning local experiments* in this chapter.

There's more...

Do you want to trigger this error message? It's simple! Open a new Terminal and run the following commands to download a few container images, which would eat up a lot of space:

```
aws ecr get-login-password --region us-east-1 | docker login
--username AWS --password-stdin 763104351884.dkr.ecr.us-east-1.
amazonaws.com
docker pull 763104351884.dkr.ecr.us-east-1.amazonaws.com/
pytorch-training:1.5.0-cpu-py3
docker pull 763104351884.dkr.ecr.us-east-1.amazonaws.com/
pytorch-inference:1.5.0-cpu-py3
docker pull 763104351884.dkr.ecr.us-east-1.amazonaws.com/mxnet-
inference:1.7.0-cpu-py36-ubuntu16.04
docker pull 763104351884.dkr.ecr.us-east-1.amazonaws.com/mxnet-
inference-eia:1.4.1-cpu-py36-ubuntu16.04
```

Continue using `docker pull` on different container images other than the container image we want to pull when we use the `fit()` function. If you need a list of container images and their container image URI values, refer to the list of available images here: `https://github.com/aws/deep-learning-containers/blob/master/available_images.md`.

You may stop pulling image containers once you receive an error message similar to the following:

```
failed to register layer: Error processing tar file(exit
status 1): write /usr/local/lib/python3.6/site-packages/...:
no space left on device
```

At this point, you may return to the Jupyter notebook running the experiment using SageMaker local mode. Run the cell again involving the `fit()` function call and see if that triggers the error.

```
-/anaconda3/envs/tensorflow_p36/lib/python3.6/site-packages/sagemaker/local/image.py in _pull_image(image)
   1089        logger.info("docker command: %s", pull_image_command)
   1090
-> 1091        subprocess.check_output(pull_image_command.split())
   1092        logger.info("image pulled: %s", image)

-/anaconda3/envs/tensorflow_p36/lib/python3.6/subprocess.py in check_output(timeout, *popenargs, **kwargs)
    354
    355        return run(*popenargs, stdout=PIPE, timeout=timeout, check=True,
--> 356              **kwargs).stdout
    357
    358

-/anaconda3/envs/tensorflow_p36/lib/python3.6/subprocess.py in run(input, timeout, check, *popenargs, **kwargs)
    436        if check and retcode:
    437            raise CalledProcessError(retcode, process.args,
--> 438                              output=stdout, stderr=stderr)
    439        return CompletedProcess(process.args, retcode, stdout, stderr)
    440

CalledProcessError: Command '['docker', 'pull', '763104351884.dkr.ecr.us-east-1.amazonaws.com/tensorflow-training:2.
1.0-cpu-py3']' returned non-zero exit status 1.
```

Figure 3.67 – CalledProcessError due to a disk space issue

In *Figure 3.67*, we can see that we were able to trigger the disk space issue error by pulling a lot of container images. Of course, there are different ways to fill up the available space but this one should work just fine!

Important note

If you are not getting the error involving `docker pull` and you are getting a `docker compose` error instead, this means that the training image is already downloaded. The workaround for this is to specify a different `framework_version` when initializing the `Estimator` object. This will force pulling a new Docker container image, which will consequently trigger the desired issue and error message.

Now that you have triggered this error message, you may now proceed with executing the instructions in the *How to do it* section in this recipe to debug and solve this disk space issue.

Debugging container execution issues when using local mode

If you have encountered an issue similar to what is shown in *Figure 3.68* when calling the `fit()` function with **SageMaker local mode**, this is the recipe for you!

```
~/anaconda3/envs/tensorflow_p36/lib/python3.6/site-packages/sagemaker/local/entities.py in start(self, input_data_con
fig, output_data_config, hyperparameters, job_name)
    219
    220            self.model_artifacts = self.container.train(
--> 221                input_data_config, output_data_config, hyperparameters, job_name
    222            )
    223            self.end_time = datetime.datetime.now()

~/anaconda3/envs/tensorflow_p36/lib/python3.6/site-packages/sagemaker/local/image.py in train(self, input_data_confi
g, output_data_config, hyperparameters, job_name)
    241            # which contains the exit code and append the command line to it.
    242            msg = "Failed to run: %s, %s" % (compose_command, str(e))
--> 243            raise RuntimeError(msg)
    244        finally:
    245            artifacts = self.retrieve_artifacts(compose_data, output_data_config, job_name)

RuntimeError: Failed to run: ['docker-compose', '-f', '/tmp/tmplzcruyz6/docker-compose.yaml', 'up', '--build', '--abo
rt-on-container-exit'], Process exited with code: 1
```

Figure 3.68 – Error running the fit() function

In *Figure 3.68*, we can see that we encountered issues when we executed the `fit()` function. Sometimes, the error message includes the following log message towards the end of the debug information:

```
RuntimeError: Failed to run: ['docker-compose', '-f', '/
tmp/abcdefghij12345/docker-compose.yaml', 'up', '--build',
'--abort-on-container-exit'], Process exited with code: 1
```

In some cases, the root cause of the errors is not really displayed back to the user, which makes this issue hard to debug for some machine learning practitioners. Do not worry as this recipe will prove useful in debugging these types of issues!

> **Tip**
>
> If everything went smoothly when running the recipes in this chapter, feel free to check the *There's more...* section for instructions on how to replicate this issue. Once this issue has been replicated, you may proceed with following the instructions in the *How to do it* section.

Getting ready

The prerequisite of this recipe is a `RunTimeError` triggered when using the `fit()` function. You may try to trigger this error yourself by checking the *There's more...* section before proceeding with the steps in the *How to do it* section.

How to do it...

The steps in this recipe focus on identifying the container ID and using the `docker logs` command to help us debug what went wrong inside the container:

Use the `docker container ls -a` command to list down all the containers. Take note of the container ID value:

```
!docker container ls -a
```

We should get a set of results similar to what is shown in *Figure 3.69*.

```
CONTAINER ID     IMAGE                                                                              COMMAND
CREATED          STATUS                         PORTS          NAMES
3a9f16d7ce0e     763104351884.dkr.ecr.us-east-1.amazonaws.com/tensorflow-training:2.1.0-cpu-py3     "train"
46 seconds ago   Exited (1) 39 seconds ago                     ss25108nrl-algo-1-lo166
89aee80b0f3e     763104351884.dkr.ecr.us-east-1.amazonaws.com/tensorflow-training:2.1.0-cpu-py3     "train"
9 minutes ago    Exited (1) 9 minutes ago                      cxdzhkhp2j-algo-1-jm3di
710b04ae0288     763104351884.dkr.ecr.us-east-1.amazonaws.com/tensorflow-training:2.1.0-cpu-py3     "train"
47 minutes ago   Exited (137) 46 minutes ago                   hld9bz6x3j-algo-1-9viei
49d8fdf4ae37     763104351884.dkr.ecr.us-east-1.amazonaws.com/tensorflow-training:2.1.0-cpu-py3     "train"
49 minutes ago   Exited (137) 49 minutes ago                   ymwawl29n2-algo-1-1r7n8
1eb422ec8c17     763104351884.dkr.ecr.us-east-1.amazonaws.com/tensorflow-training:2.1.0-cpu-py3     "train"
52 minutes ago   Exited (0) 52 minutes ago                     15sa3k5il2-algo-1-83zyl
165e36341d42     763104351884.dkr.ecr.us-east-1.amazonaws.com/tensorflow-training:2.1.0-cpu-py3     "train"
About an hour ago  Exited (137) About an hour ago              hpz09q7k5o-algo-1-t1fa7
```

Figure 3.69 Result of running docker container ls -a

Alternatively, we can get the container ID simply by inspecting the logs after calling the `fit()` function.

```
estimator.fit({'train': train_input, 'validation': val_input})
```
```
Creating ss25108nrl-algo-1-lo166 ...
Creating ss25108nrl-algo-1-lo166 ... done
Attaching to ss25108nrl-algo-1-lo166
ss25108nrl-algo-1-lo166 | 2021-08-28 23:35:23,367 sagemaker-containers INFO     Imported framework sagemaker_tensorfl
ow_container.training
ss25108nrl-algo-1-lo166 | 2021-08-28 23:35:23,376 sagemaker-containers INFO     No GPUs detected (normal if no gpus i
nstalled)
```

Figure 3.70 – Getting the container ID

You may choose either of the two preceding options in getting the container ID. In our example in *Figure 3.70*, the container ID is `ss25108nrl-algo-1-lo166`. Of course in your case, you will get a different container ID as this is randomly generated every time a container is run.

Use the `docker logs` command to inspect the logs of the target container. If we were to use the container ID from the example from the previous step, we would replace `"<insert container id here>"` with `"ss25108nrl-algo-1-lo166"`:

```
CONTAINER_ID="<insert container id here>"
docker logs $CONTAINER_ID
```

Note that the container ID changes each time the `fit()` function is executed so make sure to get its latest value from the logs in your Jupyter notebook.

You may also run the following block of code to get the logs of the last container run:

```
%%bash
LAST_CONTAINER_ID=$(docker container ls -a -q | head -n 1)
docker logs $LAST_CONTAINER_ID
```

What happened here? The `docker container ls -a -q` command will list all the container IDs while the `head -n 1` command will return the first entry it gets. As the results of `docker container ls` are sorted from the latest to the oldest container executed, these commands will return the ID of the last container executed. Of course, this block of code needs to be executed right after the `fit()` function fails as we are inspecting the logs of the last container executed. We should get an output similar to what is shown in *Figure 3.71*.

```
Invoking script with the following command:

/usr/bin/python3 tensorflow_script_broken.py --model_dir s3://sagemaker-cookbook-bucket/chapter03/output/tensorflow/t
ensorflow-training-2021-04-08-23-40-54-844/model


Traceback (most recent call last):
  File "tensorflow_script_broken.py", line 11, in <module>
    from numpy.random import seeds
ImportError: cannot import name 'seeds'
2021-04-08 23:41:02,315 sagemaker-containers ERROR    ExecuteUserScriptError:
Command "/usr/bin/python3 tensorflow_script_broken.py --model_dir s3://sagemaker-cookbook-bucket/chapter03/output/ten
sorflow/tensorflow-training-2021-04-08-23-40-54-844/model"
```

Figure 3.71 – Logs of the container

We can see in *Figure 3.71* that we accidentally misspelled `seed` as `seeds` in the line from `numpy.random import seeds`. Of course, this was not an accident! Check the *There's more...* section on how to replicate and trigger this issue.

> **Tip**
>
> Sometimes, an issue gets resolved by running `!sudo service docker restart`. You may try that first before proceeding with the other debugging steps.

Now, let's see how this works!

How it works...

In this recipe, we used the `docker logs` command to help us debug a failed container run. When the `fit()` function is called while in **local mode**, the training container image is pulled from the ECR repository and then run with the appropriate parameters.

This would invoke the `train` script inside the container and run the code inside the `train` script. After the script has completed its execution, some of the info and error logs are returned back to the user. In most cases, these error logs are helpful but in some cases, the error messages printed in the Jupyter notebook cells may not help us at all.

There are many reasons why the `fit()` function may fail. Here are some of the possible causes and reasons:

- Disk space issues
- Permission issues
- Custom script bugs and issues
- Imported library or package issues
- Container bugs and issues

When dealing with disk space issues, we can simply follow the steps in the recipe *Debugging disk space issues when using local mode*. For the other issues, this would involve opening a Terminal tab and running a few bash commands so that we can inspect the logs from the last execution. As the `entry_point` script is run inside the container, this means that we can use the `docker logs` command to inspect the logs of both running and previously run containers.

Lastly, there will be cases when the root cause of the issue is the bugs introduced in the deep learning framework or container image. One example of this would be for a training job to be working just fine with the **PyTorch** deep learning container version `1.5.0` and then suddenly failing when version `1.6.0` is used. Of course, this depends on the context of the issue but knowing this would help you know which configuration options to change in order to debug the issues faster.

There's more...

Do you want to trigger this error message? It's simple!

The first step involves copying the originally working `entrypoint` script file used during training and inference.

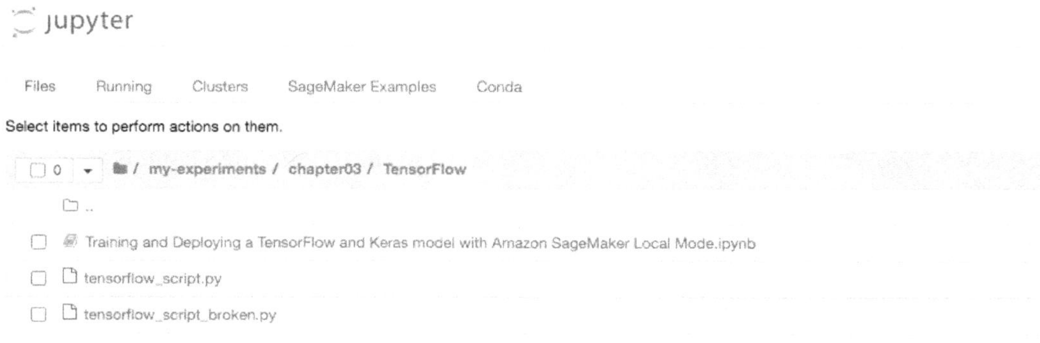

⊂ Jupyter

| Files | Running | Clusters | SageMaker Examples | Conda |

Select items to perform actions on them.

- ☐ 0 ▾ ▮ / my-experiments / chapter03 / TensorFlow
  - ☐ ..
- ☐ 🔊 Training and Deploying a TensorFlow and Keras model with Amazon SageMaker Local Mode.ipynb
- ☐ ☐ tensorflow_script.py
- ☐ ☐ tensorflow_script_broken.py

Figure 3.72 – Copying the entrypoint script file to preserve the working version

The second step involves intentionally causing an error within the source code of the new `entrypoint` file. In this case, we can simply replace `from numpy.random import seed` with `from numpy.random import seeds`. Given that there is no `seeds` function inside `numpy.random`, then this should cause an error.

The third step involves updating the relevant block of code in the Jupyter notebook when initializing the estimator. We will just have to update the `entry_point` parameter value to point to the new file:

```
estimator = TensorFlow(entry_point='tensorflow_script_broken.
py',
                       output_path=s3_output_location,
                       role=role,
                       sesion=sagemaker_session,
                       instance_count=1,
                       instance_type='local',
                       framework_version='2.1.0',
                       py_version='py3')
```

Finally, when we call the `fit()` function, we should get an error message similar to what is shown in *Figure 3.73*.

```
~/anaconda3/envs/tensorflow_p36/lib/python3.6/site-packages/sagemaker/local/entities.py in start(self, input_data_con
fig, output_data_config, hyperparameters, job_name)
    219
    220            self.model_artifacts = self.container.train(
--> 221                input_data_config, output_data_config, hyperparameters, job_name
    222            )
    223            self.end_time = datetime.datetime.now()

~/anaconda3/envs/tensorflow_p36/lib/python3.6/site-packages/sagemaker/local/image.py in train(self, input_data_confi
g, output_data_config, hyperparameters, job_name)
    241                # which contains the exit code and append the command line to it.
    242                msg = "Failed to run: %s, %s" % (compose_command, str(e))
--> 243                raise RuntimeError(msg)
    244        finally:
    245            artifacts = self.retrieve_artifacts(compose_data, output_data_config, job_name)

RuntimeError: Failed to run: ['docker-compose', '-f', '/tmp/tmplzcruyz6/docker-compose.yaml', 'up', '--build', '--abo
rt-on-container-exit'], Process exited with code: 1
```

Figure 3.73 – Failed to run command (RuntimeError)

Now that you have triggered this error message, you may now proceed with executing the instructions in the *How to do it...* section of this recipe to debug and solve this disk space issue.

4
Preparing, Processing, and Analyzing the Data

Before we can start training our machine learning model, we have to prepare, process, and transform our data into a structure and format that the algorithm can work on. There are different techniques and services we can use to handle our different data processing and analysis requirements. The recipes in this chapter focus on key SageMaker capabilities, algorithms, and features when performing these tasks. These include **SageMaker Processing** for our managed data processing and transformation requirements, support for invoking deployed SageMaker machine learning models with **Amazon Athena** to analyze our data with SQL statements, the built-in **Principal Component Analysis (PCA)** algorithm for performing dimensionality reduction, and the built-in **KMeans** algorithm for performing cluster analysis.

We will start with a gentle introduction to Amazon Athena and we will use it to help us process and analyze our large datasets and files in S3 using SQL syntax. We will also invoke a deployed **Random Cut Forest** (**RCF**) model using SQL statements to detect anomalies in our synthetic dataset. If this is the first time you have heard of Amazon Athena, it is a fully managed service that helps us analyze data in Amazon S3 using SQL syntax. We will also show how to convert data in CSV format to `protobuf recordIO` format. Most SageMaker algorithms work best with this format as it allows us to take advantage of **Pipe mode**. We will train a **K-Nearest Neighbors** (**KNN**) model with the generated `protobuf recordIO` file. Finally, we will use SageMaker Processing to help us automate and abstract processing jobs from the infrastructure. SageMaker Processing can be used to help us run scripts within managed infrastructure outside of our **SageMaker notebook instance**. These scripts may include tasks that process data and evaluate models using small or large datasets.

We will cover the following recipes in this chapter:

- Generating a synthetic dataset for anomaly detection experiments
- Training and deploying an RCF model
- Invoking machine learning models with Amazon Athena using SQL queries
- Analyzing data with Amazon Athena in Python
- Generating a synthetic dataset for analysis and transformation
- Performing dimensionality reduction with the built-in PCA algorithm
- Performing cluster analysis with the built-in KMeans algorithm
- Converting CSV data into `protobuf recordIO` format
- Training a KNN model using the `protobuf recordIO` training input type
- Preparing the SageMaker Processing prerequisites using the AWS CLI
- Managed data processing with SageMaker Processing in Python
- Managed data processing with SageMaker Processing in R

These recipes will prove useful to data scientists and machine learning practitioners as a major portion of the work done in experiments involves data processing, transformation, and analysis. If you are looking for recipes on pre-training and post-training bias detection with **SageMaker Clarify**, feel free to check *Chapter 7, Working with SageMaker Feature Store, SageMaker Clarify, and SageMaker Model Monitor*.

Technical requirements

To execute the recipes in the chapter, make sure you have the following:

- A running Amazon SageMaker notebook instance (for example, `ml.t2.large`)
- An Amazon S3 bucket

If you do not have these prerequisites ready yet, feel free to check the *Launching an Amazon SageMaker notebook instance* and *Preparing the Amazon S3 bucket and the training dataset for the linear regression experiment* recipes in *Chapter 1, Getting Started with Machine Learning Using Amazon SageMaker.*

As the recipes in this chapter involve a bit of code, we have made the scripts and notebooks available in this repository: `https://github.com/PacktPublishing/Machine-Learning-with-Amazon-SageMaker-Cookbook/tree/master/Chapter04`.

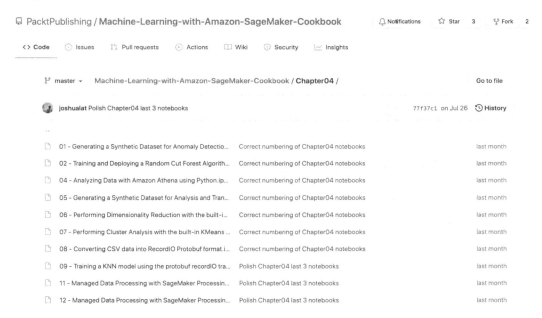

Figure 4.1 – Machine-Learning-with-Amazon-SageMaker-Cookbook GitHub repository

As seen in *Figure 4.1*, we have the source code for the scripts and notebooks for the recipes in this chapter organized inside the `Chapter04` directory.

Check out the following link to see the relevant Code in Action video:

`https://bit.ly/3he32PO`

Generating a synthetic dataset for anomaly detection experiments

In this recipe, we will generate a synthetic dataset that contains outliers or anomalies. This will enable us to perform anomaly detection experiments using algorithms such as the **Random Cut Forest** (RCF). If this is your first time hearing about anomaly detection, it is the identification of outliers or records that differ significantly from the rest of the records of the dataset. What's the RCF algorithm? The RCF algorithm is an unsupervised algorithm used for detecting these anomalies in the dataset.

After we have generated the synthetic dataset in this recipe, we will use the generated dataset to train and deploy an RCF model and trigger this model within an Amazon Athena query in the *Invoking machine learning models with Amazon Athena using SQL queries* recipe. This will enable us to tag anomalies in our dataset during the data preparation and analysis phase.

> **Tip**
> Since we will show the steps on how to generate a synthetic dataset in this recipe, you will have the opportunity to tweak this recipe later on to fit your needs. You can decide to generate more records to test the scaling capabilities of Amazon Athena and see how your queries perform when dealing with large datasets.

Getting ready

The following is the prerequisite for this recipe:

- A new Jupyter notebook using the `conda_python3` kernel within a SageMaker notebook instance

How to do it...

The first set of steps in this recipe involves setting up and preparing a few prerequisites before we generate the synthetic dataset:

1. Navigate to the `my-experiments/chapter04` directory inside your SageMaker notebook instance. Feel free to create this directory if it does not exist yet.

2. Create a new notebook using the `conda_python3` kernel inside the `my-experiments/chapter04` directory and name it with the name of this recipe (that is, `Generating a Synthetic Dataset for Anomaly Detection Experiments`). Open this notebook for editing as we will update this file with the code in the next couple of steps.

3. Start the notebook by importing a few prerequisites to generate random number and string values:

```
import random
from string import ascii_uppercase
from random import randint, choice
```

4. Define three functions, called `generate_normal_point()`, `generate_abnormal_point()`, and `generate_random_string()`:

```
def generate_normal_point():
    return randint(0, 10)

def generate_abnormal_point():
    return randint(70, 80)

def generate_random_string():
    letters = ascii_uppercase
    return ''.join(
        choice(letters) for i in range(10)
    )
```

5. Define the `normal_or_abnormal()` function:

```
def normal_or_abnormal():
    tmp = randint(0, 20)

    if tmp == 20:
        return "abnormal"
    else:
        return "normal"
```

Now that we have all the prerequisites and functions that we need, we will run a few blocks of code to generate the dataset.

6. Generate 1,000 points with a small set of numbers tagged as abnormal:

```
list_of_points = []

for _ in range (0,1000):
    point_type = normal_or_abnormal()

    point_value = 0
    string_value = generate_random_string()

    if point_type == "normal":
        point_value = generate_normal_point()
    else:
        point_value = generate_abnormal_point()

    point = {
        "label": string_value,
        "value": point_value
    }
    list_of_points.append(point)
```

The preceding block of code simply generates a record 1,000 times using a `for` loop. Out of these 1,000 records, a certain percentage (such as 5%) would be abnormal points with values significantly larger than the normal points.

7. Inspect the `list_of_points` list. All values between `70` and `80` are tagged as abnormal:

```
list_of_points
```

This should yield a set of results similar to what is shown in *Figure 4.2*:

```
[{'label': 'ZVMLWQXCGF', 'value': 10},
 {'label': 'IGYEJYTDZQ', 'value': 1},
 {'label': 'ZXGCWFNKVI', 'value': 0},
 {'label': 'KGWZAYPMUX', 'value': 2},
 {'label': 'KJHPHJPQFZ', 'value': 3},
 {'label': 'PQBCGKKSVG', 'value': 8},
 {'label': 'WOPTTXLLNV', 'value': 3},
 {'label': 'LAASUTDWXV', 'value': 8},
 {'label': 'QFMQWXDEUR', 'value': 9},
 {'label': 'BAYLBAQTWL', 'value': 5},
 {'label': 'LCYRCAZXOK', 'value': 2},
 {'label': 'JPIIJBVZCG', 'value': 5},
 {'label': 'IGBYILNJJQ', 'value': 9},
 {'label': 'YVMRVUWEMY', 'value': 7},
 {'label': 'SDUNEGGDRC', 'value': 8},
 {'label': 'SPLQISZBPS', 'value': 10},
 {'label': 'QVQEIUWUAP', 'value': 6},
 {'label': 'OYOHNOPISD', 'value': 5},
 {'label': 'PZJQMPOOYJ', 'value': 71},
```

Figure 4.2 – Label and value pairs inside the list_of_points list

In *Figure 4.2*, we can see the string label and integer value pairs inside the `list_of_points` list. We should have significantly more normal points between 0 and 10 than abnormal points between 70 and 80.

8. Render a quick plot of the list of points generated:

```python
import pandas as pd
import matplotlib.pyplot as plt
plt.rcParams['figure.figsize'] = [15, 5]
pd.DataFrame(list_of_points).plot()
```

This should render a plot similar to what is shown in *Figure 4.3*:

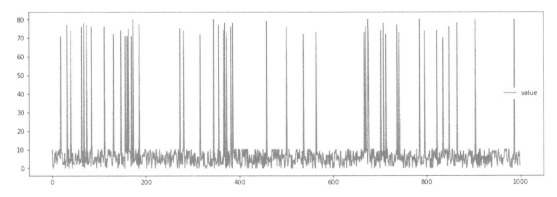

Figure 4.3 – Plot of the synthetic dataset containing outliers

In *Figure 4.3*, we have a plot of the synthetic dataset showing the normal data points and the outliers. The long, vertical lines indicate the presence of outliers as these abnormal data points have a value between 70 and 80. Note that the normal data points have a value between 0 and 10.

In the final set of steps, we focus our efforts on uploading this dataset to an Amazon S3 bucket.

9. Prepare the directory named s3_files, which we will use to store the generated points in JSON format:

```
!rm -rf s3_files
!mkdir -p s3_files
```

10. Define the save_json_file() function. This function accepts an element of list_of_points and creates a JSON file inside the directory:

```
import json

def save_json_file(point):
    label = point['label']
    filename = "s3_files/" + label + '.json'
    with open(filename, 'w') as file:
        json.dump(point, file)
        print(f"Saved {label}!")
```

> **Note**
>
> When working with file paths, it is recommended to use os.path.join() to concatenate paths. Note that we have taken a shortcut here to simplify this recipe a bit.

11. Use the save_json_file() function for each of the points generated inside list_of_points:

```
for point in list_of_points:
    save_json_file(point)
```

At this point, we should have a total of 1,000 files inside the s3_files directory.

12. Create a new S3 bucket to store the Athena results. Note that this should be different from the S3 bucket we created in *Chapter 1, Getting Started with Machine Learning Using Amazon SageMaker*:

```
bucket_name = "<insert S3 bucket name here>"
!aws s3 mb s3://{bucket_name}
```

This should yield a log message similar to `make_bucket: <bucket name>`.

> **Important Note**
>
> Take note of the S3 bucket name here as we will use this bucket name in the *Invoking machine learning models with Amazon Athena using SQL queries* recipe.

13. Copy and upload the files inside the `s3_files` directory to the S3 bucket we just created:

```
!aws s3 cp s3_files/ s3://{bucket_name}/ --recursive
```

This should yield some log messages similar to what is shown in *Figure 4.4*:

```
upload: s3_files/ADUGCFEYWU.json to s3://sagemaker-cookbook-anomaly-detection-data-bucket/ADUGCFEYWU.json
upload: s3_files/ACORCGUENY.json to s3://sagemaker-cookbook-anomaly-detection-data-bucket/ACORCGUENY.json
upload: s3_files/AECUYQSMLR.json to s3://sagemaker-cookbook-anomaly-detection-data-bucket/AECUYQSMLR.json
upload: s3_files/ADPSERBSZR.json to s3://sagemaker-cookbook-anomaly-detection-data-bucket/ADPSERBSZR.json
upload: s3_files/AHQGYKKMTS.json to s3://sagemaker-cookbook-anomaly-detection-data-bucket/AHQGYKKMTS.json
upload: s3_files/AEPVNPINSE.json to s3://sagemaker-cookbook-anomaly-detection-data-bucket/AEPVNPINSE.json
upload: s3_files/AHNZKMXVOE.json to s3://sagemaker-cookbook-anomaly-detection-data-bucket/AHNZKMXVOE.json
upload: s3_files/ALOTJURVEK.json to s3://sagemaker-cookbook-anomaly-detection-data-bucket/ALOTJURVEK.json
upload: s3_files/ALJDPCEZML.json to s3://sagemaker-cookbook-anomaly-detection-data-bucket/ALJDPCEZML.json
upload: s3_files/AOGVSRCPFU.json to s3://sagemaker-cookbook-anomaly-detection-data-bucket/AOGVSRCPFU.json
upload: s3_files/AGCBUFOWUD.json to s3://sagemaker-cookbook-anomaly-detection-data-bucket/AGCBUFOWUD.json
upload: s3_files/ALUFUVWJMM.json to s3://sagemaker-cookbook-anomaly-detection-data-bucket/ALUFUVWJMM.json
upload: s3_files/AESBBDYCPY.json to s3://sagemaker-cookbook-anomaly-detection-data-bucket/AESBBDYCPY.json
```

Figure 4.4 – Uploading the files to the S3 bucket

We can see in *Figure 4.4* that the files stored in the `s3_files` directory are being uploaded one at a time to the S3 bucket we just created.

14. Finally, use the `%store` magic to store the `list_of_points` variable for later use:

```
%store list_of_points
```

This should yield a log message similar to `Stored 'list_of_points' (list)`.

Now, let's see how this works!

How it works...

In this recipe, we have generated a synthetic dataset that involves a simple label-value pair. Around 5% of this synthetic dataset are outliers and we have intentionally made the outlier values much higher than the "normal" values.

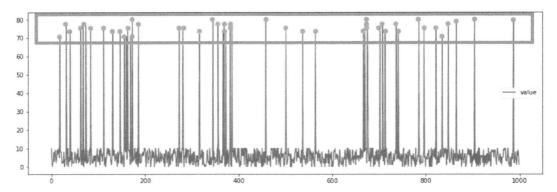

Figure 4.5 – Synthetic dataset with outlier values

We can see in *Figure 4.5* that the outliers have values in the 70 to 80 range, while the "normal" points have values in the 0 to 10 range. This is expected as the generate_normal_point() and generate_abnormal_point() functions dictate what the highest and lowest possible values are for these randomly generated numbers.

Why are we doing this? We will use this dataset to train and deploy an RCF model and demonstrate how a deployed model can be triggered within an Amazon Athena query. We will do this later in the *Invoking machine learning models with Amazon Athena using SQL queries* recipe. This will enable us to tag anomalies in our dataset during the data preparation and analysis phase.

> **Note**
>
> This feature of Amazon Athena has been in preview mode for quite some time and now that AWS has recently announced the general availability of this capability, we have decided to include these recipes here! Feel free to check this page for more information: https://aws.amazon.com/about-aws/whats-new/2021/04/announcing-general-availability-of-amazon-athena-ml-powered-by-amazon-sagemaker/.

Training and deploying an RCF model

In this recipe, we will train and deploy an RCF model using the **SageMaker Python SDK**. The RCF algorithm is an unsupervised algorithm used for detecting anomalies in a dataset. It associates each record with an anomaly score value, with higher anomaly score values associated with records that may potentially be tagged as outliers or anomalies.

After we have trained and deployed an RCF model in this recipe, we will trigger this model within an Amazon Athena SQL query in the *Invoking machine learning models with Amazon Athena using SQL queries* recipe. This will enable us to tag anomalies in our dataset during the data preparation and analysis phase.

Getting ready...

This recipe continues from *Generating a synthetic dataset for anomaly detection experiments*.

How to do it...

The next set of steps focus on using the dataset we generated in the previous recipe to prepare the RCF model:

1. Navigate to the `my-experiments/chapter04` directory inside your SageMaker notebook instance. Feel free to create this directory if it does not exist yet.

2. Create a new notebook using the `conda_python3` kernel inside the `my-experiments/chapter04` directory and name it with the name of this recipe (that is, `Training and Deploying a Random Cut Forest Model`). Open this notebook for editing as we will update this file with the code in the next couple of steps.

3. Start the notebook by using the `%store` magic command to load the stored list of values for `list_of_points`:

```
%store -r list_of_points
list_of_points
```

We should get a list of dictionaries similar to what is shown in *Figure 4.6*:

```
[{'label': 'ZVMLWQXCGF', 'value': 10},
 {'label': 'IGYEJYTDZQ', 'value': 1},
 {'label': 'ZXGCWFNKVI', 'value': 0},
 {'label': 'KGWZAYPMUX', 'value': 2},
 {'label': 'KJHPHJPQFZ', 'value': 3},
 {'label': 'PQBCGKKSVG', 'value': 8},
 {'label': 'WOPTTXLLNV', 'value': 3},
 {'label': 'LAASUTDWXV', 'value': 8},
 {'label': 'QFMQWXDEUR', 'value': 9},
 {'label': 'BAYLBAQTWL', 'value': 5},
 {'label': 'LCYRCAZXOK', 'value': 2},
 {'label': 'JPIIJBVZCG', 'value': 5},
 {'label': 'IGBYILNJJQ', 'value': 9},
 {'label': 'YVMRVUWEMY', 'value': 7},
 {'label': 'SDUNEGGDRC', 'value': 8},
 {'label': 'SPLQISZBPS', 'value': 10},
```

Figure 4.6 – list_of_points

As we can see in *Figure 4.6*, it should contain the exact same set of values as the `list_of_points` list from the *Generating a synthetic dataset for anomaly detection experiments* recipe.

4. Extract the values using the `map()` function and store them in a list:

```
point_values = list(map(lambda x: x["value"], list_of_
points))
point_values
```

We should get a list of results similar to `[10, 1, 0, 2, 3, 8, 3, … ]`.

5. Use numpy to reshape the list:

```
import numpy as np
np_array = np.array(point_values)
np_array = np_array.reshape(-1,1)
np_array
```

We should get a set of results with a structure similar to `array([[10],[1],[0], [2],…])`.

6. Initialize and prepare a few prerequisites such as `role_arn` and `session` for the SageMaker experiments:

```
import sagemaker
from sagemaker import get_execution_role
role_arn = get_execution_role()
session = sagemaker.Session()
```

7. Initialize the `RandomCutForest` estimator:

```
from sagemaker import RandomCutForest
estimator = RandomCutForest(
    role_arn,
    instance_count=1,
    instance_type='ml.m5.xlarge',
    sagemaker_session=session)
```

8. Run the `fit()` function to start preparing the model:

```
record_set_input = estimator.record_set(np_array)
estimator.fit(record_set_input)
```

This should yield a set of logs similar to what is shown in *Figure 4.7*:

```
Defaulting to the only supported framework/algorithm version: 1. Ignoring framework/algorithm version: 1.
Defaulting to the only supported framework/algorithm version: 1. Ignoring framework/algorithm version: 1.

2021-04-18 05:37:15 Starting - Starting the training job...
2021-04-18 05:37:17 Starting - Launching requested ML instancesProfilerReport-1618724234: InProgress
......
2021-04-18 05:38:42 Starting - Preparing the instances for training......
2021-04-18 05:39:43 Downloading - Downloading input data
2021-04-18 05:39:43 Training - Downloading the training image...
2021-04-18 05:40:13 Uploading - Uploading generated training model
2021-04-18 05:40:13 Completed - Training job completed
Docker entrypoint called with argument(s): train
Running default environment configuration script
[04/18/2021 05:40:03 INFO 140512099010368] Reading default configuration from /opt/amazon/lib/python3.7/site-package
s/algorithm/resources/default-conf.json: {'num_samples_per_tree': 256, 'num_trees': 100, 'force_dense': 'true', 'eval
_metrics': ['accuracy', 'precision_recall_fscore'], 'epochs': 1, 'mini_batch_size': 1000, '_log_level': 'info', '_kvs
tore': 'dist_async', '_num_kv_servers': 'auto', '_num_gpus': 'auto', '_tuning_objective_metric': '', '_ftp_port': 899
9}
```

Figure 4.7 – Logs after using fit() to prepare the RCF model

It should take a few minutes before we have the model ready. Note that we are not using labeled data during this step as the RCF algorithm is unsupervised.

9. Use the `deploy()` function to deploy the model to an inference endpoint. Here we specify the `endpoint_name` value as we will use this endpoint name later in the SQL query triggering this machine learning endpoint:

```
predictor = estimator.deploy(
    initial_instance_count=1,
    instance_type="ml.m5.xlarge",
    endpoint_name="sagemaker-cookbook-rcf")
```

10. Call the `predict()` function to get the anomaly scores of each of the points passed to the inference endpoint:

```
results = predictor.predict(np_array)
results
```

This should yield a set of results similar to what is shown in *Figure 4.8*:

```
[label {
    key: "score"
    value {
      float32_tensor {
        values: 1.3119031190872192
      }
    }
  },
  label {
    key: "score"
    value {
      float32_tensor {
        values: 1.2045546770095825
      }
    }
  },
  label {
    key: "score"
```

Figure 4.8 – Anomaly score results

As seen in *Figure 4.8*, the inference endpoint returns the corresponding anomaly score to each of the data points passed as the payload to the `predict()` function. The higher the value, the higher the chance that the data point is an outlier or anomaly.

Important Note

Do not delete the inference endpoint until you have completed the *Analyzing data with Amazon Athena in Python*.

Now, let's see how this works!

How it works...

In this recipe, we used the synthetic dataset from *Generating a synthetic dataset for anomaly detection experiments* to prepare and deploy the RCF model. When using the RCF algorithm, the data points are associated with an anomaly score and anomalies are associated with higher scores.

Using the RCF algorithm is pretty straightforward. All we need to do is pass a set of numerical values inside a reshaped list as the payload to the `fit()` function to prepare the model for a few minutes. After we have deployed the model to the inference endpoint using the `deploy()` function, we can pass the different values as the payload to the `predict()` function to get the corresponding anomaly score values.

See also

If you are looking for examples of using RCF models to detect anomalies in real datasets and more complex examples, feel free to check some of the notebooks in the `aws/amazon-sagemaker-examples` GitHub repository along with notebooks shared in AWS webinar videos:

- Using an RCF model to detect anomalies in the Numenta Anomaly Benchmark NYC Taxi dataset: `https://github.com/aws/amazon-sagemaker-examples/blob/35e2faf7d1cc48ccedf0b2ede1da9987a18727a5/introduction_to_amazon_algorithms/random_cut_forest/random_cut_forest.ipynb`

- Training and deploying an RCF model for fraud detection: `https://github.com/awslabs/fraud-detection-using-machine-learning/blob/master/source/notebooks/sagemaker_fraud_detection.ipynb`

Now, let's take a closer look at how to invoke the RCF model we deployed in this recipe in Amazon Athena with SQL statements in the next recipe!

Invoking machine learning models with Amazon Athena using SQL queries

Amazon Athena is a **serverless** interactive query service that helps us analyze data in Amazon S3 using SQL syntax. As it is a serverless service, machine learning practitioners no longer need to manage any infrastructure, so we can focus on the work that needs to be done. If you have used or heard of Amazon Athena before, you must be aware that this solution can easily scale and support **big data** requirements. Amazon Athena also supports a variety of data formats (such as CSV and text files), columnar formats (such as Parquet and ORC), and compressed data formats (such as Snappy and GZIP).

> **Note**
>
> Of course, this is a simplified description of what serverless is all about. Feel free to check `https://aws.amazon.com/serverless/` for more information.

In this recipe, we will use Amazon Athena to analyze our dataset stored in Amazon S3 using SQL statements. We will make use of a deployed machine learning model within our query to detect anomalies in our data.

Getting ready

This recipe continues from *Training and deploying an RCF model.*

How to do it...

The first set of steps focuses on making sure that we are using the Athena engine version 2:

1. Navigate to the Athena console. If you see a notification similar to what is shown in *Figure 4.9,* click the **Edit workgroup page** link.

Figure 4.9 – Athena console showing the New Athena query engine available notification

If you don't see the notification message in *Figure 4.9*, you should be able to navigate to the list of workgroups by clicking the **Workgroup : primary** tab in the navigation bar, as shown in the following screenshot:

Figure 4.10 – Navigation menu

This should redirect us to the workgroups list page showing the existing workgroups. Next, click the **View details** button, as seen in *Figure 4.11*:

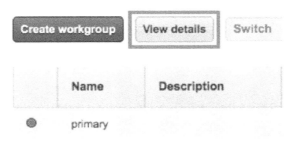

Figure 4.11 – List of existing Athena workgroups

Finally, click the **Edit workgroup** button, as seen in *Figure 4.12*:

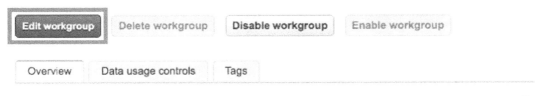

Figure 4.12 – Location of the Edit workgroup button on the workgroup details page

As we can see in *Figure 4.12*, we have the details of a specific workgroup on the workgroup details page. We also have the option to edit, disable, or delete the workgroup.

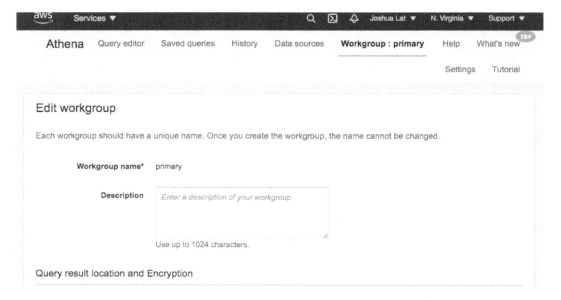

Figure 4.13 – Edit workgroup page

Once we are on the **Edit workgroup** page, we can proceed with the next step.

2. On the **Edit workgroup** page, specify the **Query result location** path where we will store the query results.

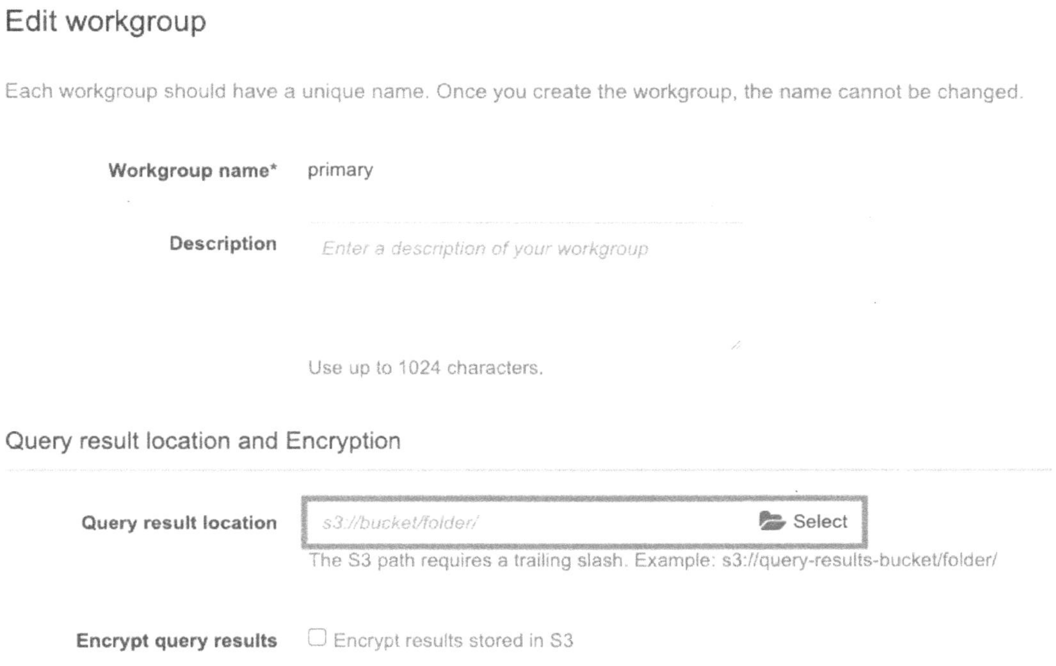

Edit workgroup

Each workgroup should have a unique name. Once you create the workgroup, the name cannot be changed.

Workgroup name*   primary

Description   *Enter a description of your workgroup*

Use up to 1024 characters.

Query result location and Encryption

Query result location   s3://bucket/folder/   📂 Select

The S3 path requires a trailing slash. Example: s3://query-results-bucket/folder/

Encrypt query results   ☐ Encrypt results stored in S3

Figure 4.14 – Selecting a query result location

Refer to *Figure 4.14* to see which field to update for this step. Clicking the **Select** button will open a popup where we can select the target S3 bucket and the folder.

3. Next, scroll down to the **Query engine version** section and make sure that the **Athena engine version 2 (recommended)** option is selected.

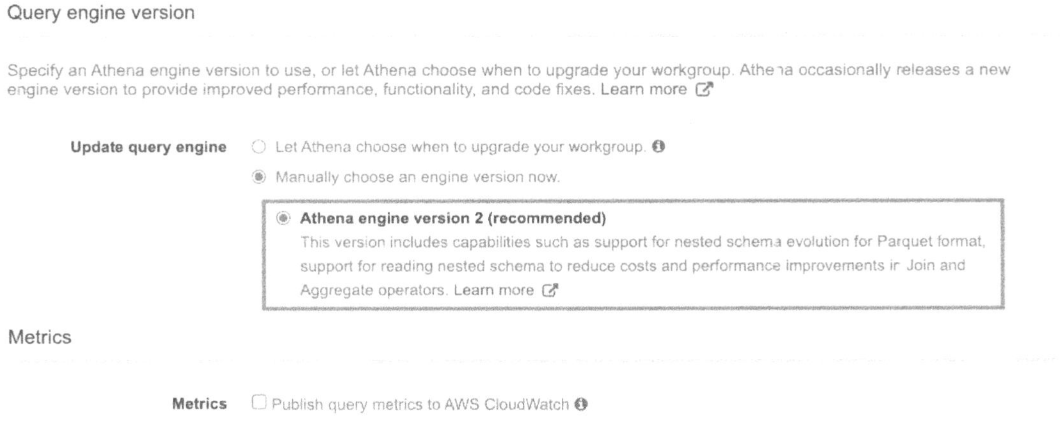

Query engine version

Specify an Athena engine version to use, or let Athena choose when to upgrade your workgroup. Athena occasionally releases a new engine version to provide improved performance, functionality, and code fixes. Learn more ☑

Update query engine   ○ Let Athena choose when to upgrade your workgroup. ❶

⦿ Manually choose an engine version now.

⦿ Athena engine version 2 (recommended)
This version includes capabilities such as support for nested schema evolution for Parquet format, support for reading nested schema to reduce costs and performance improvements in Join and Aggregate operators. Learn more ☑

Metrics

Metrics   ☐ Publish query metrics to AWS CloudWatch ❶

Figure 4.15 – Using Athena engine version 2

As in *Figure 4.15*, make sure that you have selected **Manually choose an engine version now** as well.

4. Scroll down to the end of the page, and then click the **Save** button.

> **Tip**
> If you encounter issues working with workgroups, feel free to check the tips shared on this page: `https://docs.aws.amazon.com/athena/latest/ug/workgroups-troubleshooting.html`.

5. After we have been redirected back to the workgroup details page, click **Query editor** in the navigation bar.

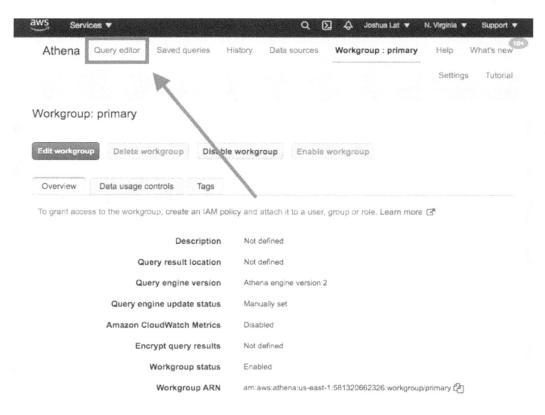

Figure 4.16 – Workgroup details page

As we can see in *Figure 4.16*, the **Query editor** tab should be located in the upper-left corner of the page (beside **Saved queries**).

6. On the left-hand side of the **Query editor** page, look for the **Data source** pane and click the **Connect data source** link.

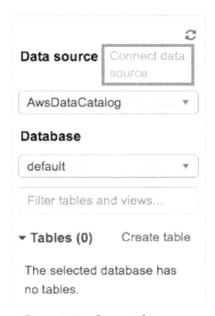

Figure 4.17 – Connect data source

Note that the **Data source** pane in *Figure 4.17* is located in the left-hand corner of the **Query editor** page. We should see the **Connect data source** link just below the refresh icon.

7. Make sure that the **Query data in Amazon S3** and **AWS Glue Data Catalog** options are selected. Click the **Next** button.

Connect data source

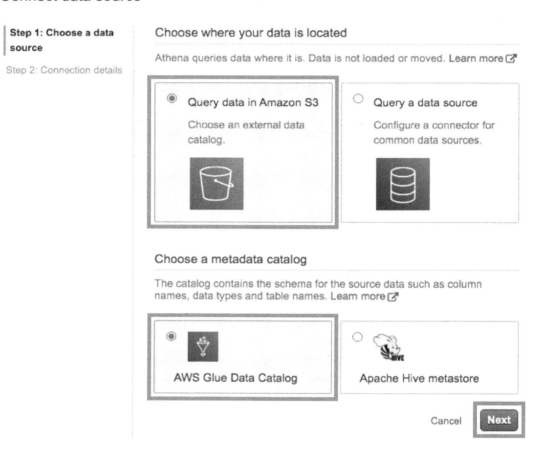

Figure 4.18 – Choose a data source

In *Figure 4.18*, we have selected **Query data in Amazon S3** as our dataset is inside an S3 bucket.

8. Choose **AWS Glue Data Catalog in this account**. Next, choose **Create a table using the Athena table wizard**. Click the **Continue to add table** button.

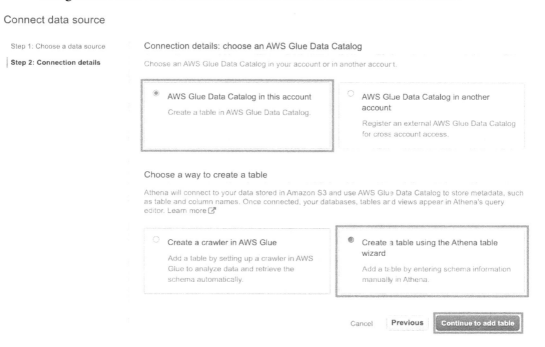

Figure 4.19 – Connection details

Clicking the **Continue to add table** button, as seen in *Figure 4.19*, would then have us proceed with creating a new database and adding a table in that database in the next set of steps.

9. Choose **Create a new database** for **Database**. Specify `cookbook_athena_db` for the database name. For the **Table Name** field, specify `athena_table`. Specify the S3 bucket name (that is, `s3://<bucket name>`) where the JSON files in the *Generating a synthetic dataset for anomaly detection experiments* recipe were uploaded in the **Location of Input Data Set** field. Click the **Next** button afterward.

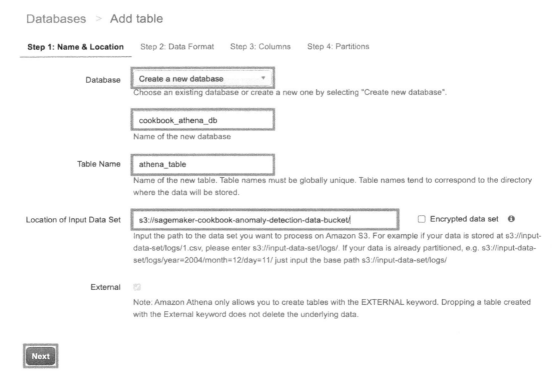

Figure 4.20 – Add table form

Make sure to check that the field values specified are the same as what we have in *Figure 4.20*.

10. Under **Step 2: Data Format**, select **JSON** and then click **Next**.

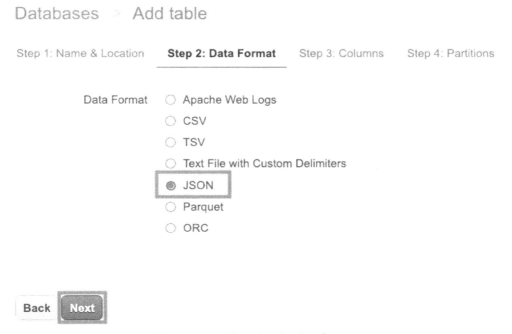

Figure 4.21 – Choosing the data format

In *Figure 4.21*, we can see that multiple data formats are supported. These include Apache web logs, CSV, and TSV. As we have uploaded JSON files inside the S3 bucket, we will choose **JSON** from this option group.

11. Under **Step 3: Columns**, specify two columns—**Column Name**: `label` and **Column type**: `string` and **Column Name**: `value`, and **Column type**: `int`. Click **Next**.

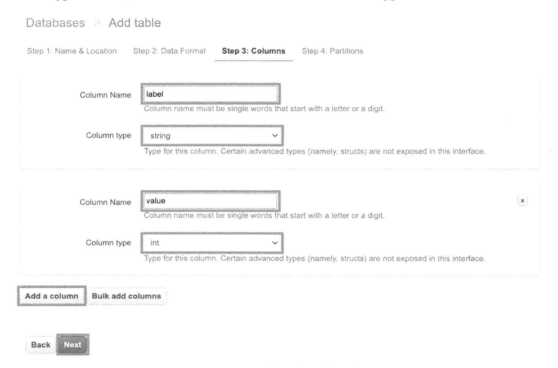

Figure 4.22 – Specifying the table columns

Make sure to check that the field values specified are the same as what we have in *Figure 4.22*.

12. Under **Step 4: Partitions**, click the **Create table** button.

Databases   Add table

Step 1: Name & Location   Step 2: Data Format   Step 3: Columns   **Step 4: Partitions**

Configure Partitions (Optional)

Partitions are a way to group specific information together. Partition are virtual columns. In case of partitioned tables, subdirectories are created under the table's data directory for each unique value of a partition column. In case the table is partitioned on multiple columns, then nested subdirectories are created based on the order of partition columns in the table definition. Learn more.

Back   Create table

Figure 4.23 – Adding a partition

Note that we will not add any partitions in this step, as seen in *Figure 4.23*. We simply click the **Create table** button to complete the table creation process.

Tip
For more information on how partitions can be used to improve query performance and reduce cost, feel free to check `https://docs.aws.amazon.com/athena/latest/ug/partitions.html`.

13. Beside the **New query 1** or **New query 2** tab, click the + icon to create a new tab where we can run a new query.

Figure 4.24 – Creating a new tab

This should open a new tab, as shown in *Figure 4.25*:

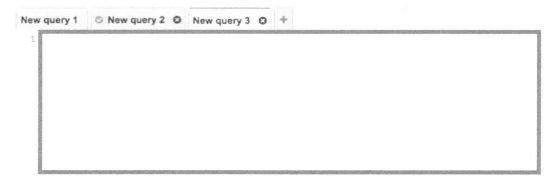

Figure 4.25 – New tab

Click the text area inside the new tab as we have in *Figure 4.25*.

14. Type the following SQL statement in the query editor pane. Click the **Run query** button to execute the query:

```
SELECT label, value FROM cookbook_athena_db.athena_table
LIMIT 10;
```

If everything is configured and set up correctly, we should get a set of results similar to what is shown in *Figure 4.26*:

Figure 4.26 – Query results

In *Figure 4.26*, we can see the query results shown at the bottom of the query editor pane. Note that each of the items in the query results corresponds to a JSON file uploaded in the S3 bucket we prepared earlier.

15. Next, update the query in the text area with the following SQL query. Click the **Run query** button to execute the query. Note that the SageMaker inference endpoint with the name `sagemaker-cookbook-rcf` must be running for this query to execute:

```
USING EXTERNAL FUNCTION detect_anomaly(value INT)
    RETURNS DOUBLE
    SAGEMAKER 'sagemaker-cookbook-rcf'
SELECT label, value, detect_anomaly(value) AS anomaly_
score
    FROM cookbook_athena_db.athena_table
```

We should get results similar to what is shown in *Figure 4.27*:

Figure 4.27 – Query results returning an anomaly_score value

In *Figure 4.27*, we can see that we were able to use the deployed RCF model within the SQL statement. Note that we are not limited to just using a deployed RCF model inside an Athena SQL query as we can use the other deployed models here as well. Feel free to add "LIMIT 1000" at the end of the SQL statement if you encounter a ThrottlingException.

Let's see how this works!

How it works...

In this recipe, we have generated and stored sample files inside our S3 bucket for our dummy data source. Then, we used SQL statements to query the data stored in the S3 bucket using Amazon Athena.

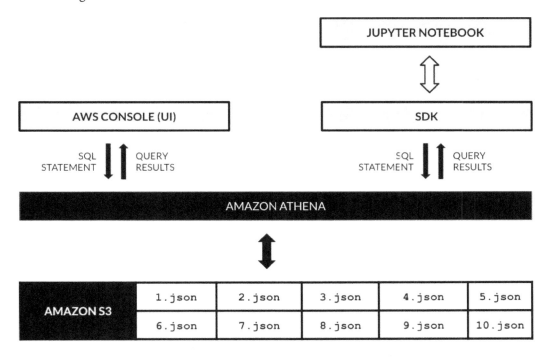

Figure 4.28 – Analyzing data stored in S3 with SQL statements using Amazon Athena

In *Figure 4.28*, we can see that Amazon Athena helps us query and analyze data inside the files stored in our S3 buckets using SQL statements. Take note that one of the advantages of using Amazon Athena is that it is a serverless query service—there are no servers to manage. In addition to this, the pricing is also proportional to its usage. This makes it one of the most practical solutions to use, especially when dealing with big data requirements. In addition to not worrying about having to manage infrastructure yourself, you will not have to worry about the costs of underutilized or idle resources. Using a similar setup as in this recipe, Amazon Athena can easily scale and process much larger datasets and still respond within a few seconds. Of course, there are tweaks, configurations, and best practices (for example, partitioning and compression) that we can perform to tune the performance of this setup.

We have also made use of a deployed RCF model within the SQL query to detect anomalies within our dataset. Note that we are not limited to just performing anomaly detection here and we can use other different types of deployed models, such as using **XGBoost** or **Factorization Machines**, for example.

Analyzing data with Amazon Athena in Python

Amazon Athena is a query service from AWS to query data stored in Amazon S3 using SQL syntax. In the previous recipe, we ran a couple of SQL queries using the Athena console (UI). Of course, when working with machine learning and machine learning engineering tasks, we want this performed through a script so that we have the opportunity to automate certain steps of the process.

In this recipe, we will use the **boto3 Python SDK** to programmatically run Amazon Athena SQL queries. Once we have completed this recipe, we will have the JSON data stored in our S3 bucket loaded, queried, and transformed into a tabular format using Amazon Athena using Python. We will perform two queries in this recipe—a simple SELECT query and a query that invokes a deployed machine learning model in Amazon SageMaker.

Getting ready

This recipe continues from *Invoking machine learning models with Amazon Athena using SQL queries*.

How to do it...

The next couple of steps focus on using the Python SDK to programmatically run Athena queries without using the UI. Make sure that you are running the next set of commands using an empty notebook using the conda_python3 kernel:

1. Navigate to the my-experiments/chapter04 directory inside your SageMaker notebook instance. Feel free to create this directory if it does not exist yet.

2. Create a new notebook using the conda_python3 kernel inside the my-experiments/chapter04 directory and name it with the name of this recipe (that is, Analyzing Data with Amazon Athena in Python). Open this notebook for editing as we will update this file with the code in the next couple of steps.

3. Start the notebook by importing boto3 and initializing the Athena client:

```
import boto3
athena = boto3.client('athena', region_name='us-east-1')
```

4. Specify the value for `athena_results_bucket`. Note that this is a different S3 bucket and we will use this bucket to store the results of the Athena queries:

```
athena_results_bucket = "<insert S3 bucket name here>"
!aws s3 mb s3://{athena_results_bucket}
```

This should yield a log message similar to `make_bucket: <bucket name>`.

5. Initialize the variable values for `query`, `database`, and `results_bucket`. Make sure to replace the database name, table name, and S3 target bucket in the following variables with the values used in the *Invoking machine learning models with Amazon Athena in SQL queries* recipe:

```
query = "SELECT label, value FROM cookbook_athena_
db.athena_table;"
database = "cookbook_athena_db"
results_bucket = "s3://" + athena_results_bucket
```

It is important to note that the preceding block of code is composed only of three lines (in case the block of code gets rendered as four or more lines because of the length of the statements).

6. Define the `execute_athena_query()` function, which uses the `start_query_execution()` function from the boto3 SDK to run the query specified in the parameter:

```
def execute_athena_query(query, database, results_
bucket):
    response = athena.start_query_execution(
        QueryString = query,
        QueryExecutionContext = {
            'Database' : database
        },
        ResultConfiguration = {
            'OutputLocation': results_bucket
        }
    )
    return response['QueryExecutionId']
```

7. Define the `get_output_path()` function, which uses the `get_query_execution()` function to load the details of the Athena query using a specified execution ID and returns the output location where the Athena query results are stored:

```
def get_output_path(execution_id):
    query_details = athena.get_query_execution(
        QueryExecutionId = execution_id
    )
    execution = query_details['QueryExecution']
    configuration = execution['ResultConfiguration']
    return configuration['OutputLocation']
```

8. Call the `execute_athena_query()` and `get_output_path()` functions. Note that the Athena query may run for a few seconds so we may need to wait for 3-5 seconds before the output CSV file becomes available in the S3 output path:

```
execution_id = execute_athena_query(query, database,
results_bucket)
output_path = get_output_path(execution_id)
output_path
```

We should get a value similar to `'s3://<bucket name>/64957fbb-b873-48ec-91aa-7377343da412.csv'`.

9. Create the `tmp` directory if it does not exist yet:

```
!mkdir -p tmp
```

10. Download the CSV file containing the results of the query to the `tmp` directory created in the previous step:

```
!aws s3 cp {output_path} tmp/output.csv
```

This should yield a log message similar to `download: s3://<bucket name>/97d9da6f-6426-46a1-b775-c96580fd29f4.csv to tmp/output.csv`.

11. Load and inspect the contents of `tmp/output.csv` using the `read_csv()` function:

```
import pandas as pd
pd.read_csv("tmp/output.csv")
```

We should get a set of results similar to what is shown in *Figure 4.29*:

	label	value
0	GCWKINKQYM	6
1	LTBDZXYYZB	78
2	UIRHTTIVJQ	10
3	QFIKGEMAYH	4
4	OSUQRHPDQX	3
...	...	...
995	QCWDWGUFQL	7
996	SCOWIGYIHW	10
997	ZUFCRCVYMD	4
998	FIWVSQWNJS	9
999	OWCDZSRNTF	2

1000 rows × 2 columns

Figure 4.29 – Contents of the output.csv file

In *Figure 4.29*, we can see the label and value pairs inside the output.csv file. Each record here should map properly to the records we generated earlier in the *Generating a synthetic dataset for anomaly detection experiments* recipe.

12. Set the query string value to the Athena SQL query that triggers the existing SageMaker inference endpoint. This is the same query we used in the console in the *Invoking machine learning models with Amazon Athena in SQL queries* recipe:

```
query = """
USING EXTERNAL FUNCTION detect_anomaly(value INT)
    RETURNS DOUBLE
    SAGEMAKER 'sagemaker-cookbook-rcf'
SELECT label, value, detect_anomaly(value) AS anomaly_
score
    FROM cookbook_athena_db.athena_table
"""
```

13. Call the `execute_athena_query()` and `get_output_path()` functions again but this time using the query that triggers the machine learning endpoint. Note that the Athena query may run for a few seconds so we may need to wait for 3-5 seconds before the output CSV file becomes available in the S3 output path:

```
execution_id = execute_athena_query(query,
    database, results_bucket)
output_path = get_output_path(execution_id)
output_path
```

This should yield an output similar to `'s3://<bucket name>/d457328e-b456-4d11-a012-6ea26a22ceb9.csv'`.

14. Download the CSV file containing the results of the query to the `tmp` directory:

```
!aws s3 cp {output_path} tmp/output.csv
```

This should yield a log message similar to `download: s3://<bucket name>/d457328e-b456-4d11-a012-6ea26a22ceb9.csv to tmp/output.csv`.

15. Load and inspect the contents of `tmp/output.csv`:

```
df = pd.read_csv("tmp/output.csv")
df
```

This should give us a `DataFrame` with a structure similar to what is shown in *Figure 4.30*:

	label	value	anomaly_score
0	TWQNHWXFHX	3	0.931371
1	DAVLHEUSFA	10	1.311903
2	DGOPPHCDLB	10	1.311903
3	THNNUOYJVZ	6	0.828076
4	FVHAEGAHGQ	10	1.311903
...	...	...	...
995	WRMIRAXDUP	10	1.311903
996	QWNYXWMTNZ	5	0.848408
997	RRBZBPZEOW	3	0.931371
998	WUYBRZQEXF	76	2.603616
999	EZZOWESLKD	5	0.848408

1000 rows × 3 columns

Figure 4.30 – Query results including anomaly_score

In *Figure 4.30*, we can see that each record now has a corresponding anomaly_score value.

16. Check the number of records with an anomaly_score value greater than 2:

```
len(df[df.anomaly_score > 2])
```

We should get a value equal or close to 47. Note that we just selected 2 as an arbitrary number. Given that higher anomaly_score values indicate a higher risk of a data point being an outlier, choosing 2 is a good start. Note that other machine learning practitioners and data scientists may use a formal definition and formula for an outlier but we will skip that here in this recipe.

At this point, we can now delete the endpoint we deployed in the *Training and deploying an RCF model* recipe. Now, let's see how this recipe works!

How it works...

In this recipe, we have used boto3 to programmatically run Amazon Athena SQL queries inside our Python notebook. As discussed in the previous recipe, Amazon Athena helps us analyze data inside the files stored in our S3 buckets without us having to worry about server and infrastructure management.

The steps performed in this recipe can be summarized by the following diagram:

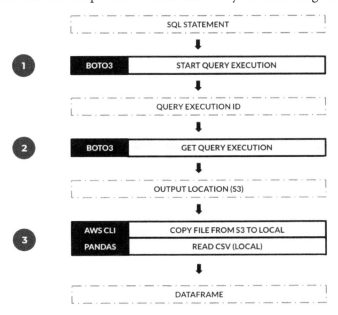

Figure 4.31 – Using boto3 and pandas to perform a query on data stored in S3 using SQL statements

In *Figure 4.31*, we can see that there are a couple of steps involved when using boto3 and Amazon Athena to perform SQL queries in order to get the results inside a `pandas` DataFrame. Once we have the results in a `pandas` DataFrame, the next steps may include processing, transforming, and analyzing these results before performing the model training step.

Generating a synthetic dataset for analysis and transformation

In this recipe, we will generate a synthetic dataset that will be used in the next four recipes involving dimensionality reduction, cluster analysis, and conversion to `protobuf` `recordIO` format. We will generate one labeled version of the dataset and one unlabeled version of the dataset. This dataset will have two easily identifiable clusters, as shown in *Figure 4.32*. It will also have six columns for the labeled version of the dataset and five columns for the unlabeled version of the dataset.

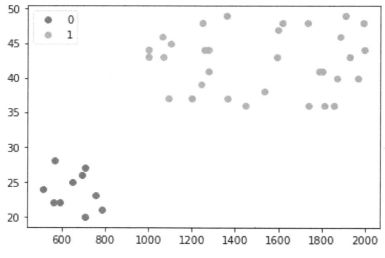

Figure 4.32 – Synthetic dataset

After we have completed this recipe, we should have a synthetic dataset similar to what is shown in *Figure 4.32*. In the *Performing dimensionality reduction with the built-in PCA algorithm* recipe, we will use the PCA algorithm to perform dimensionality reduction with this synthetic dataset. In the *Performing cluster analysis with the built-in KMeans algorithm* recipe, we will process the unlabeled version of this dataset and use the KMeans unsupervised learning algorithm to automatically detect the two clusters and assign the points to the closest cluster. We will also show how to convert the CSV data into `protobuf recordIO` format in the *Converting CSV data into protobuf recordIO format* recipe using the labeled version of this dataset.

> **Tip**
> Since we will show the steps on how to generate a synthetic dataset in this recipe, we will have the opportunity to tweak this recipe later on to fit our needs. We will discuss a few different ways to tweak it to test the performance and behavior of different algorithms in the *How it works…* section of this recipe.

Getting ready

The following is the prerequisite for this recipe:

- A new Jupyter notebook using the `conda_python3` kernel inside a SageMaker notebook instance

How to do it...

The first set of steps focuses on generating random x and y integer values:

1. Navigate to the `my-experiments/chapter04` directory inside your SageMaker notebook instance. Feel free to create this directory if it does not exist yet.

2. Create a new notebook using the `conda_python3` kernel inside the `my-experiments/chapter04` directory and name it with the name of this recipe (that is, `Generating a Synthetic Dataset for Analysis and Transformation`). Open this notebook for editing as we will update this file with the code in the next couple of steps.

3. Use the `seed` functions from `numpy` and `random` so that we can generate the same set of random numbers to make our experiment reproducible:

```
import random
from numpy.random import seed as np_seed
from random import randint
```

```
np_seed(42)
random.seed(42)
```

4. Define the `generate_x_value()` and `generate_y_value()` functions:

```
def generate_x_value():
    return randint(500,2000)

def generate_y_value():
    return randint(20,50)
```

With these functions, we are expected to get x values between 500 and 2000 and we are expected to get y values between 20 and 50.

5. Using the `generate_x_value()` function, we generate 100 random values and store them inside the `x_values` list:

```
x_values = []

for _ in range(0, 100):
    x_values.append(generate_x_value())

x_values[0:5]
```

We would get values similar to [1809, 728, 551, 1063, 1001].

6. Similarly, we generate 100 values using the `generate_y_value()` function and store these values inside the `y_values` list:

```
y_values = []

for _ in range(0, 100):
    y_values.append(generate_y_value())

y_values[0:5]
```

We would get values similar to [27, 41, 30, 46, 44].

The next set of steps focuses on generating derived values from the original x and y values we have prepared in the previous set of steps. We will use a couple of one-line statements using `list()`, `map()`, and lambda anonymous functions to generate a list containing derived values for x and y.

7. We then generate a derived list of values from the `x_values` list and store them in `x2_values`:

```
x2_values = list(map(lambda x: x * 2 + 7000, x_values))
x2_values[0:5]
```

We would get values similar to `[10618, 8456, 8102, 9126, 9002]`.

8. In a similar fashion, we generate a derived list of values and store them in the `x3_values` list:

```
x3_values = list(map(lambda x: x * 3 - 20, x_values))
x3_values[0:5]
```

We would get values similar to `[5407, 2164, 1633, 3159, 2983]`.

9. We also generate a derived list of values from `y_values` and store them in the `y2_values` list:

```
y2_values = list(map(lambda y: y * 2 + 1000, y_values))
y2_values[0:5]
```

We would get values similar to `[1054, 1082, 1060, 1092, 1088]`.

10. Now that we have all the column values inside the lists, we combine them all inside a single DataFrame:

```
import pandas as pd

df = pd.DataFrame({
    "x": x_values,
    "x2": x2_values,
    "x3": x3_values,
    "y": y_values,
    "y2": y2_values
})
```

11. Let's inspect what the DataFrame looks like:

```
df
```

We should get a `DataFrame` similar to what is shown in *Figure 4.33*:

	x	x2	x3	y	y2
0	1809	10618	5407	27	1054
1	728	8456	2164	41	1082
2	551	8102	1633	30	1060
3	1063	9126	3169	46	1092
4	1001	9002	2983	44	1088
...	...	...	...	...	...
95	1277	9554	3811	44	1088
96	1052	9104	3136	25	1050
97	1810	10620	5410	36	1072
98	1909	10818	5707	49	1098
99	1640	10280	4900	23	1046

100 rows × 5 columns

Figure 4.33 – DataFrame with the values from x_values, x2_values, x3_values, y_values, and y2_values

In *Figure 4.33*, we can see the DataFrame with five columns—x, x2, x3, y, and y2. The x2 and x3 columns were derived from x and the y2 column was derived from the y values.

12. Next, generate the values for the `label` column. If the x values are greater than `1000` and the y values are greater than `35`, we specify a value of `1` for the `label` column. Otherwise, we specify a value of `0`:

```
df["label"] = (df.x > 1000) & (df.y > 35)
df['label'] = df['label'].apply(lambda x: 1 if x else 0)
df
```

We should get a DataFrame similar to what is shown in *Figure 4.34*:

	x	x2	x3	y	y2	label
0	1809	10618	5407	27	1054	0
1	728	8456	2164	41	1082	0
2	551	8102	1633	30	1060	0
3	1063	9126	3169	46	1092	1
4	1001	9002	2983	44	1088	1
...	...	...	...	...	...	...
95	1277	9554	3811	44	1088	1
96	1052	9104	3136	25	1050	0
97	1810	10620	5410	36	1072	1
98	1909	10818	5707	49	1098	1
99	1640	10280	4900	23	1046	0

100 rows × 6 columns

Figure 4.34 – DataFrame with the label column values

In *Figure 4.34*, we can see the new label column containing the values 1 and 0 depending on the x and y values.

Now, we will focus our efforts on deleting some records in the dataset so that we will have two distinct clusters of points left.

13. We then generate the values for the keep column. As the name suggests, we will decide whether to keep the record or not based on the value in the keep column:

```
df["keep"] = ((df.x > 1000) & (df.y > 35)) | ((df.x <
800) & (df.y < 30))
df
```

We should get a DataFrame similar to what is shown in *Figure 4.35*:

	x	x2	x3	y	y2	label	keep
0	1809	10618	5407	27	1054	0	False
1	728	8456	2164	41	1082	0	False
2	551	8102	1633	30	1060	0	False
3	1063	9126	3169	46	1092	1	True
4	1001	9002	2983	44	1088	1	True
...	...	...	...	...	...	...	...
95	1277	9554	3811	44	1088	1	True
96	1052	9104	3136	25	1050	0	False
97	1810	10620	5410	36	1072	1	True
98	1909	10818	5707	49	1098	1	True
99	1640	10280	4900	23	1046	0	False

100 rows × 7 columns

Figure 4.35 – DataFrame with the new keep column

We can see in *Figure 4.35* the new `keep` column containing the `True` and `False` values. As we will be performing cluster analysis with this dataset, we will intentionally delete values in the "middle" so that the trimmed dataset will have two clusters—the first cluster having x values greater than `1000` and y values greater than `35` and the second cluster having x values less than `800` and y values less than `30`.

14. Next, select and continue with the records with the `keep` column value equal to `True`:

```
df = df[df.keep]
df.head()
```

We should get a DataFrame similar to what is shown in *Figure 4.36*:

	x	x2	x3	y	y2	label	keep
3	1063	9126	3169	46	1092	1	True
4	1001	9002	2983	44	1088	1	True
6	785	8570	2335	21	1042	0	True
7	709	8418	2107	27	1054	0	True
8	1885	10770	5635	46	1092	1	True

Figure 4.36 – Trimmed dataset

Although we are not able to see the whole trimmed dataset in *Figure 4.36*, we should notice that we are missing the 0, 1, 2, and 5 indexes in the initial list of records using the head() function. That's because the records with these indexes have been filtered out as their value in the keep column is equal to False.

15. Now that we have trimmed the dataset, let's delete the keep column using the del keyword:

```
del df["keep"]
```

16. With everything ready, generate the scatterplot showing the labeled clusters using the following lines of code:

```
import matplotlib.pyplot as plt

groups = df.groupby("label")
for name, group in groups:
    plt.plot(group["x"],
             group["y"],
             marker="o",
             linestyle="",
             label=name)

plt.legend()
```

The preceding block of code made use of the `groupby()` function to group the points using the value of the `label` column. This should generate a scatterplot diagram similar to what is shown in *Figure 4.37*:

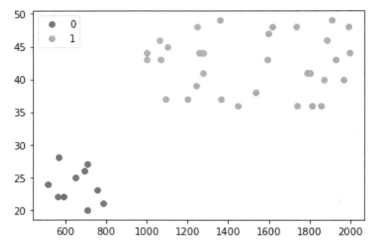

Figure 4.37 – matplotlib scatterplot diagram showing the labeled clusters

In *Figure 4.37*, we can see that there are two clusters corresponding to each of the labeled groups. The first cluster with the label `0` is found in the lower-left portion of the plot and the second cluster with the label `1` is found in the upper-right portion of the plot.

17. Generate the `tmp` directory if it does not exist yet:

```
!mkdir -p tmp
```

18. Use the `to_csv()` function to store the labeled DataFrame in a CSV file:

```
df.to_csv("tmp/synthetic.all.labeled.csv")
```

We will use this CSV file later in the *Converting CSV data into protobuf recordIO format* recipe.

19. Store the labeled DataFrame using the `%store` magic command as well:

```
labeled_df = df
%store labeled_df
```

We would get a success message equal or similar to `Stored 'labeled_df'` (`DataFrame`).

In the last set of steps, we will work toward preparing and storing the unlabeled version of our synthetic dataset.

20. Delete the `label` column using the `del` keyword:

```
del df["label"]
df.head()
```

We should get a DataFrame similar to what is shown in *Figure 4.38*:

	x	x2	x3	y	y2
3	1063	9126	3169	46	1092
4	1001	9002	2983	44	1088
6	785	8570	2335	21	1042
7	709	8418	2107	27	1054
8	1885	10770	5635	46	1092

Figure 4.38 – DataFrame without the label column

We can see in *Figure 4.38* the DataFrame without the `label` column. Where will we use this? After we have performed the scaling step, we will perform an unsupervised cluster analysis with this unlabeled DataFrame in the *Performing cluster analysis with the built-in KMeans algorithm* recipe.

21. Use `MinMaxScaler` from `scikit-learn` to scale the values in the unlabeled DataFrame:

```
from sklearn.preprocessing import MinMaxScaler


scaler = MinMaxScaler()
scaled_array = scaler.fit_transform(df.astype(float))
normalized_df = pd.DataFrame(scaled_array)
normalized_df.columns = df.columns
normalized_df.index = df.index


normalized_df.head()
```

We should get a DataFrame similar to what is shown in *Figure 4.39*:

	x	x2	x3	y	y2
3	0.371371	0.371371	0.371371	0.896552	0.896552
4	0.329507	0.329507	0.329507	0.827586	0.827586
6	0.183660	0.183660	0.183660	0.034483	0.034483
7	0.132343	0.132343	0.132343	0.241379	0.241379
8	0.926401	0.926401	0.926401	0.896552	0.896552

Figure 4.39 – Scaled DataFrame

As seen in *Figure 4.39*, our values have been scaled down to a set of values between 0 and 1. Note that we have the option to specify the value for the `feature_range` parameter for `MinMaxScaler` but we will stick to the default value of (0, 1).

22. Use the `to_csv()` function to save the values of the normalized unlabeled DataFrame to a CSV file:

```
normalized_df.to_csv("tmp/synthetic.all_normalized.
unlabeled.csv")
```

23. Finally, use the `%store` magic command to store the values of the `unlabeled_normalized_df` variable:

```
unlabeled_normalized_df = normalized_df
%store unlabeled_normalized_df
```

We would get a success message equal or similar to `Stored 'unlabeled_normalized_df' (DataFrame)`.

Let's see how this works!

How it works...

In this recipe, we have generated a synthetic dataset that will be used in the recipes involving **dimension reduction**, **cluster analysis**, and serialization to `protobuf` `recordIO` format. As a machine learning practitioner, it is important that you are familiar with these recipes on data preparation and analysis as your data will most likely be transformed a couple of times before it is passed as input to the model training step.

What are the properties of this dataset? The data points in this dataset are grouped into two clusters intentionally so that it will be easier for us to know whether our cluster analysis recipe works or not. We have also added a couple of derived columns that would help us know whether our dimension reduction recipe works as well.

What are some of the tweaks we can perform on this recipe to come up with variations of this dataset?

- We can generate (significantly) more records and test the performance of the algorithms used in the succeeding recipes. Some algorithms and solutions work well with a relatively small dataset size but may struggle when working with larger datasets.

- We can add (significantly) more columns to the generated dataset. For example, instead of having just 5 columns, we can generate 100 columns instead. We can test and compare the performance of an algorithm on a dataset with just 5 columns against the performance of the same algorithm on a dataset with 100 columns.

- We can also increase the number of clusters generated from 2 to 10 and see how the clustering algorithms behave when working with a higher number of clusters.

With this in mind, let's proceed with the recipes that make use of this synthetic dataset!

Performing dimensionality reduction with the built-in PCA algorithm

In this recipe, we will demonstrate how to use the built-in PCA algorithm to perform dimensionality reduction on a synthetic dataset. Dimensionality reduction involves bringing down the number of columns of a dataset to a smaller number of essential columns. If you're wondering why this is important, it's because some algorithms perform better and faster when dealing with fewer dimensions!

We will use the PCA algorithm on the unlabeled dataset from the *Generating a synthetic dataset for analysis and transformation* recipe and reduce the number of columns of that dataset from five to two. By using PCA, we will also notice that the resulting values are different from any of the row values from the original dataset.

Getting ready

This recipe continues from *Generating a synthetic dataset for analysis and transformation*.

How to do it...

The next set of steps focuses on using the unlabeled dataset we generated in the previous recipe to prepare the PCA model we will use for dimension reduction:

1. Navigate to the `my-experiments/chapter04` directory inside your SageMaker notebook instance. Feel free to create this directory if it does not exist yet.

2. Create a new notebook using the `conda_python3` kernel inside the `my-experiments/chapter04` directory and name it with the name of this recipe (that is, `Performing Dimensionality Reduction with the built-in PCA Algorithm`). Open this notebook for editing as we will update this file with the code in the next couple of steps.

3. Start the notebook by using the `%store` magic command to load the values of `unlabeled_normalized_df`:

   ```
   %store -r unlabeled_normalized_df
   ```

   If you can remember, we stored this value in the *Generating a synthetic dataset for analysis and transformation* recipe. We will use the values in this DataFrame and perform dimension reduction with it using the PCA model.

4. Import and prepare a few prerequisites for our SageMaker experiment. These include `session` and `role`, which we will pass as parameter values to the `PCA` estimator in a later step:

   ```
   import sagemaker
   from sagemaker import get_execution_role

   session = sagemaker.Session()
   role = get_execution_role()
   ```

5. We initialize a `PCA` estimator object using the parameter values prepared in the previous step as well as the values for the `instance_count`, `instance_type`, and `num_components` parameters:

   ```
   from sagemaker import PCA

   estimator = PCA(
       role=role,
       instance_count=1,
       instance_type='ml.c5.xlarge',
   ```

```
        num_components=2,
        sagemaker_session=session)
```

The num_components key in this estimator maps to the num_components hyperparameter for the PCA algorithm. This value corresponds to the number of principal components to compute. This means that if the hyperparameter value for num_components is 2, then if we input a record with 5-6 components or columns, then we will compute for 2 values, which would represent this record instead.

6. Create a record set using the record_set() function. The record_set() function accepts a NumPy ndarray object, uploads it to S3, and returns a Record object:

```
        data_np = unlabeled_normalized_df.values.
        astype('float32')
        record_set = estimator.record_set(data_np)
```

7. We pass the record set to the estimator's fit() function:

```
        estimator.fit(record_set)
```

We should get a similar set of logs after running the preceding line of code.

```
2021-04-18 07:33:48 Starting - Starting the training job...ProfilerReport-1618731227: InProgr
ess
...
2021-04-18 07:34:31 Starting - Launching requested ML instances......
2021-04-18 07:35:45 Starting - Preparing the instances for training...........
2021-04-18 07:37:32 Downloading - Downloading input data
2021-04-18 07:37:32 Training - Downloading the training image..Docker entrypoint called with
argument(s): train
Running default environment configuration script
[04/18/2021 07:37:57 INFO 140097614255936] Reading default configuration from /opt/amazon/li
b/python3.7/site-packages/algorithm/resources/default-conf.json: {'algorithm_mode': 'regula
r', 'subtract_mean': 'true', 'extra_components': '-1', 'force_dense': 'true', 'epochs': 1, '_
log_level': 'info', '_kvstore': 'dist_sync', '_num_kv_servers': 'auto', '_num_gpus': 'auto'}
[04/18/2021 07:37:57 INFO 140097614255936] Merging with provided configuration from /opt/ml/i
```

Figure 4.40 – Logs after calling the fit() function of the PCA estimator

In *Figure 4.40*, we can see the logs after calling the fit() function. This step should take a few minutes to complete.

8. Use the deploy() function to deploy the model to an inference endpoint:

```
        predictor = estimator.deploy(
            initial_instance_count=1,
            instance_type='ml.t2.medium')
```

9. Call the `predict()` function to perform dimension reduction with the PCA model:

```
results = predictor.predict(data_np)
results
```

This will yield the following structure of results similar to what is shown in *Figure 4.41*. We should have the same number of elements in the `data_np` array and in the list returned by the inference endpoint.

```
[label {
    key: "projection"
    value {
        float32_tensor {
            values: -0.48789161443710327
            values: 0.011455059051513672
        }
    }
},
label {
    key: "projection"
    value {
        float32_tensor {
            values: -0.4503028988838196
            values: 0.12702858448028564
        }
    }
},
```

Figure 4.41 – Results after using the predict() function

As seen in *Figure 4.41*, each corresponding projection object contains two values. Since we specified the value of 2 for the `num_components` hyperparameter when we were initializing the `PCA` estimator. Let's access the values of the first result from the `results` list:

```
results[0].label['projection'].float32_tensor.values
```

We should get a set of values with a structure similar to `[-0.48789161443710327, 0.011455059051513672]`. What this means is that the first record in `data_np` with the column values for $x, x2, x3, y,$ and $y2$ have been reduced to just two values. The same goes for the other elements of the input data.

10. Define the `extract_values()` function. This function accepts an element of the results list and returns the two components of that element:

```
def extract_values(item):
    projection = item.label['projection']
    pair = projection.float32_tensor.values
    x = pair[0]
    y = pair[1]

    return {
        "x": x,
        "y": y
    }
```

11. Next, run the following block of code, which uses the `extract_values()` function we defined in the previous step to extract the component values from the list of results. After running this, the `new_xs` and `new_ys` lists should contain the values of the two new components:

```
new_xs = []
new_ys = []
for result in results:
    x_and_y = extract_values(result)
    new_xs.append(x_and_y["x"])
    new_ys.append(x_and_y["y"])
```

12. We check the first five values of `new_xs`:

```
new_xs[0:5]
```

We should get a set of values with a structure similar to
```
[-0.48789161443710327, -0.4503028988838196,
0.31829965114593506, 0.028231695294380188,
0.06837558001279831].
```

13. Then, we check the first five values of `new_ys`:

```
new_ys[0:5]
```

We should get a set of values with a structure similar to
```
[0.011455059051513672, 0.12702858448028564,
0.9820671081542969, 0.885252058506012, -0.772600531578064].
```

14. Create a `pandas` DataFrame using the following lines of code:

```
import pandas as pd

new_df = pd.DataFrame({
    "new_x": new_xs,
    "new_y": new_ys
})

new_df
```

We should get a DataFrame with two columns, as shown in *Figure 4.42*:

	new_x	new_y
0	-0.487892	0.011455
1	-0.450303	0.127029
2	0.318300	0.982067
3	0.028232	0.885252
4	0.068376	-0.772601
5	-0.108989	0.994387
6	0.126941	1.167509
7	0.055823	0.930639

Figure 4.42 – DataFrame containing the reduced component values

As seen in *Figure 4.42*, we have the two components derived from the initial five components of the scaled dataset. Take note that the x and y columns here are different from the x and y columns from the original dataset. There may be no direct mapping between the original set of components with the new set of components after performing dimension reduction with PCA.

15. Let's see a scatterplot of the points by running the following code:

```
new_df.plot.scatter(x="new_x", y="new_y")
```

We should get a scatterplot diagram similar to the one in *Figure 4.43*:

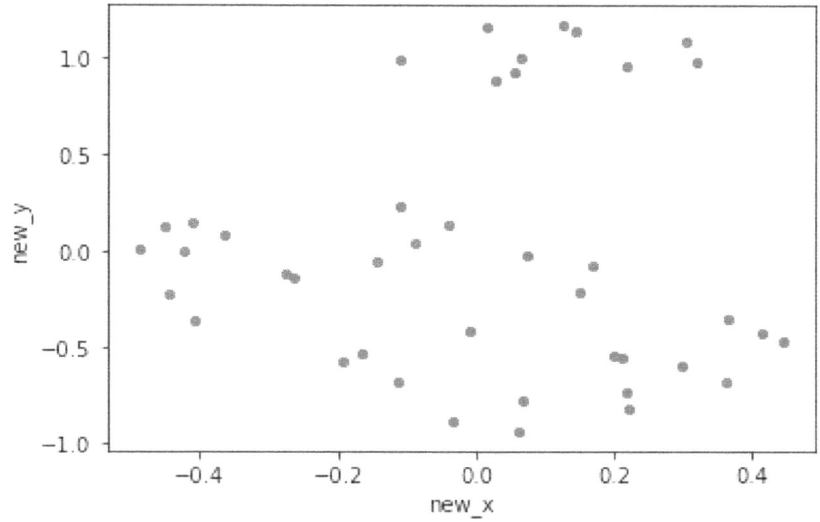

Figure 4.43 – Scatterplot of points after performing dimension reduction

In *Figure 4.43*, we see a slightly tilted version of the scatterplot generated from the original set of points passed to the PCA model. We can still see that there are two clusters with one cluster significantly larger than the other one.

At this point, we can delete the endpoint deployed in this recipe. Now, let's see how this works!

How it works...

The PCA algorithm helps us perform dimension reduction in our datasets. Why perform dimensionality reduction? The more features we work with, the higher the chance a trained model becomes more complex and the more samples we need during training. With more features and more data, the longer the training times will be and the higher the chance of the trained model overfitting. That said, we can make use of the PCA algorithm to perform dimension reduction in our dataset before passing the data as input to the training step for another model (for example, XGBoost).

> **Tip**
> For more information on the PCA algorithm, feel free to check the following link: `https://docs.aws.amazon.com/sagemaker/latest/dg/pca.html`.

See also

If you are looking for examples of using PCA models to perform dimension reduction with real datasets and more complex examples, feel free to check some of the notebooks in the `aws/amazon-sagemaker-examples` GitHub repository:

- Analyzing the MNIST dataset with the PCA algorithm: `https://github.com/aws/amazon-sagemaker-examples/blob/master/introduction_to_amazon_algorithms/pca_mnist/pca_mnist.ipynb`

- Population segmentation using the PCA and KMeans clustering algorithms: `https://github.com/aws/amazon-sagemaker-examples/blob/master/introduction_to_applying_machine_learning/US-census_population_segmentation_PCA_Kmeans/sagemaker-countycensusclustering.ipynb`

Now, let's take a closer look at the KMeans algorithm in the next recipe!

Performing cluster analysis with the built-in KMeans algorithm

In this recipe, we will demonstrate how to use the KMeans algorithm to perform cluster analysis with the synthetic dataset. **Cluster analysis** involves identifying subgroups of records within the dataset that exhibit similar properties. This helps solve different problems and requirements related to market segmentation, fraud detection, and document analysis.

Getting ready

This recipe continues from *Generating a synthetic dataset for analysis and transformation*.

How to do it...

The next set of steps focus on using the unlabeled dataset we generated in the *Generating a synthetic dataset for analysis and transformation* recipe to prepare the KMeans model we will use for cluster analysis:

1. Navigate to the `my-experiments/chapter04` directory inside your SageMaker notebook instance. Feel free to create this directory if it does not exist yet.

2. Create a new notebook using the `conda_python3` kernel inside the `my-experiments/chapter04` directory and name it with the name of this recipe (that is, `Performing Cluster Analysis with the built-in KMeans Algorithm`). Open this notebook for editing as we will update this file with the code in the next couple of steps.

3. Start the notebook by using the `%store` magic command to load the stored value for `unlabeled_normalized_df`:

    ```
    %store -r unlabeled_normalized_df
    unlabeled_normalized_df.head()
    ```

    We should get a DataFrame similar to what is shown in *Figure 4.44*:

	x	x2	x3	y	y2
3	0.371371	0.371371	0.371371	0.896552	0.896552
4	0.329507	0.329507	0.329507	0.827586	0.827586
6	0.183660	0.183660	0.183660	0.034483	0.034483
7	0.132343	0.132343	0.132343	0.241379	0.241379
8	0.926401	0.926401	0.926401	0.896552	0.896552

    Figure 4.44 – Unlabeled normalized DataFrame of values

    We should get the same set of values as when we generated this `DataFrame` in the *Generating a synthetic dataset for analysis and transformation* recipe.

4. Import and prepare a few prerequisites for our SageMaker experiment. These include `session` and `role`, which we will pass as parameter values to the `KMeans` estimator in a later step:

    ```
    import sagemaker
    from sagemaker import get_execution_role
    ```

```
session = sagemaker.Session()
role = get_execution_role()
```

5. We initialize a KMeans estimator object using the parameter values prepared in the previous step, as well as the values for the instance_count, instance_type, and k parameters:

```
from sagemaker import KMeans

estimator = KMeans(
    role=role,
    instance_count=1,
    instance_type='ml.c4.xlarge',
    k=2)
```

The k key in this estimator maps to the number of clusters for the KMeans algorithm. This means that if the hyperparameter value for k is 2, then we would group the points in the dataset into two clusters using this unsupervised machine learning algorithm.

6. Create a record set using the record_set() function on the unlabeled DataFrame values. The record_set() function accepts a NumPy ndarray object, uploads it to S3, and returns a Record object:

```
data_np = unlabeled_normalized_df.values.
astype('float32')
record_set = kmeans.record_set(data_np)
estimator.fit(record_set)
```

This should yield a set of logs similar to what is shown in *Figure 4.45*:

```
2021-04-18 08:02:17 Starting - Starting the training job...
2021-04-18 08:02:18 Starting - Launching requested ML instancesProfilerReport-1618732936: InP
rogress
......
2021-04-18 08:03:44 Starting - Preparing the instances for training........
2021-04-18 08:05:10 Downloading - Downloading input data...
2021-04-18 08:05:44 Training - Downloading the training image..Docker entrypoint called with
argument(s): train
Running default environment configuration script
[04/18/2021 08:05:55 INFO 140342932858240] Reading default configuration from /opt/amazon/li
b/python3.6/site-packages/algorithm/resources/default-input.json: {'init_method': 'random',
'mini_batch_size': '5000', 'epochs': '1', 'extra_center_factor': 'auto', 'local_lloyd_max_ite
r': '300', 'local_lloyd_tol': '0.0001', 'local_lloyd_init_method': 'kmeans++', 'local_lloyd_n
um_trials': 'auto', 'half_life_time_size': '0', 'eval_metrics': '["msd"]', 'force_dense': 'tr
```

Figure 4.45 – Logs after using the fit() function

This should take a few minutes to complete, so feel free to grab a cup of coffee or tea!

7. Use the `deploy()` function to deploy the KMeans model to an inference endpoint:

```
predictor = estimator.deploy(
    initial_instance_count=1,
    instance_type='ml.t2.medium')
```

8. Call the `predict()` function to perform cluster analysis with the KMeans model:

```
results = predictor.predict(data_np)
results
```

This should yield a set of results similar to what is shown in *Figure 4.46*:

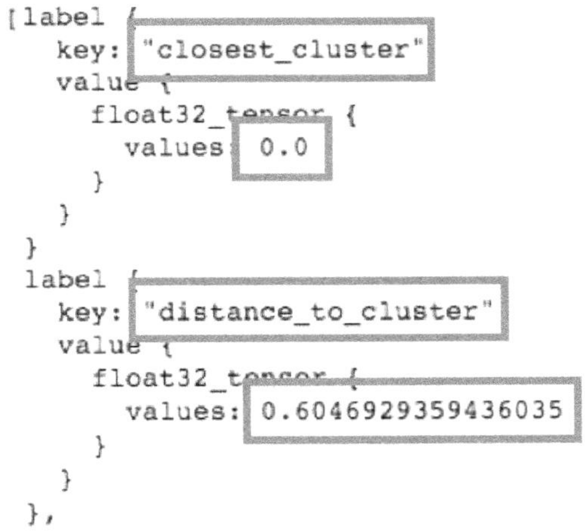

```
[label {
    key: "closest_cluster"
    value {
        float32_tensor {
            values 0.0
        }
    }
}
label {
    key: "distance_to_cluster"
    value {
        float32_tensor {
            values: 0.6046929359436035
        }
    }
},
```

Figure 4.46 – Results after using the predict() function on a deployed KMeans model

In *Figure 4.46*, we can see that for each data point we pass in the `predict()` function, we get two values—`closest_cluster` and `distance_to_cluster`.

9. Define the `extract_values()` function. This function accepts an element of the results list and returns the `closest_cluster` and `distance_to_cluster` values:

```
def extract_values(item):
    closest_cluster = item.label['closest_cluster']
    cc_value = int(closest_cluster.float32_tensor.
values[0])
    distance_to_cluster = item.label['distance_to_
cluster']
```

```
    dtc_value = distance_to_cluster.float32_tensor.
values[0]

    return {
        "closest_cluster": cc_value,
        "distance_to_cluster": dtc_value
    }
```

10. Next, run the following block of code, which uses the `extract_values()` function we defined in the previous step to extract the `closest_cluster` and `distance_to_cluster` values from the response data. After running this, the `closest_cluster_list` and `distance_to_cluster_list` lists should contain the appropriate values:

```
closest_cluster_list = []
distance_to_cluster_list = []
for result in results:
    cv = extract_values(result)
    closest_cluster_list.append(cv["closest_cluster"])
    distance_to_cluster_list.append(cv["distance_to_
cluster"])
```

11. Inspect the first six elements of `closest_cluster_list`:

```
closest_cluster_list[0:6]
```

We should get a set of results similar to [0, 0, 1, 1, 0, 1].

12. Update the `unlabeled_normalized_df` DataFrame with the corresponding values for `closest_cluster` and `distance_to_cluster` of each record:

```
df = unlabeled_normalized_df
df = df.assign(closest_cluster=closest_cluster_list)
df = df.assign(distance_to_cluster=distance_to_cluster_
list)
df.head()
```

We should see a DataFrame similar to what is shown in *Figure 4.47*:

	x	x2	x3	y	y2	closest_cluster	distance_to_cluster
3	0.371371	0.371371	0.371371	0.896552	0.896552	0	0.604693
4	0.329507	0.329507	0.329507	0.827586	0.827586	0	0.657982
6	0.183660	0.183660	0.183660	0.034483	0.034483	1	0.231180
7	0.132343	0.132343	0.132343	0.241379	0.241379	1	0.167338
8	0.926401	0.926401	0.926401	0.896552	0.896552	0	0.412965

Figure 4.47 – DataFrame with the closest_cluster and distance_to_cluster columns

We can see in *Figure 4.47* that our df DataFrame now has two new columns—closest_cluster and distance_to_cluster.

13. Let's see a scatterplot of the cluster of points by running the following code:

```
import matplotlib.pyplot as plt
groups = df.groupby("closest_cluster")
for name, group in groups:
    plt.plot(group["x"],
             group["y"],
             marker="o",
             linestyle="",
             label=name)

plt.legend()
```

The preceding block of code made use of the `groupby()` function to group the points based on the `closest_cluster` value. This should render a scatterplot chart similar to what is shown in *Figure 4.48*:

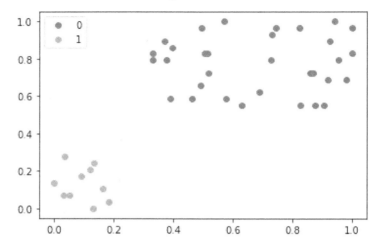

Figure 4.48 – Scatterplot of points showing the two clusters

We can see in *Figure 4.48* that the **KMeans** model has properly grouped the data points together into two clusters. We can try comparing the `closest_cluster` values with the original `label` values from the *Generating a synthetic dataset for analysis and transformation* recipe but we will leave that to you as an exercise.

At this point, we can delete the endpoint deployed in this recipe. Now, let's see how this works!

How it works...

In this recipe, we used the KMeans algorithm to help us perform cluster analysis on our synthetic dataset. How does this work? The KMeans algorithm divides the dataset into k subgroups with k equal to the number of desired clusters. The data points within these subgroups are automatically grouped together based on similarity by this unsupervised learning algorithm.

The deployed KMeans model returns two values for each of the data points passed as the payload to the `predict()` function:

- `closest_cluster`: Group or cluster that a data point is predicted to belong to

- `distance_to_cluster`: Euclidean distance of the data point values to the center of the cluster

With this cluster analysis step, we may easily perform automated unsupervised tagging of data points with class labels even without labeled training data. Another application of cluster analysis is the detection of anomalies as the anomalies can easily be detected, especially if they are too far from the cluster centers.

See also

If you are looking for examples of training and deploying KMeans clustering models in SageMaker using real datasets, feel free to check some of the notebooks in the `aws/amazon-sagemaker-examples` GitHub repository:

- KMeans clustering on the MNIST dataset: `https://github.com/aws/amazon-sagemaker-examples/blob/master/sagemaker-python-sdk/1P_kmeans_highlevel/kmeans_mnist.ipynb`

- Population segmentation using the PCA and KMeans clustering algorithms: `https://github.com/aws/amazon-sagemaker-examples/blob/master/introduction_to_applying_machine_learning/US-census_population_segmentation_PCA_Kmeans/sagemaker-countycensusclustering.ipynb`

- Deploying a pre-trained KMeans clustering model in Amazon SageMaker: `https://github.com/aws/amazon-sagemaker-examples/blob/master/advanced_functionality/kmeans_bring_your_own_model/kmeans_bring_your_own_model.ipynb`

Now, let's take a closer look at the `protobuf recordIO` format in the next recipe!

Converting CSV data into protobuf recordIO format

In this recipe, we will convert and serialize the synthetic data stored in CSV format into the `protobuf recordIO` format. With the data serialized into the `protobuf recordIO` format, we can take advantage of Pipe mode, where training start times will be faster as the training job streams data directly from the S3 bucket source. That said, the SageMaker algorithms may perform much better with this training file format.

Getting ready

This recipe continues from *Generating a synthetic dataset for analysis and transformation*.

How to do it...

In the first few steps of this recipe, we will focus on scaling and transforming the synthetic labeled dataset into a set of values between 0 and 1 using `MinMaxScaler` from `sklearn`:

1. Navigate to the `my-experiments/chapter04` directory inside your SageMaker notebook instance. Feel free to create this directory if it does not exist yet.

2. Create a new notebook using the `conda_python3` kernel inside the `my-experiments/chapter04` directory and name it with the name of this recipe (that is, `Converting CSV data into protobuf recordIO format`). Open this notebook for editing as we will update this file with the code in the next couple of steps.

3. Start the notebook by using the `%store` magic command to read the value of `labeled_df`. Remember that we stored this value in the *Generating a synthetic dataset for analysis and transformation* recipe:

```
%store -r labeled_df
labeled_df.head()
```

This should show the first five elements of `labeled_df`. We should see a DataFrame similar to what is shown in *Figure 4.49*:

	x	x2	x3	y	y2	label
3	1063	9126	3169	46	1092	1
4	1001	9002	2983	44	1088	1
6	785	8570	2335	21	1042	0
7	709	8418	2107	27	1054	0
8	1885	10770	5635	46	1092	1

Figure 4.49 – labeled_df

We have in *Figure 4.49* the labeled DataFrame containing the label value of each record.

4. Use `MinMaxScaler` from `scikit-learn` to scale the values in the labeled DataFrame:

```
import pandas as pd
from sklearn.preprocessing import MinMaxScaler
```

```
scaler = MinMaxScaler()
scaled_values = scaler.fit_transform(labeled_
df.astype(float))
normalized_df = pd.DataFrame(scaled_values)
normalized_df.columns = labeled_df.columns
normalized_df.index = labeled_df.index


normalized_df.head()
```

This should show the first five elements of `normalized_df`. This should show a DataFrame similar to what is shown in *Figure 4.50*:

	x	x2	x3	y	y2	label
3	0.371371	0.371371	0.371371	0.896552	0.896552	1.0
4	0.329507	0.329507	0.329507	0.827586	0.827586	1.0
6	0.183660	0.183660	0.183660	0.034483	0.034483	0.0
7	0.132343	0.132343	0.132343	0.241379	0.241379	0.0
8	0.926401	0.926401	0.926401	0.896552	0.896552	1.0

Figure 4.50 – normalized_df

We have in *Figure 4.50* the normalized labeled DataFrame containing values scaled and transformed to fit the `(0, 1)` range.

5. Check the shape of the normalized DataFrame:

```
normalized_df.shape
```

This yields a tuple equal or similar to `(42, 6)`.

In the next set of steps, we will perform the actual conversion and serialization to the `protobuf recordIO` format using the `write_numpy_to_dense_tensor()` function from the SageMaker Python SDK.

6. Perform the train-test split using the `train_test_split()` function on the normalized DataFrame:

```
from sklearn.model_selection import train_test_split


y = normalized_df["label"].values
X = normalized_df[["x", "x2", "x3", "y", 'y2"]].values
```

```
X_train, X_test, y_train, y_test = train_test_split(X, y,
test_size=0.2, random_state=0)
```

7. Store the values of X_train and y_train in train_np and label_np, respectively:

```
train_np = X_train
label_np = y_train
```

8. Use the write_numpy_to_dense_tensor() function from the SageMaker Python SDK to convert the NumPy array to protobuf recordIO format:

```
import io
from sagemaker.amazon.common import \
write_numpy_to_dense_tensor

buf = io.BytesIO()
write_numpy_to_dense_tensor(buf, train_np, label_np)
buf.seek(0)
```

9. Create the tmp directory if it does not exist yet:

```
!mkdir -p tmp
```

10. Define the save_bytesio() function, which saves the protobuf recordIO data into a file:

```
def save_bytesio(filename, buf):
    with open("tmp/" + filename, "wb") as file:
        file.write(buf.getbuffer())
        print(f"Successfully saved {filename}")
```

11. Use the save_bytesio() function to save the data into the tmp/train.io file:

```
save_bytesio("train.io", buf)
```

This should yield a log message similar to Successfully saved train.io.

12. Use the %store magic command to store the value of the buf variable:

```
%store buf
```

This should yield a log message similar to Stored 'buf' (BytesIO).

13. In a similar fashion, use the `%store` magic command to store the values for X_train, X_test, y_train, and y_test:

```
%store X_train
%store X_test
%store y_train
%store y_test
```

This should yield a log message similar to `Stored 'X_train' (ndarray)  Stored 'X_test' (ndarray)  Stored 'y_train' (ndarray)  Stored 'y_test' (ndarray)`. We will use these values in the *Training a KNN model using the protobuf recordIO training input type* recipe.

Now, let's see how this works!

How it works...

In this recipe, we converted and serialized tabular CSV data into `protobuf recordIO` format. As this format allows SageMaker to stream data directly with Pipe mode, training start times would be faster as we would not need to wait for all data to be downloaded before the training step starts. This means that most SageMaker algorithms work better in general when using this data format during training.

Note that instead of us trying to create our own converter or serializer, we used the `write_numpy_to_dense_tensor()` function from the **SageMaker Python SDK** to convert NumPy array values to `protobuf recordIO` format. Feel free to check the implementation of the `write_numpy_to_dense_tensor()` function here: `https://github.com/aws/sagemaker-python-sdk/blob/master/src/sagemaker/amazon/common.py`.

Training a KNN model using the protobuf recordIO training input type

In this recipe, we will train two k-Nearest Neighbors (KNN) models with the SageMaker Python SDK—one with the training input data using `record_set()` with the NumPy array of values as the parameter and another using the `protobuf recordIO` training input file generated from the *Converting CSV data into protobuf recordIO format* recipe.

Once we have completed this recipe, we will have a better understanding of some key differences when using the different types of estimators and when using the different training input types.

Getting ready

This recipe continues from *Converting CSV data into protobuf recordIO format*.

How to do it...

The first few steps in this recipe focus on loading and using the values of X_train and y_train to train and deploy a KNN model. We will use the KNN estimator class along with the record_set() function to serialize and process the X_train and y_train values before using the fit() function to start the training job:

1. Navigate to the my-experiments/chapter04 directory inside your SageMaker notebook instance. Feel free to create this directory if it does not exist yet.

2. Create a new notebook using the conda_python3 kernel inside the my-experiments/chapter04 directory and name it with the name of this recipe (that is, Training a KNN model using the protobuf recordIO training input type). Open this notebook for editing as we will update this file with the code in the next couple of steps.

3. Start the notebook by using the %store magic command to retrieve the values of X_train, y_train, X_test, and y_test. Remember that we stored these values in the *Converting CSV data into protobuf recordIO format* recipe:

```
%store -r X_train
%store -r y_train
%store -r X_test
%store -r y_test
```

4. Import and prepare a few prerequisites for our SageMaker experiment. These include session and role, which we will pass as parameter values to the KNN estimator in a later step:

```
import sagemaker
from sagemaker import get_execution_role
session = sagemaker.Session()
role = get_execution_role()
```

5. We initialize a KNN estimator object using the parameter values prepared in the previous step, as well as the values for the `instance_count`, `instance_type`, `sample_size`, k, `feature_dim`, and `predictor_type` parameters:

```
from sagemaker import KNN

estimator1 = KNN(
    role=role,
    instance_count=1,
    instance_type='ml.c5.xlarge',
    sample_size=50,
    k=3,
    feature_dim=5,
    predictor_type="classifier",
    sagemaker_session=session)
```

6. Create a record set using the `record_set()` function on the unlabeled DataFrame values. The `record_set()` function accepts a NumPy `ndarray` object, uploads the data to S3, and returns a `Record` object:

```
record_set = estimator1.record_set(train=X_train,
labels=y_train)
estimator1.fit(record_set)
```

This should yield a set of logs similar to what is shown in *Figure 4.51*:

```
2021-04-18 09:15:52 Starting - Starting the training job...
2021-04-18 09:16:15 Starting - Launching requested ML instancesProfilerReport-1618737351: InP
rogress
......
2021-04-18 09:17:22 Starting - Preparing the instances for training.........
2021-04-18 09:18:35 Downloading - Downloading input data
2021-04-18 09:18:35 Training - Downloading the training image.....Docker entrypoint called wi
th argument(s): train
Running default environment configuration script
[04/18/2021 09:19:33 INFO 140449387971776] Reading default configuration from /opt/amazon/li
b/python2.7/site-packages/algorithm/resources/default-conf.json: {u'index_metric': u'L2', u'_
tuning_objective_metric': u'', u'_num_gpus': u'auto', u'_log_level': u'info', u'feature_dim':
u'auto', u'faiss_index_ivf_nlists': u'auto', u'epochs': u'1', u'index_type': u'faiss.Flat',
```

Figure 4.51 – Logs after running the fit() function

The training job should complete in a few minutes. Once the training job has completed, feel free to proceed with the deployment step.

7. Use the `deploy()` function to deploy the KNN model to an inference endpoint:

```
predictor1 = estimator1.deploy(
    initial_instance_count=1,
    instance_type='ml.t2.medium')
```

8. Inspect the x values in `X_test` before using the `predict()` function:

```
X_test
```

We should get a set of values similar to `array([0.86968265, 0.86968265, 0.86968265, 0.72413793, 0.72413793])`.

9. Use the `predict()` function to get the corresponding predicted labels for each of the values in `X_test`:

```
results1 = predictor1.predict(X_test)
results1[0:3]
```

This should return a set of results with a structure similar to what is shown in *Figure 4.52*:

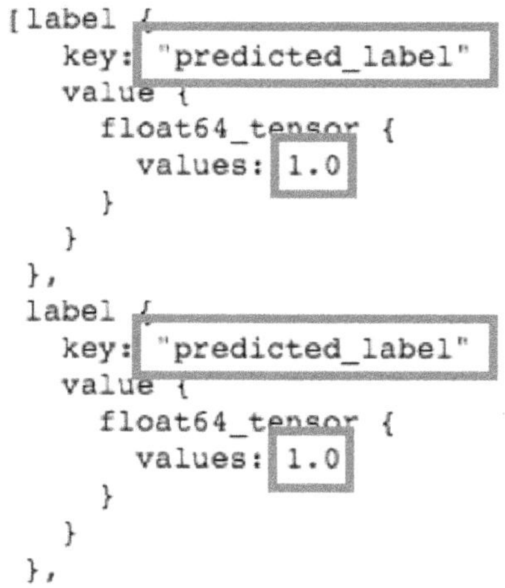

```
[label {
    key: "predicted_label"
    value {
        float64_tensor {
            values: 1.0
        }
    }
},
label {
    key: "predicted_label"
    value {
        float64_tensor {
            values: 1.0
        }
    }
},
```

Figure 4.52 – Results after using the predict() function

In *Figure 4.52*, we can see that the `predict()` function returned the predicted label for each of the x values passed as the payload.

The second half of this recipe focuses on using the `protobuf recordIO` file we generated in the *Converting CSV data into protobuf recordIO format* recipe to train and deploy a second KNN model. We will use the `Estimator` class along with `TrainingInput()` with the `content_type` parameter value set to `application/x-recordio-protobuf` before using the `fit()` function to start the training job.

10. Specify the S3 bucket and the prefix. Make sure to replace the value of `"<insert s3 bucket name here>"` with the name of the bucket we created in the *Preparing the Amazon S3 bucket and the training dataset for the linear regression experiment* recipe in *Chapter 1, Getting Started with Machine Learning Using Amazon SageMaker*:

    ```
    s3_bucket = "<insert s3 bucket name here>"
    prefix = "chapter04/knn"
    ```

11. Upload the generated `train.io` file to S3. Remember that this file was generated in the *Converting CSV data into protobuf recordIO format* recipe:

    ```
    !aws s3 cp tmp/train.io s3://{s3_bucket}/{prefix}/input/
    train.io
    ```

    This should yield a log message similar to `upload: tmp/train.io to s3://<bucket name>/chapter04/knn/input/train.io`.

12. Create a `TrainingInput` channel configuration object with the `content_type` parameter set to `application/x-recordio-protobuf`:

    ```python
    from sagemaker.inputs import TrainingInput

    train_path = f"s3://{s3_bucket}/{prefix}/input/train.io"
    train = TrainingInput(
        train_path,
        content_type="application/x-recordio-protobuf")
    ```

13. We initialize a second estimator object using the `Estimator` class:

    ```python
    from sagemaker.estimator import Estimator

    estimator2 = Estimator(
        image_uri=estimator1.training_image_uri(),
        role=role,
        instance_count=1,
    ```

```
            instance_type='ml.c5.xlarge',
            sagemaker_session=session)
```

Note that when using the `Estimator` class, we also pass the algorithm training container image URI as the value to the `image_uri` parameter.

14. Set the `hyperparameters` using the `set_hyperparameters()` function:

```
    estimator2.set_hyperparameters(
        sample_size=50,
        k=3,
        predictor_type="classifier")
```

15. Use the `fit()` function to start the training job. Note that in this step, we pass the S3 path of the training dataset (wrapped with a `TrainingInput` object) with the `protobuf recordIO` training input format:

```
    estimator2.fit({"train": train})
```

This should yield a set of logs similar to what is shown in *Figure 4.53*:

```
2021-04-18 10:52:15 Starting - Starting the training job...
2021-04-18 10:52:38 Starting - Launching requested ML instancesProfilerReport-1618743135: InP
rogress
......
2021-04-18 10:53:44 Starting - Preparing the instances for training......
2021-04-18 10:54:39 Downloading - Downloading input data
2021-04-18 10:54:39 Training - Downloading the training image...
2021-04-18 10:55:16 Uploading - Uploading generated training model.Docker entrypoint called w
ith argument(s): train
Running default environment configuration script
[04/18/2021 10:55:13 INFO 140499858621824] Reading default configuration from /opt/amazon/li
b/python2.7/site-packages/algorithm/resources/default-conf.json: {u'index_metric': u'L2', u'_
tuning_objective_metric': u'', u'_num_gpus': u'auto', u'_log_level': u'info', u'feature_dim':
u'auto', u'faiss_index_ivf_nlists': u'auto', u'epochs': u'1', u'index_type': u'faiss.Flat',
u'_faiss_index_nprobe': u'5', u'_kvstore': u'dist_async', u'_num_kv_servers': u'1', u'mini_ba
tch_size': u'5000'}
```

Figure 4.53 – Logs after running the fit() function

The training job should complete in a few minutes. Once the training job has completed, feel free to proceed with the next set of steps.

16. Deploy the trained KNN model using the `deploy()` function. As this step will provision a machine learning instance and deploy the model to the instance, this step may take 5-10 minutes to complete:

```
    predictor2 = estimator2.deploy(
        initial_instance_count=1,
        instance_type='ml.t2.medium')
```

17. Update `serializer` and `deserializer` of `predictor2`:

```
import sagemaker
from sagemaker.serializers import CSVSerializer
from sagemaker.deserializers import JSONDeserializer
predictor2.serializer = CSVSerializer()
predictor2.deserializer = JSONDeserializer()
```

18. Use the `predict()` function to predict the corresponding labels for each of the values in `X_test`:

```
results2 = predictor2.predict(X_test)
results2["predictions"][0:3]
```

We should get a set of values with a structure similar to `[{'predicted_label': 1.0}, {'predicted_label': 1.0}, {'predicted_label': 0.0}]`.

19. Delete the two inference endpoints using the `delete_endpoint()` function:

```
predictor1.delete_endpoint()
predictor2.delete_endpoint()
```

This should delete both endpoints.

Let's see how this works!

How it works...

In this recipe, we have demonstrated a few different ways to initialize an estimator and perform training jobs using different training input formats:

- With an algorithm-specific estimator class with the training input data using `record_set()` with the NumPy array of values as the parameter

- With an `Estimator` class with the `protobuf recordIO` training input file as the training input data as the parameter to the `fit()` function call

Take note of the key differences in this recipe when running the training jobs as these differences apply to other built-in algorithms as well.

As discussed in the *Converting CSV data into protobuf recordIO format* recipe, we will benefit from faster training start times if the training data is serialized into the `protobuf recordIO` format. When using this format, we can make use of Pipe mode in SageMaker, which enables data streaming while the training job is running. This means that less volume disk space is needed and training jobs will start (and finish) faster as well.

Preparing the SageMaker Processing prerequisites using the AWS CLI

One of the most important steps in the machine learning process involves the preparation, processing, and transformation of the data before the actual training step. After the training step, the data needs to be analyzed and may need to be processed further before and during the evaluation step. Amazon SageMaker Processing is one of the most powerful options for fulfilling these types of requirements.

If you have a custom data processing script (for example, a data transformation script), your data is stored in an Amazon S3 bucket, or you are planning to run this script in an isolated managed environment that can easily be configured to handle larger datasets for production workloads at a later stage, then the next three recipes are for you!

> **Tip**
> Technically, you can use Amazon SageMaker Processing for any processing requirement that involves using a managed service to handle the infrastructure component and a custom script that performs a specific action.

In this recipe, we will prepare the prerequisites for the following SageMaker Processing recipes—`dataset.processing.csv` containing the data we will load and process, the S3 bucket where the file(s) will be stored, and optionally, the **Amazon ECR** repository. Once we have completed this recipe, we will be able to use Python and R to configure, launch, and monitor SageMaker Processing jobs.

Getting ready

The following is the prerequisite of this recipe:

- A running SageMaker notebook instance where we will run our commands and create directory files

How to do it...

The first set of steps focuses on setting up the directory structure and creating a dummy `dataset.processing.csv` file inside the `SageMaker/my-experiments/chapter04/tmp` directory:

1. If you do not have an existing running terminal in your SageMaker notebook instance, create one by clicking the **New** button and selecting **Terminal** from the drop-down list of options.

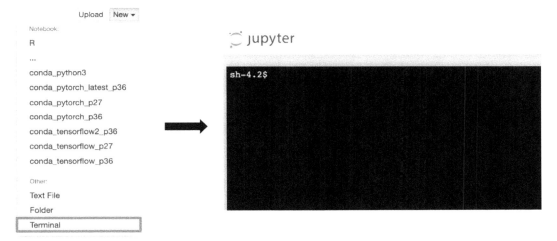

Figure 4.54 – Creating a new terminal

Note that the **Terminal** option is located near the bottom of the dropdown, as seen in *Figure 4.54*.

2. Using Bash commands, navigate to the target directory where you will create your Python and R notebooks, CSV file(s), and `Dockerfile`(s):

```
cd SageMaker/my-experiments
mkdir -p chapter04/tmp
cd chapter04/tmp
```

3. Create an empty file named `dataset.processing.csv`:

```
touch dataset.processing.csv
```

4. Run the following block of code to set some dummy values inside the `dataset.processing.csv` file. This sample `dataset.processing.csv` file is an arbitrary CSV file with three columns, `label`, `a`, and `b`:

```
cat > dataset.processing.csv <<EOF
label,a,b
one,1,2
two,3,4
EOF
```

5. Check the contents of the `dataset.processing.csv` file:

```
cat dataset.processing.csv
```

This would yield a similar output as shown in *Figure 4.55*:

```
sh-4.2$ cat dataset.processing.csv
label,a,b
one,1,2
two,3,4
```

Figure 4.55 – Contents of the dummy dataset.processing.csv file

We can proceed with the next set of steps if the `dataset.processing.csv` file is a valid CSV file.

The next set of steps focuses on creating the Amazon ECR repository for the custom container image we will build in the *Managed data processing with SageMaker Processing in R* recipe.

6. Authenticate with Amazon ECR using the following command:

```
ACCOUNT_ID=$(aws sts get-caller-identity | jq -r
".Account")
aws ecr get-login-password --region us-east-1 | docker
login --username AWS --password-stdin $ACCOUNT_ID.dkr.
ecr.us-east-1.amazonaws.com
```

This should give us the `Login Succeeded` message.

> **Important Note**
>
> Note that we have made the assumption that your repository will be created in the `us-east-1` region. Feel free to modify the region in the command if needed. This applies to all the commands in this chapter.

7. Create an Amazon ECR repository using the following command. Make sure to specify your own ECR repository name (for example, `sagemaker-processing-r`):

```
ECR_REPO_NAME="<insert name here>"
aws ecr create-repository --repository-name $ECR_REPO_
NAME --region us-east-1 --image-tag-mutability IMMUTABLE
```

This should yield a set of values similar to what is shown in *Figure 4.56*:

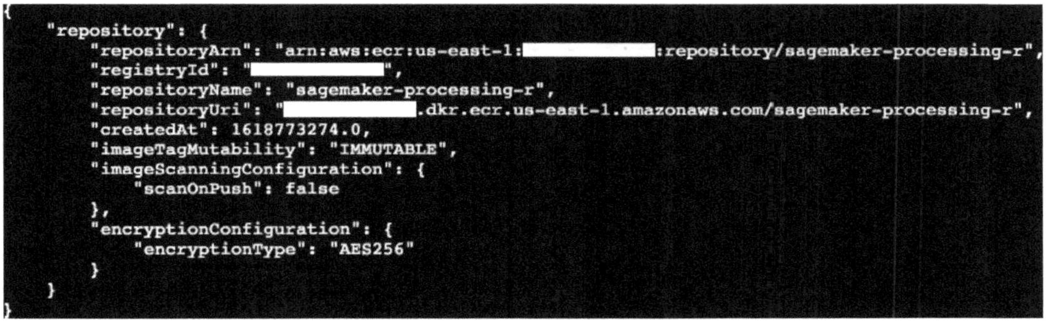

```
{
    "repository": {
        "repositoryArn": "arn:aws:ecr:us-east-1:          :repository/sagemaker-processing-r",
        "registryId": "          ",
        "repositoryName": "sagemaker-processing-r",
        "repositoryUri": "          .dkr.ecr.us-east-1.amazonaws.com/sagemaker-processing-r",
        "createdAt": 1618773274.0,
        "imageTagMutability": "IMMUTABLE",
        "imageScanningConfiguration": {
            "scanOnPush": false
        },
        "encryptionConfiguration": {
            "encryptionType": "AES256"
        }
    }
}
```

Figure 4.56 – Amazon ECR repository successfully created

In *Figure 4.56*, we can see that we have successfully created the Amazon ECR repository. Take note of the `repositoryUri` and `repositoryName` values as we will use them later in the *Managed data processing with SageMaker Processing in R* recipe.

Now, let's see how this works!

How it works...

In this recipe, we have prepared a few prerequisites for the next couple of recipes involving SageMaker Processing in Python and R. For the Python recipe using **script mode**, the prerequisites include the following:

- S3 bucket (input and output)
- Dummy dataset/file(s)

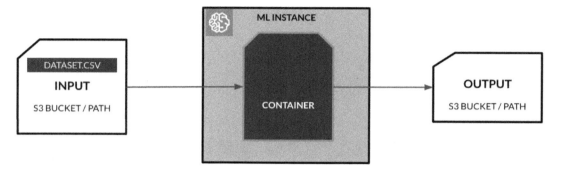

Figure 4.57 – Getting the SageMaker Processing job prerequisites ready

In *Figure 4.57*, we can see that the S3 bucket(s) we have prepared in this recipe will be used as the input and output sources when running the SageMaker Processing jobs in the next two recipes.

On the other hand, for the R recipe using a custom container image, the prerequisites include the following:

- S3 bucket (input and output)
- Dummy dataset/file(s)
- ECR repository

Depending on your needs and requirements, you will choose one of these two approaches. We will see in the next two recipes these two approaches in action.

Managed data processing with SageMaker Processing in Python

In the previous recipe, we prepared a few prerequisites, including preparing the dummy dataset within a specified directory, for the SageMaker Processing job we will run in this recipe. Now, we will create a Python script and use SageMaker Processing to run the custom Python script inside a managed environment. This managed environment is automatically created, configured, and destroyed when the processing job is launched and executed. If you are working on a requirement that is similar to one of the following, then this recipe is for you:

- Normalizing numerical features with `sklearn` (`scikit-learn`)
- Text preprocessing with `nltk` (**Natural Language Toolkit**)
- Automated feature engineering with `pandas`
- Performing post-training processing and evaluation steps

Once we have completed this recipe, we will have the custom Python script executed inside an isolated and managed SageMaker Processing environment and the output of the processing job stored in an S3 bucket.

> **Important Note**
>
> Note that in this recipe, we were able to run a SageMaker Processing job using Python without having to build and prepare a custom container image. In the *Managed data processing with SageMaker Processing in R* recipe, we will build and use a custom container image to run a SageMaker Processing job using the R language.

Getting ready

The following is the prerequisite for this recipe:

- This recipe continues from the *Preparing the SageMaker Processing prerequisites using the AWS CLI* recipe.

How to do it...

This recipe is composed of two parts: writing the processing.py script and using SKLearnProcessor from the **SageMaker Python SDK** to run the processing.py script. The first set of steps focuses on preparing the processing.py script inside the my-experiments/chapter04 directory:

1. Navigate to the my-experiments/chapter04 directory inside your SageMaker notebook instance. Feel free to create this directory if it does not exist yet.

2. Create an empty text file inside the my-experiments/chapter04 directory and name it processing.py. Open this file for editing as we will update this file with the code in the next couple of steps.

> **Important Note**
>
> Note that this recipe assumes that both the processing.py file and the corresponding Jupyter Notebook file are inside the my-experiments/chapter04 directory. If we were to change the location of these files, we would need to update the paths in this recipe as well. For now, let's stick with this assumption so that we can get the SageMaker Processing job running!

3. Start the script by importing pandas, argparse, os, subprocess, importlib, and sys.executable:

```
import pandas as pd
import argparse
import os
import subprocess
import importlib
from sys import executable
```

4. Prepare the function that installs and loads a specified package using pip and subprocess.call:

```
def install_and_load(target):
    sequence = [executable, "-m",
                "pip", "install", target]
    subprocess.call(sequence)

    print(f'[+] Successfully installed {target}')
    return importlib.import_module(target)
```

5. Prepare the function that uninstalls a specified package. This will be used while testing the script locally before using the SageMaker Python SDK to run the script inside the managed environment:

```python
def uninstall(target):
    sequence = [
        executable, "-m",
        "pip", "uninstall", "-y", target]

    subprocess.call(sequence)
    print(f'[+] Successfully uninstalled {target}')
```

6. Define the `process_args()` function, which returns the arguments when the script is run. If no arguments are specified, the default values will be used instead:

```python
def process_args():
    parser = argparse.ArgumentParser()
    parser.add_argument(
        '--sample-argument',
        type=int, default=1)
    arguments, _ = parser.parse_known_args()

    return arguments
```

7. Define the `load_input()` function, which reads and returns the content of a CSV file using its filename:

```python
def load_input(input_target):
    df = pd.read_csv(input_target)
    return df
```

8. Define the `save_output()` function, which generates a sample file in a specified location:

```python
def save_output(output_target):
    with open(output_target, 'w') as writer:
        writer.write("sample output\n")
```

9. Generate the `main()` function, which makes use of the functions created in this recipe:

```
def main():
    args = process_args()
    print(args)

    plt = install_and_load('matplotlib')
    print(plt)
    uninstall('matplotlib')

    path = "/opt/ml/processing/input/dataset.processing.csv"
    sample_input = load_input(path)
    print(sample_input)

    save_output("/opt/ml/processing/output/output.csv")
    print('[+] DONE')
```

We performed the following actions inside the `main()` function:

- Read the arguments using the `process_args()` function
- Performed a test installation of a library using the `install_and_load()` function
- Uninstalled the library using the `uninstall()` function
- Loaded a dummy CSV file using the `load_input()` function
- Generated a dummy CSV file using the `save_output()` function

Note

When working with file paths, it is recommended to use `os.path.join()` to concatenate paths. Note that we have taken a shortcut here to simplify this recipe a bit. At the same time, we have hardcoded the name of the file we will load. In a more realistic example, we would use several statements to list the contents of a directory and load the relevant files in that directory.

---

10. Finally, wrap up the script file by adding the following two lines, which run the `main` function when the script is executed:

```
if __name__ == "__main__":
    main()
```

When building on top of this recipe, take note that the script must not use hardcoded values. We took a shortcut in this recipe to get the `processing.py` file as short as possible. Feel free to use the utility functions we have used to manage the paths and the files in the *Preparing and testing the train script in Python* recipe in *Chapter 2, Building and Using Your Own Algorithm Container Image*.

> **Tip**
>
> Feel free to check whether the `processing.py` file you have just coded is correct by checking a working copy from the Machine-Learning-with-Amazon-SageMaker-Cookbook repository: `https://github.com/PacktPublishing/Machine-Learning-with-Amazon-SageMaker-Cookbook/blob/master/Chapter04/processing.py`.

The next few steps focus on running the Python script using `SKLearnProcessor` from the SageMaker Python SDK. Make sure that you're running the next set of commands using the empty notebook using the `conda_python3` kernel.

11. Prepare and import the required libraries and prerequisites:

```
import boto3
import sagemaker
from sagemaker import get_execution_role
from sagemaker.sklearn.processing import SKLearnProcessor
role = get_execution_role()
```

12. Initialize `SKLearnProcessor`:

```
sklearn_processor = SKLearnProcessor(
    framework_version='0.20.0',
    role=role,
    instance_count=1,
    instance_type='ml.m5.large')
```

13. Initialize `pinput1` and `poutput1`, which are `ProcessingInput` and `ProcessingOutput` objects that will be used in the next step:

```
from sagemaker.processing import ProcessingInput,
ProcessingOutput
source = 'tmp/dataset.processing.csv'
pinput1 = ProcessingInput(
    source=source, destination='/opt/ml/processing/
input')
poutput1 = ProcessingOutput(
    source='/opt/ml/processing/output')
```

14. Use the `run()` API with the specified parameters shown here:

```
sklearn_processor.run(
    code='processing.py',
    arguments = ['--sample-argument', '3'],
    inputs=[pinput1],
    outputs=[poutput1]
)
```

> **Important Note**
>
> As mentioned earlier, this recipe assumes that the notebook running these lines of Python code is in the same directory as the `processing.py` file. If the `processing.py` file is inside another directory, make sure to update the path specified in the `code` parameter value.

15. Store the S3 output path inside the destination variable:

```
latest_job = sklearn_processor.latest_job
destination = latest_job.outputs[0].destination
```

16. Copy the `output.csv` file generated by the script from the S3 target path to the SageMaker notebook instance directory where the Python notebook is located. Take note that there is a space and a dot in the following snippet:

```
!aws s3 cp "{destination}/output.csv" tmp/output.csv
```

17. Inspect the `output.csv` file:

```
!cat tmp/output.csv
```

The `output.csv` file should contain the `output` string value.

Now, let's see how this works!

How it works...

In this recipe, we have used the **script mode** option with **SageMaker Processing** to run our Python script. When using Python, we may no longer need to build and use a custom container as we can use built-in classes from SageMaker Python SDK such as `SKLearnProcessor` to run the script "directly" using script mode. Behind the scenes, script mode makes use of a pre-built container image prepared by the AWS team.

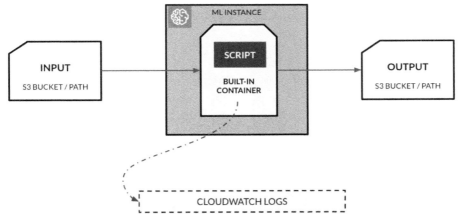

Figure 4.58 – Using script mode with SageMaker Processing

In *Figure 4.58*, we can see that SageMaker Processing automatically runs our custom script inside a built-in container inside a machine learning instance provisioned by SageMaker, and the logs generated by the script inside the custom container are pushed to CloudWatch Logs. Before the script is executed, the files located in the specified input S3 path are loaded, copied, and mounted to the container. After the script is executed, the files generated by the custom script stored in the `/opt/ml/processing/output` directory are automatically copied to the specified output S3 path.

The example in this recipe involves executing some basic print and save statements. In a more realistic scenario, SageMaker Processing can be used in one of (but not limited to) the following use cases:

- Feature engineering and data preparation of large datasets
- Data visualization
- File type conversion
- Model evaluation

Given that we are using a built-in container image of SageMaker Processing with this approach, we may encounter scenarios where we need to use a Python library (for example, `matplotlib`) that is not installed in the built-in container. In this recipe, we have demonstrated how to handle this workaround using the `install_and_load()` function we have defined in the script. This workaround makes use of the `subprocess.call()` function to run the `pip install` Bash command in a separate process.

There's more...

To manage and lower the cost of running experiments and processing jobs, the SageMaker notebook instance where you run your Jupyter notebooks containing your experiment code should ideally be using smaller instance types (for example, `t2.medium`) as it will be running most of the time. On the other hand, larger instance types (for example, `ml.m5.xlarge`) should be used when processing large datasets and these are expected to run for a limited amount of time only (for example, 5-15 minutes).

This is one of the advantages of SageMaker Processing—we only pay for the billable time the resources are running. SageMaker Processing automatically creates the resources when the processing job is initiated and deletes the resources after the processing job has completed.

See also

If you are looking for examples on how to use SageMaker Processing using real datasets and more complex examples, feel free to check some of the notebooks in the `aws/amazon-sagemaker-examples` GitHub repository:

- Feature transformation using SageMaker Processing and SparkML: `https://github.com/aws/amazon-sagemaker-examples/blob/master/sagemaker_processing/feature_transformation_with_sagemaker_processing/feature_transformation_with_sagemaker_processing.ipynb`

- Distributed data processing with SageMaker Processing and PySpark: `https://github.com/aws/amazon-sagemaker-examples/blob/master/sagemaker_processing/spark_distributed_data_processing/sagemaker-spark-processing.ipynb`

Now, let's take a closer look at how to use SageMaker Processing to run R scripts in the next recipe!

Managed data processing with SageMaker Processing in R

In the *Preparing the SageMaker Processing prerequisites using the AWS CLI* recipe, we prepared a few prerequisites including the dummy dataset we will use in our SageMaker Processing job and the ECR repository where we will store the custom container image we will prepare in this recipe.

Now, we will create an R script, build a custom R container image, and use SageMaker Processing to run the R script inside a managed environment that is automatically created, configured, and destroyed when the processing job is launched and executed. If you are working on a requirement that is similar to one of the following, then this recipe is for you:

- Normalizing numerical features with the `normalr` package

- Text preprocessing with the `tm` (text mining) package

- Automated feature engineering with the `dplyr` package

- Performing post-training processing and evaluation steps

Once we have completed this recipe, we will have the custom R script executed inside an isolated and managed SageMaker Processing environment and the output of the processing job stored in an S3 bucket.

> **Important Note**
> Note that in this recipe, we are building and using a custom container image to run a SageMaker Processing job using the R language, while in the *Managed data processing with SageMaker Processing in Python* recipe, we were able to run a SageMaker Processing job using Python without having to build and prepare a custom container image. Technically, we can also build a custom container image when using Python, but we used a pre-built container image already made available to us by AWS and the SageMaker team.

Getting ready

The following is the prerequisite for this recipe:

- This recipe continues from the *Preparing the SageMaker Processing prerequisites using the AWS CLI* recipe.

How to do it...

This recipe involves writing the `processing.r` script, preparing the Dockerfile, building the Docker container image and pushing it to the ECR repository, and using `ScriptProcessor` to run the R script inside the custom container. The first few steps focus on getting us a barebones `processing.r` script inside the `my-experiments/chapter04` directory:

1. Navigate to the `my-experiments/chapter04` directory inside your SageMaker notebook instance. Feel free to create this directory if it does not exist yet.

2. Create an empty text file inside the `my-experiments/chapter04` directory and name it `processing.r`. Open this file for editing as we will update this file with the code in the next couple of steps.

    > **Important Note**
    >
    > Note that this recipe assumes that the `processing.r` file, the Dockerfile, and the corresponding Jupyter Notebook (`ipynb`) file are inside the `my-experiments/chapter04` directory. If we were to change the location of these files, we would need to update the paths in this recipe as well. For now, let's stick with this assumption so that we can get the SageMaker Processing job running!

3. Start the `processing.r` script by loading the `readr` and `argparse` libraries. Take note that the `argparse` package needs to be installed separately and may not be part of the base R installation. In this case, given that this script will run inside the custom container, we will have to ensure that the dependencies are installed inside the custom container image so that the script runs without issues:

    ```
    library(readr)
    library("argparse")
    ```

4. Add the following lines of code to load the arguments passed to the script from the SDK. If no arguments are passed, the default value(s) specified will be used instead:

```
parser <- ArgumentParser()
parser$add_argument("--sample-argument", default=1L)
args <- parser$parse_args()
print(args)
```

5. Load and print the contents the sample `dataset.processing.csv` file by adding the following lines of code. Let's also create an `output.csv` file containing the `output` string value:

```
filename <- "/opt/ml/processing/input/dataset.processing.
csv"
df <- read_csv(filename)
print(df)
cat("output",
    file="/opt/ml/processing/output/output.csv",
    sep="\n")
```

When building on top of this recipe, take note that the script must not use hardcoded values. We took a shortcut in this recipe to make the `processing.r` file as short as possible. Feel free to use the utility functions we have used to manage the paths and the files in the *Preparing and testing the train script in R* recipe in *Chapter 2, Building and Using Your Own Algorithm Container Image*.

> **Tip**
>
> Feel free to check whether the `processing.r` file you have just coded is correct by checking a working copy from the *Machine-Learning-with-Amazon-SageMaker-Cookbook* repository: `https://github.com/PacktPublishing/Machine-Learning-with-Amazon-SageMaker-Cookbook/blob/master/Chapter04/processing.r`.

Now that we have prepared the `processing.r` script file, we will now focus our efforts on generating the Dockerfile for the custom R container image.

6. Make sure you are inside the `my-experiments/chapter04` directory within the SageMaker notebook instance. Create an empty text file inside the `my-experiments/chapter04` directory and name it `Dockerfile`. Open this file for editing as we will update this file with the code in the next couple of steps.

7. Start the `Dockerfile` text file by adding the following line:

```
FROM rocker/tidyverse:latest
```

8. Add the following lines to install the `argparse` library during the container build step:

```
RUN install2.r --error \
    argparse
```

9. Finally, specify the entry point using the `ENTRYPOINT` directive:

```
ENTRYPOINT ["Rscript"]
```

When building the container image in the next set of steps, we will make use of the `rocker/tidyverse` container image as the base image. This container image has the base R installation and packages we need to get the SageMaker Processing job running.

> **Tip**
> Feel free to check whether the Dockerfile you have just coded is correct by checking a working copy from the *Machine-Learning-with-Amazon-SageMaker-Cookbook* repository: `https://github.com/PacktPublishing/Machine-Learning-with-Amazon-SageMaker-Cookbook/blob/master/Chapter04/Dockerfile`

The next few steps focus on building and pushing the custom container image to an existing ECR repository.

10. Navigate back to the `my-experiments/chapter04` directory in the Jupyter Notebook tab. Create a new terminal by clicking the **New** button and selecting **Terminal** from the list of drop-down options.

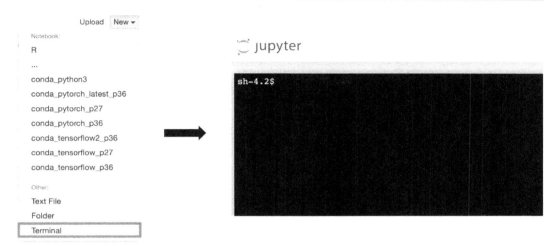

Figure 4.59 – Creating a new terminal

We should find the **New** button beside the **Upload** button located in the right corner of the Jupyter Notebook page. Note that the **Terminal** option should be near the bottom of the drop-down list as we can see in *Figure 4.59*. Once we have the **Terminal** tab available, we will run the commands in the next set of steps in this terminal.

11. Navigate to the directory containing the Dockerfile and the `processing.r` script by running the following code:

```
cd SageMaker/my-experiments/chapter04
```

12. Run the following lines of code to authenticate with Amazon ECR. Make sure to specify the correct value for the ECR URI of the repository by replacing `<ECR URI>` in the following code snippet. The Amazon ECR URI should have a format similar to `1234567890.dkr.ecr.us-east-1.amazonaws.com/sagemaker-processing-r`. Remember that we created this ECR repository in the *Preparing the SageMaker Processing prerequisites using the AWS CLI* recipe:

```
ECR_URI="<insert ECR URI here>"
aws ecr get-login-password --region us-east-1 | docker login --username AWS --password-stdin $ECR_URI
```

This should yield a message similar to what is shown in *Figure 4.60*:

```
WARNING! Your password will be stored unencrypted in /home/ec2-user/.docker/config.json.
Configure a credential helper to remove this warning. See
https://docs.docker.com/engine/reference/commandline/login/#credentials-store

Login Succeeded
```

Figure 4.60 – Authenticating with Amazon ECR

In *Figure 4.60*, we can see that we have successfully authenticated with Amazon ECR. If you encounter issues running the previous set of commands, make sure that the region and the ECR URI arguments are correct, then try again.

13. Build and push the Docker container image by running the following command. Note that there is a space and a dot after $TAG:

```
TAG=1
docker build -t $ECR_URI:$TAG .
```

This should start the build process as seen in *Figure 4.61*:

```
Sending build context to Docker daemon  2.815MB
Step 1/3 : FROM rocker/tidyverse:latest
latest: Pulling from rocker/tidyverse
a70d879fa598: Pull complete
c4394a92d1f8: Pull complete
10e6159c56c0: Pull complete
833bebb909e4: Pull complete
9bceb8b50184: Pull complete
e925a327b30d: Pull complete
01c3425e0ade: Pull complete
aa899054de42: Pull complete
Digest: sha256:f9671fa9329160cc57f76ec822896467f420f4b8d55bf3a811293b9f94
283a3e
Status: Downloaded newer image for rocker/tidyverse:latest
 ---> 5f4c0516e1a2
Step 2/3 : RUN install2.r --error       argparse
 ---> Running in 190caa01d5e3
also installing the dependency 'findpython'

trying URL 'https://packagemanager.rstudio.com/all/__linux__/focal/latest
/src/contrib/findpython_1.0.7.tar.gz'
Content type 'binary/octet-stream' length 20373 bytes (19 KB)
==================================================
downloaded 19 KB

trying URL 'https://packagemanager.rstudio.com/all/__linux__/focal/latest
/src/contrib/argparse_2.0.3.tar.gz'
Content type 'binary/octet-stream' length 141423 bytes (138 KB)
==================================================
downloaded 138 KB

* installing *binary* package 'findpython' ...
* DONE (findpython)
* installing *binary* package 'argparse' ...
* DONE (argparse)
```

Figure 4.61 – Running the docker build command

In *Figure 4.61*, we can see that the build process has been triggered after running the `docker build` command. After pulling the base container image, the `argparse` package will be installed.

> Tip
> The build process should take a few minutes, so feel free to grab a cup of coffee or tea!

14. Once the Docker container image has successfully been built, let's run the `docker push` command:

```
docker push $ECR_URI:$TAG
```

This yields a set of logs similar to what is shown in *Figure 4.62*:

```
4b1cbfe7b765: Pushed
c54f6eed973b: Pushed
975f6564f91e: Pushed
904f4d5a6efd: Pushed
8e5e03028587: Pushed
9fbdaeca625e: Pushed
346be19f13b0: Pushed
935f303ebf75: Pushed
0e64bafdc7ee: Pushed
1: digest: sha256:adaf2daf8190a5210a0a41cfd4328800ad37d22529f0787426da338
01c4ddfb9 size: 2211
```

Figure 4.62 – Logs after running the docker push command

The next couple of steps focus on running the R script inside the custom R container using `ScriptProcessor` from the SDK.

15. Navigate to the `my-experiments/chapter04` directory inside your SageMaker notebook instance again. Create a new notebook using the R kernel inside `my-experiments/chapter04`. Open the notebook (`ipynb`) file as we will update this file with the statements and the lines of code in the next couple of steps. Feel free to rename the notebook with the title of this recipe—`Managed Data Processing with SageMaker Processing in R`.

16. Load the `reticulate` library:

```
library('reticulate')
```

17. Perform the required imports for the SageMaker Processing job:

```
sagemaker <- import('sagemaker')
boto3 <- import('boto3')
role <- sagemaker$get_execution_role()
```

18. Prepare the other prerequisites for the `ScriptProcessor` job. Make sure to replace the value of `<ECR Repo URI>` with the Amazon ECR URI of the ECR repository created in the *Preparing the SageMaker Processing prerequisites using the AWS CLI* recipe. It should have a format similar to `1234567890.dkr.ecr.us-east-1.amazonaws.com/sagemaker-processing-r`. In addition to this, make sure to replace the value of `<TAG>` with the value equal to the latest tag value used when building and pushing the Docker container image:

```
processing_repository_uri <- "<ECR Repo URI>:<TAG>"

session <- boto3$session$Session()
sagemaker_session <- sagemaker$Session(boto_
session=session)
```

19. Initialize `ScriptProcessor` and specify the parameters as shown:

```
ScriptProcessor <- sagemaker$processing$ScriptProcessor
script_processor <- ScriptProcessor(
    command=list('Rscript'),
    image_uri=processing_repository_uri,
    role=role,
    sagemaker_session=sagemaker_session,
    instance_count=1L,
    instance_type='ml.c5.large')
```

20. Initialize `pinput1` and `poutput1`, which are `ProcessingInput` and `ProcessingOutput` objects that will be used in the next step:

```
ProcessingInput <- sagemaker$processing$ProcessingInput
ProcessingOutput <- sagemaker$processing$ProcessingOutput

source <- 'tmp/dataset.processing.csv'
pinput1 <- ProcessingInput(
    source=source, destination='/opt/ml/processing/
input')
poutput1 <- ProcessingOutput(
    source='/opt/ml/processing/output')
```

21. Use the `run()` function using the specified parameters, as shown in the following block of code. You may decide to set `wait=FALSE` to proceed with the next steps but take note that the SageMaker Processing job might take around 5-10 minutes to complete:

```
script_processor$run(code='processing.r',
                     inputs=list(pinput1),
                     outputs=list(poutput1),
                     arguments=list('--sample-
argument','3'),
                     wait=TRUE)
```

> **Important Note**
>
> As mentioned earlier, this recipe assumes that the notebook running these lines of R code is in the same directory as the `processing.r` file. If the `processing.r` file is inside another directory, make sure to update the path specified in the `code` parameter value.

22. Prepare the `cmd` function, which will be used to run Bash commands inside the R notebook:

```
cmd <- function(bash_command) {
    print(bash_command)
    output <- system(bash_command, intern=TRUE)
    last_line = ""

    for (line in output) {
        cat(line)
        cat("\n")
        last_line = line
    }
    return(last_line)
}
```

23. Install the `awslogs` utility using `pip`:

```
cmd('pip install awslogs')
```

24. Use the `awslogs` utility to read the generated logs from CloudWatch Logs. Modify this as needed:

```
cmd("awslogs get /aws/sagemaker/ProcessingJobs -s1h
--aws-region=us-east-1")
```

> **Important Note**
> Make sure that the SageMaker execution role attached to the notebook instance has the `CloudWatchLogsReadOnlyAccess` policy attached.

Running the `aws logs` command generates logs similar to what is shown in *Figure 4.63*:

```
/aws/sagemaker/ProcessingJobs r-processing-2020-11-30-16-26-43-013/algo-1
-1606753788 $sample_argument
/aws/sagemaker/ProcessingJobs r-processing-2020-11-30-16-26-43-013/algo-1
-1606753788 [1] "3"
/aws/sagemaker/ProcessingJobs r-processing-2020-11-30-16-26-43-013/algo-1
-1606753788 — Column specification ———————————————————

/aws/sagemaker/ProcessingJobs r-processing-2020-11-30-16-26-43-013/algo-1
-1606753788 cols(
  label = col_character(),
  a = col_double(),
  b = col_double()
/aws/sagemaker/ProcessingJobs r-processing-2020-11-30-16-26-43-013/algo-1
-1606753788 )
/aws/sagemaker/ProcessingJobs r-processing-2020-11-30-16-26-43-013/algo-1
-1606753788 # A tibble: 2 x 3
  label     a      b
  <chr> <dbl> <dbl>
/aws/sagemaker/ProcessingJobs r-processing-2020-11-30-16-26-43-013/algo-1
-1606753788 1 one       1      2
/aws/sagemaker/ProcessingJobs r-processing-2020-11-30-16-26-43-013/algo-1
-1606753788 2 two       3      4
```

Figure 4.63 – Processing job logs using awslogs

Figure 4.63 shows us the logs generated by our custom script in the processing job. As we can see from the logs, the R script was able to get the argument value 3 from the arguments list provided using the SDK; the input CSV file was loaded into a DataFrame and then printed into the logs

The last set of steps in this recipe focuses on retrieving the output file generated by the SageMaker Processing job. If you can remember, we added a line of code in the `processing.r` script file that stores the `output` string value in the `/opt/ml/processing/output/output.csv` file. SageMaker will pick this up inside the running container and upload this to the Amazon S3 destination path.

25. Next, retrieve the output path of the latest processing job:

```
latest_job <- script_processor$latest_job
destination <- latest_job$outputs[[1]]$destination
```

26. Download the output CSV file:

```
csv_path <- paste0(destination, "/output.csv")
command <- paste("aws s3 cp", csv_path, "tmp/output.
processing.r.csv")
cmd(command)
```

27. Finally, use `read.csv()` to read the output file contents:

```
read.csv("tmp/output.processing.r.csv", header=FALSE)
[[1]]
```

We should get the `output` string as the result of the preceding line of code.

Now, let's see how this works!

How it works...

In this recipe, we have built and used a custom container image with SageMaker Processing to run our R script. When using Python, we may no longer need to build and use a custom container as we can use built-in classes from the SageMaker Python SDK such as the SKLearnProcessor class to run the script directly with script mode. Behind the scenes, script mode makes use of a pre-built container image. When using the ScriptProcessor class with the custom container image option, there is a bit more flexibility as the custom container image can include pre-installed libraries and tools specified and configured by the user.

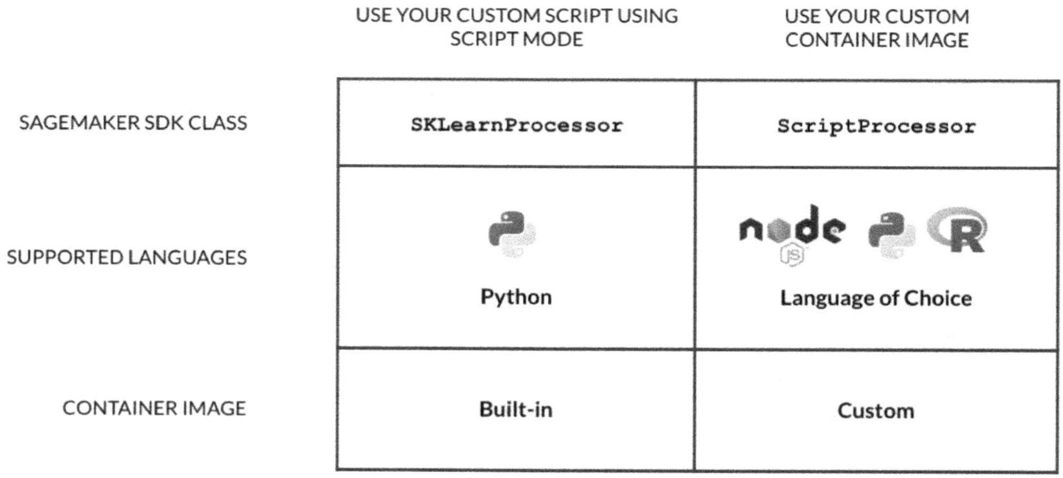

	USE YOUR CUSTOM SCRIPT USING SCRIPT MODE	USE YOUR CUSTOM CONTAINER IMAGE
SAGEMAKER SDK CLASS	SKLearnProcessor	ScriptProcessor
SUPPORTED LANGUAGES	Python	Language of Choice
CONTAINER IMAGE	Built-in	Custom

Figure 4.64 – Choosing between script mode and using your own container image

We can see in *Figure 4.64* that when we need to use R or other languages in SageMaker Processing, we need to build and use our own container image. If you want to use Python and you want to use your own custom container image, then that is possible as well using the ScriptProcessor class.

> **Important Note**
>
> Take note that the base class of SKLearnProcessor is the ScriptProcessor class. Most of their parameters are the same (for example, sagemaker_session, role, and base_job_name); what makes the SKLearnProcessor class different is that it does not accept the image_uri parameter value as it is using a pre-built container provided by AWS.

In the script we have prepared in this recipe, we just had the script print a couple of dummy values just to show where we can see the script output after the processing job has completed.

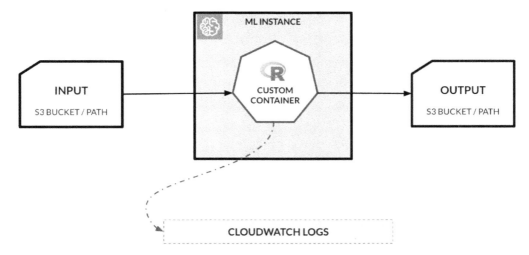

Figure 4.65 – Using a custom container image with SageMaker Processing

In *Figure 4.65*, we can see that the logs generated by the script inside the custom container are pushed to CloudWatch Logs. Toward the end of this recipe, we were able to pull these logs using the `awslogs` command-line tool.

In a more realistic example, we would use different R packages inside the script. Make sure of the following:

- The R packages are properly installed in the container through the Dockerfile or by using a Docker container image that already includes these R packages.

- The packages are loaded in the script using the `library()` function.

Finally, it is best for the R script to be developed and tested outside of the container image and SageMaker Processing first to reduce any roadblocks and delays due to the number of steps required to build and run the container.

5
Effectively Managing Machine Learning Experiments

In the previous chapter, we worked on several recipes that focused on preparing and processing the data before passing it as input to the training jobs. In this chapter, we will focus on different solutions and capabilities to help us manage **machine learning** (**ML**) experiments in **Amazon SageMaker**.

Once we have performed a certain number of ML experiments, we will realize that not all experiments succeed, and it takes a bit of trial and error to build high-quality ML models. This is somewhat similar to software development where bugs in the code need to be detected as early as possible to prevent these bugs from accidentally getting deployed into a production environment. Debugging ML experiments is generally much harder compared to debugging issues in software code since we would need a specialized tool that inspects and monitors changes in the values of parameters, metrics, and other variables in the experiment and performs a specified action when specific rules or conditions are met. This tool would also be extremely useful when working with clusters of ML instances when running distributed training jobs. The good news is that **SageMaker Debugger** is that specialized tool that we are looking for!

In addition to using the right tools for debugging ML experiments, it is critical that data science and ML engineering teams actively keep track of datasets, hyperparameters, and other input and output values of multiple ML experiments as well. This would allow data scientists and ML practitioners to easily access previous ML experiments without having to spend a lot of time trying to search for a needle in a haystack. That said, we have more good news for you—**SageMaker Experiments** can help us keep track of and manage this growing number of entities.

In this chapter, we will demonstrate how to use **SageMaker Debugger** to detect issues in our training jobs. After that, we will use SageMaker Experiments to manage and track multiple experiments at the same time. We will inspect the logs and entities produced while using these capabilities during our sample ML experiments as well. Once we have become quite familiar with the services, tools, and techniques discussed in this chapter, we will be able to manage and debug multiple ML experiments with ease.

We will cover the following recipes in this chapter:

- Synthetic data generation for classification problems
- Identifying issues with SageMaker Debugger
- Inspecting SageMaker Debugger logs and results
- Running and managing multiple experiments with SageMaker Experiments
- Experiment analytics with SageMaker Experiments
- Inspecting experiments, trials, and trial components with SageMaker Experiments

As you get to work with more ML experiments using SageMaker, you will find the recipes in this chapter very useful as they will help you manage and audit both running or completed ML experiments.

Technical requirements

To execute the recipes in the chapter, make sure you have the following:

- A running Amazon SageMaker notebook instance (for example, `ml.t2.large`)
- An Amazon S3 bucket

If you do not have these prerequisites ready yet, feel free to check the *Launching an Amazon SageMaker notebook instance* and *Preparing the Amazon S3 bucket and the training dataset for the linear regression experiment* recipes in *Chapter 1, Getting Started with Machine Learning Using Amazon SageMaker*.

As the recipes in this chapter involve a bit of code, we have made these scripts and notebooks available in this repository: `https://github.com/PacktPublishing/Machine-Learning-with-Amazon-SageMaker-Cookbook/tree/master/Chapter05`.

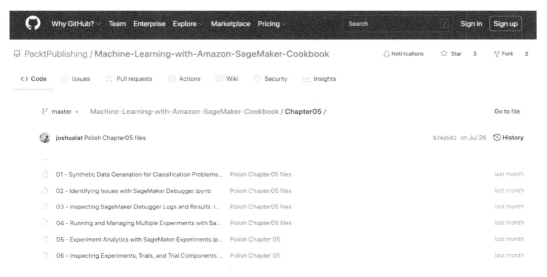

Figure 5.1 – Machine-Learning-with-Amazon-SageMaker-Cookbook GitHub repository

As seen in *Figure 5.1*, we have the source code for the scripts and notebooks for the recipes in this chapter organized inside the `Chapter05` directory. Before starting on each of the recipes in this chapter, make sure that the `my-experiments/chapter05` directory is ready. If it has not yet been created, please do so now as this keeps things organized as we go through each of the recipes in this book.

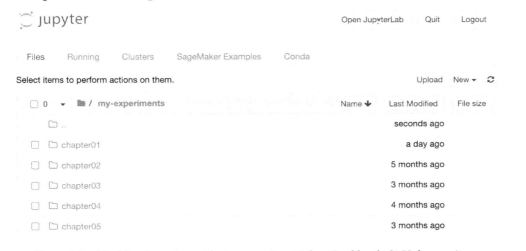

Figure 5.2 – Machine-Learning-with-Amazon-SageMaker-Cookbook GitHub repository

In *Figure 5.2*, we can see the **New** button beside the **Upload** button in the upper left-hand corner of the page. To create a new directory inside the `my-experiments` directory, click the **New** button and select **Folder** in the list of drop-down options to create a new directory. We will create multiple Jupyter notebooks using the `conda_python3` kernel inside this directory in this chapter.

> **Note**
>
> As we are not using *local mode* in the recipes in this chapter, we can technically work on the recipes here inside SageMaker Studio notebooks. If you are wondering when we will start using **SageMaker Studio** in this book, do not worry as we will properly introduce and set it up in *Chapter 6, Automated Machine Learning in Amazon SageMaker*.

That said, let's start working on the recipes of this chapter!

Check out the following link to see the relevant Code in Action video:

`https://bit.ly/2YAwHMl`

Synthetic data generation for classification problems

In this recipe, we will generate a synthetic dataset using scikit-learn. This dataset will serve as a dummy dataset for the classification problems in this chapter. This dataset has only three columns—`label`, `a`, and `b`. In *Figure 5.3*, we have a scatterplot diagram of the dataset showing the two groups of points grouped by their `label` values:

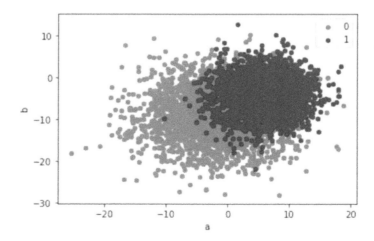

Figure 5.3 – Synthetic dataset for binary classification problems

We will divide this dataset into training, validation, and test datasets with a train-test split and upload these to an Amazon S3 bucket. Once we have them ready, we can run ML experiments while working with SageMaker Debugger and SageMaker Experiments in the following recipes in this chapter.

> **Tip**
> Since we will show the steps on how to generate a synthetic dataset in this recipe, we will have the opportunity to tweak this recipe later on to fit our needs. We can decide to make this dataset generate more records for `label = 0` and use SageMaker Debugger to detect whether the generated dataset suffers from the class imbalance problem. We will talk about class imbalance along with other ML bias issues in the *Detecting pretraining bias with SageMaker Clarify* recipe in *Chapter 7, Working with SageMaker Feature Store, SageMaker Clarify, and SageMaker Model Monitor*. In the following recipes in this chapter, we will use the SageMaker Debugger `LossNotDecreasing` rule to monitor and detect whether the decrease in the target metric of a running training job (with the dataset generated in this recipe) is less than the threshold we have specified.

Getting ready

The following is the prerequisite for this recipe:

- A running Amazon SageMaker notebook instance (for example, `ml.t2.large`)

How to do it...

The first set of steps in this recipe focuses on setting up a DataFrame containing the `x` and `y` values of the generated synthetic dataset:

1. Navigate to the `my-experiments/chapter05` directory inside your SageMaker notebook instance. Feel free to create this directory if it does not exist yet.

2. Create a new notebook using the `conda_python3` kernel inside the `my-experiments/chapter05` directory and name it with the name of this recipe (that is, `Synthetic Data Generation for Classification Problems`). Open this notebook for editing as we will update this file with the code in the next couple of steps.

3. Use the `make_blobs()` function from `scikit-learn` to generate a dataset that can be used for classification problems:

```
from sklearn.datasets import make_blobs
X, y = make_blobs(n_samples=5000, centers=2,
                  cluster_std=[6, 4], n_features=2,
                  random_state=40)
```

4. Using the X and y variables from the previous step, create a `pandas` DataFrame with three columns: `label`, `a`, and `b`:

```
import pandas as pd
all_dataset = pd.DataFrame(
    dict(label=y, a=X[:,0], b=X[:,1]))


print(all_dataset)
```

Figure 5.4 shows us what the output looks like after running the previous block of code:

```
      label           a           b
0         1   11.253262   -5.355250
1         0    5.155165   -1.044914
2         0    4.023015   -6.368246
3         0   -6.094907   -7.146572
4         1    0.839381  -10.651923
...     ...         ...         ...
4995      0    0.194730   -8.277397
4996      0    4.966191   -7.437370
4997      0   -5.376702   -7.324231
4998      1    6.371769   -1.671036
4999      0    2.474043  -13.858433

[5000 rows x 3 columns]
```

Figure 5.4 – Dataset containing all the records after using the make_blobs() function

In *Figure 5.4*, we can see that in the `all_dataset` DataFrame, we have three columns—`label`, containing 1s and 0s, and `a` and `b`, containing floating-point values.

5. Generate a scatterplot of `all_dataset` with the following lines of code:

```
from matplotlib import pyplot

colors = {0:'red', 1:'blue'}
fig, ax = pyplot.subplots()
grouped = all_dataset.groupby('label')

for key, group in grouped:
    group.plot(ax=ax, kind='scatter',
               x='a', y='b',
               label=key,
               color=colors[key])

pyplot.show()
```

This should generate a scatterplot similar to what is shown in *Figure 5.5*:

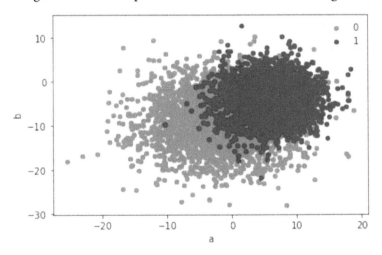

Figure 5.5 – Synthetic dataset for classification problems

In *Figure 5.5*, we have the scatterplot of `all_dataset` generated using `matplotlib`.

Now that we have the synthetic dataset generated, we will perform the train-test split before uploading the resulting training, validation, and test datasets to an Amazon S3 bucket.

6. Perform the train-test split step and get the training, validation, and test datasets:

```
from sklearn.model_selection import train_test_split

train_val, test = train_test_split(all_dataset,
                                    test_size=0.2,
                                    random_state=0)
training, validation = train_test_split(
    train_val,
    test_size=0.25,
    random_state=0)
```

This should give us 3,000 records for the `training` dataset, 1,000 records for the `validation` dataset, and 1,000 records for the `test` dataset.

7. Inspect the training dataset by running the next line of code:

```
training
```

This should show a DataFrame of values similar to what is shown in *Figure 5.6*:

	label	a	b
2869	1	5.495257	-7.298110
4588	1	11.538980	1.776162
4423	0	-7.362446	-0.777593
963	1	8.311903	-6.773622
3018	0	0.617189	-5.809337
...	...	...	...
3	0	-6.094907	-7.146572
3661	0	0.587368	-10.858990
3911	1	-2.410089	-8.436436
3588	0	-4.745941	-1.500113
2035	1	4.567330	-3.688829

3000 rows × 3 columns

Figure 5.6 – Training dataset with 3,000 records

Figure 5.6 shows us a quick view of the training dataset. Note that this dataset has 3,000 records, which is 60% of the total number of records in `all_dataset`.

8. Inspect the validation dataset by running the next line of code:

```
validation
```

This should give us a DataFrame of values similar to what is shown in *Figure 5.7*:

	label	a	b
1827	0	-1.272569	0.388111
3688	1	7.930955	-2.510119
3591	0	-2.982887	-6.956733
1840	0	-10.263374	-10.968153
1979	1	5.839357	-3.068643
...	...	...	...
2492	0	-1.694507	-3.569015
4835	1	5.447907	-5.371618
3793	1	1.906472	-7.107010
2709	0	9.129038	-9.253943
745	0	-0.572530	-2.297837

1000 rows × 3 columns

Figure 5.7 – Validation dataset with 1,000 records

Figure 5.7 shows us a quick view of the validation dataset. Take note that this dataset has 1,000 records, which is 20% of the total number of records.

9. Inspect the test dataset by running the next line of code:

```
test
```

This should give us a DataFrame similar to what is shown in *Figure 5.8*:

	label	a	b
398	1	2.698427	-0.811067
3833	0	-7.546137	-7.444046
4836	0	-6.850566	-15.322851
4572	1	8.338603	-5.083704
636	0	1.083114	-8.722642
...	...	...	...
4228	0	12.950594	-15.304722
2367	0	-1.767492	-1.240460
788	0	3.271917	-7.494765
1452	1	5.960743	-8.790329
3265	0	-5.680270	-14.951039

1000 rows × 3 columns

Figure 5.8 – Test dataset with 1,000 records

Figure 5.8 shows us a quick view of the test dataset. Take note that this dataset has 1,000 records, which is 20% of the total number of records in `all_dataset`.

10. Generate a temporary directory called `tmp` using the `mkdir` Bash command:

```
!mkdir -p tmp
```

11. Next, use the `to_csv()` function to generate a CSV file from the `training`, `validation`, and `test` DataFrames:

```
training.to_csv('tmp/training_data.csv', header=False,
index=False)
validation.to_csv('tmp/validation_data.csv',
header=False, index=False)
test.to_csv('tmp/test_data.csv', header=False,
index=False)
```

12. Use the **AWS CLI** Bash commands to copy the CSV files to our target S3 path. Make sure to replace the value of "<insert S3 bucket name here>" with the name of the bucket we have created in the *Preparing the Amazon S3 bucket and the training dataset for the linear regression experiment* recipe from *Chapter 1, Getting Started with Machine Learning Using Amazon SageMaker*:

```
s3_bucket = "<insert S3 bucket name here>"
prefix = "chapter05"
path = f"s3://{s3_bucket}/{prefix}/input"
!aws s3 cp tmp/training_data.csv {path}/training_data.csv
!aws s3 cp tmp/validation_data.csv {path}/validation_
data.csv
!aws s3 cp tmp/test_data.csv {path}/test_data.csv
```

This should upload the CSV files of the training, validation, and test datasets into the S3 bucket.

Let's see how this works!

How it works...

In this recipe, we have used the make_blobs() function to generate a synthetic dataset for classification problems. Let's describe a few options we can use when calling this function:

- n_samples: Number of samples

- cluster_std: Standard deviation of the clusters

- n_features: Number of features

- random_state: Random state value that makes the operation reproducible

The dataset we have produced has three columns—label, a, and b. Our goal when building a binary classifier is to use the a and b values of a record to predict the label value.

> **Note**
>
> In real-life datasets and examples, a and b can easily map to one or more predictor variables. One example of this would be using the properties of a breast mass image (such as perimeter_mean, area_mean, concavity_worst) to detect whether a tumor is benign or malignant.

After generating the dataset containing 5,000 records, we have used the `train_test_split()` function twice to obtain the training, validation, and test sets. We get a *60-20-20* split for these datasets. We then uploaded these datasets into the target S3 bucket. These will be used later when we work on the *Identifying issues with SageMaker Debugger* and *Running and managing multiple experiments with SageMaker Experiments* recipes.

If you are curious which algorithms we can use for training a classifier using this dataset, here are a few algorithms:

- **XGBoost** with `objective='binary:logistic'`
- **Linear Learner** with `predictor_type='binary_classifier'`
- **Factorization Machines** with `predictor_type='binary_classifier'`
- A custom algorithm (for example, **support vector machines**) using a custom container image

Out of these options, we will use XGBoost in the succeeding recipes to solve our binary classification problem.

There's more...

What are some of the tweaks we can perform on this recipe to come up with variations of this dataset?

- We can generate (significantly) more records and test the performance of the algorithms used in the succeeding recipes. Issues during training may or may not appear depending on the size of the dataset, so this is something you can experiment with after we have finished this chapter.

- We can add (significantly) more columns to the generated dataset. For example, instead of having just 2 predictor columns, we can generate 40 predictor columns by modifying the parameter values passed in the `make_blobs()` function. Again, some issues during training may appear because of this change and we want to check whether certain SageMaker Debugger rules can be used to detect these issues.

With these in mind, let's proceed with the recipes that make use of this synthetic dataset!

Identifying issues with SageMaker Debugger

Amazon SageMaker Debugger is one of the more powerful capabilities of Amazon SageMaker that can help us manage our ML experiments. With SageMaker Debugger, we can automatically detect issues and profile training jobs using **Debugger rules**. We are then able to eliminate these issues and bottlenecks, which would help improve training time and significantly reduce costs. SageMaker Debugger can also be used to monitor the hardware resource usage of training jobs. This feature can help significantly reduce costs as we are able to profile training jobs, detect issues caused by hardware resource usage early, and optimize training time and resource usage. **SageMaker Debugger** supports ML frameworks and algorithms such as **XGBoost**, **PyTorch**, **TensorFlow**, and **MXNet**.

There are several built-in Debugger rules to choose from. These include (but are not limited to) the `VanishingGradient`, `PoorWeightInitialization`, `ExplodingTensor`, `DeadRelu`, and `LossNotDecreasing` rules. In most cases, we can configure SageMaker Debugger with these rules without requiring any changes in our custom training scripts.

In this recipe, we will use the SageMaker Debugger `LossNotDecreasing` rule to monitor and detect whether the decrease in the target metric of a training job using the XGBoost built-in algorithm is less than the threshold we have specified.

> **Tip**
> Using SageMaker Debugger is free! Generally, we only pay for the compute and storage resources based on their usage. This means that we will pay for the **SageMaker Processing** jobs launched (for a few minutes) while using SageMaker Debugger. For more information, refer to the pricing page here: `https://aws.amazon.com/sagemaker/pricing/`.

Getting ready

The following is the prerequisite for this recipe:

- This recipe continues from *Synthetic data generation for classification problems*.

How to do it...

The following instructions show us how to use SageMaker Debugger to detect issues with our training jobs with a few lines of code:

1. Navigate to the `my-experiments/chapter05` directory inside your SageMaker notebook instance. Feel free to create this directory if it does not exist yet.

2. Create a new notebook using the `conda_python3` kernel inside the `my-experiments/chapter05` directory and name it with the name of this recipe (that is, `Identifying Issues with SageMaker Debugger`). Open this notebook for editing as we will update the file with the code in the next couple of steps.

3. Import and prepare a few prerequisites such as `session`, `role_arn`, `boto3`, and the `get_execution_role()` function using the following lines of code:

```
import sagemaker
import boto3
from sagemaker import get_execution_role
role_arn = get_execution_role()
session = sagemaker.Session()
```

4. Specify the S3 location of the training, validation, and test CSV files. Make sure to replace the value of `"<insert S3 bucket name here>"` with the name of the bucket we created in the *Preparing the Amazon S3 bucket and the training dataset for the linear regression experiment* recipe from *Chapter 1, Getting Started with Machine Learning Using Amazon SageMaker*:

```
s3_bucket = '<insert S3 bucket name here>'
prefix = "chapter05"
path = f"s3://{s3_bucket}/{prefix}/input"

training_path = f"{path}/training_data.csv"
validation_path = f"{path}/validation_data.csv"
```

Note that the `s3_bucket` value should be the same as the value used in the *Synthetic dataset generation for classification problems* recipe in this chapter.

5. Use the `retrieve()` function to get the ECR repository URI of the XGBoost container image:

```
from sagemaker.image_uris import retrieve
container = retrieve('xgboost',
                     boto3.Session().region_name,
                     version="0.90-2")
```

6. Import and initialize the prerequisites to use SageMaker Debugger in our experiment:

```
from sagemaker.debugger import rule_configs
from sagemaker.debugger import Rule
from sagemaker.debugger import DebuggerHookConfig
from sagemaker.debugger import CollectionConfig

save_interval = 2
prefix = "debugger"
bucket_path = 's3://{}/{}'.format(s3_bucket, prefix)
```

7. Initialize and configure the `CollectionConfig` and `DebuggerHookConfig` objects:

```
metrics_collection_config = CollectionConfig(
    name="metrics",
    parameters={
        "save_interval": str(save_interval)
    })

debugger_hook_config = DebuggerHookConfig(
    s3_output_path=bucket_path,
    collection_configs=[metrics_collection_config]
)
```

Here, we specify the data to collect along with the other configuration parameters.

> **Note**
>
> For a comprehensive list of built-in collections available with SageMaker Debugger, feel free to check `https://github.com/awslabs/sagemaker-debugger/blob/master/docs/api.md#built-in-collections`.

8. Prepare the SageMaker Debugger rule using the following lines of code:

```
loss_not_decreasing_rule = Rule.sagemaker(
    rule_configs.loss_not_decreasing(),
    rule_parameters={
        "collection_names": "metrics",
        "diff_percent": "5",
        "num_steps": "2",
    },
)
rules = [loss_not_decreasing_rule]
```

9. Next, initialize the `Estimator` function:

```
estimator = sagemaker.estimator.Estimator(
    role=role_arn,
    instance_count=1,
    instance_type='ml.m5.xlarge',
    image_uri=container,
    debugger_hook_config=debugger_hook_config,
    rules=rules,
    sagemaker_session=session)
```

Note that we are using XGBoost as a built-in algorithm in this recipe. When using XGBoost as a built-in algorithm, we will not use the `XGBoost` class from the SageMaker Python SDK as that is used when using the open source XGBoost algorithm.

> **Note**
>
> We can use XGBoost as a built-in algorithm or use the open source XGBoost algorithm, which involves writing a custom training script. For more information about the XGBoost built-in algorithm, feel free to check the following link for more information: `https://sagemaker.readthedocs.io/en/stable/frameworks/xgboost/using_xgboost.html`.

10. Specify the hyperparameters using the `set_hyperparameters()` function:

```
estimator.set_hyperparameters(max_depth=16,
                              objective='binary:logistic',
                              num_round=10000)
```

11. Prepare the inputs for the training job with `TrainingInput`:

```
from sagemaker.inputs import TrainingInput
s3_input_training = TrainingInput(
    training_path, content_type="text/csv")
s3_input_validation = TrainingInput(
    validation_path, content_type="text/csv")
```

12. Use the `fit()` function to start the training step. Take note that the `wait` argument is set to `False`, which allows us to run the next steps while the training jobs are running:

```
estimator.fit({'train': s3_input_training,
               'validation': s3_input_validation},
              wait=False)
```

13. Import, initialize, and load a few prerequisites for printing the rule job summary:

```
import time
job_name = estimator.latest_training_job.name
client = \
estimator.sagemaker_session.sagemaker_client
print("Job Name:", job_name)
EVALUATION_STOP_STATES = [
    "Stopped", "IssuesFound",
    "NoIssuesFound", "Error"
]
```

14. Define the `display_rule_job_summary()` function. This function will help display the details of the SageMaker Processing job running the SageMaker Debugger rule logic:

```
def display_rule_job_summary(rule_job_summary):
    break_after_this = False

    for rule_job in rule_job_summary:
        rj = rule_job
        rule_name = rj["RuleConfigurationName"]
        es = rj["RuleEvaluationStatus"]
        evaluation_status = es
        print("Rule [{}]: {}".format(
            rule_name, evaluation_status))

        if evaluation_status == 'IssuesFound':
            summary = rule_job_summary[0]
            sd = summary['StatusDetails']
            status_details = sd
            print("{}".format(status_details))

        stopped = es in EVALUATION_STOP_STATES
        np = 'ProfilerReport' not in rule_name
        not_profiler = np

        if stopped and not_profiler:
            break_after_this = True

    return break_after_this
```

We will use this function in the next step to help us see whether a violation of the `LossNotDecreasing` rule has been detected by the SageMaker Debugger job.

15. Run the following lines of code to display the status of the training and processing jobs. The `describe_training_job()` function accepts the training job name and returns the details and metadata associated with the training job. Inside this loop, we also make use of the `display_rule_job_summary()` function to help us extract and print a quick summary of the rule job and the current status of the training job:

```
for _ in range(200):
    description = client.describe_training_job(
        TrainingJobName=job_name
    )
    ts = description["TrainingJobStatus"]
    training_job_status = ts
    print("\nTraining job Status: {}".format(
        training_job_status))
    latest_job = estimator.latest_training_job
    rule_job_summary = latest_job.rule_job_summary()
    break_after_this = display_rule_job_summary(
        rule_job_summary
    )
    if break_after_this:
        break

    time.sleep(10)
```

Figure 5.9 shows us a certain portion of the `print` logs after running the previous block of code:

```
Job Name: sagemaker-xgboost-2020-12-17-19-10-41-877

Training job Status: InProgress
Rule [LossNotDecreasing]: InProgress
Rule [ProfilerReport-1608232241]: InProgress

Training job Status: InProgress
Rule [LossNotDecreasing]: InProgress
Rule [ProfilerReport-1608232241]: InProgress

Training job Status: InProgress
Rule [LossNotDecreasing]: InProgress
Rule [ProfilerReport-1608232241]: InProgress

Training job Status: InProgress
Rule [LossNotDecreasing]: InProgress
Rule [ProfilerReport-1608232241]: InProgress
```

Figure 5.9 – Logs showing the status of the training and processing jobs

We can see in *Figure 5.9* the logs generated by the loop in the preceding block of code. This log summary includes the training job status and the `LossNotDecreasing` rule job status.

> **Important Note**
>
> This step may take more than 30 minutes to complete, so feel free to read the *How it works…* and *There's more…* sections of this recipe while waiting. You may also proceed to work on the succeeding recipes in this chapter as well!

After 30 minutes or so, we should see that a violation of the `LossNotDecreasing` rule is detected toward the end of the job summary logs.

16. Store the debugger artifact's S3 path using `%store` magic. We will use this later in the first half of the *Inspecting SageMaker Debugger logs and results* recipe:

```
e = estimator
ap = e.latest_job_debugger_artifacts_path()
artifacts_path = ap
%store artifacts_path
```

17. Similarly, store the rule job summary dictionary value using `%store` magic as well. We will use this in the second half of the *Inspecting SageMaker Debugger logs and results* recipe:

```
rjs = e.latest_training_job.rule_job_summary()
rule_job_summary = rjs
%store rule_job_summary
```

We will use the value of `rule_job_summary` later in the *Inspecting SageMaker Debugger logs and results* recipe.

> **Note**
>
> The ML instances provisioned behind the scenes by SageMaker training and SageMaker Processing jobs are automatically deleted when these jobs are completed. This means that we do not have to turn off running resources manually similar to what we performed in the other chapters. In terms of costs, we will only pay for the time the jobs are running. For more information on this, feel free to check the pricing calculator along with the pricing examples provided here: https://aws.amazon.com/sagemaker/pricing/.

Now, let's see how this works!

How it works...

In this recipe, we have used SageMaker Debugger to help us detect issues with our training job. It works by collecting debug data during the training step and checking whether certain rule conditions are met while the debug data is being processed and analyzed.

Let's define a few things so that we can understand how SageMaker Debugger works:

- **Debugger Hooks**: Capture the debug data (for example, `train-error`, `validation-error`) during the training step. When initializing `Estimator`, we can specify `DebuggerHookConfig` with one or more `CollectionConfig` objects.

- **Debugger rules**: Code that detects whether certain conditions are met. When the training step is executed, a rules processing container will automatically process the debug data and identify whether issues are found. Take note that we can specify as many rules as we want when initializing `Estimator`.

Once the Debugger Hook and Debugger rule configurations have been prepared and passed as parameter values when initializing the `Estimator` object, all we need to do is call the `fit()` function to start the training job.

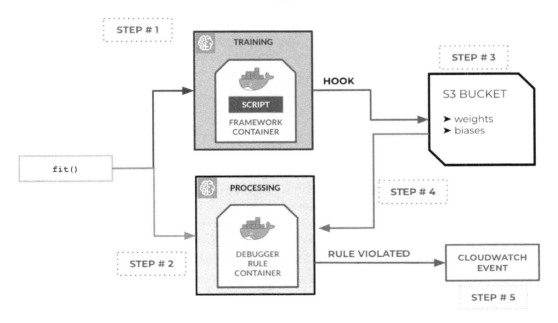

Figure 5.10 – What happens behind the scenes when using SageMaker Debugger

Refer to *Figure 5.10* for a quick visual of what happens behind the scenes when using SageMaker Debugger. Let's discuss each step in detail to see how this works:

1. When the estimator's `fit()` function is called, one or more ML training instances are launched to start a training job.

2. SageMaker will start the SageMaker Processing jobs using the Debugger rule containers to inspect and monitor the training job. For more information on SageMaker Processing jobs, feel free to check the *Managed data processing with SageMaker Processing in Python* recipe from *Chapter 4, Preparing, Processing, and Analyzing the Data*.

3. Based on how `DebuggerHookConfig` is configured, certain values such as losses, weights, and gradients are extracted and stored in S3 after a specified number of steps for each interval.

4. Once the data from the previous step is available in S3, the SageMaker Processing jobs from *step 2* will run the configured rule logic and check whether a rule is violated.

5. A **CloudWatch event** is triggered.

Note that we can work on a quick integration using AWS Lambda to help us react to the CloudWatch events triggered by SageMaker Debugger and stop the running training job. For more information on how to accomplish this, feel free to check `https://github.com/aws/amazon-sagemaker-examples/blob/master/sagemaker-debugger/tensorflow_action_on_rule/tf-mnist-stop-training-job.ipynb`.

There's more...

You might be wondering why there is a second rule job running with the `ProfilerReport` prefix even if we have only specified the `LossNotDecreasing` rule when initializing the `Estimator` object. That's because SageMaker Debugger runs the `ProfilerReport` rule by default to generate a profiling report. The generated profiling report starts with the training job summary and the system usage statistics of each worker node.

Framework metrics summary

Rules summary

The following table shows a profiling summary of the Debugger built-in rules. The table is sorted by the rules that triggered the most frequently. During your training job, the GPUMemoryIncrease rule was the most frequently triggered. It processed 0 datapoints and was triggered 0 times.

	Description	Recommendation	Number of times rule triggered	Number of datapoints	Rule parameters
GPUMemoryIncrease	Measures the average GPU memory footprint and triggers if there is a large increase.	Choose a larger instance type with more memory if footprint is close to maximum available memory.	0	0	increase:5 patience:1000 window:10
CPUBottleneck	Checks if the CPU utilization is high and the GPU utilization is low. It might indicate CPU bottlenecks, where the GPUs are waiting for data to arrive from the CPUs. The rule evaluates the CPU and GPU utilization rates, and triggers the issue if the time spent on the CPU bottlenecks exceeds a threshold percent of the total training time. The default threshold is 50 percent.	Consider increasing the number of data loaders or applying data pre-fetching.	0	91	threshold:50 cpu_threshold:90 gpu_threshold:10 patience:1000
LowGPUUtilization	Checks if the GPU utilization is low or fluctuating. This can happen due to bottlenecks, blocking calls for synchronizations, or a small batch size.	Check if there are bottlenecks, minimize blocking calls, change distributed training strategy, or increase the batch size.	0	0	threshold_p95:70 threshold_p5:10 window:500 patience:1000
Dataloader	Checks how many data loaders are running in parallel and whether the total number is equal the number of available CPU cores. The rule triggers if number is much smaller or larger than the number of available cores. If too small, it might lead to low GPU utilization. If too large, it might impact other compute intensive operations on CPU.	Change the number of data loader processes.	0	0	min_threshold:70 max_threshold:200

Figure 5.11 – Profiler report rules summary

It would also provide a profiling summary of the Debugger built-in rules similar to what is shown in *Figure 5.11*. Feel free to check `https://docs.aws.amazon.com/sagemaker/latest/dg/debugger-profiling-report.html` for more information.

See also

If we want to create and use a custom rule, we can use the `Rule.custom()` function from the SageMaker Python SDK. Creating a custom rule is out of the scope of this book so we will provide some references and examples here instead:

- Training TensorFlow Keras models with SageMaker Debugger custom rules: `https://sagemaker-examples.readthedocs.io/en/latest/sagemaker-debugger/tensorflow_keras_custom_rule/tf-keras-custom-rule.html`

- Running iterative model pruning with a SageMaker Debugger custom rule: `https://github.com/aws/amazon-sagemaker-examples/blob/master/sagemaker-debugger/pytorch_iterative_model_pruning/iterative_model_pruning_alexnet.ipynb`

For more information on creating custom debugger rules for training job analysis, feel free to check `https://docs.aws.amazon.com/sagemaker/latest/dg/debugger-custom-rules.html` as well.

Inspecting SageMaker Debugger logs and results

In the previous recipe, we used SageMaker Debugger to help us detect a violation of a rule while the training job is running. In this recipe, we will use the `smdebug` library, the AWS CLI tool, and the `awslogs` command-line tool to inspect the files and logs generated by SageMaker Debugger. This will be very useful when debugging the results of SageMaker Debugger itself.

Sometimes, if we have a misconfigured setup with SageMaker Debugger, we may end up getting the `NoIssuesFound` result even if certain `DebuggerRule` conditions have been met already. That said, it is important that we know how to debug our setup and where to look for the files generated by SageMaker Debugger.

Getting ready

The following is the prerequisite for this recipe:

- This recipe continues from *Identifying issues with SageMaker Debugger*.

How to do it...

The following instructions show us the steps to inspect the files and logs generated by SageMaker Debugger processing jobs:

1. Navigate to the `my-experiments/chapter05` directory inside your SageMaker notebook instance. Feel free to create this directory if it does not exist yet.

2. Create a new notebook using the `conda_python3` kernel inside the `my-experiments/chapter05` directory and name it with the name of this recipe (that is, `Inspecting SageMaker Debugger Logs and Results`). Open this notebook for editing as we will update this file with the code in the next couple of steps.

3. Use the %store magic command to load the value for artifacts_path. Note that we stored this value in IPython's database in the *Inspecting SageMaker Debugger logs and results* recipe:

```
%store -r artifacts_path
```

4. Install smdebug with pip install:

```
!pip install smdebug
```

5. Import create_trial and run the create_trial() function with the path of the artifacts generated by SageMaker Debugger:

```
from smdebug.trials import create_trial
trial = create_trial(artifacts_path)
```

This trial object lets us query for the tensors of a training job as we will see in the next set of steps.

6. Use the tensor_names() function to list the available tensors of the trial:

```
trial.tensor_names()
```

Running the previous line of code will return a list containing the train-error and validation-error string values.

7. Prepare the target_path variable using the bucket_name and prefix_name attributes of the trial object. The target_path variable points to the S3 path containing the debug output files:

```
target_path = f"s3://{trial.bucket_name}/{trial.prefix_
name}"
```

8. Use the AWS CLI to show all the files inside the target S3 path:

```
s3_contents = !aws s3 ls {target_path} --recursive
s3_contents
```

This should yield a set of results similar to what is shown in *Figure 5.12*:

```
Out[11]:  ['2021-08-29 18:33:20          5504 debugger/sagemaker-xgboost-2021-08-29-18-29-09-733/debug-out
          put/collections/000000000/worker_0_collections.json',
          '2021-08-29 18:33:20           222 debugger/sagemaker-xgboost-2021-08-29-18-29-09-733/debug-out
          put/events/000000000000/000000000000_worker_0.tfevents',
          '2021-08-29 18:33:20           226 debugger/sagemaker-xgboost-2021-08-29-18-29-09-733/debug-out
          put/events/000000000002/000000000002_worker_0.tfevents',
          '2021-08-29 18:33:20           226 debugger/sagemaker-xgboost-2021-08-29-18-29-09-733/debug-out
          put/events/000000000004/000000000004_worker_0.tfevents',
          '2021-08-29 18:33:20           226 debugger/sagemaker-xgboost-2021-08-29-18-29-09-733/debug-out
          put/events/000000000006/000000000006_worker_0.tfevents',
          '2021-08-29 18:33:20           226 debugger/sagemaker-xgboost-2021-08-29-18-29-09-733/debug-out
          put/events/000000000008/000000000008_worker_0.tfevents',
          '2021-08-29 18:33:20           228 debugger/sagemaker-xgboost-2021-08-29-18-29-09-733/debug-out
          put/events/000000000010/000000000010_worker_0.tfevents',
          '2021-08-29 18:33:20           228 debugger/sagemaker-xgboost-2021-08-29-18-29-09-733/debug-out
          put/events/000000000012/000000000012_worker_0.tfevents',
          '2021-08-29 18:33:20           228 debugger/sagemaker-xgboost-2021-08-29-18-29-09-733/debug-out
          put/events/000000000014/000000000014_worker_0.tfevents',
          '2021-08-29 18:33:20           228 debugger/sagemaker-xgboost-2021-08-29-18-29-09-733/debug-out
          put/events/000000000016/000000000016_worker_0.tfevents',
```

Figure 5.12 – SageMaker Debugger output files

Figure 5.12 shows us some of the files inside the target S3 path after using the `aws s3 ls` command.

9. Copy the path of one of the files in the previous step and store it inside the `chosen_fullpath` variable:

```
chosen_path = s3_contents[-1].split(" ")[-1]
chosen_fullpath = f"s3://{trial.bucket_name}/{chosen_
path}"
```

The `chosen_fullpath` variable should contain the S3 path pointing to a JSON file.

10. Create the `tmp` directory if it does not exist yet:

```
!mkdir -p tmp
```

11. Use the AWS CLI to copy the file from S3 to the `tmp` directory:

```
!aws s3 cp {chosen_fullpath} tmp/worker_0.json
```

12. Inspect the contents of the file we just downloaded from S3:

```
!cat tmp/worker_0.json
```

We should get a JSON value similar to what is shown in *Figure 5.13*:

```
{"meta": {"mode": "GLOBAL", "mode_step": 6998, "event_file_name": "events/000000006998/000000
006998_worker_0.tfevents"}, "tensor_payload": [{"tensorname": "train-error", "start_idx": 0,
"length": 112}, {"tensorname": "validation-error", "start_idx": 112, "length": 122}]}
```

Figure 5.13 – JSON value inside worker_0.json

We have in *Figure 5.13* the value of the JSON file we downloaded from the S3 bucket containing the logs generated by SageMaker Debugger.

13. Run the following lines of code to print the `train-error` and `validation-error` tensor values:

```
for i in [2, 4, 6, 8, 10, 12, 14, 16]:
    traint = trial.tensor("train-error")
    train_error = traint.value(i)[0]
    train_error = "{0:.4f}".format(train_error)
    valt = trial.tensor("validation-error")
    validation_error = valt.value(i)[0]
    validation_error = "{0:.4f}".format(validation_error)
    print(f"STEP {i}: [TRAIN ERROR]={train_error} " +
        f"[VALIDATION ERROR]={validation_error}")
```

In the preceding block of code, we have iterated over the [2, 4, 6, 8, 10, 12, 14, 16] list since we configured the Debugger Hook to collect the metric values every two steps. *Figure 5.14* shows us what the output looks like after running the previous block of code:

```
STEP 2:  [TRAIN ERROR]=0.1053 [VALIDATION ERROR]={validation_error}
STEP 4:  [TRAIN ERROR]=0.0893 [VALIDATION ERROR]={validation_error}
STEP 6:  [TRAIN ERROR]=0.0800 [VALIDATION ERROR]={validation_error}
STEP 8:  [TRAIN ERROR]=0.0670 [VALIDATION ERROR]={validation_error}
STEP 10: [TRAIN ERROR]=0.0633 [VALIDATION ERROR]={validation_error}
STEP 12: [TRAIN ERROR]=0.0613 [VALIDATION ERROR]={validation_error}
STEP 14: [TRAIN ERROR]=0.0567 [VALIDATION ERROR]={validation_error}
STEP 16: [TRAIN ERROR]=0.0540 [VALIDATION ERROR]={validation_error}
```

Figure 5.14 – Training and validation error values

We have in *Figure 5.14* the training error and validation error values collected by SageMaker Debugger. If you are wondering how we are collecting metric values every two steps, remember that in the *Identifying issues with SageMaker Debugger* recipe, we specified the value 2 for `save_interval` inside the `parameters` dictionary while initializing the `CollectionConfig` object.

14. Use the `%store` magic command to load the value of `rule_job_summary`:

```
%store -r rule_job_summary
```

Remember that we stored this value using `%store` magic in the IPython database in the *Inspecting SageMaker Debugger logs and results* recipe.

15. Get the specific rule job summary for the `LossNotDecreasing` rule:

```
def lfx(r):
    return r["RuleConfigurationName"] ==
"LossNotDecreasing"


loss_not_decreasing_summary = list(filter(
    lfx,
    rule_job_summary))[0]


loss_not_decreasing_summary
```

The previous block of code helps us extract the rule job summary for the `LossNotDecreasing` rule as there are two rule jobs that are executed after we have called the `fit()` function, as mentioned in the previous recipe. This should yield a dictionary similar to what is shown in *Figure 5.15*:

```
{'RuleConfigurationName': 'LossNotDecreasing',
 'RuleEvaluationJobArn': 'arn:aws:sagemaker:us-east-1:                :processing-job/sagemaker-x
gboost-2021-08--lossnotdecreasing-864c9cee',
 'RuleEvaluationStatus': 'IssuesFound',
 'StatusDetails': 'RuleEvaluationConditionMet: Evaluation of the rule LossNotDecreasing at st
ep 8 resulted in the condition being met\n',
 'LastModifiedTime': datetime.datetime(2021, 8, 29, 18, 39, 21, 517000, tzinfo=tzlocal())}
```

Figure 5.15 – Rule job summary for the LossNotDecreasing rule

In *Figure 5.15*, we can see that the `RuleEvaluationStatus` key in the dictionary maps to the `IssuesFound` value. We can see more details in the value mapped to the `StatusDetails` key.

16. Get and inspect the rule evaluation job ARN:

```
summary = loss_not_decreasing_summary
rule_evaluation_job_arn = summary['RuleEvaluationJobArn']
rule_evaluation_job_arn
```

This should give us a string value similar to `'arn:aws:sagemaker:us-east-1:1234567890:processing-job/sagemaker-xgboost-2021-04--lossnotdecreasing-ee3df9a4'`.

17. Load the processing job using the `ProcessingJob.from_processing_arn()` function with the evaluation job ARN obtained in the previous step:

```
import sagemaker
from sagemaker.processing import ProcessingJob

session = sagemaker.Session()
processing_job = ProcessingJob.from_processing_arn(
    sagemaker_session=session,
    processing_job_arn=rule_evaluation_job_arn)

processing_job
```

18. Next, set and prepare the `region`, `group`, and `prefix` variable values:

```
region = "us-east-1"
group = "/aws/sagemaker/ProcessingJobs"
prefix = processing_job.job_name
```

19. Install the `awslogs` command-line tool using `pip`:

```
!pip install awslogs
```

20. Run the following Bash command to load the logs generated by the processing job:

```
!awslogs get {group} --log-stream-name-prefix {prefix}
-s3h --aws-region {region}
```

This should yield a set of logs similar to what is shown in *Figure 5.16*:

```
/aws/sagemaker/ProcessingJobs sagemaker-xgboost-2021-08--LossNotDecreasing-864c9cee/algo-1-16
30262234 [2021-08-29 18:39:00.308 ip-10-2-223-179.ec2.internal:1 INFO loss_decrease.py:269] 1
loss is not decreasing over the last 2 steps at step 8
/aws/sagemaker/ProcessingJobs sagemaker-xgboost-2021-08--LossNotDecreasing-864c9cee/algo-1-16
30262234 [2021-08-29 18:39:00.308 ip-10-2-223-179.ec2.internal:1 INFO action.py:81] Invoking
actions
/aws/sagemaker/ProcessingJobs sagemaker-xgboost-2021-08--LossNotDecreasing-864c9cee/algo-1-16
30262234 Exception during rule evaluation: RuleEvaluationConditionMet: Evaluation of the rule
LossNotDecreasing at step 8 resulted in the condition being met
/aws/sagemaker/ProcessingJobs job-YCSQUFQDNB-ProfilerReport-1630261951-468d25e4/algo-1-163026
2298 [2021-08-29 18:39:05.126 ip-10-0-178-156.ec2.internal:1 INFO utils.py:27] RULE_JOB_STOP_
SIGNAL_FILENAME: /opt/ml/processing/input/profiler/signals/ProfilerReport-1630261951
/aws/sagemaker/ProcessingJobs job-YCSQUFQDNB-ProfilerReport-1630261951-468d25e4/algo-1-163026
2298 [2021-08-29 18:39:05.606 ip-10-0-178-156.ec2.internal:1 INFO profiler_trial.py:67] Waiti
ng for profiler data.
/aws/sagemaker/ProcessingJobs job-YCSQUFQDNB-ProfilerReport-1630261951-468d25e4/algo-1-163026
2298 [2021-08-29 18:39:15.616 ip-10-0-178-156.ec2.internal:1 INFO profiler_trial.py:37] Outpu
t files of ProfilerTrial will be saved to /opt/ml/processing/output/rule
```

Figure 5.16 – Logs generated after running the awslogs CLI tool

Figure 5.16 shows us a portion of the logs generated after running the `awslogs` command-line tool. As we can see in the second line in the logs shown in *Figure 5.16*, the condition for the `LossNotDecreasing` rule that we specified in `Estimator` before running the training job was met.

Now, let's see how this works!

How it works...

In this recipe, we have used `smdebug` to inspect the properties of the entities created with SageMaker Debugger. Let's quickly take a closer look at some of the relevant terms and concepts when using the `smdebug` library:

- **Step**: A unit of work performed by a training job for one batch

- **Collection**: A collection or group of tensors (matrix-like container or structure)

- **Hook**: Keeps track of the collections and generates the relevant output files after each step depending on the configuration specified

- **Trial**: Enables the querying and analysis of the tensors of a training job

Now, let's check the lines of code in one of the steps from the *Identifying issues with SageMaker Debugger* recipe:

```
metrics_collection_config = CollectionConfig(
    name="metrics",
    parameters={
        "save_interval": str(save_interval)
```

```
    })

debugger_hook_config = DebuggerHookConfig(
    s3_output_path=bucket_path,
    collection_configs=[metrics_collection_config]
)
```

Here, we have initialized `DebuggerHookConfig`, which we have used as an argument when initializing `Estimator`. With `DebuggerHookConfig`, we can specify and configure how the debugging information is emitted and collected. This `DebuggerHookConfig` configures SageMaker Debugger to do the following:

- Save the debug data (for example, `train-error`) every two intervals (`save_interval = 2`).

- Store the debug data in the Amazon S3 `bucket_path` we have specified.

When we run the training job, the debug data is stored in the target S3 path. If we want to extract the debug data from these logs, we use the `trial.tensor()` function from the `smdebug` library, and then use the `value()` function to get the value at a specified index. Finally, if one or more of the rule conditions are met, the `rule_job_summary()` function should reflect the appropriate status and description of the issues.

There's more...

Note that we were only scratching the surface of what the `smdebug` library can do in the *How to do it...* section. Once we have collected metric data into the S3 bucket, we can use the `smdebug` library to create a system reader object using the following lines of code. Feel free to replace the `"<insert training job name here>"` string with the name of the XGBoost training job we executed in the *Identifying issues with SageMaker Debugger* recipe:

```
# set up the TrainingJob object from the job name
from smdebug.profiler.analysis.notebook_utils.training_job
import TrainingJob
training_job_name = "<insert training job name here>"
training_job = TrainingJob(training_job_name, "us-east-1")

# prepare a system reader object
training_job.wait_for_sys_profiling_data_to_be_available()
reader = training_job.get_systems_metrics_reader()
```

Once we have the system reader object, we can compute and plot histograms using the `MetricsHistogram` class from the `smdebug` library:

```
from smdebug.profiler.analysis.notebook_utils.metrics_histogram
import MetricsHistogram
```

```
endtime = reader.get_timestamp_of_latest_available_file()
```

```
metrics_histogram = MetricsHistogram(reader)
metrics_histogram.plot(
    starttime=0,
    endtime=endtime
)
```

This should give us a set of charts similar to what is shown in *Figure 5.17*:

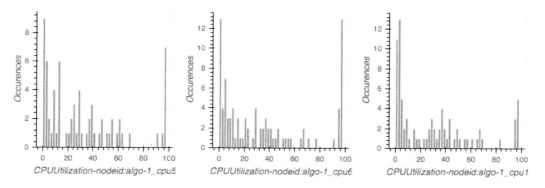

Figure 5.17 – Generated plots of the metrics histograms

Note that if you were to execute these blocks of code yourself, you would get more than 28 charts since we did not specify a specific set of dimensions and events when using the `plot()` function.

See also

Feel free to check some of the references and links here as well:

- Analyze system and framework metrics using the `smdebug` library: https://docs.aws.amazon.com/sagemaker/latest/dg/debugger-analyze-data.html

- SageMaker Debugger on SageMaker Studio: `https://docs.aws.amazon.com/sagemaker/latest/dg/debugger-on-studio.html`
- SageMaker Debugger programming model for analysis: `https://github.com/awslabs/sagemaker-debugger/blob/master/docs/analysis.md`

Now, let's proceed with the next recipe!

Running and managing multiple experiments with SageMaker Experiments

Managing a single machine learning (ML) experiment is easy. When we are dealing with a single ML experiment, it is easy to locate and audit the input and output artifacts, configuration parameters, hyperparameter values, and all the other relevant metadata and details related to this single ML experiment. Things get a bit trickier when we have to deal with multiple ML experiments as well as when retrieving information on experiments and training jobs performed in the past.

In this recipe, we will run and track multiple experiments using **SageMaker Experiments**. Each experiment trial corresponds to a specific combination of hyperparameters that we will use for the training job. We will use the XGBoost built-in algorithm to help us train and build a classifier using the synthetic dataset we generated in the *Synthetic data generation for classification problems* recipe. While setting up the experiment, we will make use of the classes and functions from the smexperiments library to help us record the relevant information while we are configuring and running each experiment.

> **Tip**
> Using SageMaker Experiments is free! Generally, we only pay for the compute and storage resources based on their usage. This means that we will only have to pay for the cost of running training jobs, running SageMaker notebook instances, and the costs associated with storing files inside the S3 bucket. For more information on this topic, refer to the pricing page here: `https://aws.amazon.com/sagemaker/pricing/`.

Getting ready

The following is the prerequisite for this recipe:

- This recipe continues from *Synthetic data generation for classification problems.*

How to do it...

The first steps in this recipe focus on setting up the prerequisites for the experiments:

1. Navigate to the `my-experiments/chapter05` directory inside your SageMaker notebook instance. Feel free to create this directory if it does not exist yet.

2. Create a new notebook using the `conda_python3` kernel inside the `my-experiments/chapter05` directory and name it with the name of this recipe (that is, `Running and Managing Multiple Experiments with SageMaker Experiments`). Open this notebook for editing as we will update this file with the code in the next couple of steps.

3. Import a few prerequisites of the training job (such as `session`, `role`, `sagemaker_session`, and `sagemaker_client`) by running the following lines of code:

```
import sagemaker, boto3
session = boto3.Session()
sagemaker_session = sagemaker.Session(boto_
session=session)
sagemaker_client = session.client('sagemaker')
role = sagemaker.get_execution_role()
```

4. Next, import a few more prerequisite libraries and utilities, such as `time`, `numpy`, `pandas`, `itertools`, and `pprint`:

```
import time, os, sys
import numpy as np
import pandas as pd
import itertools
from pprint import pprint
```

5. Install the `sagemaker-experiments` package:

```
!pip install sagemaker-experiments
```

6. Import `Experiment`, `TrialComponent`, `Trial`, and `Tracker` from `smexperiments`:

```
from smexperiments.experiment import Experiment
from smexperiments.trial import Trial
from smexperiments.trial_component import TrialComponent
from smexperiments.tracker import Tracker
```

7. Define the `generate_random_string()` function:

```
import random
import string

def generate_random_string():
    list_of_chars = random.choices(
        string.ascii_uppercase,
        k=10)
    return ''.join(list_of_chars)
```

8. Create an `Experiment` instance using the `create()` function:

```
label = generate_random_string()
training_experiment = Experiment.create(
    experiment_name = f"experiment-{label}",
    description     = "Experiment Description",
    sagemaker_boto_client=sagemaker_client)
```

9. Prepare the `hyperparam_options` dictionary:

```
hyperparam_options = {
    'max_depth': [2, 8],
    'eta': [0.2],
    'gamma': [3, 4],
    'min_child_weight': [6],
    'subsample': [0.4],
    'num_round': [10, 20],
    'objective': ['binary:logistic']
}
```

> **Important Note**
>
> Note that increasing the number of options specified in the `hyperparam_options` dictionary increases the total number of combinations of the hyperparameters used by the experiment to configure the training jobs. That said, this will increase the number of training jobs executed along with the total duration of the entire experiment. As we are paying for the amount of time the ML training instances are running, the cost may increase proportionally as well. In this recipe, we have a total of `2 x 1 x 2 x 1 x 2 x 1` combinations of hyperparameters. If each experiment takes approximately 5 minutes to complete, we will be paying for `8 x 5` minutes of training time.

10. Define the `prepare_hyperparam_variations()` function. This function accepts a list of hyperparameter options and generates and returns a list of dictionaries with all the variations of the hyperparameter configuration values:

```
def prepare_hyperparam_variations(options):
    names, values = zip(*options.items())
    return [dict(zip(names, value))
            for value in itertools.product(*values)]
```

11. Use the function from the previous step to generate a list of hyperparameter configurations:

```
hyperparam_variations = prepare_hyperparam_variations(
    hyperparam_options
)
hyperparam_variations
```

This should yield an array of dictionaries similar to what is shown in *Figure 5.18*:

```
[{'max_depth': 2,
  'eta': 0.2,
  'gamma': 3,
  'min_child_weight': 6,
  'subsample': 0.4,
  'num_round': 10,
  'objective': 'binary:logistic'},
 {'max_depth': 2,
  'eta': 0.2,
  'gamma': 3,
  'min_child_weight': 6,
  'subsample': 0.4,
  'num_round': 20,
  'objective': 'binary:logistic'},
 {'max_depth': 2,
  'eta': 0.2,
  'gamma': 4,
  'min_child_weight': 6,
  'subsample': 0.4,
  'num_round': 10,
  'objective': 'binary:logistic'},
```

Figure 5.18 – Hyperparameter variations (showing 3 of 8 variations)

In *Figure 5.18*, we can see a list of the different hyperparameter configuration variations after using the `prepare_hyperparam_variations()` function.

12. Specify the S3 bucket and the location of the training, validation, and test datasets. Make sure to replace the value of `"<insert S3 bucket name here>"` with the name of the bucket we created in the *Preparing the Amazon S3 bucket and the training dataset for the linear regression experiment* recipe in *Chapter 1, Getting Started with Machine Learning Using Amazon SageMaker*:

```
s3_bucket = '<insert S3 bucket name here>'
prefix = "chapter05"
path = f"s3://{s3_bucket}/{prefix}/input"
training_path = f"{path}/training_data.csv"
validation_path = f"{path}/validation_data.csv"
output_path = f"s3://{s3_bucket}/{prefix}/output/"
```

Note that the `s3_bucket` value should be the same as the value used in the *Synthetic dataset generation for classification problems* recipe in this chapter.

13. Use the `retrieve()` function to get the ECR container image URI for XGBoost:

```
from sagemaker.image_uris import retrieve
container = retrieve('xgboost',
                     boto3.Session().region_name,
                     version="0.90-2")
container
```

This should give us a string value similar to `683313688378.dkr.ecr.us-east-1.amazonaws.com/sagemaker-xgboost:0.90-2-cpu-py3`.

> **Note**
>
> We can use XGBoost as a built-in algorithm or use the open source XGBoost algorithm, which involves writing a custom training script. Feel free to check the following link for more information on using XGBoost with SageMaker: `https://sagemaker.readthedocs.io/en/stable/frameworks/xgboost/using_xgboost.html`.

14. Use `TrainingInput` to specify the training channel configuration for the training and validation inputs:

```
from sagemaker.inputs import TrainingInput
s3_input_training = TrainingInput(
    training_path, content_type="text/csv")
s3_input_validation = TrainingInput(
    validation_path, content_type="text/csv")
```

Now that we have the prerequisites ready, we will proceed with getting our experiments running in the next set of steps.

15. Create a tracker called `experiment_tracker` and record the values of `training_path`, `validation_path`, and `hyperparam_options` in the tracker using `log_input()`:

```
ex_name = training_experiment.experiment_name
with Tracker.create(
    display_name="xgboost-experiment-display-name",
    artifact_bucket=s3_bucket,
    artifact_prefix=ex_name,
```

```
        sagemaker_boto_client=sagemaker_client
) as experiment_tracker:
    experiment_tracker.log_input(
        name="training-input",
        media_type="s3/uri",
        value=training_path)

    experiment_tracker.log_input(
        name="validation-input",
        media_type="s3/uri",
        value=validation_path)

    experiment_tracker.log_parameters(
        hyperparam_options)
```

16. Define the `track_and_generate_config()` function. This function has three parts:

- Creating a `Trial` tracker for a specific training job and recording the hyperparameters of the training job using `log_parameters()`

- Creating a `Trial` object and adding the trial components using the `add_trial_component()` function

- Returning the experiment configuration dictionary containing the `ExperimentName`, `TrialName`, and `TrialComponentDisplayName` keys:

```
def track_and_generate_config(
    experiment_tracker,
    experiment_name,
    job_name,
    random_string,
    hyperparameters):

    tdn = f"trial-metadata-{random_string}"
    tracker_display_name = tdn
    print(f"{label} Create Tracker: {tdn}")
```

```python
# Step # 1: Creating a Trial Tracker
prefix = f"{experiment_name}/{job_name}"
with Tracker.create(
    display_name=tracker_display_name,
    artifact_bucket=s3_bucket,
    artifact_prefix=prefix,
    sagemaker_boto_client=sagemaker_client
) as trial_tracker:
    trial_tracker.log_parameters(
        hyperparameters
    )

trial_name = f'trial-{random_string}'
print(f"Create Trial: {trial_name}")


# Step # 2A: Creating a Trial object
trial = Trial.create(
    trial_name=trial_name,
    experiment_name=experiment_name,
    sagemaker_boto_client=sagemaker_client)


# Step # 2B: Adding the trial components
trial.add_trial_component(
    experiment_tracker.trial_component)
time.sleep(1)
trial.add_trial_component(
    trial_tracker.trial_component)


# Step # 3: Returning the config dictionary
return {
    "ExperimentName": experiment_name,
    "TrialName": trial.trial_name,
    "TrialComponentDisplayName": job_name
}
```

17. Run the following block of code to loop through each of the hyperparameter configuration variations and trigger a training job for each one. This block of code has three parts—setting up the variable values for iteration, label, random_string, and job_name, tracking the experiment and generating the experiment configuration dictionary using the track_and_generate_config() function we have prepared in the previous step, and running the training job with the experiment configuration specified as a parameter to the fit() function call:

```
import time
experiment_name = training_experiment.experiment_name


for index, hyperparameters in enumerate(
    hyperparam_variations
):
    # Step # 1: Setting up the variable values
    iteration = index + 1
    print(f"Iteration # {iteration}")
    label = f"[Iteration # {iteration}]"
    random_string = generate_random_string()
    job_name = f"job-{random_string}"


    # Step # 2: Tracking trials and trial
    # components and generating the
    # experiment configuration
    print(f"{label} Track and Generate Config")
    experiment_config = track_and_generate_config(
        experiment_tracker=experiment_tracker,
        experiment_name=experiment_name,
        job_name=job_name,
        random_string=random_string,
        hyperparameters=hyperparameters)


    time.sleep(1)
    # Step # 3: Running the training job
    print(f"{label} Initialize Estimator")
    estimator = sagemaker.estimator.Estimator(
        container,
```

```
        role,
        instance_count=1,
        instance_type='ml.m5.large',
        output_path=output_path,
        hyperparameters=hyperparameters,
        enable_sagemaker_metrics = True,
        sagemaker_session=sagemaker_session
    )

    print(f"{label} Call fit() function")
    data = {'train': s3_input_training,
            'validation': s3_input_validation}
    exc = experiment_config
    estimator.fit(data,
                  job_name = job_name,
                  wait=False,
                  experiment_config=exc)
```

As you will notice in the last part of the preceding code block, we have passed the dictionary we have generated using track_and_generate_config() as the value to the experiment_config parameter when calling the fit() function. As seen in a previous step, the experiment_config parameter expects the experiment management configuration in the form of a dictionary with three optional keys: "ExperimentName", "TrialName", and "TrialComponentDisplayName".

> **Note**
>
> If you are using a fairly new AWS account, you may encounter the ResourceLimitExceeded error (or similar) when launching ML instances using the fit() and deploy() functions. To resolve this, open the AWS Support Center and create a case. For more information, feel free to visit this link: https://aws.amazon.com/premiumsupport/knowledge-center/resourcelimitexceeded-sagemaker/.

Figure 5.19 shows us the sample log output after running the previous block of code:

```
Iteration # 1
[Iteration # 1] Track and Generate Config
[Iteration # 1] Create Tracker: trial-metadata-XMMFUVHMTC
Create Trial: trial-XMMFUVHMTC

Prepare Experiment Configuration
[Iteration # 1] Initialize Estimator
[Iteration # 1] Call fit() function
Iteration # 2
[Iteration # 2] Track and Generate Config
[Iteration # 2] Create Tracker: trial-metadata-HXSGRKNWMP
Create Trial: trial-HXSGRKNWMP
```

Figure 5.19 – A certain portion of the log output while running the loop that runs and tracks multiple experiments using the XGBoost built-in algorithm

We have in *Figure 5.19* the logs generated by the loop in the preceding code block. Note that we are running several training jobs using different configurations of hyperparameter values and attaching a `Trial` to each training job.

> **Important Note**
>
> This step may take more than 30 minutes to complete, so feel free to read the *How it works...* and *There's more...* sections of this recipe while waiting. You may also proceed to work on the following recipes in this chapter as well!

18. Finally, we use `%store` magic to save the value of `experiment_name`:

```
%store experiment_name
```

We will use the `experiment_name` variable value later in the *Experiment analytics with SageMaker Experiments* recipe.

Now, let's see how this works!

How it works...

In this recipe, we have used the `sagemaker-experiments` library to help us track and organize different variables, values, and artifacts in our ML experiments.

Let's define a few important terms:

- **Experiment**: Corresponds to the ML experiment, which aims to achieve a certain goal
- **Trial**: Corresponds to a training iteration or training job
- **Trial components**: Corresponds to parameters, metadata, or artifacts

Using the `smexperiments` library is fairly straightforward. We create an `Experiment` resource and link the `Trial` and `TrialComponent` objects when running the training jobs. You may think of this like a tree with the `Experiment` resource as the root and the `Trial` objects as the first level of children under the `Experiment` resource. The same goes for the `TrialComponent` objects under each `Trial`.

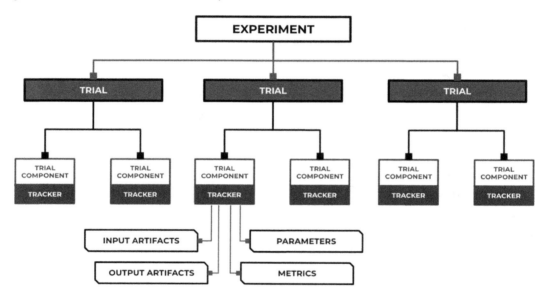

Figure 5.20 – How Experiment, Trial, TrialComponent, and Tracker resources are connected

In *Figure 5.20*, we can see that each `Experiment` can have one or more trials and each `Trial` can have one or more trial components. We can also see that the metrics, parameters, and input and output artifact information and details are linked to the `TrialComponent` objects.

In the next recipe, we will see how to analyze and process the training results of multiple experiments with a few lines of code.

There's more...

When dealing with multiple experiments, model lineage, search, and tracking requirements, there are two other related capabilities that come to mind—**SageMaker ML Lineage Tracking** and **SageMaker Search**.

SageMaker ML Lineage Tracking helps keep the history of an ML model. This includes the configuration parameters, the hyperparameters set, and the data and algorithm used to train the model. This may also include workflow metadata to help us locate, reproduce, and compare models. Lineage associations are tracked and recorded automatically in most cases and we can generate a quick `DataFrame` containing the lineage associations of a SageMaker resource (for example, training job, processing job, or model package) using the following lines of code:

```
from sagemaker.session import Session
from sagemaker.lineage.visualizer import LineageTableVisualizer
session = Session()
viz = LineageTableVisualizer(session)
viz.show(training_job_name="<insert job name here>")
```

This would give us a DataFrame similar to what is shown in *Figure 5.21*:

	Name/Source	Direction	Type	Association Type	Lineage Type
0	s3://...cket/chapter05/input/validation_data.csv	Input	DataSet	ContributedTo	artifact
1	s3://...bucket/chapter05/input/training_data.csv	Input	DataSet	ContributedTo	artifact
2	68331...aws.com/sagemaker-xgboost:0.90-2-cpu-py3	Input	Image	ContributedTo	artifact
3	s3://...1-05-18-21-07-20-154/output/model.tar.gz	Output	Model	Produced	artifact

Figure 5.21 – DataFrame containing the lineage associations of a training job

In *Figure 5.21*, we can see the connected resources to the specified training job—the location of the training and validation CSV files, the XGBoost algorithm container image, and the location of the output artifact produced by the training job. We will not discuss the details of SageMaker ML Lineage Tracking in this book, so feel free to check this link for more information on this topic: https://docs.aws.amazon.com/sagemaker/latest/dg/lineage-tracking.html.

On the other hand, SageMaker Search helps us locate SageMaker resources such as training jobs, experiments, endpoints, and model packages. One practical example of using SageMaker Search involves searching and stopping all running training jobs of an experiment. This means that we will not need to wait for the training jobs to finish before trying again. We can use the following script to search for and stop all running training jobs:

```
sagemaker = boto3.client(service_name='sagemaker')

def stop_all_running_training_jobs():
    search_params={
        "MaxResults": 50,
        "Resource": "TrainingJob",
        "SearchExpression": {
            "Filters": [
                {
                    "Name": "TrainingJobStatus",
                    "Operator": "Equals",
                    "Value": "InProgress"
                }
            ]
        }
    }

    results = sagemaker.search(**search_params)
    count = len(results['Results'])

    for result in results['Results']:
        job_name = result['TrainingJob']['TrainingJobName']
        stop_training_job(job_name)
```

This function first searches all training jobs with `TrainingJobStatus` equal to `InProgress` and then stops these training jobs using the `stop_training_job()` function. The `stop_training_job()` function (not shown in the preceding code block) simply makes use of the `sagemaker.stop_training_job()` function to stop a training job given the `TrainingJobName`. With this, we will be able to stop running training jobs right away so that we will not have to wait for them to complete before trying again.

> **Note**
>
> Note that some lines of code have been omitted here for brevity. You may check a more complete version of the code here: `https://gist.github.com/joshualat/c420ff645730034387c39b62f5d3296a`.

For more information on SageMaker Search, feel free to check the following link for more information: `https://docs.aws.amazon.com/sagemaker/latest/dg/search.html`.

Now, let's proceed with the next recipe!

Experiment analytics with SageMaker Experiments

In the previous recipe, we ran multiple training jobs and used SageMaker Experiments to keep track of the parameters, input and output artifacts, metric values, and other metadata with the `Experiment`, `Trial`, `TrialComponent`, and `Tracker` resources using the `smexperiments` library.

In this recipe, we will use `ExperimentAnalytics` from `sagemaker.analytics` to load and analyze the DataFrame containing the details of the previous experiments we have performed and tracked using SageMaker Experiments. This allows us to inspect and analyze the results of the training jobs with just a few lines of code.

Getting ready

The following is the prerequisite for this recipe:

- This recipe continues from *Running and managing multiple experiments with SageMaker Experiments*.

How to do it...

The first set of steps in this recipe focuses on preparing and loading the prerequisites:

1. Navigate to the `my-experiments/chapter05` directory inside your SageMaker notebook instance. Feel free to create this directory if it does not exist yet.

2. Create a new notebook using the `conda_python3` kernel inside the `my-experiments/chapter05` directory and name it with the name of this recipe (that is, `Experiment Analytics with SageMaker Experiments`). Open this notebook for editing as we will update this file with the code in the next couple of steps.

3. Use the `%store` magic command to load the value for `experiment_name`. Remember that we used the `%store` magic command to save this value in the *Running and managing multiple experiments with SageMaker Experiments* recipe:

```
%store -r experiment_name
```

4. Initialize the `ExperimentAnalytics` object:

```
from sagemaker.analytics import ExperimentAnalytics
import sagemaker, boto3

session = boto3.Session()
sagemaker_session = sagemaker.Session(boto_
session=session)

experiment_analytics = ExperimentAnalytics(
    sagemaker_session=sagemaker_session,
    experiment_name=experiment_name,
)
```

5. Load the `DataFrame` using the `dataframe()` function and store it inside the `experiment_details_df` variable:

```
experiment_details_df = experiment_analytics.dataframe()
```

6. Use the following lines of code to display the contents of the `DataFrame`:

```
import pandas as pd
from IPython.display import display
pd.options.display.max_columns = None
display(experiment_details_df)
```

In *Figure 5.22*, we can see a table containing all the values of the `experiment_details_df` DataFrame after using the `display()` function:

	TrialComponentName	DisplayName	SourceArn	SageMaker.ImageUri	SageMaker.InstanceCount	SageMaker.InstanceType	SageMaker.VolumeS
0	job-YCSQUFQDNB-aws-training-job	job-YCSQUFQDNB	arn:aws:sagemaker:us-east-1:581320662326:train...	683313688378.dkr.ecr.us-east-1.amazonaws.com/s...	1.0	ml.m5.large	
1	job-BQTPUSOKIT-aws-training-job	job-BQTPUSOKIT	arn:aws:sagemaker:us-east-1:581320662326:train...	683313688378.dkr.ecr.us-east-1.amazonaws.com/s...	1.0	ml.m5.large	
2	job-EXGXPANEUM-aws-training-job	job-EXGXPANEUM	arn:aws:sagemaker:us-east-1:581320662326:train...	683313688378.dkr.ecr.us-east-1.amazonaws.com/s...	1.0	ml.m5.large	
3	job-DZBTVPMIWJ-aws-training-job	job-DZBTVPMIWJ	arn:aws:sagemaker:us-east-1:581320662326:train...	683313688378.dkr.ecr.us-east-1.amazonaws.com/s...	1.0	ml.m5.large	
4	job-ZCNPCSSGBW-aws-training-job	job-ZCNPCSSGBW	arn:aws:sagemaker:us-east-1:581320662326:train...	683313688378.dkr.ecr.us-east-1.amazonaws.com/s...	1.0	ml.m5.large	
5	job-RLXREGUJLI-aws-training-job	job-RLXREGUJLI	arn:aws:sagemaker:us-east-1:581320662326:train...	683313688378.dkr.ecr.us-east-1.amazonaws.com/s...	1.0	ml.m5.large	
6	job-FWRNDEPVCR-aws-training-job	job-FWRNDEPVCR	arn:aws:sagemaker:us-east-1:581320662326:train...	683313688378.dkr.ecr.us-east-1.amazonaws.com/s...	1.0	ml.m5.large	
7	job-SYGXCDZBAB-aws-training-job	job-SYGXCDZBAB	arn:aws:sagemaker:us-east-1:581320662326:train...	683313688378.dkr.ecr.us-east-1.amazonaws.com/s...	1.0	ml.m5.large	
8	TrialComponent-2021-08-29-183229-mrcg	trial-metadata-YCSQUFQDNB	NaN	NaN	NaN	NaN	
9	TrialComponent-2021-08-29-183226-vhyo	trial-metadata-BQTPUSOKIT	NaN	NaN	NaN	NaN	

Figure 5.22 – Table containing all the values of the DataFrame

We have in *Figure 5.22* the DataFrame containing the parameters, hyperparameters, and metric values of the different training jobs we have executed in our experiment. Note that if we loaded this DataFrame using `experiment_analytics.dataframe()` before the training jobs had completed, we may not have some of the columns and column values yet.

7. To make sure that the columns are ready before proceeding with the next step, we run the following block of code:

```
from time import sleep

metric = "validation:error - Avg"
while metric not in experiment_details_df:
    eddf = experiment_analytics.dataframe()
    experiment_details_df = eddf
    print("Not yet ready. Sleeping for 10 seconds")
```

```
    sleep(10)

print("Ready")
```

This block of code will loop indefinitely until the specified metric column is detected in the DataFrame returned using the `dataframe()` function. In this case, we are waiting for `"validation:error - Avg"` to be present before we proceed with the next step.

8. Prepare a new DataFrame with a selected set of columns:

```
target_fields = [
    "TrialComponentName",
    "DisplayName",
    "eta",
    "gamma",
    "max_depth",
    "min_child_weight",
    "num_round",
    "objective",
    "subsample",
    "validation:error - Avg",
    "train:error - Avg",
    "Trials",
    "Experiments",
]

eddf = experiment_details_df
experiment_summary_df = eddf[target_fields]
```

9. Use the `display()` function to show the contents of the new DataFrame:

```
display(experiment_summary_df)
```

Figure 5.23 shows us the output after running the previous line of code:

	TrialComponentName	DisplayName	eta	gamma	max_depth	min_child_weight	num_round	objective	subsample	validation:error - Avg	train:error - Avg
0	job-YCSQUFQDNB-aws-training-job	job-YCSQUFQDNB	0.2	4	8	6	20	binary:logistic	0.4	0.16455	0.174900
1	job-BQTPUSOKIT-aws-training-job	job-BQTPUSOKIT	0.2	4	8	6	10	binary:logistic	0.4	0.16560	0.177700
2	job-EXGXPANEUM-aws-training-job	job-EXGXPANEUM	0.2	3	8	6	20	binary:logistic	0.4	0.16240	0.176367
3	job-DZBTVPMIWJ-aws-training-job	job-DZBTVPMIWJ	0.2	3	8	6	10	binary:logistic	0.4	0.16440	0.179233
4	job-ZCNPCSSGBW-aws-training-job	job-ZCNPCSSGBW	0.2	3	2	6	20	binary:logistic	0.4	0.16680	0.185383
5	job-RLXREGUJLI-aws-training-job	job-RLXREGUJLI	0.2	4	2	6	10	binary:logistic	0.4	0.16770	0.190333
6	job-FWRNDEPVCR-aws-training-job	job-FWRNDEPVCR	0.2	4	2	6	20	binary:logistic	0.4	0.16655	0.185033
7	job-SYGXCDZBAB-aws-training-job	job-SYGXCDZBAB	0.2	3	2	6	10	binary:logistic	0.4	0.16770	0.190333
8	TrialComponent-2021-08-29-183229-mrcg	trial-metadata-YCSQUFQDNB	0.2	4	8	6	20	binary:logistic	0.4	NaN	NaN
9	TrialComponent-2021-08-29-183226-vhyo	trial-metadata-BQTPUSOKIT	0.2	4	8	6	10	binary:logistic	0.4	NaN	NaN

Figure 5.23 – The updated DataFrame

We have in *Figure 5.23* the DataFrame containing only the selected columns.

10. While some experiments are still running, we may see some rows with NaN values. In this step, we remove the rows with the NaN values:

```
import math


def is_not_nan(num):
    return not math.isnan(num)


def remove_nan_rows(df):
    return df[df['train:error - Avg'].map(
        is_not_nan
    )]


experiment_summary_df = remove_nan_rows(
    experiment_summary_df
)
```

11. Sort the rows based on the value of the training error:

```
sorted_df = experiment_summary_df.sort_values(
    'train:error - Avg', ascending=True
)
```

12. Finally, prepare the DataFrame containing only the display name of the training job and the corresponding training error:

```
final_df = sorted_df[["DisplayName", "train:error - Avg"]]
final_df
```

This should yield a DataFrame similar to what is shown in *Figure 5.24*:

	DisplayName	train:error - Avg
0	job-YCSQUFQDNB	0.174900
2	job-EXGXPANEUM	0.176367
1	job-BQTPUSOKIT	0.177700
3	job-DZBTVPMIWJ	0.179233
6	job-FWRNDEPVCR	0.185033
4	job-ZCNPCSSGBW	0.185383
5	job-RLXREGUJLI	0.190333
7	job-SYGXCDZBAB	0.190333

Figure 5.24 – The final_df DataFrame containing the display name and the average training error

In *Figure 5.24*, we can see the training jobs with the average training error values sorted in ascending order.

13. Use the `plot()` function to generate the horizontal bar chart:

```
final_df.plot(kind='barh', x="DisplayName", fontsize=8)
```

This should yield a plot similar to what is shown in *Figure 5.25*

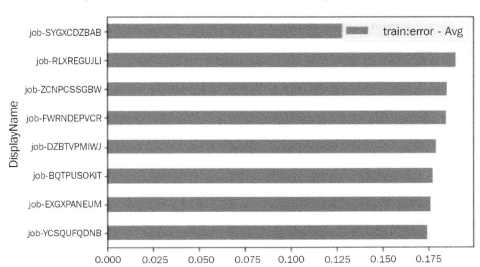

Figure 5.25 – Horizontal bar chart showing the average training error values of each job

Figure 5.25 shows us the horizontal bar chart generated after using the plot() function. Note that this is just one example chart we can generate, so feel free to play around with the DataFrame returned by ExperimentAnalytics.

Now, let's see how this works!

How it works...

In this recipe, we have performed a sample simplified end-to-end process of generating a quick report using the data we have loaded from experiment_analytics. dataframe().

We have performed the following data manipulation steps:

- Removed the NaN values
- Limited the columns to a few columns we would focus on
- Sorted the rows based on a column value

Take note that this is just an example of what we can do once we have the DataFrame containing the results and details of the experiments we have tracked with SageMaker Experiments.

`AnalyticsMetricsBase` is the base class of `ExperimentAnalytics`. In addition to the `dataframe()` function, we can also use the `export_csv()` function to export and save the analytics DataFrame values into a file.

There are actually a few classes that inherit from the `AnalyticsMetricsBase` class—`HyperparameterTuningJobAnalytics`, `TrainingJobAnalytics`, and `ExperimentAnalytics`. These classes focus on providing immediate access to data on hyperparameter tuning jobs, training jobs, and experiments, respectively. For more information, you may check the `analytics.py` file in the SageMaker Python SDK GitHub repository here: `https://github.com/aws/sagemaker-python-sdk/blob/master/src/sagemaker/analytics.py`.

Inspecting experiments, trials, and trial components with SageMaker Experiments

In this recipe, we will use the `sagemaker-experiments` library to inspect the different experiments, trials, and trial components created and linked with SageMaker Experiments. Being able to understand what we can do with the `sagemaker-experiments` library will allow us to manage our experiments better and inspect the properties of the entities created during the process.

Getting ready

The following is the prerequisite for this recipe:

- This recipe continues from *Running and managing multiple experiments with SageMaker Experiments*.

How to do it...

The first steps in this recipe focus on preparing the prerequisites and making sure that the `sagemaker-experiments` library is installed and loaded:

1. Navigate to the `my-experiments/chapter05` directory inside your SageMaker notebook instance. Feel free to create this directory if it does not exist yet.

2. Create a new notebook using the `conda_python3` kernel inside the `my-experiments/chapter05` directory and name it with the name of this recipe (that is, `Inspecting Experiments, Trials, and Trial Components with SageMaker Experiments`). Open this notebook for editing as we will update this file with the code in the next couple of steps.

3. Install `sagemaker-experiments` using `pip install`:

```
!pip install sagemaker-experiments
```

4. Import a few prerequisites from the `smexperiments` library:

```
from smexperiments.experiment import Experiment
from smexperiments.trial import Trial
from smexperiments.trial_component import TrialComponent
from smexperiments.tracker import Tracker
```

Now that we have the prerequisites loaded, we will work on loading the `Experiment`, `Trial`, and `TrialComponent` objects in the next set of steps.

5. Use `Experiment.list()` to load all the experiments stored as `ExperimentSummary` objects:

```
for experiment_summary in Experiment.list():
    print(experiment_summary)
```

Figure 5.26 shows us what the output looks like after running the previous code block:

```
ExperimentSummary(experiment_name='experiment-FMKPXRYRJH',experiment_arn='arn:aws:sagemaker:u
s-east-1:           :experiment/experiment-fmkpxryrjh',display_name='experiment-FMKPXRYRJH',
creation_time=datetime.datetime(2021, 8, 29, 18, 32, 10, 623000, tzinfo=tzlocal()),last_modif
ied_time=datetime.datetime(2021, 8, 29, 18, 38, 58, tzinfo=tzlocal()))
ExperimentSummary(experiment_name='sagemaker-autopilot-demo-aws-auto-ml-job',experiment_arn
='arn:aws:sagemaker:us-east-1:           :experiment/sagemaker-autopilot-demo-aws-auto-ml-jo
b',display_name='sagemaker-autopilot-demo-aws-auto-ml-job',experiment_source={'SourceArn': 'a
rn:aws:sagemaker:us-east-1:581320662326:automl-job/sagemaker-autopilot-demo', 'SourceType':
'SageMakerAutoMLJob'},creation_time=datetime.datetime(2021, 8, 8, 20, 51, 14, 837000, tzinfo=
tzlocal()),last_modified_time=datetime.datetime(2021, 8, 21, 18, 48, 801000, tzinfo=tzloca
l()))
ExperimentSummary(experiment_name='automl-2021-07-25-21-12-45-471-aws-auto-ml-job',experiment
_arn='arn:aws:sagemaker:us-east-1:           :experiment/automl-2021-07-25-21-12-45-471-aws-
auto-ml-job',display_name='automl-2021-07-25-21-12-45-471-aws-auto-ml-job',experiment_source=
{'SourceArn': 'arn:aws:sagemaker:us-east-1:581320662326:automl-job/automl-2021-07-25-21-12-45
-471', 'SourceType': 'SageMakerAutoMLJob'},creation_time=datetime.datetime(2021, 7, 25, 21, 1
2, 48, 360000, tzinfo=tzlocal()),last_modified_time=datetime.datetime(2021, 7, 25, 21, 40, 4
1, 528000, tzinfo=tzlocal()))
```

Figure 5.26 – Sample output after using Experiment.list()

We have in *Figure 5.26* the list of `ExperimentSummary` objects. Note that `ExperimentSummary` objects are different from `Experiment` objects as `ExperimentSummary` objects contain less information. `ExperimentSummary` objects have the `experiment_name` value, which can be used to load the `Experiment` object with `Experiment.load()`.

6. Load the experiment using `Experiment.load()`. The `load()` function accepts the experiment name as the argument and returns an `Experiment` object instead of an `ExperimentSummary` object:

```
experiment = Experiment.load(
    experiment_summary.experiment_name
)
experiment
```

> **Important Note**
> Take note of the differences between the `Experiment` object and the `ExperimentSummary` object as the `Experiment` object includes more properties than the `ExperimentSummary` object.

Figure 5.27 shows us what the output looks like after running the previous code block:

```
Out[8]: Experiment(sagemaker_boto_client=<botocore.client.SageMaker object at 0x7f74fa360390>,experim
ent_name='experiment-NPZFAITKHA',experiment_arn='arn:aws:sagemaker:us-east-1:        :exp
eriment/experiment-npzfaitkha',display_name='experiment-NPZFAITKHA',description='Experiment D
escription',creation_time=datetime.datetime(2021, 4, 24, 21, 40, 20, 404000, tzinfo=tzlocal
()),created_by={},last_modified_time=datetime.datetime(2021, 4, 24, 21, 45, 26, 163000, tzinf
o=tzlocal()),last_modified_by={},response_metadata={'RequestId': '89b44e0d-573a-495b-8401-41f
6788b2939', 'HTTPStatusCode': 200, 'HTTPHeaders': {'x-amzn-requestid': '89b44e0d-573a-495b-84
01-41f6788b2939', 'content-type': 'application/x-amz-json-1.1', 'content-length': '314', 'dat
e': 'Sun, 29 Aug 2021 18:39:07 GMT'}, 'RetryAttempts': 0})
```

Figure 5.27 – Sample output [Experiment object] after using the Experiment.load() function

As you can see in *Figure 5.27*, the loaded `Experiment` object contains a few more properties such as `description` and `response_metadata`.

7. Inspect the functions by using `dir()`:

```
dir(experiment)
```

This should give us a list of valid attributes and functions of the `experiment` object, such as `create_trial`, `delete_all`, `display_name`, `experiment_name`, and `experiment_arn`.

8. List the `Trials` of the `Experiment` using the `list_trials()` function:

```
experiment_trials = list(
    experiment.list_trials()
)
experiment_trials
```

This should yield a set of results similar to what is shown in *Figure 5.28*:

```
Out[11]:   [TrialSummary(trial_name='trial-RJNRCIKGOR',trial_arn='arn:aws:sagemaker:us-east-1:
             :experiment-trial/trial-rjnrcikgor',display_name='trial-RJNRCIKGOR',creation_time=datetime.
           datetime(2021, 4, 24, 21, 40, 38, 930000, tzinfo=tzlocal()),last_modified_time=datetime.datet
           ime(2021, 4, 24, 21, 45, 26, 163000, tzinfo=tzlocal())),
            TrialSummary(trial_name='trial-QBXPLDRXZY',trial_arn='arn:aws:sagemaker:us-east-1:
             :experiment-trial/trial-qbxpldrxzy',display_name='trial-QBXPLDRXZY',creation_time=datetime.
           datetime(2021, 4, 24, 21, 40, 36, 414000, tzinfo=tzlocal()),last_modified_time=datetime.datet
           ime(2021, 4, 24, 21, 45, 24, 322000, tzinfo=tzlocal())),
            TrialSummary(trial_name='trial-DZJUVJIHXM',trial_arn='arn:aws:sagemaker:us-east-1:
             :experiment-trial/trial-dzjuvjihxm',display_name='trial-DZJUVJIHXM ',creation_time=datetime.
           datetime(2021, 4, 24, 21, 40, 33, 881000, tzinfo=tzlocal()),last_modified_time=datetime.datet
           ime(2021, 4, 24, 21, 45, 22, 462000, tzinfo=tzlocal())),
            TrialSummary(trial_name='trial-LGPLDSCFNA',trial_arn='arn:aws:sagemaker:us-east-1:
             :experiment-trial/trial-lgpldscfna',display_name='trial-LGPLDSCFNA',creation_time=datetime.
           datetime(2021, 4, 24, 21, 40, 31, 265000, tzinfo=tzlocal()),last_modified_time=datetime.datet
           ime(2021, 4, 24, 21, 45, 0, 180000, tzinfo=tzlocal())),
```

Figure 5.28 – List of TrialSummary objects after using the list_trials() function

In *Figure 5.28*, we can see a list of `TrialSummary` objects corresponding to each `Trial` linked to the `Experiment` object. Similar to the `ExperimentSummary` object, `TrialSummary` objects contain less information than their corresponding `Trial` objects.

9. Load the corresponding `Trial` object using `Trial.load()`:

    ```
    trial = Trial.load(experiment_trials[0].trial_name)
    ```

    > **Important Note**
    >
    > Take note of the differences between the `Trial` and `TrialSummary` objects as the `Trial` object includes more properties than the `TrialSummary` object.

10. Check the properties and functions of the `Trial` object using the `dir()` function:

    ```
    dir(trial)
    ```

    This should give us a list of valid attributes and functions (methods) of the `trial` object, such as `add_trial_component`, `delete_all`, `list_trial_components`, `trial_name`, and `trial_arn`.

11. Load the trial components linked to the trial as a list of `TrialComponentSummary` objects using the `list_trial_components()` function:

```
trial_component_summary_list = list(
    trial.list_trial_components()
)
trial_component_summary_list
```

This should yield a set of results similar to what is shown in *Figure 5.29*:

```
Out[17]:  [TrialComponentSummary(trial_component_name='job-RJNRCIKGOR-aws-training-job',trial_component
          _arn='arn:aws:sagemaker:us-east-1:             :experiment-trial-component/job-rjnrcikgor-aws-
          training-job',display_name='job-RJNRCIKGOR',trial_component_source={'SourceArn': 'arn:aws:sag
          emaker:us-east-1:581320662326:training-job/job-rjnrcikgor', 'SourceType': 'SageMakerTrainingJ
          ob'},status=TrialComponentStatus(primary_status='Completed',message='Status: Completed, secon
          dary status: Completed, failure reason: .'),creation_time=datetime.datetime(2021, 4, 24, 21,
          40, 41, 629000, tzinfo=tzlocal()),created_by={},last_modified_time=datetime.datetime(2021, 4,
          24, 21, 45, 26, 163000, tzinfo=tzlocal()),last_modified_by={}),
          TrialComponentSummary(trial_component_name='TrialComponent-2021-04-24-214038-svqp',trial_com
          ponent_arn='arn:aws:sagemaker:us-east-1:            :experiment-trial-component/trialcomponen
          t-2021-04-24-214038-svqp',display_name='trial-metadata-RJNRCIKGOR',status=TrialComponentStatu
          s(primary_status='Completed',message=None),start_time=datetime.datetime(2021, 4, 24, 21, 40,
          38, tzinfo=tzlocal()),end_time=datetime.datetime(2021, 4, 24, 21, 40, 38, tzinfo=tzlocal()),c
          reation_time=datetime.datetime(2021, 4, 24, 21, 40, 38, 829000, tzinfo=tzlocal()),created_by=
          {},last_modified_time=datetime.datetime(2021, 4, 24, 21, 40, 38, 892000, tzinfo=tzlocal()),la
          st_modified_by={}),
          TrialComponentSummary(trial_component_name='TrialComponent-2021-04-24-214020-ucgr',trial_com
          ponent_arn='arn:aws:sagemaker:us-east-1:            :experiment-trial-component/trialcomponen
          t-2021-04-24-214020-ucgr',display_name='xgboost-experiment-display-name',status=TrialComponen
          tStatus(primary_status='Completed',message=None),start_time=datetime.datetime(2021, 4, 24, 2
          1, 40, 20, tzinfo=tzlocal()),end_time=datetime.datetime(2021, 4, 24, 21, 40, 20, tzinfo=tzloc
          al()),creation_time=datetime.datetime(2021, 4, 24, 21, 40, 20, 622000, tzinfo=tzlocal()),crea
          ted_by={},last_modified_time=datetime.datetime(2021, 4, 24, 21, 40, 20, 686000, tzinfo=tzloca
          l()),last_modified_by={})]
```

Figure 5.29 – List of TrialComponentSummary objects after using the list_trial_components() function

In *Figure 5.29*, we can see a list of `TrialComponentSummary` objects corresponding to each `TrialComponent` linked to the `Trial`. Similar to the `ExperimentSummary` object, `TrialComponentSumary` objects contain less information than their corresponding `TrialComponent` objects.

12. Load the corresponding `Trial` object using `TrialComponent.load()`:

```
tcs = trial_component_summary_list[0]
trial_component_summary = tcs
tc_name = tcs.trial_component_name


trial_component = TrialComponent.load(
    trial_component_name=tc_name)
```

13. List the properties and functions using the `dir()` function:

```
dir(trial_component)
```

This should give us a list of valid attributes of the `trial_component` object, such as `input_artifacts`, `output_artifacts`, `parameters`, `trial_component_name`, and `trial_component_arn`.

14. Inspect the `input_artifacts` value of `trial_component`:

```
trial_component.input_artifacts
```

This should give us a dictionary of values similar to what is shown in *Figure 5.30*:

```
Out[21]: {'train': TrialComponentArtifact(value='s3://sagemaker-cookbook-bucket/chapter05/input/traini
         ng_data.csv',media_type='text/csv'),
          'validation': TrialComponentArtifact(value='s3://sagemaker-cookbook-bucket/chapter05/input/v
         alidation_data.csv',media_type='text/csv')}
```

Figure 5.30 – trial_component.input_artifacts

We have in *Figure 5.30* a dictionary with the `train` and `validation` key-value pairs containing the S3 paths pointing to the `training_data.csv` and `validation_csv` datasets, respectively.

15. Inspect the `output_artifacts` value of `trial_component`:

```
trial_component.output_artifacts
```

Here is the output:

```
Out[22]: {'SageMaker.ModelArtifact': TrialComponentArtifact(value='s3://sagemaker-cookbook-bucket/chap
         ter05/output/job-RJNRCIKGOR/output/model.tar.gz',media_type=None)}
```

Figure 5.31 – trial_component.output_artifacts

We have in *Figure 5.31* a dictionary with the `SageMaker.ModelArtifact` key and the value set to a `TrialComponentArtifact` object with the S3 path pointing to the `model.tar.gz` file.

16. Inspect the `parameters` value of `trial_component`:

```
trial_component.parameters
```

This should give us a dictionary of values similar to what is shown in *Figure 5.32*:

```
{'SageMaker.ImageUri': '683313688378.dkr.ecr.us-east-1.amazonaws.com/sagemaker-xgboost:0.90-2
-cpu-py3',
 'SageMaker.InstanceCount': 1.0,
 'SageMaker.InstanceType': 'ml.m5.large',
 'SageMaker.VolumeSizeInGB': 30.0,
 'eta': 0.2,
 'gamma': 4.0,
 'max_depth': 8.0,
 'min_child_weight': 6.0,
 'num_round': 20.0,
 'objective': 'binary:logistic',
 'subsample': 0.4}
```

Figure 5.32 – trial_component.parameters

We have in *Figure 5.32* a dictionary with the key-value pairs for training job configuration, including `SageMaker.ImageURI` and `SageMaker.InstanceCount`, and hyperparameters such as `eta`, `gamma`, and `max_depth`.

17. Inspect the `metrics` value of `trial_component`:

```
trial_component.metrics
```

This should give us a list of objects similar to what is shown in *Figure 5.33*:

```
Out[24]: [TrialComponentMetricSummary(metric_name='validation:error',source_arn='arn:aws:sagemaker:us-
         east-1:           :training-job/job-rjnrcikgor',time_stamp=datetime.datetime(2021, 4, 24, 2
         1, 44, 10, 430000, tzinfo=tzlocal()),max=0.172,min=0.16,last=0.165,count=20,avg=0.16455000000
         000006,std_dev=0.0025021043774769865),
         TrialComponentMetricSummary(metric_name='train:error',source_arn='arn:aws:sagemaker:us-east-
         1:           :training-job/job-rjnrcikgor',time_stamp=datetime.datetime(2021, 4, 24, 21, 44,
         10, 430000, tzinfo=tzlocal()),max=0.183667,min=0.169,last=0.169667,count=20,avg=0.17490009999
         999998,std_dev=0.0043837282966611105)]
```

Figure 5.33 – trial_component.metrics

We have in *Figure 5.33* two `TrialComponentMetricSummary` objects within a list—one for the `train:error` metric and the other for the `validation:error` metric.

Now, let's see how this works!

How it works...

When using the `smexperiments` library, it has to be noted that the `Experiment`, `Trial`, and `TrialComponent` classes and objects have their own corresponding `Summary` counterparts with a minimum number of properties and attributes.

As seen in the *How to do it...* section of this recipe, we will get a list of
`ExperimentSummary` objects after using the `Experiment.list()` function. We
then have to use the `Experiment.load()` function to get the complete `Experiment`
object, which contains more properties including `description` and `response_`
`metadata`. In a similar fashion, we will get a list of `TrialSummary` objects linked to the
`Experiment` object after using the `experiment.list_trials()` function. To get
the complete `Trial` object, we would then need to use the `Trial.load()` function by
passing the `trial_name` value of the `TrialSummary` object. Finally, once we have the
`Trial` object, we simply use the `trial.list_trial_components()` function and
the `TrialComponent.load()` function to get the `TrialComponent` objects. These
`TrialComponent` objects contain the metric values, location, and details of the input
and output artifacts, hyperparameters specified at the start of the training jobs, and other
properties tracked while the experiment is running.

There's more...

We can actually search, inspect, and compare the different attributes and properties of
`Experiment`, `Trial`, and `TrialComponent` entities inside SageMaker Studio.

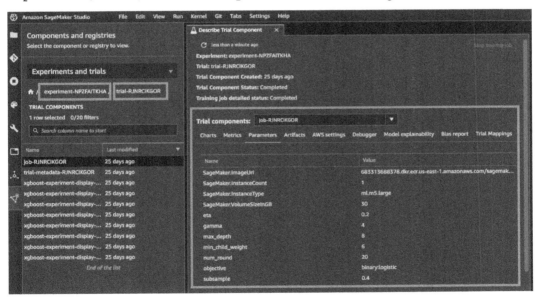

Figure 5.34 – Inspecting Trial components in SageMaker Studio

We can see in *Figure 5.34* the different parameters tracked inside the selected `Trial` component. These parameters include the training configuration parameters as well as the hyperparameter values specified before the training job has started.

See also

We will not be able to discuss the other features and solutions with SageMaker Experiments, so we will provide links to relevant references and examples instead:

- Cleaning up SageMaker Experiment resources: `https://docs.aws.amazon.com/sagemaker/latest/dg/experiments-cleanup.html`

- Streamline modeling with SageMaker Studio and SageMaker Experiments: `https://github.com/aws-samples/modeling-with-amazon-sagemaker-experiments/blob/dad91afb3ed4a0f0c0dff00cc978825972eb2936/modeling-with-amazon-sagemaker-experiments.ipynb`

Note that we have not formally introduced **SageMaker Studio** yet at this point. Do not worry as we will do this in the very next chapter!

6
Automated Machine Learning in Amazon SageMaker

Automated machine learning (AutoML) is the process of automating different aspects of the machine learning pipeline to help build and deploy high-quality models in a short period of time. This works by automating different phases of the machine learning process, such as feature engineering, architecture search, and hyperparameter optimization. Initially, tools for AutoML focused more on automating the time-consuming hyperparameter optimization tasks by looking for an optimal set of hyperparameters for a model. These past couple of years, however, AutoML has expanded to include the automation of other parts of the pipeline, including data cleaning, feature selection, model selection, and more:

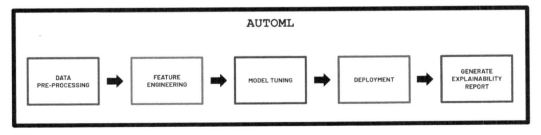

Figure 6.1 – Different phases and types of tasks that are automated through AutoML

The preceding image shows a simplified diagram showing the different phases and types of tasks that can be automated with AutoML. By just providing a dataset and a few configuration parameters, we will be able to build and deploy a high-quality model without having to write a single line of code. In this chapter, we will take a closer look at using AutoML in SageMaker using **SageMaker Autopilot**. This capability allows machine learning practitioners to easily train and tune models, even with limited machine learning experience or expertise. While easily being accessible to beginners, **SageMaker Autopilot** allows the process to be transparent and repeatable. This enables experts to understand, modify, and replace different parts of the automated machine learning process.

We will start by introducing **SageMaker Studio**, where we will create and manage our first **SageMaker Autopilot** experiment. We will also have a quick look at the model explainability report that's generated toward the end of the Autopilot experiment. Later, toward the end of this chapter, we will look at how to use and configure the **Automatic Model Tuning** capability of **Amazon SageMaker**, which allows us to search for the optimal set of hyperparameter values for our model.

We will cover the following recipes in this chapter:

- Onboarding to **SageMaker Studio**
- Generating a synthetic dataset with additional columns containing random values
- Creating and monitoring a **SageMaker Autopilot** experiment in **SageMaker Studio** (console)
- Creating and monitoring a **SageMaker Autopilot** experiment using the **SageMaker Python SDK**
- Inspecting the **SageMaker Autopilot** experiment's results and artifacts
- Performing **Automatic Model Tuning** with the SageMaker **XGBoost** built-in algorithm
- Analyzing the **Automatic Model Tuning** job results

Once we have completed the recipes in this chapter, we will have a better understanding of how to configure, manage, and monitor the automated machine learning and automatic model tuning capabilities of **Amazon SageMaker**. Note that we will take a quick look at the model explainability report that's generated during the Autopilot experiment, but we will discuss this in detail in the *Enabling ML explainability with SageMaker Clarify* recipe of *Chapter 7, Working with SageMaker Feature Store, SageMaker Clarify, and SageMaker Model Monitor*.

Now, let's proceed with the recipes in this chapter!

Technical requirements

To execute the recipes in the chapter, make sure you have the following:

- An Amazon S3 bucket.

- Permission to manage the **Amazon SageMaker** and **Amazon S3** resources if you're using an **AWS IAM** user with a custom URL. If you are using the root account, then you should be able to proceed with the recipes in this chapter. However, it is recommended that you sign in as an AWS IAM user instead of using the root account in most cases. For more information, feel free to take a look at `https://docs.aws.amazon.com/IAM/latest/UserGuide/best-practices.html`.

If you do not have these prerequisites ready yet, feel free to check out the *Preparing the Amazon S3 bucket and the training dataset for the linear regression experiment* recipe of *Chapter 1, Getting Started with Machine Learning Using Amazon SageMaker*.

As the recipes in this chapter involve a bit of code, we have made these notebooks available in this book's GitHub repository: `https://github.com/PacktPublishing/Machine-Learning-with-Amazon-SageMaker-Cookbook/tree/master/Chapter06`. Before starting each of the recipes in this chapter, make sure that the `my-experiments/chapter06` directory is ready. If it has not been created yet, please do so now as this keeps things organized as we go through each of the recipes in this book.

Check out the following link to see the relevant Code in Action video:

`https://bit.ly/3tslmcy`

Onboarding to SageMaker Studio

Amazon SageMaker Studio is a fully integrated environment for machine learning. With SageMaker Studio, we can easily use the other features and capabilities of SageMaker in this environment, such as **SageMaker Autopilot**, **SageMaker Debugger**, and **SageMaker Experiments**, using its intuitive user interface. In this recipe, we will set up SageMaker Studio so that we can use its different features and integrations in the succeeding recipes. We will assume that this is your first time using SageMaker Studio, so we will work on setting up the execution roles and other prerequisites in this recipe as well.

In the *Creating and monitoring a SageMaker Autopilot experiment in SageMaker Studio (console)* recipe, we will see how easy it is to create an Autopilot experiment using **SageMaker Studio** without having to write a single line of code. For now, we will be focusing on getting our environment set up so that we can use it in the subsequent recipes.

> **Note**
>
> Using **SageMaker Studio** is free! Generally, we only pay for the compute and storage resources based on their usage. This means that we will pay for **SageMaker Processing** and the training jobs that are launched (for a few minutes) while using the different options and features in **SageMaker Studio**. For more information, please refer to the notes on the Amazon SageMaker Studio pricing page (`https://docs.aws.amazon.com/sagemaker/latest/dg/studio-pricing.html`), along with the details in the pricing page (`https://aws.amazon.com/sagemaker/pricing/`).

Getting ready

For this recipe, you need the necessary permissions to manage the **Amazon SageMaker** and **Amazon S3** resources if you're using an **AWS IAM** user with a custom URL.

How to do it...

The steps in this recipe focus on getting **SageMaker Studio** set up. Let's get started:

1. In the search bar, type `sagemaker` and choose **Amazon SageMaker** from the search results:

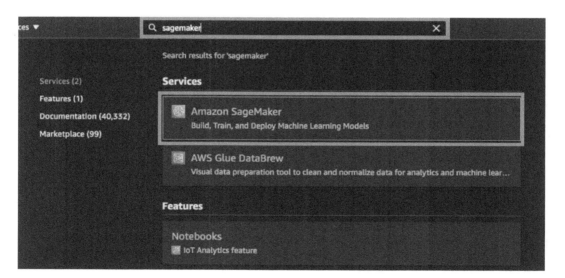

Figure 6.2 – Navigating to the SageMaker console

Clicking the **Amazon SageMaker** option from the list of search results should direct us to the SageMaker console.

2. In the SageMaker console, locate and click the **Amazon SageMaker Studio** link at the top left, as shown here:

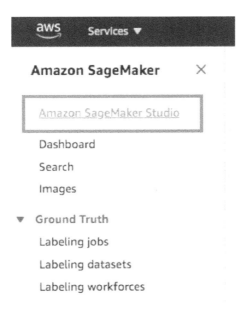

Figure 6.3 – Locating Amazon SageMaker Studio in the navigation pane

The sidebar helps us navigate to the different sections of the SageMaker console. Clicking **Amazon SageMaker Studio** should redirect us to a page showing the **Amazon SageMaker Studio** dashboard.

3. If this is your first time using **SageMaker Studio** in a specific region, you should see a screen similar to the following. Select **Standard Setup** and then choose **AWS Identity and Access Management (IAM)** as your **Authentication method**:

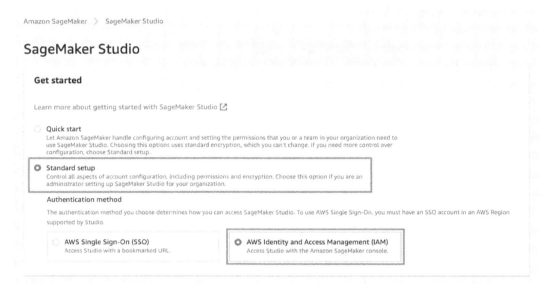

Figure 6.4 – Quick start versus Standard setup

As we can see, we have two options – **Quick start** and **Standard Setup**. Choosing **Standard Setup** gives us more flexibility and control over the configuration.

4. Under **Permission**, select **Create a new role**. This should open the **Create an IAM role** dialog box, which allows us to create an IAM role without having to leave the page:

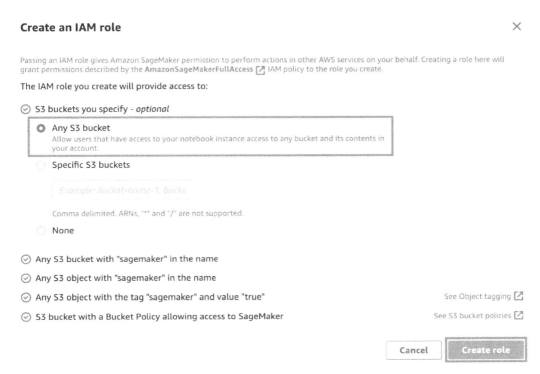

Figure 6.5 – Create an IAM role

As we can see, we will be using the default configuration specified. Then, click the **Create role** button. We should see a message stating `Success! You created an IAM role.`

> **Note**
>
> We may also decide to use an existing role instead of creating a new one.

5. Under **Network and storage**, select the existing default VPC and choose **Public internet Only**:

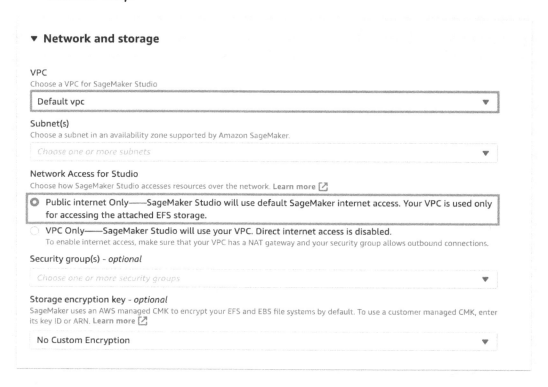

Figure 6.6 – Network and storage configuration

Here, we have two options for network access – **Public internet Only** and **VPC Only**. Choosing the **VPC Only** option will give us a more secure network configuration, but choosing the **Public internet Only** option should do the trick for now as we will be dealing with sample datasets and experiments.

6. Leave the other default configuration as-is and click the **Submit** button. We should see a **Preparing SageMaker Studio** notification, similar to what is shown in the following screenshot:

> **Preparing SageMaker Studio**
> We are configuring resources needed by Studio. This is a one-time configuration and may take a few minutes.

Figure 6.7 – Preparing SageMaker Studio notification message

We will see this notification message for about 3-5 minutes while the resources are being prepared and configured by SageMaker. After a while, we should see a message stating **SageMaker Studio is ready**.

7. In **SageMaker Studio Control Panel**, click **Add user**:

SageMaker Studio Control Panel

Choose your user name, then choose Open Studio to get started   Add user

Q Search users   < 1 >   ⊚

User name   ▽   Last modified   ▽   Created   ▽

No users

To add a user, choose Add user and enter a user name.

Figure 6.8 – Locating the Add user button within SageMaker Studio Control Panel

As we can see, the **Add user** button is located at the top right of the panel.

8. Clicking **Add user** will open the **Add user** form, as shown in the following screenshot. Specify the preferred username. Then, under **Execution role**, choose **Create a new role**:

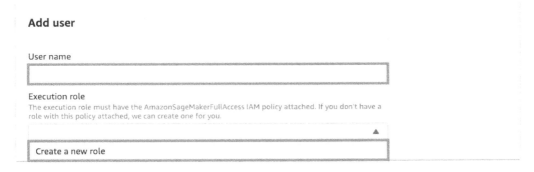

Amazon SageMaker > SageMaker Studio > Control Panel > Add user profile

SageMaker Studio Control Panel

Add user

User name

Execution role
The execution role must have the AmazonSageMakerFullAccess IAM policy attached. If you don't have a role with this policy attached, we can create one for you.

▲

Create a new role

Figure 6.9 – Add user form

Here, we can see the **Add user** form, which allows us to create a new user to access SageMaker Studio.

9. When prompted to create an IAM role, select **Any S3 bucket**:

Figure 6.10 – Creating an IAM role for the user

Here, we have the **Create an IAM role** dialog box, similar to what we had when we created the IAM role for the SageMaker Studio execution role.

10. We should see a success message after completing the previous step. Click the **Submit** button afterward.

11. Once the user has been created, click **Open Studio** under **SageMaker Studio Control Panel**:

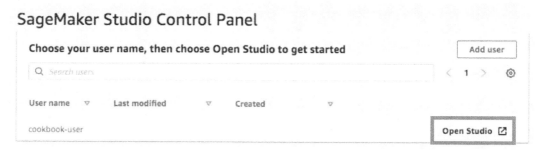

Figure 6.11 – Open Studio button

Here, we can see the **Open Studio** button at the right-hand side of the screen. After clicking this button, a new tab will open. Wait a few minutes for the **SageMaker Studio** interface to load:

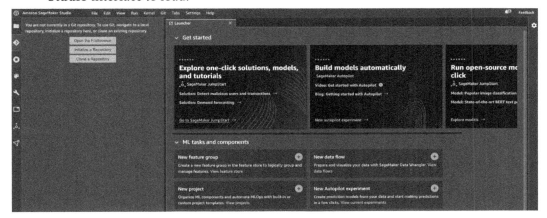

Figure 6.12 – SageMaker Studio

Here, we can see SageMaker Studio. We will discuss the different parts of **SageMaker Studio** in the *How it works…* section.

12. From the left sidebar, select the second icon, which will open the different options to manage GitHub repositories. Click the **Clone a Repository** button:

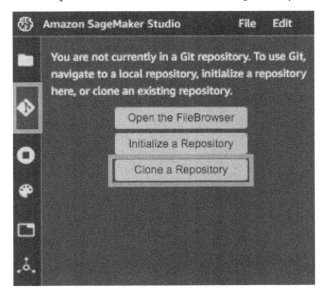

Figure 6.13 – Different options for managing GitHub repositories

Here, we can see that there are three buttons:

- **Open the File Browser**, which simply navigates us back to the file browser.

- **Initialize a Repository**, which makes the current directory a GitHub repository,

- **Clone a Repository**, which clones a repository into the current directory.

13. In the **Clone a repo** dialog box, specify the Clone URI as `https://github.com/PacktPublishing/Machine-Learning-with-Amazon-SageMaker-Cookbook.git` in the text box. Click the **CLONE** button afterward:

Figure 6.14 – Clone a repo

Here, we have the **Clone a repo** dialog box. In this step, we are cloning the repository that contains this book's notebooks, source code, and files.

14. Right-click on the empty space below the **Machine-Learning-with-Amazon-SageMaker-Cookbook** directory and choose **New Folder** from the context menu:

Figure 6.15 – Context menu

Here, we have the context menu, which pops up after right-clicking.

15. Rename the new directory my-experiments:

Figure 6.16 – File browser

Here, we can see that there are two directories:

- The Amazon-SageMaker-Cookbook directory, which contains the repository notebooks and files that were cloned from this book's GitHub repository.

- The my-experiments directory we have created.

With that, we are ready to work on the next set of recipes within SageMaker Studio. Now, let's see how this works!

How it works...

In this recipe, we set up our **SageMaker Studio** environment. This recipe is composed of three parts:

- Using the standard setup for **SageMaker Studio** onboarding

- Creating a user

- Cloning the **Amazon-SageMaker-Cookbook** GitHub repository files:

Figure 6.17 – Parts of SageMaker Studio

Let's discuss the different parts of the **SageMaker Studio** interface. As shown in the preceding screenshot, the **left sidebar** provides the different options and tools to help us manage our resources – files, terminals, kernels, GitHub repositories, SageMaker Jumpstart, and SageMaker components and registries. The **main work area** focuses on the notebooks and Terminals we will use to manage our experiments. We can have multiple tabs inside the main work area. We also have the **Settings** pane (not shown in *Figure 6.17*) at the right of the screen. Here, we can adjust the properties of tables and charts when working on the different resources (for example, trial components).

Let's look at some of the differences when using **SageMaker Studio** and **SageMaker Notebook instances**:

- In **SageMaker Studio**, we no longer need to launch instances manually before running a notebook as these are automatically managed and created behind the scenes.

- Notebooks inside **SageMaker Studio** can be viewed even if the instances are turned off.

- Notebooks inside SageMaker Studio can be shared with other users. For more information on this topic, check out `https://docs.aws.amazon.com/sagemaker/latest/dg/notebooks-sharing.html`.

- Notebooks inside **SageMaker Studio** do not support **local mode**. On the other hand, we can use **local mode** inside the notebooks running in **SageMaker Notebook instances**.

In addition to these, several capabilities and features of SageMaker can be used and managed inside SageMaker Studio without us having to write a single line of code. These capabilities and features include **SageMaker Autopilot**, **SageMaker Experiments**, **SageMaker Feature Store**, **SageMaker Pipelines**, **SageMaker Clarify**, **SageMaker Debugger**, **SageMaker Model Monitor**, and more.

There's more...

Another capability of **Amazon SageMaker**, called **SageMaker JumpStart**, can be accessed by clicking the **New Launcher** option under the list of options inside the **File** menu, and then clicking **Go to SageMaker JumpStart**, as shown in the following screenshot. We can also use the JumpStart icon (the second icon from the bottom) in the left sidebar to browse the JumpStart page:

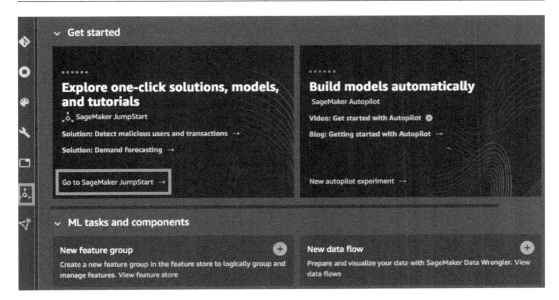

Figure 6.18 – Accessing SageMaker JumpStart

We will be redirected to a page containing the different solutions, models, notebooks, and other references, as shown in the following screenshot:

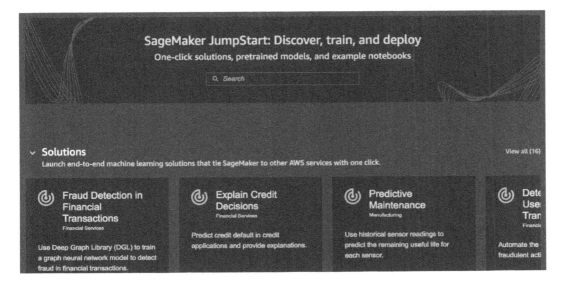

Figure 6.19 – SageMaker JumpStart

SageMaker JumpStart speeds up how we prepare machine learning experiments and workflows by providing a catalog of different solutions, models, sample notebooks, and resources to the users inside **SageMaker Studio**. For more information on **SageMaker JumpStart**, feel free to check out `https://docs.aws.amazon.com/sagemaker/latest/dg/studio-jumpstart.html`.

In addition to **SageMaker JumpStart**, we have **SageMaker Data Wrangler**, where we can import, process, transform, and analyze our data. This feature allows us to perform data preparation and feature engineering without writing a single line of code.

> **Note**
>
> Once you have used **SageMaker Data Wrangler** a bit, you will probably be aware that it supports custom transforms and formulas as well. These allow us to use PySpark, pandas, and PySpark SQL code within **SageMaker Data Wrangler** for more flexibility.

The following screenshot shows the **Export data flow** tab of **SageMaker Data Wrangler**. Here, we can export the flows and transformations we have prepared into notebooks and code, which allows us to customize and integrate these with any existing pipelines:

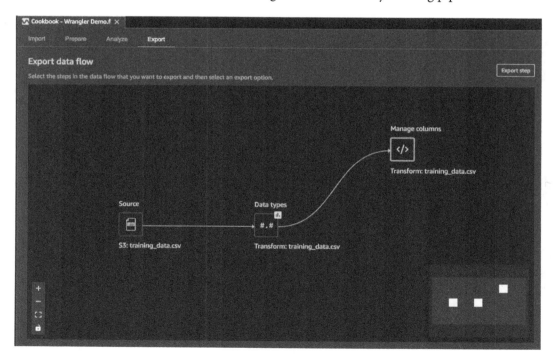

Figure 6.20 – SageMaker Data Wrangler

At a high level, we can do the following with this service:

- Import data from Amazon S3 and other data sources

- Create and manage data flows that perform different steps and transformations

- Transform data

- Analyze data using a variety of analysis and visualization tools

- Export existing flows into notebooks or code files

> **Note**
>
> For more information on this topic, please check out the following link:
> `https://docs.aws.amazon.com/sagemaker/latest/dg/`
> `data-wrangler.html`.

Generating a synthetic dataset with additional columns containing random values

In this recipe, we will generate a synthetic dataset using scikit-learn. This dataset will serve as a dummy dataset for the experiments in this chapter:

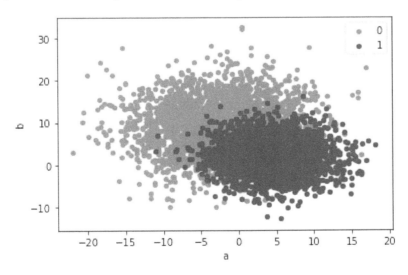

Figure 6.21 – Scatterplot of the synthetic dataset for classification problems

Just by looking at the preceding scatterplot, we can infer that we are generating a synthetic dataset for a **binary classification problem**. In addition to the primary predictor columns, a and b, that were generated by the make_blobs() function of scikit-learn, the dataset will include two columns, c and d, that contain random values that show us what the generated model explainability report looks like with these additional columns. This model explainability report will be generated in the *Creating and monitoring a SageMaker Autopilot experiment in SageMaker Studio (console)* recipe.

> **Tip**
> Since we will show the steps for how to generate a synthetic dataset in this recipe, we will have the opportunity to tweak this recipe later on so that it fits our needs. We can decide to make this dataset have (significantly) more records or more columns to see how these tweaks affect the results, as well as the configuration, of the automated machine learning and automated hyperparameter optimization jobs.

Getting ready

This recipe continues from the *Onboarding to SageMaker Studio* recipe.

How to do it...

We will be running the following set of steps inside a notebook running in **SageMaker Studio**. The first few steps will focus on setting up a few prerequisites before we generate the synthetic dataset. Let's get started:

1. Navigate to the my-experiments/chapter06 directory inside **SageMaker Studio**. Feel free to create this directory if it does not exist yet:

Figure 6.22 – The my-experiments/chapter06 directory

Here, we have the my-experiments/chapter06 directory. Double-click the chapter06 directory before proceeding.

2. Create a new notebook using the `Python 3 (Data Science)` kernel inside the `my-experiments/chapter06` directory and rename it with the name of this recipe (that is, `Generating a synthetic dataset with additional columns containing random values`):

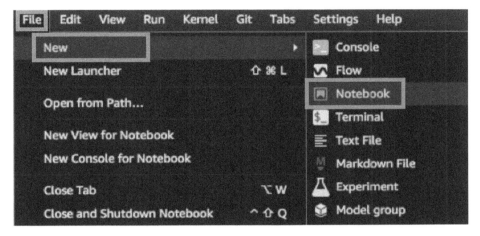

Figure 6.23 – Creating a new notebook

Here, we can see how to create a new **Notebook** from the **File** menu. When prompted for the kernel to use, choose `Python 3 (Data Science)`.

> **Note**
>
> It may take 3-5 minutes for the kernel to start if this is our first Notebook in **SageMaker Studio**. We will need to wait for the instance where the Notebook will run to be ready. After working on the recipes of this chapter, feel free to turn off the running instances in the **Running Terminals and Kernels** pane in the left sidebar of **SageMaker Studio**. This will help reduce and manage the costs that are incurred when running notebooks. For more information, feel free to check out the following link: `https://docs.aws.amazon.com/sagemaker/latest/dg/notebooks-usage-metering.html`.

3. Use the `make_blobs()` function from `scikit-learn` to generate the synthetic dataset. Note that we set the number of samples to 5,000 in the following lines of code:

```
from sklearn.datasets import make_blobs
X, y = make_blobs(n_samples=5000, centers=2,
                  cluster_std=[6, 4], n_features=2,
                  random_state=42)
```

4. Verify the number of samples in the synthetic dataset. We will use the `n_samples` variable in the next step when generating the random values for the two additional columns:

```
n_samples = len(X)
```

5. Generate random values for the two additional columns we will include in the dataset. In the following block of code, we are using the `random.randint()` function from **NumPy** to generate the random values with a length equivalent to `n_samples`:

```
import numpy as np
r1 = np.random.randint(low=-100, high=100,
                       size=(n_samples,)).astype(int)
r2 = np.random.randint(low=-100, high=100,
                       size=(n_samples,)).astype(int)
```

6. Prepare a `DataFrame` called `all_dataset`. It will contain the target column named `label` and four predictor columns named `a`, `b`, `c`, and `d`:

```
import pandas as pd
all_dataset = pd.DataFrame(
    dict(label=y, a=X[:,0], b=X[:,1], c=r1, d=r2))
print(all_dataset)
```

This should give us a `DataFrame` of values, similar to what's shown in the following screenshot:

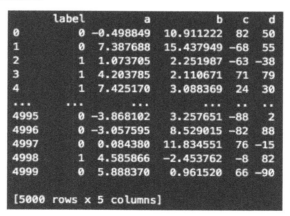

	label	a	b	c	d
0	0	-0.498849	10.911222	82	50
1	0	7.387688	15.437949	-68	55
2	1	1.073705	2.251987	-63	-38
3	1	4.203785	2.110671	71	79
4	1	7.425170	3.088369	24	30
...	...	...	...	..	..
4995	0	-3.868102	3.257651	-88	2
4996	0	-3.057595	8.529015	-82	88
4997	0	0.084380	11.834551	76	-15
4998	1	4.585866	-2.453762	-8	82
4999	0	5.888370	0.961520	66	-90

[5000 rows x 5 columns]

Figure 6.24 – The all_dataset DataFrame

Here, we can see `all_dataset`, which contains five columns – `label`, `a`, `b`, `c`, and `d`.

7. Generate a plot using the following lines of code. Here, we grouped the points by their label values using the `groupby()` function:

```
from matplotlib import pyplot
colors = {0:'red', 1:'blue'}
fig, ax = pyplot.subplots()
grouped = all_dataset.groupby('label')

for key, group in grouped:
    group.plot(ax=ax, kind='scatter',
                x='a', y='b', label=key,
color=colors[key])

pyplot.show()
```

Here, we can see the scatterplot that was generated using Matplotlib:

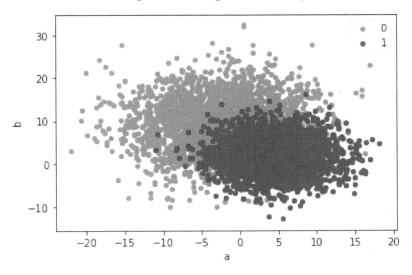

Figure 6.25 – Scatterplot of the synthetic dataset for classification problems

This is the scatterplot of `all_dataset` and shows the points grouped using the label column value.

Now that we have generated the dataset, we will need to divide it into training, validation, and test sets before uploading these to an S3 bucket:

8. Use the `train_test_split()` function twice to get the training, validation, and test datasets:

    ```
    from sklearn.model_selection import train_test_split
    train_val, test = train_test_split(
        all_dataset,
        test_size=0.2,
        random_state=0)
    training, validation = train_test_split(
        train_val,
        test_size=0.25,
        random_state=0)
    ```

9. Generate the `tmp` directory using the `mkdir` Bash command if it does not exist yet:

    ```
    !mkdir -p tmp
    ```

10. Specify the S3 bucket's name and the prefix where the data will be stored. Make sure that you replace the value of `"<insert bucket name here>"` with the name of the bucket we created in the *Preparing the Amazon S3 bucket and the training dataset for the linear regression experiment* recipe of *Chapter 1, Getting Started with Machine Learning Using Amazon SageMaker*:

    ```
    s3_bucket_name = "<insert bucket name here>"
    prefix = "chapter06/input"
    ```

11. Use the `to_csv()` function to save the contents of the training, validation, and test DataFrames to the CSV files. Note that we have set the value of `header` to `True` for this step:

    ```
    training.to_csv('tmp/training_data.csv', header=True,
    index=False)
    validation.to_csv('tmp/validation_data.csv', header=True,
    index=False)
    test.to_csv('tmp/test_data.csv', header=True,
    index=False)
    ```

12. Use the `aws s3 cp` command to copy the CSV files to the S3 destination. After performing this step, you can verify, using the AWS Console S3 UI, if the files were successfully uploaded to the target S3 bucket and path:

```
!aws s3 cp tmp/training_data.csv s3://{s3_bucket_name}/
{prefix}/training_data.csv
!aws s3 cp tmp/validation_data.csv s3://{s3_bucket_name}/
{prefix}/validation_data.csv
!aws s3 cp tmp/test_data.csv s3://{s3_bucket_name}/
{prefix}/test_data.csv
```

13. Use the `to_csv()` function again to generate 3 CSV files from the training, validation, and test DataFrames. This time, note that the filenames are different and that the `header` parameter value is set to `False`:

```
training.to_csv('tmp/training_data_no_header.csv',
header=False, index=False)
validation.to_csv('tmp/validation_data_no_header.csv',
header=False, index=False)
test.to_csv('tmp/test_data_no_header.csv', header=False,
index=False)
```

14. Finally, use the `aws s3 cp` command to copy the CSV files to the S3 destination:

```
!aws s3 cp tmp/training_data_no_header.csv s3://{s3_
bucket_name}/{prefix}/training_data_no_header.csv
!aws s3 cp tmp/validation_data_no_header.csv s3://{s3_
bucket_name}/{prefix}/validation_data_no_header.csv
!aws s3 cp tmp/test_data_no_header.csv s3://{s3_bucket_
name}/{prefix}/test_data_no_header.csv
```

15. Use the `%store` magic to store the values of `s3_bucket_name` and `prefix`:

```
%store s3_bucket_name
%store prefix
```

We will use these values in the succeeding recipes so that we can easily specify and provide the S3 paths to the CSV files we have uploaded in this recipe.

16. Finally, print the S3 path of the CSV file containing the training data:

```
f"s3://{s3_bucket_name}/{prefix}/training_data.csv"
```

Copy the output after running this cell (without the quotes) to your clipboard; we will be using this in the *Creating and monitoring a SageMaker Autopilot experiment in SageMaker Studio (console)* recipe.

Now, let's see how this works!

How it works...

In this recipe, we have generated a synthetic dataset that can be used for classification problems. This recipe involved three major steps – synthetic data generation using scikit-learn and NumPy, data visualization using Matplotlib, and uploading the CSV files containing the datasets to S3.

Let's look at these steps in more detail:

- **Synthetic Data Generation**: We used the make_blobs() function to generate a dataset that has three columns, including a target column. We then generated two columns containing random values and merged them with the synthetic dataset. Later in this chapter, we will see that **SageMaker Autopilot** will still come up with a model that makes use of the values of the first two columns, primarily to predict the target column value.

- **Data Visualization**: Given that columns c and d only contain random values, we simply used the values of the a and b columns, along with the target label column, in the scatterplot chart. What are our initial findings after looking at the chart? When the value of a is near 5 and the value of b is near 0, the target value (label) is most likely 1. This, of course, is an oversimplification and is just one of the remarks we can make for now.

> **Note**
>
> In real-life datasets and examples, a and b can easily map to one or more predictor variables. One example of this would be using the properties of a breast mass image (for example, perimeter_mean, area_mean, and concavity_worst) to detect if the tumor is benign or malignant. In addition to this, c and d can easily map to one or more predictor variables that may not necessarily contribute significantly to the target variable.

- **Uploading the CSV Files**: Finally, we uploaded two versions of the CSV files: one version without the header included and another with the header included. When we run the Autopilot experiment later, we will use the training CSV file with the header included. In some of the other recipes in this chapter, we will use the test dataset without the header included.

There's more...

Let's look at some of the tweaks we can perform on this recipe to come up with variations of this dataset:

- We can generate (significantly) more records and more predictor columns, and then test how these tweaks affect the configuration of the steps we'll perform during the automated machine learning experiment.

- We can also change the machine learning problem type to a **multiclass classification** problem by modifying the parameter values for `centers` and `cluster_std` when calling the `make_blobs()` function.

- We can also change the machine learning problem type to a **regression** problem by replacing the target column and the steps used to generate the target column.

We will most likely see different model families being used instead when running the AutoML jobs. In addition to this, we may see a few differences in the notebooks that are generated by the Autopilot experiments once we have made these tweaks and rerun the experiment. With these in mind, let's make use of this synthetic dataset!

Creating and monitoring a SageMaker Autopilot experiment in SageMaker Studio (console)

In this recipe, we will use **Amazon SageMaker Autopilot** to perform **AutoML** using the synthetic dataset we generated previously. We will simply pass the training dataset CSV file, along with a few configuration parameter values. These should be enough to get our Autopilot experiment running.

With **SageMaker Autopilot**, the different steps of the machine learning process are performed automatically. These include preprocessing, feature engineering, and model tuning. The cool thing here is that even if the process is completely automated, we still have the option to see what's happening behind the scenes using the generated notebooks. This allows seasoned professionals to modify the generated machine learning code when needed.

Getting ready

This recipe continues from any of the recipes that come after the *Generating a synthetic dataset with additional columns containing random values* recipe.

How to do it...

The steps in this recipe focus on using the SageMaker Studio console (UI) to create and monitor a **SageMaker Autopilot** experiment. Let's get started:

1. Create a new Autopilot experiment by clicking the **File** menu and choosing **Experiment** under the **New** submenu:

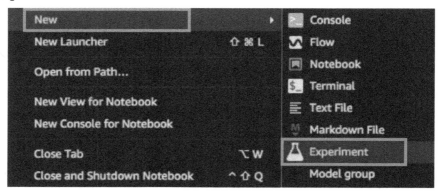

Figure 6.26 – Creating a new experiment from the File menu

Here, we can see that the **New** submenu is under the **File** menu.

2. Specify the **Experiment name** value. Feel free to specify your preferred experiment name. You can set the experiment name so that it follows the convention of `sagemaker-cookbook-autopilot`:

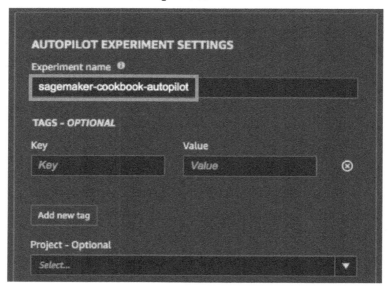

Figure 6.27 – AUTOPILOT EXPERIMENT SETTINGS

Here, we have the **AUTOPILOT EXPERIMENT SETTINGS** form, where we can specify the configuration of our new experiment.

3. Under **Connect your Data**, select **Enter S3 bucket location** and then set the S3 bucket address to the location of the training data. Make sure that you replace the value of `<S3 bucket name>` with the name of the bucket we created in the *Preparing the Amazon S3 bucket and the training dataset for the linear regression experiment* recipe of *Chapter 1, Getting Started with Machine Learning Using Amazon SageMaker*:

   `s3://<S3 bucket name>/chapter06/input/training_data.csv`

   Note that the S3 bucket name that's specified here must be the same as the bucket name that was specified in the *Generating a synthetic dataset with additional columns containing random values* recipe of this chapter.

4. Under **Output data location (S3 bucket)**, select **Enter S3 bucket location**:

   `s3://<s3 bucket name>/chapter06/output`

   Make sure that you replace the value of `<S3 bucket name>` with the name of the bucket we used in the previous step:

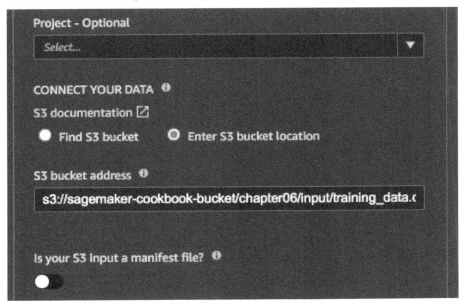

Figure 6.28 – CONNECT YOUR DATA

Here, we have the **CONNECT YOUR DATA** form, with the fields updated with the preferred values.

5. Create and specify `label` in the **Target** field. Make sure that you click **Create "label"** after typing `label`, as shown in the following screenshot:

Figure 6.29 – Specifying the Target field value

This target column should match the `label` column in the dataset we generated in the *Generating a synthetic dataset with additional columns containing random values* recipe.

6. When presented with the option to auto-deploy the model, just leave the setting as **On**.

7. Under **ADVANCED SETTINGS**, specify the following values for the following fields:

-   **Max trial runtime in seconds**: `3600`

-   **Max job runtime in seconds**: `600`

-   **Max candidates**: `12` (as shown in the following screenshot):

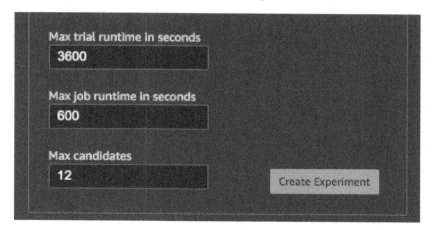

Figure 6.30 – Create Experiment button

This will make sure that the Autopilot experiment will only take about an hour to complete.

8. Finally, click the **Create Experiment** button. Clicking this button will start the Autopilot experiment:

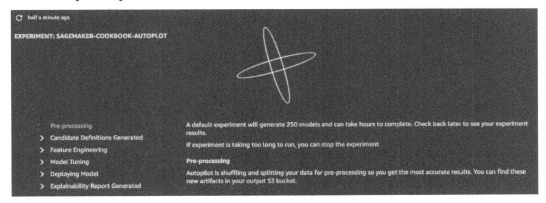

Figure 6.31 – Autopilot experiment has started

The preceding screenshot shows what our screen should look like for the first few minutes after initiating the Autopilot experiment. While waiting, let's proceed so that we can inspect some of the processing jobs that have started and the notebooks that have been generated by the Autopilot experiment.

> **Important note**
> Make sure that you delete the inference endpoint that's generated after the experiment has been completed. You will be charged for the amount of time the inference endpoint has been running for.

9. In the AWS Console, navigate to the **SageMaker Processing Jobs** page to see the processing jobs that were automatically started by the Autopilot experiment:

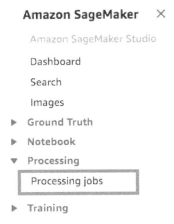

Figure 6.32 – Navigating to the Processing jobs page

This should redirect us to the **Processing jobs** page, where we can see the processing jobs that have been started by the Autopilot experiment:

Figure 6.33 – Processing jobs

Here, we can see the **SageMaker Processing** jobs that have been started by the Autopilot experiment. Feel free to check out the jobs that have been executed, along with the processing images that have been used in these jobs. Once the experiment has finished, we will see a variety of **SageMaker Processing** jobs, including data splitter, pipeline recommender, and model explainability report generator jobs.

10. After about 10 to 15 minutes, the Data Exploration and Candidate Generation Notebooks should have been generated. Once the two buttons shown in the following screenshot appear in the **Autopilot experiment** tab, click the **Open data exploration notebook** button:

Figure 6.34 – Open data exploration notebook button

This should open the **Amazon SageMaker Autopilot Data Exploration** notebook, similar to what is shown in the following screenshot:

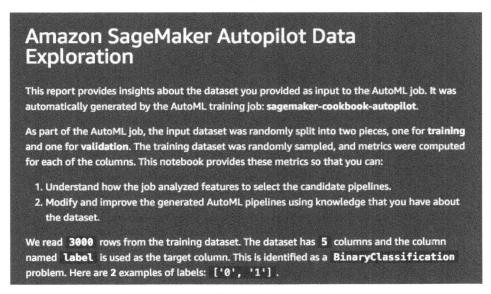

Figure 6.35 – Autopilot Data Exploration Notebook

We will take a closer look at this notebook in the *Inspecting the SageMaker Autopilot experiment results and artifacts* recipe.

11. In the **Autopilot experiment** tab, click the **Open candidate generation notebook** button:

Figure 6.36 – Open candidate generation notebook button

This should open the **Amazon SageMaker Autopilot Candidate Definition Notebook** page, as shown here:

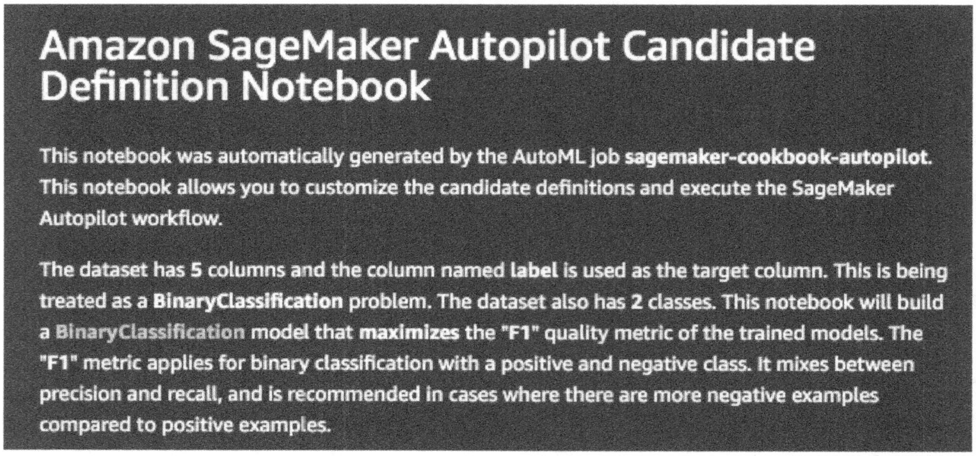

Amazon SageMaker Autopilot Candidate Definition Notebook
This notebook was automatically generated by the AutoML job sagemaker-cookbook-autopilot. This notebook allows you to customize the candidate definitions and execute the SageMaker Autopilot workflow.
The dataset has 5 columns and the column named label is used as the target column. This is being treated as a BinaryClassification problem. The dataset also has 2 classes. This notebook will build a BinaryClassification model that maximizes the "F1" quality metric of the trained models. The "F1" metric applies for binary classification with a positive and negative class. It mixes between precision and recall, and is recommended in cases where there are more negative examples compared to positive examples.

Figure 6.37 – Autopilot Candidate Definition Notebook

We will take a closer look at this notebook in the *Inspecting the SageMaker Autopilot experiment results and artifacts* recipe.

12. After a few more minutes, the best model will be deployed. This should be reflected in the quick progress report, as shown in the following screenshot:

less than 5 seconds ago
EXPERIMENT: SAGEMAKER-COOKBOOK-AUTOPLOT
Problem type: BinaryClassification
Open candidate generation notebook
Open data exploration notebook
Pre-processing
Candidate Definitions Generated
Feature Engineering
Model Tuning
Deploying Model
Explainability Report Generated
A default experiment will generate 250 models and can take hours to complete. Check back later to see your experiment results.
If experiment is taking too long to run, you can stop the experiment
Explainability Report Generated
Autopilot is generating feature importance for the best model.

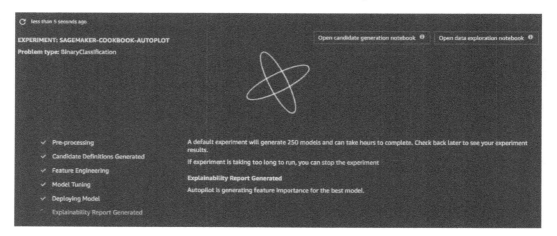

Figure 6.38 – Model has been deployed

At this point, the explainability report is being generated. We will see what this looks like and what information this provides later. We should also see a **SageMaker Processing** job running in the SageMaker console that is generating this report.

13. After a few more minutes, the Autopilot experiment should be completed and we will be able to inspect the results. Right-click on the row of the "best" tuning job trial and choose **Open in model details** from the list of options in the context menu, as shown here:

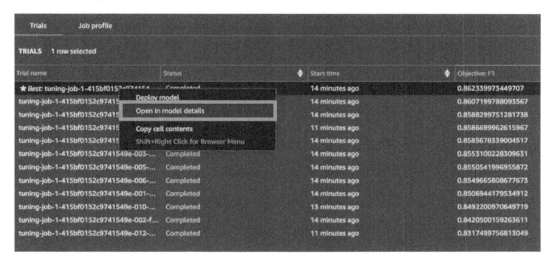

Figure 6.39 – Inspecting the results of the best tuning job trial

Here, we can see that the target objective metric (F1 score) value is approximately `0.86234` for the best model produced by the Autopilot experiment. This objective metric value corresponds to the `validation:f1 metric` value that's computed automatically while the training jobs are running. We should see the details of the model, including an excerpt from the model explainability report, as shown in the following screenshot:

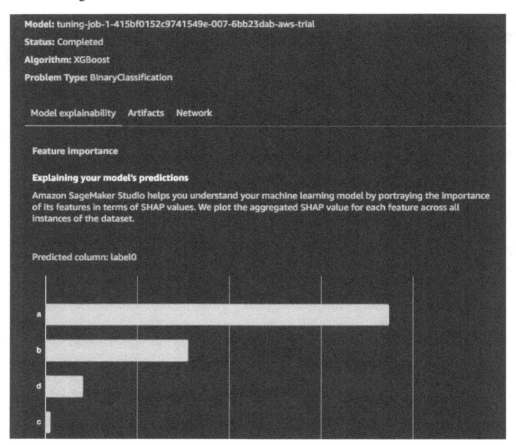

Figure 6.40 – Model explainability report

Here, we can see that columns a and b are considered to be more important than columns c and d. This is expected as columns c and d only contain random values that are not expected to contribute to the predicted label value. We can also see that **XGBoost** is the algorithm that is being used to produce the best model. Also, SageMaker Autopilot automatically detected that we are dealing with a **binary classification problem**, without us having to specify it when initiating the Autopilot experiment.

> **Note**
>
> We will take a closer look at model explainability and feature importance in the *Enabling ML explainability with SageMaker Clarify* recipe of *Chapter 7, Working with SageMaker Feature Store, SageMaker Clarify, and SageMaker Model Monitor.*

14. Finally, scroll down to the bottom of the page to see the **Metrics** and **Parameters** properties of the best model:

Metrics

Name	Minimum	Maximum	Standard Deviation	Average	Final value
ObjectiveMetric	0	0	0	0	0.862339973449707
train:f1	0	0	0	0	0.8763099908828735
validation:error	0	0	0	0	0.13732999563217163
validation:f1	0	0	0	0	0.862339973449707
train:error	0	0	0	0	0.12342000007629395

Figure 6.41 – Metrics

Feel free to check out the other details presented on the **Model Details** page. This includes the input and output artifacts that were used and produced in the experiment, along with the network configuration settings.

> **Important note**
>
> Make sure that you delete the endpoint after running this experiment. Navigate to the **Endpoints** page in the SageMaker console and delete the endpoint that was created during the experiment.

Let's see how this works!

How it works...

In this recipe, we used **SageMaker Autopilot** to process the dataset we generated previously, and then automatically performed the different stages of the machine learning process to come up with the best model.

SageMaker Autopilot automates the following stages for us:

- **Data Analysis**: Different statistics are computed and collected to help machine learning practitioners understand the data better.

- **Problem Definition**: The problem type (for example, linear regression, binary classification, or multiclass classification) is determined based on the target column in the provided dataset CSV file.

- **Dataset Schema Detection**: The data type of each of the columns/variables in the provided dataset CSV file is determined.

- **Candidate Definitions Generation**: Multiple candidate pipeline definitions are generated. Each candidate pipeline is composed of a data processing job and a training job.

- **Data Preprocessing and Feature Engineering**: The data is processed and prepared before the training step.

- **Algorithm Selection**: Based on the problem type, the algorithms to be used for training and tuning will be selected. Currently, SageMaker Autopilot supports the usage of **Linear Learner**, **XGBoost**, and a **multilayer perceptron (MLP)** deep learning algorithm when training and tuning models. Note that Autopilot selects the algorithm to be used for the training step.

- **Model Tuning**: Different models are trained and tuned. In this stage, the **Automatic Model Tuning** capability of SageMaker is used to find the best hyperparameters of the target model(s), which would result in the best objective metric being used.

- **Deployment**: The best model is deployed to an inference endpoint. The best model is selected based on the value of the objective metric (for example, the F1 score).

- **Explainability Report Generation**: The model explainability report is generated for the best model.

When working with automated machine learning solutions and capabilities such as **Autopilot** and **Automatic Model Tuning**, it is important to note that generally, the higher the number of training and tuning jobs used to produce models, the higher the chance that we will get a high(er) quality model. Of course, this means that we will need to wait longer as we will have to wait for each of the training jobs to complete. That said, it is about finding the right set of configuration parameters (**Max trial runtime in seconds**, **Max job runtime in seconds**, and **Max candidates**) when initializing the Autopilot experiment.

Note

If this is your first time working with an AutoML solution, you might be wondering if this will replace data scientists since we can technically perform end-to-end machine learning experiments and deployments without having to know anything about machine learning. The answer to this would be "not anytime soon" as we should think of AutoML solutions as tools and services that help assist and speed up the work of data scientists and machine learning practitioners. AutoML allows them to focus more on the complex and custom requirements of the projects by modifying the generated scripts and notebooks of the Autopilot jobs.

Now, let's perform a similar AutoML experiment programmatically.

Creating and monitoring a SageMaker Autopilot experiment using the SageMaker Python SDK

In the previous recipe, we created and monitored a **SageMaker Autopilot** experiment using the **SageMaker Studio** interface. In this recipe, we will use the **SageMaker Python SDK** to programmatically create and monitor a similar AutoML experiment. Using the **SageMaker Python SDK**, we will be able to get the properties of the **Autopilot** experiment, such as the primary and secondary status of the Autopilot job. Once certain stages of the experiment have been completed, we will also get the S3 paths where the different artifacts and files are stored. We will also use the `best_candidate()` function to inspect the properties of the "best candidate" after running the hyperparameter optimization jobs.

Getting ready

This recipe continues from any of the recipes after the *Generating a synthetic dataset with additional columns containing random values* recipe.

How to do it...

The steps in this recipe focus on using the **SageMaker Python SDK** to create and monitor a **SageMaker Autopilot** experiment. Let's get started:

1. Navigate to the `my-experiments/chapter06` directory inside **SageMaker Studio**. Feel free to create this directory if it does not exist yet:

Figure 6.42 – The my-experiments/chapter06 directory

Here, we have the `my-experiments/chapter06` directory. Double-click the `chapter06` directory before proceeding.

2. Create a new notebook using the `Python 3 (Data Science)` kernel inside the `my-experiments/chapter06` directory and rename it with the name of this recipe (that is, `Creating and monitoring a SageMaker Autopilot experiment using the SageMaker Python SDK`):

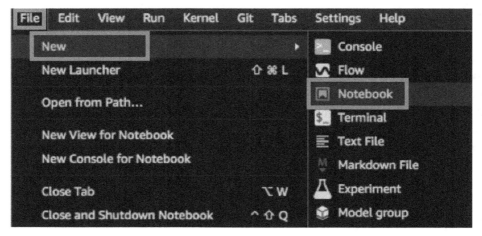

Figure 6.43 – Creating a new notebook

Here, we can see how to create a new **Notebook** from the **File** menu. When prompted for the kernel to use, choose **Python 3 (Data Science)**.

3. Use the `%store` magic to load the values of `s3_bucket_name` and `prefix`:

```
%store -r s3_bucket_name
%store -r prefix
```

Remember that we used the `%store` magic to save these variable values in the *Generating a synthetic dataset with additional columns containing random values* recipe.

4. Specify the input path, the output path, and the max number of candidates:

```
s3_data_source = \
f"s3://{s3_bucket_name}/{prefix}/training_data.csv"
output_target = f"s3://{s3_bucket_name}/{prefix}/output"
max_candidates = 25
```

5. Initialize a few prerequisites, such as the session and the role:

```
import sagemaker
session = sagemaker.Session()
role = sagemaker.get_execution_role()
```

6. Initialize the Autopilot experiment by running the following block of code:

```
from sagemaker.automl.automl import AutoML
experiment = AutoML(
    role=role,
    sagemaker_session=session,
    target_attribute_name="label",
    output_path=output_target,
    max_candidates=25,
    max_runtime_per_training_job_in_seconds=1000,
    total_job_runtime_in_seconds=6000
)
```

Note that the parameter values for `total_job_runtime_in_seconds`, `max_candidates`, and `max_runtime_per_training_job_in_seconds` dictate how long the **Autopilot** experiment will last. Once one of these values has been reached, the Autopilot experiment will end as well.

7. Use the `fit()` function to start the Autopilot job:

```
experiment.fit(inputs=s3_data_source, logs=False,
wait=False)
```

Here, we specified `wait=False` when calling the `fit()` function so that we can proceed with the next set of steps.

8. Import the `pprint` and `sleep` functions:

```
from pprint import pprint
from time import sleep
```

Feel free to use the `pprint()` function to inspect the response objects and see what other properties and attributes are available.

9. Use the following lines of code to check the status and wait for the experiment to complete:

```
%%time
status = "InProgress"

while status == "InProgress":
    response = experiment.describe_auto_ml_job()
    status = response['AutoMLJobStatus']
    secondary_status =
response['AutoMLJobSecondaryStatus']
    print(f"{status} - {secondary_status}")
    sleep(15)
```

This should give us a set of logs similar to what is shown in the following screenshot:

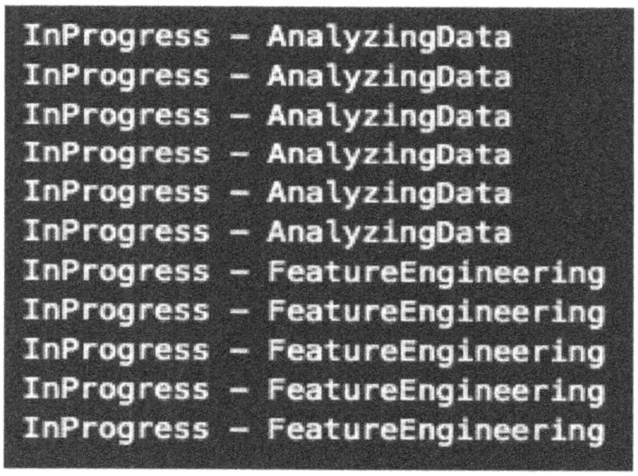

```
InProgress - AnalyzingData
InProgress - AnalyzingData
InProgress - AnalyzingData
InProgress - AnalyzingData
InProgress - AnalyzingData
InProgress - AnalyzingData
InProgress - FeatureEngineering
InProgress - FeatureEngineering
InProgress - FeatureEngineering
InProgress - FeatureEngineering
InProgress - FeatureEngineering
```

Figure 6.44 – Logs showing the primary and secondary statuses of the Autopilot job

As this may take about 40 minutes to complete, feel free to proceed with the next steps while waiting. Here, we should see the primary and secondary status text change in the following sequence: InProgress - Starting, InProgress - AnalyzingData, InProgress - FeatureEngineering, InProgress - ModelTuning, InProgress - GeneratingExplainabilityReport, and Completed - Completed.

10. Navigate to the **Components and registries** pane by clicking the last icon in the left sidebar, as shown here:

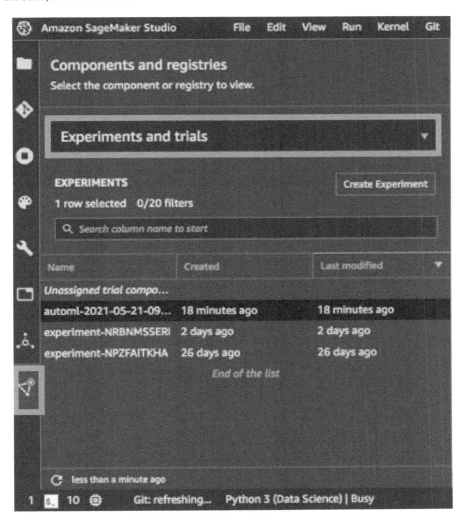

Figure 6.45 – Locating the Autopilot experiment

From the list of options in the dropdown, choose **Experiments and trials** to show all the experiments.

11. Right-click on the Autopilot experiment that we initiated using the **SageMaker Python SDK**:

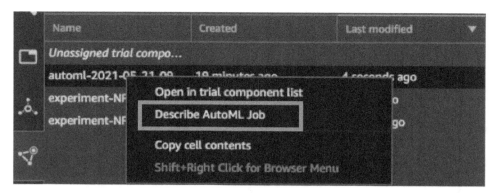

Figure 6.46 Describe AutoML Job

After that, choose **Describe AutoML Job** from the list of options in the context menu:

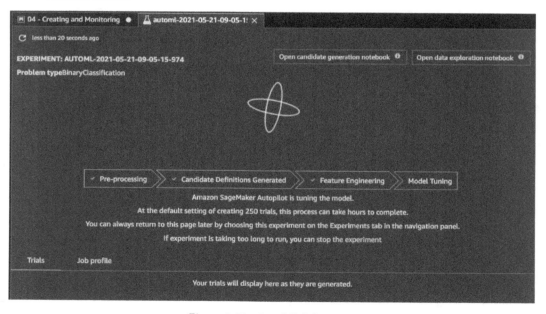

Figure 6.47 – AutoML Job screen

We should see a view similar to the one shown in the preceding screenshot. This screen should reflect the same status that we had when running the while loop previously.

12. Once the Autopilot job has finished running, we can compute the total by running the following block of code:

```
delta = response['EndTime'] - response['CreationTime']
total_minutes = int(delta.total_seconds() / 60)
total_minutes
```

This should yield a value close to 35.

13. Let's check the S3 location of the generated output artifacts:

```
artifacts = response['AutoMLJobArtifacts']
artifacts
```

This should give us a dictionary similar to the following:

{'CandidateDefinitionNotebookLocation': 's3://sagemaker-cookbook-bucket/chapter06/input/output/automl-2021-05-21-11-24-23-889/sagemaker-automl-candidates/automl-2021-05-21-11-24-23-889-pr-1-87a141a5b40e4650abd19edc6f3/notebooks/SageMakerAutopilotCandidateDefinitionNotebook.ipynb', 'DataExplorationNotebookLocation': 's3://sagemaker-cookbook-bucket/chapter06/input/output/automl-2021-05-21-11-24-23-889/sagemaker-automl-candidates/automl-2021-05-21-11-24-23-889-pr-1-87a141a5b40e4650abd19edc6f3/notebooks/SageMakerAutopilotDataExplorationNotebook.ipynb'}

Figure 6.48 – Dictionary containing the locations of the artifacts that were generated

Here, we have the S3 paths of the notebooks generated by **SageMaker Autopilot**.

14. Next, copy the generated artifacts using the following lines of code:

```
for s3_path in list(artifacts.values()):
    !aws s3 cp {s3_path} tmp/.
```

15. Get the best candidate using the best_candidate() function:

```
best = experiment.best_candidate()
best
```

This should give us a dictionary with a structure similar to the following:

```
{'CandidateName': 'tuning-job-1-b98ef1f7b388435894-014-34973fd5'
 'FinalAutoMLJobObjectiveMetric': {'MetricName': 'validation:f1'
 'Value': 0.8728799819946289},
 'ObjectiveStatus': 'Succeeded',
 'CandidateSteps': [{'CandidateStepType': 'AWS::SageMaker::ProcessingJob',
  'CandidateStepArn': 'arn:aws:sagemaker:us-east-1:581320662326:processing-job/db-1-8b89b
3b8f06d4191a92d0ffc866b2e9a36b12c8ef7b64db1aac22abb19',
  'CandidateStepName': 'db-1-8b89b3b8f06d4191a92d0ffc866b2e9a36b12c8ef7b64db1aac22abb1
9'},
  {'CandidateStepType': 'AWS::SageMaker::TrainingJob',
   'CandidateStepArn': 'arn:aws:sagemaker:us-east-1:581320662326:training-job/automl-202-d
pp3-1-cdad32b7e086418fa8ed8322c63e1ed383986dc609444',
   'CandidateStepName': 'automl-202-dpp3-1-cdad32b7e086418fa8ed8322c63e1ed383986dc60944
4'},
  {'CandidateStepType': 'AWS::SageMaker::TransformJob',
   'CandidateStepArn': 'arn:aws:sagemaker:us-east-1:581320662326:transform-job/automl-202-
dpp3-rpb-1-9fdf994cd19046e2b14e9547c44f23304fa8b06c9',
   'CandidateStepName': 'automl-202-dpp3-rpb-1-9fdf994cd19046e2b14e9547c44f23304fa8b06c
9'},
  {'CandidateStepType': 'AWS::SageMaker::TrainingJob',
   'CandidateStepArn': 'arn:aws:sagemaker:us-east-1:581320662326:training-job/tuning-job-1
-b98ef1f7b388435894-014-34973fd5',
```

Figure 6.49 – Dictionary containing the metadata of the "best candidate"

Here, we can see the results of the `best_candidate()` function.

16. Check the objective metric value of the best candidate:

```
best['FinalAutoMLJobObjectiveMetric']
```

This should give us a dictionary of values similar to `{'MetricName': 'validation:f1', 'Value': 0.87288}`.

17. Use the `%store` magic to save the name of the Autopilot job:

```
autopilot_job_name = response['AutoMLJobName']
%store autopilot_job_name
autopilot_job_name
```

This should allow us to inspect the Autopilot experiment details later.

> **Note**
>
> If we want to deploy the best model of the Autopilot experiment, performing the deployment is as simple as using `autopilot_job.deploy(...)`. This `deploy()` function should have the same parameters as the `deploy()` function of an `Estimator` or `Model` object – `endpoint_name`, `instance_type`, and `initial_instance_count`.

Now, let's see how this works!

How it works...

In this recipe, we used the `AutoML` class from the **SageMaker Python SDK** to initialize the **SageMaker Autopilot** experiment object. As we can see, using the **SageMaker Python SDK** to run Autopilot jobs is straightforward. We simply pass a set of configuration parameters, similar to what we had in the *Creating and monitoring a SageMaker Autopilot experiment in SageMaker Studio (console)* recipe.

The primary statements we have used in this recipe to get things done involve doing the following:

1. Initialize the `AutoML` object and specify the properties of the Autopilot job:

    ```
    experiment = AutoML(...)
    ```

2. Start the experiment using the `fit()` function:

    ```
    experiment.fit(..., wait=False)
    ```

3. Inspect the progress of the AutoML job using the `describe_auto_ml_job()` function:

    ```
    experiment.describe_auto_ml_job()
    ```

The AutoML job may take some time to complete (for example, > 30 minutes). The `describe_auto_ml_job()` function would be handy in letting us know the status of the AutoML job.

There's more...

Note that we can use the `ExperimentAnalytics` class from the **SageMaker Python SDK** to help us analyze the results of the Autopilot experiment. The following is a quick example of how to obtain a `DataFrame` with the details of the processing and training jobs:

```
%store -r autopilot_job_name
experiment_name = f"{autopilot_job_name}-aws-auto-ml-job"
from sagemaker.analytics import ExperimentAnalytics
from sagemaker import Session
```

```
analytics = ExperimentAnalytics(
    sagemaker_session=Session(),
    experiment_name=experiment_name
)
analytics.dataframe()
```

This should give us a `DataFrame` with values similar to the following:

	TrialComponentName	DisplayName	SourceArn	SageMaker.ImageUri	SageMaker.InstanceCount	SageMaker.InstanceType
0	tuning-job-1-da61eaf8193b4bbba9-018-620fe364-a...	tuning-job-1-da61eaf8193b4bbba9-018-620fe364-a...	arn:aws:sagemaker:us-east-1:581320662326:train...	382416733822.dkr.ecr.us-east-1.amazonaws.com/m...	1.0	ml.m5.4xlarge
1	tuning-job-1-da61eaf8193b4bbba9-025-deeda273-a...	tuning-job-1-da61eaf8193b4bbba9-025-deeda273-a...	arn:aws:sagemaker:us-east-1:581320662326:train...	683313688378.dkr.ecr.us-east-1.amazonaws.com/s...	1.0	ml.m5.4xlarge
2	tuning-job-1-da61eaf8193b4bbba9-023-1642c50d-a...	tuning-job-1-da61eaf8193b4bbba9-023-1642c50d-a...	arn:aws:sagemaker:us-east-1:581320662326:train...	382416733822.dkr.ecr.us-east-1.amazonaws.com/L...	1.0	ml.m5.4xlarge
3	tuning-job-1-da61eaf8193b4bbba9-017-1acc82b0-a...	tuning-job-1-da61eaf8193b4bbba9-017-1acc82b0-a...	arn:aws:sagemaker:us-east-1:581320662326:train...	382416733822.dkr.ecr.us-east-1.amazonaws.com/m...	1.0	ml.m5.4xlarge
4	tuning-job-1-da61eaf8193b4bbba9-024-573a8576-a...	tuning-job-1-da61eaf8193b4bbba9-024-573a8576-a...	arn:aws:sagemaker:us-east-1:581320662326:train...	683313688378.dkr.ecr.us-east-1.amazonaws.com/s...	1.0	ml.m5.4xlarge
5	tuning-job-1-da61eaf8193b4bbba9-016-64ef1329-a...	tuning-job-1-da61eaf8193b4bbba9-016-64ef1329-a...	arn:aws:sagemaker:us-east-1:581320662326:train...	382416733822.dkr.ecr.us-east-1.amazonaws.com/m...	1.0	ml.m5.4xlarge
6	tuning-job-1-da61eaf8193b4bbba9-021-48684286-a...	tuning-job-1-da61eaf8193b4bbba9-021-48684286-a...	arn:aws:sagemaker:us-east-1:581320662326:train...	382416733822.dkr.ecr.us-east-1.amazonaws.com/L...	1.0	ml.m5.4xlarge

Figure 6.50 – DataFrame containing the experiment analytics data

For more information on this topic, feel free to check out the *Experiment analytics with SageMaker experiments* recipe of *Chapter 5, Effectively Managing Machine Learning Experiments*.

Inspecting the SageMaker Autopilot experiment's results and artifacts

In the previous recipe, we used the **SageMaker Python SDK** to launch and monitor an **Autopilot** job, as well as to deploy the best model once the AutoML job has finished running.

In this recipe, we will inspect the notebooks that were generated by a SageMaker Autopilot experiment:

- Data Exploration Notebook
- Candidate Definition Notebook

SageMaker Autopilot has generated these notebooks to help us understand what is happening inside the AutoML job. These notebooks allow data scientists and machine learning practitioners to build on top of the Autopilot experiment by modifying and customizing parts of these notebooks as they see fit.

Finally, we will take a quick look at what is stored in the S3 output path, now that the Autopilot job has finished executing.

Getting ready

This recipe continues from the *Creating and monitoring a SageMaker Autopilot experiment using the SageMaker Python SDK* recipe.

How to do it...

The next couple of steps assume that we have just completed the *Creating and monitoring a SageMaker Autopilot experiment using the SageMaker Python SDK* recipe and that the AutoML job artifacts have been copied to the tmp directory. Let's get started:

1. Inside the tmp directory, we will find two files called SageMakerAutopilotCandidateDefinitionNotebook and SageMakerAutopilotDataExplorationNotebook. Open the tmp directory. You will see the downloaded CSV files and the generated Autopilot notebooks, as shown here:

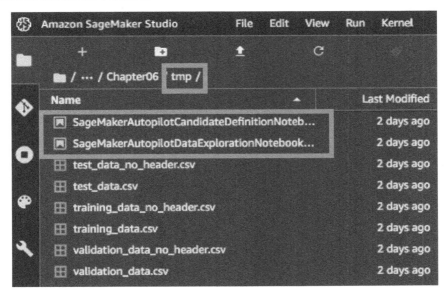

Figure 6.51 – Autopilot generated notebooks inside the tmp directory

2.   Open the Data Exploration Notebook:

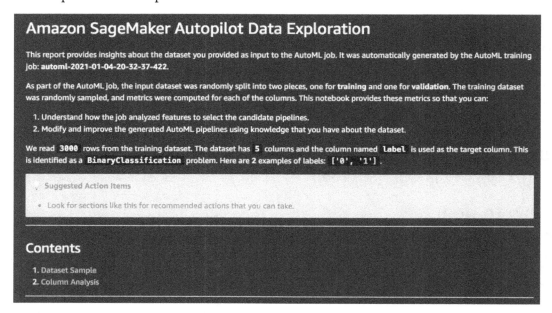

Figure 6.52 – Generated Data Exploration Notebook

Here, we can see the Data Exploration Notebook. Note that the cells in this notebook can be run like any normal notebook cells, which allows us to inspect the results after each step.

3.   Open the Candidate Definition Notebook:

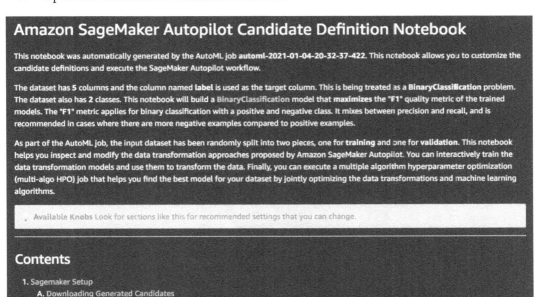

Figure 6.53 – Candidate Definition Notebook

Here, we have the Candidate Definition Notebook. Note that the cells in this notebook can be executed like any normal notebook cells, which allows us to inspect the results after each step.

4. Scroll down and read through the sections and steps in the notebook:

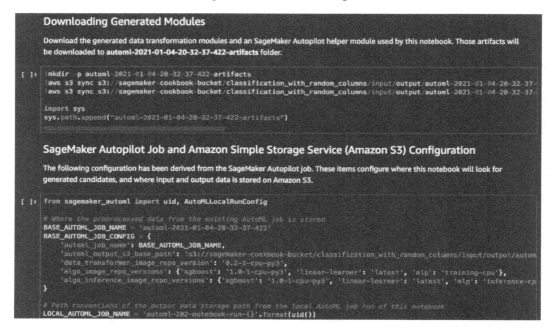

Figure 6.54 – Other steps inside Candidate Definition Notebook

Here, we can see the other steps in the Candidate Definition Notebook. The Candidate Definition Notebook is generally longer than the Data Exploration Notebook as it contains the different preprocessing steps and machine learning selection steps.

5. Scroll further down and check out the algorithms and hyperparameter range configuration. This will be used in the automated hyperparameter tuning jobs:

```python
from sagemaker.parameter import CategoricalParameter, ContinuousParameter, IntegerParameter

ALGORITHM_TUNABLE_HYPERPARAMETER_RANGES = {
    'xgboost': {
        'num_round': IntegerParameter(2, 1024, scaling_type='Logarithmic'),
        'max_depth': IntegerParameter(2, 8, scaling_type='Logarithmic'),
        'eta': ContinuousParameter(1e-3, 1.0, scaling_type='Logarithmic'),
        'gamma': ContinuousParameter(1e-6, 64.0, scaling_type='Logarithmic'),
        'min_child_weight': ContinuousParameter(1e-6, 32.0, scaling_type='Logarithmic'),
        'subsample': ContinuousParameter(0.5, 1.0, scaling_type='Linear'),
        'colsample_bytree': ContinuousParameter(0.3, 1.0, scaling_type='Linear'),
        'lambda': ContinuousParameter(1e-6, 2.0, scaling_type='Logarithmic'),
        'alpha': ContinuousParameter(1e-6, 2.0, scaling_type='Logarithmic'),
    },
    'linear-learner': {
        'wd': ContinuousParameter(1e-7, 1.0, scaling_type='Logarithmic'),
        'l1': ContinuousParameter(1e-7, 1.0, scaling_type='Logarithmic'),
        'learning_rate': ContinuousParameter(1e-5, 1.0, scaling_type='Logarithmic'),
    },
    'mlp': {
        'mini_batch_size': IntegerParameter(128, 512, scaling_type='Linear'),
        'learning_rate': ContinuousParameter(1e-6, 1e-2, scaling_type='Logarithmic'),
        'weight_decay': ContinuousParameter(1e-12, 1e-2, scaling_type='Logarithmic'),
        'dropout_prob': ContinuousParameter(0.25, 0.5, scaling_type='Linear'),
        'embedding_size_factor': ContinuousParameter(0.65, 0.95, scaling_type='Linear'),
        'network_type': CategoricalParameter(['feedforward', 'widedeep']),
        'layers': CategoricalParameter(['256', '50, 25', '100, 50', '200, 100', '256, 128', '300, 150', '200, 100, 50'])
    },
}
```

Figure 6.55 – Defined hyperparameter ranges inside Candidate Definition Notebook

Here, we can see that the defined hyperparameter ranges are inside the Candidate Definition Notebook.

6. In the AWS Console, use the search bar to navigate to the S3 console:

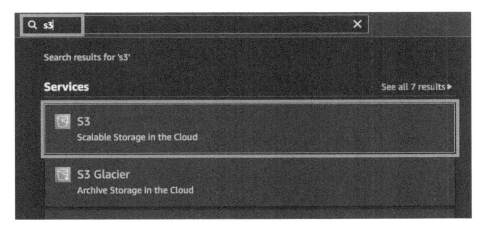

Figure 6.56 – Navigating to the S3 console

Type s3 in the search bar and select **S3 (Scalable Storage in the Cloud)**.

7. Navigate to the S3 bucket containing the output artifacts (that is, `<bucket name>/chapter06/output/...>`):

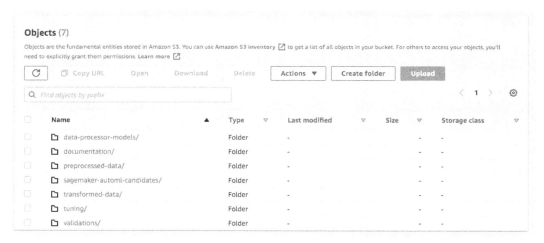

Objects (7)

Objects are the fundamental entities stored in Amazon S3. You can use Amazon S3 inventory to get a list of all objects in your bucket. For others to access your objects, you'll need to explicitly grant them permissions. Learn more

	Name	Type	Last modified	Size	Storage class
	data-processor-models/	Folder	-	-	-
	documentation/	Folder	-	-	-
	preprocessed-data/	Folder	-	-	-
	sagemaker-automl-candidates/	Folder	-	-	-
	transformed-data/	Folder	-	-	-
	tuning/	Folder	-	-	-
	validations/	Folder	-	-	-

Figure 6.57 – S3 output path containing the output artifacts produced by the Autopilot job

Here, we should find a couple of folders containing the different artifacts produced by the Autopilot job. These include `data-processor-models`, `documentation`, `preprocessed-data`, `sagemaker-automl-candidates`, `transformed-data`, `tuning`, and `validation`.

Now, let's see how this works!

How it works...

In this recipe, we took a closer look at the notebooks that were automatically generated in the Autopilot experiment. These generated notebooks help make the process as transparent as possible.

Let's discuss the contents of these notebooks in detail:

- **Data Exploration Notebook**: This notebook helps machine learning practitioners identify problems in the dataset. It is generated during the first phase of the AutoML job (that is, the analysis phase). It has two main sections:

  - **Dataset Sample**: This section shows a small sample of the dataset.

  - **Column Analysis**: This section shows the percentage of missing values, count statistics, and descriptive statistics.

- **Candidate Definition Notebook**: This notebook contains the suggested machine learning algorithm, along with the hyperparameter ranges. It also contains the suggested preprocessing step(s) before training begins. It has four main sections:

  - **SageMaker Setup**: This section focuses on downloading the generated modules and configuring the Autopilot job.

  - **Candidate Pipelines**: This section focuses on creating an `AutoMLInteractiveRunner` object that automates the steps for executing the feature engineering and tuning steps.

  - **Executing the Candidate Pipelines**: This section focuses on executing the feature transformation and multi-algorithm hyperparameter tuning steps for each candidate pipeline.

  - **Model Selection and Deployment**: This section focuses on selecting and deploying the best model, along with its corresponding feature engineering models, using the `PipelineModel` class from the **SageMaker Python SDK**.

In this recipe, we also took a quick look at the S3 bucket containing the artifacts that were generated and produced by the Autopilot job. *Figure 6.59* shows a couple of folders (prefixes) containing the different output artifacts:

- `data-processor-models/`: This folder contains the data processing models.

- `documentation/`: This folder contains the model explainability report.

- `preprocessed-data/`: This folder contains the preprocessed data, including the training and validation dataset split into chunks of CSV files (for example, `chunk_20.csv` and `chunk_21.csv`).

- `sagemaker-automl-candidates/`: This folder contains the generated notebooks (that is, `SageMakerAutopilotCandidateDefinitionNotebook. ipynb` and `SageMakerAutopilotDataExplorationNotebook.ipynb`).

- `transformed-data/`: This folder contains the corresponding transformed versions of the CSV files from `preprocessed-data/`.

- `tuning/`: This folder contains the model artifacts inside the `model.tar.gz` files that were generated by the tuning jobs.

- `validations/`: This folder contains the `output_validation` file.

For more information on this topic, feel free to check out the following link: `https:// docs.aws.amazon.com/sagemaker/latest/dg/autopilot-automate- model-development.html`.

Performing Automatic Model Tuning with the SageMaker XGBoost built-in algorithm

Hyperparameters are the properties of a machine learning algorithm that influence how the algorithm works and behaves. These properties are not learned and modified by the algorithm during the training step, and it is this key characteristic that makes it different from parameters. Hyperparameters must be specified before a training job starts while the parameters of a model are obtained when processing the training data during the training step. **Hyperparameter optimization** is the process of looking for the best configuration and combination of hyperparameter values that produce the best model.

That said, **Automatic Model Tuning** runs multiple training jobs with different hyperparameter configurations to look for the "best" version of a model.

> **Note**
> In this case, the best model is the model that yields the best objective metric. This objective metric depends on the problem being solved.

In this recipe, we will use the **Automatic Model Tuning** capability of **Amazon SageMaker** to look for the "best" **XGBoost** binary classifier model by searching for the optimal set of hyperparameter values. We will use the synthetic dataset we generated earlier in this chapter as the input dataset for the `fit()` function.

Getting ready

This recipe continues from the *Generating a synthetic dataset with additional columns containing random values* recipe.

How to do it...

The first set of steps in this recipe focus on preparing the prerequisites before running the **Automatic Model Tuning** job. Let's get started:

1. Navigate to the `my-experiments/chapter06` directory inside **SageMaker Studio**. Feel free to create this directory if it does not exist yet:

Figure 6.58 – The my-experiments/chapter06 directory

Here, we have the `my-experiments/chapter06` directory. Double-click the `chapter06` directory before proceeding

2. Create a new notebook using the `Python 3 (Data Science)` kernel inside the `my-experiments/chapter06` directory and rename it to the name of this recipe (that is, `Performing Automatic Model Tuning with the SageMaker XGBoost built-in algorithm`):

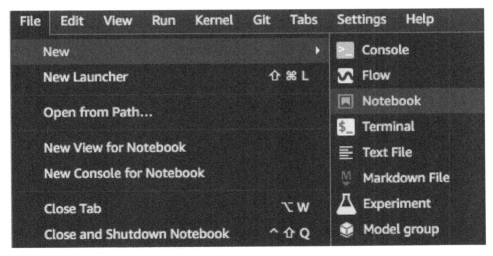

Figure 6.59 – Creating a new notebook

Here, we can see how to create a new **Notebook** from the **File** menu. When prompted for the kernel to use, choose **Python 3 (Data Science)**.

3. Import, load, and prepare a few prerequisites for the **Automatic Model Tuning** job. Similar to the previous recipes, some of the prerequisites include our `region`, boto3 `client`, SageMaker `session`, and the SageMaker execution `role`. We will also import `tuner` from the **SageMaker Python SDK** in the following block of code:

```
import sagemaker
import boto3
import numpy as np
import pandas as pd
import os
from sagemaker import tuner

region = boto3.Session().region_name
client = boto3.Session().client('sagemaker')
session = sagemaker.Session()
role = sagemaker.get_execution_role()
```

4. Use the `%store` magic to load the variable values for `s3_bucket_name` and `prefix`:

```
%store -r s3_bucket_name
%store -r prefix
```

5. Prepare the variables that will point to the S3 input and output locations to be used in this recipe:

```
training_s3_input_location = \
f"s3://{s3_bucket_name}/{prefix}/training_data_no_header.
csv"
validation_s3_input_location = \
f"s3://{s3_bucket_name}/{prefix}/validation_data_no_
header.csv"
test_s3_input_location = \
f"s3://{s3_bucket_name}/{prefix}/test_data_no_header.csv"
training_s3_output_location = \
f"s3://{s3_bucket_name}/output/"
```

6. Use the `retrieve()` function from the **SageMaker Python SDK** to get the container URI of the specified version of the **XGBoost** built-in algorithm:

```
from sagemaker.image_uris import retrieve
container = retrieve('xgboost', region, version="0.90-2")
container
```

7. Use `TrainingInput` to specify the content type of the input data as `text/csv`:

```
from sagemaker.inputs import TrainingInput
s3_input_training = TrainingInput(
    training_s3_input_location,
    content_type="text/csv")
s3_input_validation = TrainingInput(
    validation_s3_input_location,
    content_type="text/csv")
```

8. Initialize the `Estimator` object:

```
estimator = sagemaker.estimator.Estimator(
            container,
            role,
            instance_count=1,
            instance_type='ml.m5.large',
            output_path=training_s3_output_location,
            sagemaker_session=session)
```

9. Specify a few hyperparameters that we will keep constant through the different training jobs:

```
estimator.set_hyperparameters(
    eval_metric='auc',
    objective='binary:logistic',
    num_round=50)
```

10. Specify the hyperparameter range configuration for `eta`, `min_child_weight`, and `max_depth`. As you can see, we are using the `ContinuousParameter` and `IntegerParameter` classes to specify the configuration of the continuous and discrete hyperparameters, respectively:

```
hyperparameter_ranges = {
    'eta': tuner.ContinuousParameter(0, 1),
    'min_child_weight': tuner.ContinuousParameter(3, 7),
    'max_depth': tuner.IntegerParameter(2, 8)
}
```

> **Note**
> Take note that this hyperparameter configuration is just an example. Feel free to modify this `hyperparameter_ranges` configuration as you see fit.

11. Specify the value of `objective_metric_name`:

```
objective_metric_name = 'validation:auc'
```

In this recipe, we are going to specify the **area under the ROC curve (AUC)**. Generally, the higher the AUC metric value of a model, the better it performs.

12. Initialize the `HyperparameterTuner` object. We specified the estimator, objective metric name, and the hyperparameter range configuration in the previous steps as three of the arguments for initializing the `HyperparameterTuner` object:

```
hyperparameter_tuner = tuner.HyperparameterTuner(
    estimator,
    objective_metric_name,
    hyperparameter_ranges,
    max_jobs=20,
    max_parallel_jobs=3)
```

13. Use the `fit()` function to start the hyperparameter tuning job. Note that we have set the `wait` value to `False` so that we can proceed with the next steps while the tuning job is running:

```
hyperparameter_tuner.fit(
    {'train': s3_input_training,
     'validation': s3_input_validation},
```

```
        include_cls_metadata=False,
        wait=False
    )
```

14. Use the `describe_hyper_parameter_tuning_job()` function to check the status of the tuning job:

```
job_name = hyperparameter_tuner.latest_tuning_job.job_
name
response = client.describe_hyper_parameter_tuning_job(
    HyperParameterTuningJobName=job_name
)
response['HyperParameterTuningJobStatus']
```

Once the tuning job has finished running, we should get a value of `Completed`. We should get an `InProgress` value while the tuning job is running.

> **Important note**
> Feel free to use the SageMaker console to check the status and process of the training jobs that were launched by the hyperparameter tuning job. If there are errors when running these training jobs, we will be able to see the details and troubleshoot the issue in **CloudWatch Logs**.

15. Use `pprint()` to display the content of `response`:

```
from pprint import pprint
pprint(response)
```

This should give us a nested dictionary of values, similar to what is shown in the following screenshot:

```
{'CreationTime': datetime.datetime(2021, 5, 21, 14, 8, 43, 808000, tzinfo=tzlocal()),
 'HyperParameterTuningJobArn': 'arn:aws:sagemaker:us-east-1:581320662326:hyper-parameter-tuning-job/sagema
ker-xgboost-210521-1408',
 'HyperParameterTuningJobConfig': {'HyperParameterTuningJobObjective': {'MetricName': 'validation:auc',
                                                                        'Type': 'Maximize'},
                                   'ParameterRanges': {'CategoricalParameterRanges': [],
                                                       'ContinuousParameterRanges': [{'MaxValue': '1',
                                                                                      'MinValue': '0',
                                                                                      'Name': 'eta',
                                                                                      'ScalingType': 'Aut
o'},
                                                                                     {'MaxValue': '7'.
```

Figure 6.60 – Results after calling describe_hyper_parameter_tuning_job()

Here, we can see that we were able to achieve a value of `0.91098` for the `validation:auc` metric.

16. Finally, get the name of the tuning job and use the `%store` magic to save its value:

```
r = response
tuning_job_name = r['HyperParameterTuningJobName']
%store tuning_job_name
tuning_job_name
```

We will use the value of `tuning_job_name` later in the *Analyzing the Automatic Model Tuning job results* recipe.

> **Note**
> If we want to deploy the best model of the Automatic Model Tuning job, it is as simple as using `hyperparameter_tuner.deploy(...)`. This `deploy()` function should have the same parameters as the `deploy()` function of an `Estimator` or `Model` object – `endpoint_name`, `instance_type`, and `initial_instance_count`.

Now, let's see how this works!

How it works...

The **Automatic Model Tuning** capability of **Amazon SageMaker** helps us look for the best version of the model. It does this by searching for the optimized values of the hyperparameters of a model automatically:

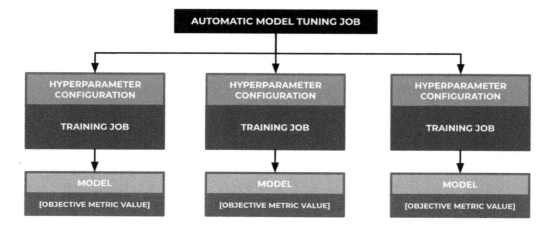

Figure 6.61 – How Automatic Model Tuning works

As shown in the preceding diagram, the **Automatic Model Tuning** capability of SageMaker involves testing different hyperparameter configuration values with multiple training jobs to come up with an optimal model. The optimal model is selected based on the objective metric values for each model that are computed while executing the training jobs. Let's discuss this process in detail:

- **Preparing the hyperparameter range configuration**: We provide the range of values for each hyperparameter we wish to tweak and optimize. In this recipe, we specified the hyperparameter range configuration for `eta`, `min_child_weight`, and `max_depth` by providing the minimum and maximum values. For each of these ranges, the scaling type can be specified to change how a specific hyperparameter range will be searched by the tuner algorithm. The scaling type can be linear, logarithmic, or reverse logarithmic. By default, it is set to **auto**.

- **Initializing the HyperparameterTuner object**: When we initialized the `HyperparameterTuner` object, we passed the `Estimator` object, the objective metric name, the hyperparameter range configuration, and a few other arguments. The objective metric is critical in knowing if the model that's been produced is better than other models that were produced by the training jobs during the model tuning process. In this recipe, the AUC is used and the higher its value, the better the model is.

- **Calling the fit() function to start the tuning job**: To start the tuning job, the `HyperparameterTuner` object's `fit()` function is called. Similar to an estimator object's `fit()` function, the `HyperparameterTuner` object's `fit()` function accepts the training data, along with other values for parameters, such as `wait` and `include_cls_metadata`.

- **Waiting for the tuning job to complete**: The automatic model tuning job takes some time to complete as multiple training jobs are configured and executed. Once all the training jobs have been completed, the best tuning job is selected and we can proceed with analyzing and deploying the model (which needs to be done separately).

At this point, we should have a good idea of how **Automatic Model Tuning** works.

There's more...

We are only scratching the surface of what we can do with the **Automatic Model Tuning** capability of **Amazon SageMaker**. Let's discuss a few of these capabilities:

- **Warm Start**: We can use this to start a hyperparameter tuning job by utilizing previous tuning jobs as the starting point. This is useful for different scenarios, including continuing stopped tuning jobs or tuning a model using new data.

- **Early Stopping**: We can use this to stop a hyperparameter tuning job when the objective metric does not have significant improvements across several iterations. Not all built-in SageMaker algorithms support early stopping. We can use the early stopping option for algorithms such as Linear Learner, XGBoost, Image Classification, and a few more.

- **Strategy**: We can use this to specify whether we will use the **Bayesian Search** or the **Random Search** option to look for the optimized values for the hyperparameters. In **Random Search**, values are randomly sampled from the hyperparameter range and then evaluated after running the training jobs that have been configured with these randomly selected hyperparameter values. Compared to the **Random Search** option, the **Bayesian Search** option makes use of evaluation results from previous iterations when identifying the next set of values. The **Bayesian Search** option, which is the default option, makes use of the "explore/exploit" approach when looking for better results as it sometimes explores hyperparameter values close to the current best-evaluated training job. Sometimes, it explores a configuration of values far from any existing attempted configuration of hyperparameter values. Feel free to check out https://docs.aws.amazon.com/sagemaker/latest/dg/automatic-model-tuning-how-it-works.html for more information.

We are also not limited to SageMaker's built-in algorithms when performing automated hyperparameter optimization. We can also tune models using the following:

- Deep learning libraries and frameworks (for example, **TensorFlow** and **PyTorch**)

- Custom container images (for example, using R)

Finally, we can perform automatic hyperparameter tuning on multiple models at the same time. Here is a sample code snippet on how to get this to work:

```
estimator_dict = { 'nn': <est1>, 'll': <est2> }
hyperparams_dict = { 'nn': { ... }, 'll': { ... }}
metrics_dict = { 'nn': '<metric1>', 'll': '<metric2>' }
metric_definitions_dict = { 'nn': [...], 'll': [...] }

hyperparameter_tuner = HyperparameterTuner.create(
    ...,
    estimator_dict=estimator_dict,
    hyperparameter_ranges_dict=hyperparams_dict,
    objective_metric_name_dict=metrics_dict,
    metric_definitions_dict=metric_definitions_dict
)
```

Here, we can see that we simply need to provide the parameter values that follow a common nested dictionary structure for `estimator_dict`, `hyperparameter_ranges_dict`, `objective_metric_name_dict`, and `metric_definitions_dict`. For a more complete example, you can check out the multi-algorithm support for the **Automatic Model Tuning** capability of SageMaker. This can be found in the **Candidate Definition Notebook** of the *Inspecting the SageMaker Autopilot experiment results and artifacts* recipe.

See also

If you are looking for examples of using the **Automatic Model Tuning** capability of SageMaker on real datasets and more complex examples, feel free to check out some of the notebooks in the `aws/amazon-sagemaker-examples` GitHub repository:

- Using **early stopping** to speed up the tuning process for a multiclass image classifier with **Automatic Model Tuning**: `https://github.com/aws/amazon-sagemaker-examples/blob/master/hyperparameter_tuning/image_classification_early_stopping/hpo_image_classification_early_stopping.ipynb`

- Performing automated hyperparameter optimization with a custom algorithm container image using R in SageMaker: `https://github.com/aws/amazon-sagemaker-examples/blob/master/hyperparameter_tuning/r_bring_your_own/tune_r_bring_your_own.ipynb`

Now, let's take a closer look at how to analyze the automatic model tuning job results!

Analyzing the Automatic Model Tuning job results

In the previous recipe, we used the **Automatic Model Tuning** capability of SageMaker to help us identify the optimal set of hyperparameter values for our model. In this recipe, we will use the `HyperparameterTuningJobAnalytics` class from the **SageMaker Python SDK** to load the properties and details of the automatic model tuning job. This will come in handy when we want to analyze and compare the properties, hyperparameters, and results of the different training jobs.

> **Tip**
> We can run this recipe even if the **Automatic Model Tuning** job has not finished yet.

Getting ready

This recipe continues from the *Performing Automatic Model Tuning with the SageMaker XGBoost built-in algorithm* recipe.

How to do it...

The following steps focus on using `HyperparameterTuningJobAnalytics` to load and inspect the results and current state of the **Automatic Model Tuning** job. Let's get started:

1. Navigate to the `my-experiments/chapter06` directory inside **SageMaker Studio**. Feel free to create this directory if it does not exist yet:

Figure 6.62 – The my-experiments/chapter06 directory

Here, we have the `my-experiments/chapter06` directory. Double-click the `chapter06` directory before proceeding.

2. Create a new notebook using the `Python 3 (Data Science)` kernel inside the `my-experiments/chapter06` directory and rename it with the name of this recipe (that is, `Analyzing the Automatic Model Tuning job results`):

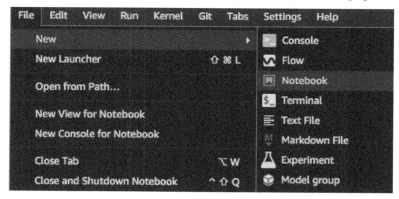

Figure 6.63 – Creating a new notebook

Here, we can see how to create a new **Notebook** from the **File** menu. When prompted for the kernel to use, choose **Python 3 (Data Science)**.

3. Use the `%store` magic to load the value of `tuning_job_name`. Remember that we stored this value earlier in the *Performing Automatic Model Tuning with the SageMaker XGBoost built-in algorithm* recipe:

```
%store -r tuning_job_name
```

4. Initialize a `HyperparameterTuningJobAnalytics` object with `tuning_job_name` as the parameter. Load the DataFrame containing the results and metrics of the automated tuning job using the `dataframe()` function:

```
import pandas as pd
import sagemaker
analytics = sagemaker.HyperparameterTuningJobAnalytics(
    tuning_job_name)
full_df = analytics.dataframe()
full_df
```

This should give us a `DataFrame` of values similar to the following:

	eta	max_depth	min_child_weight	TrainingJobName	TrainingJobStatus	FinalObjectiveValue	TrainingStartTime	Trainin
0	0.322555	3.0	6.056740	sagemaker-xgboost-210521-1408-006-88ce905a	InProgress	NaN	NaT	
1	0.350235	2.0	5.328425	sagemaker-xgboost-210521-1408-005-6eaea126	InProgress	NaN	NaT	
2	0.025056	4.0	3.962702	sagemaker-xgboost-210521-1408-004-87d73e81	InProgress	NaN	NaT	
3	0.180004	4.0	6.319783	sagemaker-xgboost-210521-1408-003-66a3e21e	Completed	0.903793	2021-05-21 14:11:05+00:00	2(14:12
4	0.726608	5.0	6.885498	sagemaker-xgboost-210521-1408-002-6eb2bd68	Completed	0.880860	2021-05-21 14:11:18+00:00	2(14:12
5	0.076452	8.0	5.140763	sagemaker-xgboost-210521-1408-001-f367ccc4	Completed	0.897652	2021-05-21 14:11:09+00:00	2(14:12

Figure 6.64 – Results after using HyperparameterTuningJobAnalytics.dataframe()

Here, we can see the parameters, hyperparameters, and the metric values associated with the training jobs that were triggered by the **Automatic Model Tuning** job. At this point, we can analyze this `DataFrame` using the different data manipulation options using the **pandas** library.

5. Sort the DataFrame by the `FinalObjectiveValue` column in descending order:

```
full_df.sort_values("FinalObjectiveValue",
ascending=False)
```

This should give us a `DataFrame` of values similar to the following:

	eta	max_depth	min_child_weight	TrainingJobName	TrainingJobStatus	FinalObjectiveValue	TrainingStartTime	Trainin
3	0.180004	4.0	6.319783	sagemaker-xgboost-210521-1408-003-66a3e21e	Completed	0.903793	2021-05-21 14:11:05+00:00	2(14:12
5	0.076452	8.0	5.140763	sagemaker-xgboost-210521-1408-001-f367ccc4	Completed	0.897652	2021-05-21 14:11:09+00:00	2(14:12
4	0.726608	5.0	6.885498	sagemaker-xgboost-210521-1408-002-6eb2bd68	Completed	0.880860	2021-05-21 14:11:18+00:00	2(14:12
0	0.322555	3.0	6.056740	sagemaker-xgboost-210521-1408-006-88ce905a	InProgress	NaN	NaT	
1	0.350235	2.0	5.328425	sagemaker-xgboost-210521-1408-005-6eaea126	InProgress	NaN	NaT	
2	0.025056	4.0	3.962702	sagemaker-xgboost-210521-1408-004-87d73e81	InProgress	NaN	NaT	

Figure 6.65 – Training job results sorted by FinalObjectiveValue

Here, the results are sorted by `FinalObjectiveValue` in decreasing order. The first row in this `DataFrame` will give us the model with the highest current `FinalObjectiveValue`.

6. Get the median value of the number of seconds the training jobs has executed for using the `median()` function:

```
full_df["TrainingElapsedTimeSeconds"].median()
```

This should return a value close to `82.0`.

Now, let's see how this works!

How it works...

In this recipe, we loaded the results and details of a **SageMaker Automatic Model Tuning** job. We expect to see a lot of training jobs being run during this process, and this recipe is one quick way of getting the properties and results of the training jobs with a few lines of code.

`HyperparameterTuningJobAnalytics` allows us to extract the properties and metrics of the training jobs we have executed while using the Automatic Model Tuning capability of **Amazon SageMaker**. As shown in *Figure 6.67*, the following columns are available in the DataFrame's hyperparameter configuration (that is, `eta`, `max_depth`, and `min_child_weight`): `TrainingJobName`, `TrainingJobStatus`, and `FinalObjectiveValue`.

In the **SageMaker Python SDK**, we have the following classes. These can be used while analyzing different jobs and experiments:

- `HyperparameterTuningJobAnalytics`
- `TrainingJobAnalytics`
- `ExperimentAnalytics`

Each of these classes allow us to gather more details about their corresponding target entity (for example, `ExperimentAnalytics` to load the details and results for **SageMaker Experiments**).

7
Working with SageMaker Feature Store, SageMaker Clarify, and SageMaker Model Monitor

In the previous chapter, we had our first look at **SageMaker Studio**, along with its automated machine learning capabilities, by using **SageMaker Autopilot** and **Automatic Model Tuning** to prepare high-quality models. In this chapter, we will focus on a few more capabilities of SageMaker that have great integration with **SageMaker Studio** – **SageMaker Feature Store**, **SageMaker Clarify**, and **SageMaker Model Monitor**. These capabilities help data scientists and machine learning practitioners handle specific but relevant requirements when working on production-level machine learning experiments and deployments.

These include using online and offline feature stores, detecting bias in the data, enabling machine learning explainability, and monitoring the deployed model. The following diagram shows how these capabilities are used in the different stages of the machine learning process:

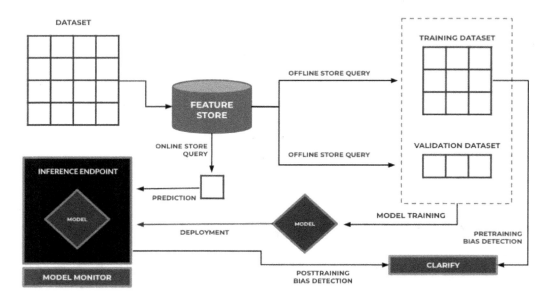

Figure 7.1 – Working with SageMaker Feature Store, Clarify, and Model Monitor

We will start by using **SageMaker Feature Store** to store and manage the ML features in our dataset. We will also take a closer look at how to use **SageMaker Clarify** in detecting pre-training and post-training bias in our data. In addition to this, we will also learn how to generate ML explainability reports using **SageMaker Clarify**. We will then learn how to use **SageMaker Model Monitor** to capture and analyze the data in our inference endpoint. Finally, we will learn how we can use **SageMaker Model Monitor** to automatically detect data quality drift.

We will cover the following recipes in this chapter:

- Generating a synthetic dataset and using **SageMaker Feature Store** for storage and management

- Querying data from the offline store of **SageMaker Feature Store** and uploading it to Amazon S3

- Detecting pre-training bias with **SageMaker Clarify**

- Detecting post-training bias with **SageMaker Clarify**

- Enabling ML explainability with **SageMaker Clarify**

- Deploying an endpoint from a model and enabling data capture with **SageMaker Model Monitor**

- Baselining and scheduled monitoring with **SageMaker Model Monitor**

Once we have completed the recipes in this chapter, we will have a better understanding of how these capabilities and features help make ML experiments and deployments successful.

Now, let's proceed with the recipes in this chapter!

Technical requirements

To execute the recipes in the chapter, make sure you have an Amazon S3 bucket, as well as permissions to manage the **Amazon SageMaker** and **Amazon S3** resources if you're using an **AWS IAM** user with a custom URL. If you are using the root account, then you should be able to proceed with the recipes in this chapter. However, it is recommended that you're signed in as an AWS IAM user instead of using the root account in most cases. For more information, feel free to take a look at `https://docs.aws.amazon.com/IAM/latest/UserGuide/best-practices.html`.

As the recipes in this chapter involve a bit of code, we have made the necessary scripts and notebooks available in this book's GitHub repository: `https://github.com/PacktPublishing/Machine-Learning-with-Amazon-SageMaker-Cookbook/tree/master/Chapter07`. Before starting each of the recipes in this chapter, make sure that the `my-experiments/chapter07` directory is ready. If it has not been created yet, please do so now as this will help keep things organized as we go through each of the recipes in this book.

Check out the following link to see the relevant Code in Action video:

`https://bit.ly/3nfR8Zg`

Generating a synthetic dataset and using SageMaker Feature Store for storage and management

In this recipe, we will generate a synthetic dataset similar to what is shown in the following screenshot. Here, we will emulate a sample scenario where a school is planning to build a binary classifier model that automatically approves or rejects candidates for a scholarship based on certain attributes, such as their scores for the math, science, and technology exams:

	approved	sex	math	science	technology	random1	random2	index	event_time
0	1	1	97	97	98	93	82	1	1.623605e+09
1	1	1	85	68	62	92	65	2	1.623605e+09
2	1	1	99	100	80	71	60	3	1.623605e+09
3	1	1	91	79	84	60	70	4	1.623605e+09
4	1	1	73	86	66	70	98	5	1.623605e+09
...	...	...	...	...	...	...	...	...	...
995	1	0	99	62	92	71	75	996	1.623605e+09
996	1	0	85	74	91	69	63	997	1.623605e+09
997	1	1	72	99	86	61	65	998	1.623605e+09
998	1	1	79	89	79	98	80	999	1.623605e+09
999	1	1	72	97	79	87	60	1000	1.623605e+09

1000 rows × 9 columns

Figure 7.2 – Synthetic dataset

This dataset will have four primary predictor columns called `sex`, `math`, `science`, and `technology`, along with two columns called `random1` and `random2` that contain random values. These columns will help us verify whether the feature importance report generated by **SageMaker Clarify** in the *Enabling ML explainability with SageMaker Clarify* recipe is working or not. In addition to these, the generated dataset will have two additional columns, called `event_time` and `index`, as we will use **SageMaker Feature Store** to ingest and manage our feature groups and processed datasets.

SageMaker Feature Store will serve as the centralized repository where the processed features are stored. These stored records will be loaded for training and inference in the succeeding recipes in this chapter. As we will see in the *Detecting pre-training bias with SageMaker Clarify* recipe, the synthetic dataset we will generate in this recipe is considered to be imbalanced as it does not have an equal number of instances of each class – the `"female"` (`sex=0`) class has significantly fewer records than the `"male"` (`sex=1`) class. In the *Detecting post-training bias with SageMaker Clarify* recipe, we will also see that with the dataset we will generate in this recipe and a model we will train using this dataset will generate a post-training bias report. This will suggest that the model has a higher chance of approving `male` applicants than `female` applicants.

Getting ready

Here are the prerequisites for this recipe:

- **SageMaker Studio** needs to be set up and ready before you work on this recipe. Feel free to check out the *Onboarding to SageMaker Studio* recipe of *Chapter 6, Automated Machine Learning in Amazon SageMaker*, if you have not set this up yet.

- Make sure that the execution role associated with **SageMaker Studio** has the necessary permissions to upload files to **Amazon S3**. You may attach the `AmazonS3FullAccess` policy to the said execution role.

How to do it...

We will be running the following set of steps in **SageMaker Studio**:

1. Create a new notebook using the `Python 3 (Data Science)` kernel inside the *my-experiments/chapter07* directory and rename it so that it's the name of this recipe (`Generating a synthetic dataset and using SageMaker Feature Store for storage and management`):

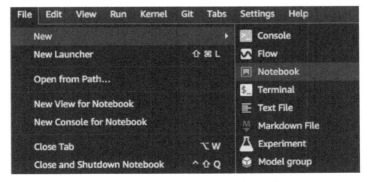

Figure 7.3 – Creating a new notebook

The preceding screenshot shows how to create a new **Notebook** from the **File** menu. When prompted for the kernel to use, choose Python 3 (Data Science).

2. Use the %load magic command to load the data_generator.py script file from the *Machine-Learning-with-Amazon-SageMaker-Cookbook* GitHub repository:

```
%load \
https://raw.githubusercontent.com/PacktPublishing/
Machine-Learning-with-Amazon-SageMaker-Cookbook/master/
Chapter07/scripts/generator.py
```

This should magically replace the original cell content with the code from the specified script file, similar to what is shown in the following screenshot:

```python
# %load https://raw.githubusercontent.com/PacktPublishing/Amazon-SageMaker-Cookbook/master/Chapter0
import random
import pandas as pd
from time import time, sleep


def log(message):
    print(f"[+] {message}\n")


def generate_random_score():
    return random.randint(60, 100)


def generate_list_of_random_scores(total_records=1000):
    return list(map(lambda x: generate_random_score(), range(total_records)))


def generate_event_time_records(num_records):
    time_value = int(round(time()))
    output = pd.Series([time_value]*num_records,
                       dtype="float64")

    return output


def main():
    log("Generating column values for math, science, technology")
    math = generate_list_of_random_scores()
    science = generate_list_of_random_scores()
    technology = generate_list_of_random_scores()
```

Figure 7.4 – Using %load to load the data_generator.py script from this book's GitHub repository

As we can see, the generator.py script file makes use of generate_list_of_random_scores() to generate scores for the math, science, technology, random1, and random2 columns. In addition to this, the script makes use of the generate_event_time_records() function to generate the values for the event_time column.

3. Run the cell containing the code from the previous step to start generating the synthetic dataset. This should generate a set of logs and a `DataFrame` of values, similar to what is shown in the following screenshot:

	approved	sex	math	science	technology	random1	random2	index
0	1	1	97	97	98	93	82	1
1	1	1	85	68	62	92	65	2
2	1	1	99	100	80	71	60	3
3	1	1	91	79	84	60	70	4
4	1	1	73	86	66	70	98	5
..	...	...	...	...	...	...	...	...
995	1	0	99	62	92	71	75	996
996	1	0	85	74	91	69	63	997
997	1	1	72	99	86	61	65	998
998	1	1	79	89	79	98	80	999
999	1	1	72	97	79	87	60	1000

	event_time
0	1.623605e+09
1	1.623605e+09
2	1.623605e+09
3	1.623605e+09
4	1.623605e+09
..	...
995	1.623605e+09
996	1.623605e+09
997	1.623605e+09
998	1.623605e+09
999	1.623605e+09

[1000 rows x 9 columns]

Figure 7.5 – DataFrame containing the generated synthetic dataset

Here, we can see that we have generated 1,000 records for the synthetic dataset. Each of the records in this dataset has values for the following fields: `approved` (target variable), `sex`, `math`, `science`, `technology`, `random1`, `random2`, `index`, and `event_time`.

4. In the next cell, import and prepare a few prerequisites, including `boto3`, `sagemaker`, and `session`:

```
import boto3
import sagemaker
from sagemaker.session import Session
region = boto3.Session().region_name
session = boto3.Session(region_name=region)
```

5. Initialize the `Session` object called `feature_store_session` as well:

```
client = session.client(
    service_name='sagemaker',
```

```
        region_name=region
    )
runtime = session.client(
    service_name='sagemaker-featurestore-runtime',
    region_name=region
)
feature_store_session = Session(
    boto_session=session,
    sagemaker_client=client,
    sagemaker_featurestore_runtime_client=runtime
)
```

6. Specify the S3 bucket name and the prefix where the data will be stored. Make sure to replace the value of `"<insert s3 bucket name here>"` with the name of the bucket we created in the *Preparing the Amazon S3 bucket and the training dataset for the linear regression experiment* recipe of *Chapter 1, Getting Started with Machine Learning Using Amazon SageMaker*:

```
s3_bucket_name = "<insert s3 bucket name here>"
prefix = "chapter07"
```

7. Initialize the execution role and the `boto3` S3 client:

```
from sagemaker import get_execution_role
role = get_execution_role()
s3_client = boto3.client('s3', region_name=region)
```

8. Specify the feature group name in the `feature_group_name` variable:

```
feature_group_name = 'cookbook-feature-group'
```

9. Next, initialize the `FeatureGroup` object and pass the `feature_group_name` and `feature_store_session` variables as the parameter values:

```
from sagemaker.feature_store.feature_group import
FeatureGroup
feature_group = FeatureGroup(
    name=feature_group_name,
    sagemaker_session=feature_store_session
)
```

10. Use the `load_feature_definitions()` function to have SageMaker automatically detect the data types for the features in the feature group from a specified `DataFrame`:

```
feature_group.load_feature_definitions(
    data_frame=all_df
)
sleep(1)
```

> **Important note**
>
> Before proceeding with the next step, make sure that the execution role associated with **SageMaker Studio** has the necessary permissions to upload files to **Amazon S3**. You may attach the `AmazonS3FullAccess` policy to the said execution role.

11. Use the `create()` function to create the feature group. Make sure to set the `enable_online_store` parameter value to `True`:

```
feature_group.create(
    s3_uri=f"s3://{s3_bucket_name}/{prefix}/input",
    record_identifier_name="index",
    event_time_feature_name="event_time",
    role_arn=role,
    enable_online_store=True
)
sleep(60)
```

Note that in this step, we have specified that the `index` column will serve as the unique identifier of a record. Later, we will see that we can get a specific record from the feature group using the `get_record()` function by specifying the `index` column. In addition to this, we specified that the `event_time` column would serve as the field that corresponds to the `created_at` or `updated_at` values of a record, which would allow us to get a specific version of a record when using the offline store. Lastly, we used the `sleep()` function to block and wait for a minute for these changes to take effect.

> **Note**
>
> One of the major differences between the online store and the offline store is that the online store stores only the latest version of the record, while the offline store contains all the records, including the previous ones. Offline stores are generally used to store months or years of feature data, which means that there may be duplicate information when performing queries for certain records. We will discuss this in more detail in the *There's more…* section of the *Querying data from the offline store of SageMaker Feature Store and uploading it to Amazon S3* recipe.

12. Inspect the feature group's status using the following code:

```
feature_group.describe()
```

This should give us a dictionary of values, similar to what is shown in the following screenshot:

```
{'FeatureGroupArn': 'arn:aws:sagemaker:us-east-1:         :feature-group/cookbook-feature-group',
 'FeatureGroupName': 'cookbook-feature-group',
 'RecordIdentifierFeatureName': 'index',
 'EventTimeFeatureName': 'event_time',
 'FeatureDefinitions': [{'FeatureName': 'approved', 'FeatureType': 'Integral'},
  {'FeatureName': 'sex', 'FeatureType': 'Integral'},
  {'FeatureName': 'math', 'FeatureType': 'Integral'},
  {'FeatureName': 'science', 'FeatureType': 'Integral'},
  {'FeatureName': 'technology', 'FeatureType': 'Integral'},
  {'FeatureName': 'random1', 'FeatureType': 'Integral'},
  {'FeatureName': 'random2', 'FeatureType': 'Integral'},
  {'FeatureName': 'index', 'FeatureType': 'Integral'},
  {'FeatureName': 'event_time', 'FeatureType': 'Fractional'}],
 'CreationTime': datetime.datetime(2021, 6, 13, 17, 15, 42, 30000, tzinfo=tzlocal()),
 'OnlineStoreConfig': {'EnableOnlineStore': True},
 'OfflineStoreConfig': {'S3StorageConfig': {'S3Uri': 's3://sagemaker-cookbook-bucket/chapter07/input',
   'ResolvedOutputS3Uri': 's3://sagemaker-cookbook-bucket/chapter07/input/581320662326/sagemaker/us-east-1/offline-store/
cookbook-feature-group-1623604542/data'},
  'DisableGlueTableCreation': False,
  'DataCatalogConfig': {'TableName': 'cookbook-feature-group-1623604542',
   'Catalog': 'AwsDataCatalog',
   'Database': 'sagemaker_featurestore'}},
 'RoleArn':                                 ',
 'FeatureGroupStatus': 'Created',
 'ResponseMetadata': {'RequestId': 'f67f3434-e47b-47cf-b32b-856b598fb3e1',
  'HTTPStatusCode': 200,
  'HTTPHeaders': {'x-amzn-requestid': 'f67f3434-e47b-47cf-b32b-856b598fb3e1',
   'content-type': 'application/x-amz-json-1.1',
   'content-length': '1614',
   'date': 'Sun, 13 Jun 2021 17:16:42 GMT'},
  'RetryAttempts': 0}}
```

Figure 7.6 – Result of feature_group.describe()

In the preceding screenshot, we can see the properties of the feature group after calling the `feature_group.describe()` function. These include the values for `FeatureDefinitions`, `OnlineStoreConfig`, and `FeatureGroupStatus`, along with other properties.

13. Inspect the feature group's status using the following code:

```
feature_group.describe().get("FeatureGroupStatus")
```

This should return a 'Created' status.

14. Now that we have prepared everything, we will use the ingest() function with the all_df DataFrame as the value of the data_frame parameter:

```
%%time
feature_group.ingest(
    data_frame=all_df, max_workers=3, wait=True
)
```

This function simply loads the records in the all_df DataFrame to the feature store. Wait about a minute or so for this step to complete. Note that when running this block of code, you may encounter issues or errors when the feature group definitions do not match the data types of the values inside the DataFrame. You may be able to solve this issue by making sure that the versions of the **SageMaker Python SDK** and other relevant libraries are updated to the latest versions, or by casting the column values of the **pandas** DataFrame to the appropriate type (for example, using the astype() function).

15. To test if the records have been ingested properly in the online store, use the get_record() function and load the last value in the DataFrame that was ingested by specifying 300 as the record identifier value:

```
runtime.get_record(
    FeatureGroupName=feature_group.name,
    RecordIdentifierValueAsString="300"
)
```

This should return a nested dictionary of values, similar to what is shown in the following screenshot:

```
{'ResponseMetadata': {'RequestId': '3f780b7f-ec25-4bb3-8381-d61a46f4c3b8',
 'HTTPStatusCode': 200,
 'HTTPHeaders': {'x-amzn-requestid': '3f780b7f-ec25-4bb3-8381-d61a46f4c3b8',
 'content-type': 'application/json',
 'content-length': '442',
 'date': 'Sun, 13 Jun 2021 17:16:48 GMT'},
 'RetryAttempts': 0},
'Record': [{'FeatureName': 'approved', 'ValueAsString': '0'},
 {'FeatureName': 'sex', 'ValueAsString': '0'},
 {'FeatureName': 'math', 'ValueAsString': '65'},
 {'FeatureName': 'science', 'ValueAsString': '61'},
 {'FeatureName': 'technology', 'ValueAsString': '86'},
 {'FeatureName': 'random1', 'ValueAsString': '91'},
 {'FeatureName': 'random2', 'ValueAsString': '68'},
 {'FeatureName': 'index', 'ValueAsString': '300'},
 {'FeatureName': 'event_time', 'ValueAsString': '1623604510.0'}]}
```

Figure 7.7 – Result after using the get_record() function

Here, we can see that we were able to retrieve the feature values of the record with an index equal to 300.

16. Use the %store magic to store the variable values for feature_group_name, s3_bucket_name, and prefix:

```
%store feature_group_name
%store s3_bucket_name
%store prefix
```

We will use these variables later in the succeeding recipes of this chapter.

Let's see how this works!

How it works...

In this recipe, we generated a synthetic dataset that can be used for classification problems. It has the following columns:

- An approved column containing the target variable.

- The sex, math, science, and technology columns, which were generated by the generate_list_of_random_scores() function.

- The random1 and random2 columns, which contain a random set of numbers to help identify the contribution of each feature to predicting the target label when using the **SHAP** values in the *Enabling ML explainability with SageMaker Clarify* recipe.

- The event_time and index columns, which we need when using **SageMaker Feature Store**. Toward the end of the generator.py script, we updated a few rows in the DataFrame to introduce bias in our dataset which gave "male" (sex=1) candidates a higher chance of getting approved over "female" (sex=0) candidates.

Once we had generated and prepared the DataFrame containing the target and predictor variable values, we used **SageMaker Feature Store** to ingest it. There are two types of stores – **online** and **offline**. The online store is generally used to load records in real time and works well in use cases that involve loading records to test the inference endpoints. The offline store, on the other hand, is useful in scenarios that involve loading a batch of records that are used during the training phase.

> **Note**
>
> As we mentioned in a note in the *How to do it...* section, one of the major differences between the online store and the offline store is that the online store stores only the latest version of the record, while the offline store may contain all the records, including the previous ones. For example, the offline store would have a copy of the previous snapshots of a customer's purchase data for each month, while the online store would only have the latest version. These properties, along with the other benefits that help machine learning practitioners save time organizing features, make the online and offline stores much better options over just purely using Amazon S3 for storage.

When dealing with feature stores, the process usually involves the following steps:

1. The feature group properties and feature definition configuration are loaded.
2. The feature group is created.
3. The data is ingested into the feature group.
4. Multiple records are loaded from the offline store for training data.
5. A record is loaded from the online store for inference test data.

In this recipe, we performed the first three steps to ingest the DataFrame values for the feature group. In the *Querying data from the offline store of SageMaker Feature Store and uploading it to Amazon S3* recipe, we will perform the fourth step and load multiple records when preparing the training, validation, and test datasets before the training phase. Finally, we will perform the fifth step in the *Deploying an endpoint from a model and enabling data capture with SageMaker Model Monitor* recipe, when loading a test record to verify if the inference endpoint is working and configured properly.

Querying data from the offline store of SageMaker Feature Store and uploading it to Amazon S3

In the previous recipe, we generated a synthetic dataset and stored it in **SageMaker Feature Store** using the ingest() function. In this recipe, we will demonstrate how to load the data from the feature group of the offline store where the data was stored in the previous recipe. As we discussed in the *Generating a synthetic dataset and using SageMaker Feature Store for storage and management* recipe, the offline store is useful for use cases that involve loading a batch of records that are used during the training phase. That said, the training, validation, and test datasets will be loaded from the offline store, exported in CSV format, and then uploaded to S3.

> **Important note**
> Note that you may need a wait for a few minutes before the offline store data is available for querying if you've just finished ingesting data into the feature group in the *Generating a synthetic dataset and using SageMaker Feature Store for storage and management* recipe.

Getting ready

This recipe continues from the *Generating a synthetic dataset and using SageMaker Feature Store for storage and management* recipe.

How to do it...

The next set of steps focus on loading the training, validation, and test datasets from **SageMaker Feature Store** offline store and storing these in S3. Let's get started:

1. Create a new notebook using the `Python 3 (Data Science)` kernel inside the `my-experiments/chapter07` directory and rename it to the name of this recipe (`Querying data from the offline store of SageMaker Feature Store and uploading it to Amazon S3`). When prompted for the kernel to use, choose `Python 3 (Data Science)`.

2. Import and prepare a few prerequisites, such as the `Session` and `FeatureGroup` classes, from the **SageMaker Python SDK** and the `boto3` `Session` object:

   ```
   import boto3
   import sagemaker
   from sagemaker.session import Session
   from sagemaker.feature_store.feature_group import
   FeatureGroup
   region = boto3.Session().region_name
   session = boto3.Session(region_name=region)
   ```

3. Next, initialize the `feature_store_session` session:

   ```
   client = session.client(
       service_name='sagemaker',
       region_name=region
   )
   runtime = session.client(
   ```

```
    service_name='sagemaker-featurestore-runtime',
    region_name=region
)
feature_store_session = Session(
    boto_session=session,
    sagemaker_client=client,
    sagemaker_featurestore_runtime_client=runtime
)
```

4. Use the %store magic to load the variable value for feature_group_name and use it to load the feature group we created in the *Generating a synthetic dataset and using SageMaker Feature Store for storage and management* recipe:

```
%store -r feature_group_name
feature_group = FeatureGroup(
    name=feature_group_name,
    sagemaker_session=feature_store_session
)
```

5. Run the following block of code to store the Athena table name in the table variable:

```
table = feature_group.athena_query().table_name
```

> **Note**
> Note that the offline store makes use of Amazon Athena to fetch data with SQL queries.

6. Get the S3 URI from the response nested dictionary value after using the describe() function of the feature_group object:

```
describe_response = feature_group.describe()
offline_config = describe_response['OfflineStoreConfig']
s3_uri = offline_config['S3StorageConfig']['S3Uri']
!aws s3 ls {s3_uri} --recursive
```

This should give us a list of Parquet files that correspond to the data we have stored in the offline feature store. The columnar Parquet format works well with **Amazon Athena**, since queries on data stored in this format generally have better performance than CSV or TSV files.

7. Use the `%store` magic to load the variable values for `s3_bucket_name` and `prefix`. Using these values, prepare the S3 output path stored in the `output_location` variable:

```
%store -r s3_bucket_name
%store -r prefix
base = f's3://{s3_bucket_name}/{prefix}'
output_location = f'{base}/query_results/'
```

In the succeeding steps, we will see that the results of the Athena queries are stored inside this output location.

8. Define the `query_data()` function, which we will use to load the DataFrame containing the results of the Athena query:

```
def query_data(query_string):
    print(f"QUERY: {query_string}\n")
    query = feature_group.athena_query()
    query.run(query_string=query_string,
              output_location=output_location)

    query.wait()

    return query.as_dataframe()
```

The next few steps involve loading the training, validation, and test datasets by querying the data using the `query_data()` function we defined in the previous step:

9. Use the `query_data()` function to load the first `600` records for the training dataset using Amazon Athena:

```
query = f"""SELECT approved, sex, math, science,
technology, random1, random2 FROM "{table}" ORDER BY
index ASC LIMIT 600"""
training_df = query_data(query)
training_df
```

This should give us a `DataFrame` similar to what is shown in the following screenshot:

	approved	sex	math	science	technology	random1	random2
0	1	1	97	97	98	93	82
1	1	1	85	68	62	92	65
2	1	1	99	100	80	71	60
3	1	1	91	79	84	60	70
4	1	1	73	86	66	70	98
...	...	...	...	...	...	...	...
595	1	1	99	86	85	98	87
596	1	1	71	97	90	86	99
597	1	1	95	86	62	69	73
598	1	1	78	71	68	72	68
599	1	1	89	82	65	97	83

600 rows × 7 columns

Figure 7.8 – Training dataset

Here, we have a `DataFrame` containing the training data.

> **Important note**
> If you do not get any results after running the previous query, please wait a few minutes and then try again.

10. Use the `query_data()` function to load the next `200` records for the validation dataset:

```
query = f"""SELECT approved, sex, math, science,
technology, random1, random2 FROM "{table}" WHERE index >
600 ORDER BY index ASC LIMIT 200"""
validation_df = query_data(query)
```

11. In a similar fashion, use the `query_data()` function to load the last `200` records for the test dataset:

```
query = f"""SELECT approved, sex, math, science,
technology, random1, random2 FROM "{table}" WHERE index >
800 ORDER BY index ASC LIMIT 200"""
test_df = query_data(query)
```

12. Create the `tmp` directory if it does not exist yet:

```
!mkdir -p tmp
```

13. Use the `to_csv()` function to generate the CSV files from the DataFrames:

```
training_df.to_csv('tmp/training_data.csv',
                   header=True,
                   index=False)
validation_df.to_csv('tmp/validation_data.csv',
                   header=True,
                   index=False)
test_df.to_csv('tmp/test_data.csv',
               header=True,
               index=False)
```

14. Upload the CSV files we generated in the previous step to S3 using the `aws s3 cp` command:

```
path = f"s3://{s3_bucket_name}/{prefix}"
training_data_path = f"{path}/input/training_data.csv"
validation_data_path = f"{path}/input/validation_data.csv"
test_data_path = f"{path}/input/test_data.csv"
!aws s3 cp tmp/training_data.csv {training_data_path}
!aws s3 cp tmp/validation_data.csv {validation_data_path}
!aws s3 cp tmp/test_data.csv {test_data_path}
```

15. Use the `to_csv()` function to generate the CSV files without the header:

```
target = 'tmp/training_data_no_header.csv'
training_df.to_csv(target,
                   header=False,
                   index=False)
target = 'tmp/validation_data_no_header.csv'
validation_df.to_csv(target,
```

```
                        header=False,
                        index=False)
    test_df.to_csv('tmp/test_data_no_header.csv',
                   header=False,
                   index=False)
```

16. Upload the CSV files we generated in the previous step to S3 using the `aws s3 cp` command:

```
training_data_path_nh = f"{path}/input/training_data_no_
header.csv"
validation_data_path_nh = f"{path}/input/validation_data_
no_header.csv"
test_data_path_nh = f"{path}/input/test_data_no_header.
csv"
!aws s3 cp tmp/training_data_no_header.csv {training_
data_path_nh}
!aws s3 cp tmp/validation_data_no_header.csv {validation_
data_path_nh}
!aws s3 cp tmp/test_data_no_header.csv {test_data_path_
nh}
```

17. Finally, use the `%store` magic to store the variable values for `training_data_path`, `validation_data_path`, `test_data_path`, `training_data_path_nh`, `validation_data_path_nh`, and `test_data_path_nh`:

```
%store training_data_path
%store validation_data_path
%store test_data_path
%store training_data_path_nh
%store validation_data_path_nh
%store test_data_path_nh
```

We will use one or more of these stored variable values in the succeeding recipes.

Let's see how this works!

How it works...

In this recipe, we loaded the records from the offline store of the SageMaker feature store using **Amazon Athena** and SQL queries to query and load the data. In the *Generating a synthetic dataset and using SageMaker Feature Store for storage and management* recipe, we ingested data to both the offline and online stores. The offline store data is stored in **Amazon S3** and the steps in this recipe simply load the data stored in the offline store. When the feature group is defined and created, the corresponding **glue data catalog** is created. This serves as the schema of the data so that we can perform queries using Amazon Athena.

The SQL queries used in this recipe involve making sure that the training, validation, and test datasets are mutually exclusive. The first 600 (60%) of the records were loaded for the training dataset, the next 200 (20%) of the records were loaded for the validation dataset, and the last 200 (20%) records were loaded for the test dataset. Note that the OFFSET keyword is not supported in **Amazon Athena**, which is why we used an alternative method to pull the records from the offline store.

There's more...

The example in this recipe involved a simplified example of how to use **SageMaker Feature Store**. There are more things we can do with the feature store, including the following:

- Joining and pulling data from two feature groups
- Loading a previous version of a dataset and removing duplicates and deleted records using an Athena query, which involves using the event_time and is_deleted columns in the WHERE clause

This recipe only involved a single layer of records that had the same event_time value. In real-world examples, a feature group may contain duplicates of a single record with the same index but with different event_time values. When analyzing and processing data, the version of a dataset at a specified time is pulled, and only the latest version of a record before the specified timestamp is loaded from the feature store.

If you need to load the latest version of a record before a certain timestamp, use the following SQL statement:

```
SELECT * FROM
    (SELECT *, row_number() OVER (PARTITION BY index
        ORDER BY event_time desc) AS row_num
```

```
        FROM <table name>
        where event_time <= timestamp '<timestamp>')
WHERE row_num = 1 and
NOT is_deleted
```

Note that the index and event_time column names are just the names/keys used in this set of recipes. These can be set when initializing the feature group before the ingestion step.

Detecting pre-training bias with SageMaker Clarify

As we deal with more real-world examples, we will start to encounter requirements that involve detecting and managing ML bias. For example, deployed machine learning models may reject applications from disfavored or underrepresented groups, since the training data used to train these models is already biased against the disfavored groups to begin with. This reduces opportunities for these disfavored groups, which then perpetuates their lack of fitness for an application. That said, once we start to realize the importance of ensuring fairness in machine learning, we will start looking for solutions that will help us handle the legal, ethical, and technical considerations as well. The good news is that **SageMaker Clarify** is there to help us detect ML bias in our data and models!

AI and ML bias may be present in specific stages in the machine learning pipeline – before, during, and after training. In this recipe, we will use **SageMaker Clarify** to help detect pre-training bias – specifically **class imbalance** – for the given dataset. After the **SageMaker Clarify** processing job has finished executing, we will be able to analyze the metric values that have been computed and returned as output by the processing job.

> **Note**
>
> A dataset is considered to be imbalanced if it does not have an equal number of instances for each class. We need to be careful when dealing with the imbalanced class problem as some of the algorithms for classification assume an equal number of samples for each class. In addition to this, even if an algorithm does not assume an equal number of samples for each of the classes, having extreme class imbalance can result in a model that may not be doing anything – if a dataset has 90% of the data for class A and the remaining 10% for class B, the model can get a seemingly acceptable 90% accuracy by just predicting class A every time. That said, failure to detect and manage this properly may result in incorrect and erroneous conclusions, especially when making predictions on the underrepresented group.

Getting ready

This recipe continues from the *Querying data from the offline store of SageMaker Feature Store and uploading it to Amazon S3* recipe.

How to do it...

The first set of steps in this recipe focus on preparing the prerequisites for the **SageMaker Clarify** processing job. Let's get started:

1. Create a new notebook using the Python 3 (Data Science) kernel inside the my-experiments/chapter07 directory and rename it to the name of this recipe (Detecting pre-training bias with SageMaker Clarify). When prompted for the kernel to use, choose Python 3 (Data Science).

2. Specify the s3_bucket_name and prefix values:

```
%store -r s3_bucket_name
%store -r prefix
%store -r training_data_path
```

3. Import, prepare, and load a few prerequisites, such as session, region, and role:

```
import sagemaker
session = sagemaker.Session()
region = session.boto_region_name
role = sagemaker.get_execution_role()
```

4. Import pandas and numpy:

```
import pandas as pd
import numpy as np
```

5. Create the tmp directory if it does not exist yet:

```
!mkdir -p tmp
```

6. Specify the s3_training_data_path and s3_output_path values:

```
s3_training_data_path = training_data_path
s3_output_path = f"s3://{s3_bucket_name}/{prefix}/output"
```

7. Use the AWS CLI to copy the file to the `tmp` directory:

```
!aws s3 cp {s3_training_data_path} tmp/training_data.csv
```

8. Load the training data using the `read_csv()` function:

```
training_data = pd.read_csv("tmp/training_data.csv")
```

Now that we have prepared the prerequisites, we can focus on getting the **SageMaker Clarify** processing job running in the next set of steps:

9. Initialize the `SageMakerClarifyProcessor` class. This class is simply a purpose-built wrapper class that makes use of the **SageMaker Processing** capability:

```
from sagemaker import clarify
processor = clarify.SageMakerClarifyProcessor(
    role=role,
    instance_count=1,
    instance_type='ml.m5.large',
    sagemaker_session=session
)
```

10. Initialize the `DataConfig` object:

```
data_config = clarify.DataConfig(
    s3_data_input_path=s3_training_data_path,
    s3_output_path=s3_output_path,
    label='approved',
    headers=training_data.columns.to_list(),
    dataset_type='text/csv'
)
```

11. Initialize the `BiasConfig` object:

```
bias_config = clarify.BiasConfig(
    label_values_or_threshold=[1],
    facet_name='sex',
)
```

If you are wondering what these configuration values mean, do not worry – we will explain how this configuration works in the *How it works…* section.

12. Use the `run_pre_training_bias()` function and wait a few minutes for the job to complete. In the following code block, we are specifying the `data_config` and the `bias_config` objects we initialized in the previous steps as arguments when calling the `run_pre_training_bias()` function:

```
%%time
processor.run_pre_training_bias(
    data_config=data_config,
    data_bias_config=bias_config,
    methods=['CI']
)
```

Note that this step may take about 5 to 10 minutes to complete. Calling `run_pre_training_bias()` would trigger a **SageMaker Clarify** processing job. This would run a container using one of the prebuilt **SageMaker Clarify** container images.

13. Once the processing job has completed, we get the output destination and can store the URI string value in the `output_uri` variable:

```
output = processor.latest_job.outputs[0]
output_destination = output.destination
```

14. Use the AWS CLI to copy the generated files to the `tmp` directory:

```
!aws s3 cp {output_destination}/ tmp/ --recursive
!ls -lahF tmp/
```

15. Inspect the content of the `analysis.json` file:

```
!cat tmp/analysis.json
```

This should give us a nested structure of values, similar to what is shown in the following screenshot:

```json
{
    "version": "1.0",
    "pre_training_bias_metrics": {
        "label": "approved",
        "facets": {
            "sex": [
                {
                    "value_or_threshold": "1",
                    "metrics": [
                        {
                            "name": "CI",
                            "description": "Class Imbalance (CI)",
                            "value": -0.5933333333333334
                        }
                    ]
                },
                {
                    "value_or_threshold": "0",
                    "metrics": [
                        {
                            "name": "CI",
                            "description": "Class Imbalance (CI)",
                            "value": 0.5933333333333334
                        }
                    ]
                }
            ]
        },
        "label_value_or_threshold": "1"
    }
}
```

Figure 7.9 – Content of the tmp/analysis.json file

Here, we can see the results of the pre-training bias job. Just by looking at the metric value for CI in the `tmp/analysis.json` file, we can infer that we have an imbalanced dataset.

Now, let's see how this works!

How it works...

SageMaker Clarify makes use of **SageMaker Processing** to run a job that assesses data. In this case, we specified that we only want to get the **class imbalance (CI)** metric from a larger list of pre-training bias metrics list. In addition to the class imbalance metric, there are other bias metrics we can compute using **SageMaker Clarify**. These include **Difference in Positive Proportions in Labels (DPL)**, **Kullback-Liebler divergence (KL)**, **Jensen-Shannon divergence (JS)**, **LP Norm (LP)**, **Total Variation Distance (TVD)**,

Komogorov-Smirnov (KS), and **Conditional Demographic Disparity in Labels (CDDL)**. These pre-training bias metrics are model-agnostic, which means we can compute on raw datasets before the training step.

So, how do we interpret what is shown in the following diagram? A positive score near 1 or -1, such as 0.5933333334, for the class imbalance metric means that the sex facet has more training samples in the dataset for a given value (for example, 1). This means that there is an imbalance between the number of males and females in the provided dataset:

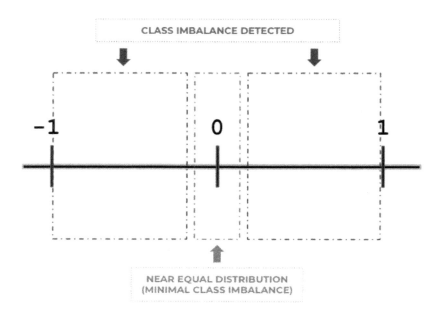

Figure 7.10 – Interpreting the class imbalance score

As we can see, the range of possible scores for the class imbalance metric is between -1 and 1. A score near 1 or -1 would mean that the dataset is imbalanced. On the other hand, a score near 0 would mean that we have a close to equal distribution of males and females in the dataset. A model that has been trained with an imbalanced dataset may have less accurate predictions on the underrepresented group.

Important note

How do we manage this **class imbalance** issue? There are different ways to handle class imbalance in a dataset. One possible approach would be to perform downsampling on the advantaged group and then use the **SageMaker Clarify** run_pre_training_bias() function on the adjusted dataset to see how the modification (for example, downsampling) has affected the CI metric score. If we get a CI metric score significantly closer to zero, this means that we have resolved the class imbalance issue that exists in the dataset. For more information on this topic, feel free to check out the whitepaper here: https://pages.awscloud.com/rs/112-TZM-766/images/Amazon.AI.Fairness.and.Explainability.Whitepaper.pdf.

There's more...

We can actually use the run_bias() function to run both pre-training and post-training analyses in the processing job at the same time:

```
clarify_processor.run_bias(
    data_config=bias_data_config,
    data_bias_config=bias_config,
    ...
    pre_training_methods=['CI'],
    post_training_methods=['DPPL', 'RD']
)
```

Of course, this approach may involve a few more configuration parameter values and prerequisites. However, it has its advantages, such as running a single processing job for both pre-training and post-training bias detection and analysis.

Now, let's take a closer look at detecting post-training bias!

Detecting post-training bias with SageMaker Clarify

In the previous recipe, we used **SageMaker Clarify** to help us detect pre-training bias in our data. In this recipe, we will use **SageMaker Clarify** to detect post-training bias in the same dataset we used in the previous recipe. In addition to this, we will train a model using this dataset and use it to compute the post-training bias metrics. Specifically, we will compute the **Difference in Positive Proportions in Predicted Labels (DPPL)** and **Recall Difference (RD)** metric values and check the results after the processing job has finished running.

> **Note**
>
> Why is this important? If the metric value for **DPPL** suggests bias against a disadvantaged group, this means that the machine learning model has a higher chance of predicting positive outcomes for the advantaged group. For example, if the advantaged group involves male applicants and the disadvantaged group involves female applicants, a machine learning model may accept more scholarship applications for the advantaged group over the disadvantaged group. Similarly, if the metric value for **RD** suggests bias against a disadvantaged group, this means that the machine learning model has a higher chance of *correctly predicting* possible outcomes for the advantaged group over the disadvantaged group. Using the same application example, the machine learning model has a higher chance of correctly accepting scholarship applications for applicants older than 30 years. Lastly, note that **DPPL** and **RD** measure different things. RD measures the differences in the **True Positive Rate** between two groups, which may mean higher false negatives for another group.

Getting ready

This recipe continues from the *Detecting pre-training bias with SageMaker Clarify* recipe.

How to do it...

The first set of steps in this recipe focus on preparing the model using the **SageMaker Python SDK**. Note that before we can run the post-training bias job, we need to get the model ready. Let's get started:

1. Create a new notebook using the `Python 3 (Data Science)` kernel inside the `my-experiments/chapter07` directory and rename it to the name of this recipe (`Detecting post-training bias with SageMaker Clarify`). When prompted for the kernel to use, choose `Python 3 (Data Science)`.

2. Use the `%store` magic to load the variable values for `s3_bucket_name`, `prefix`, and `training_data_path`:

```
%store -r s3_bucket_name
%store -r prefix
%store -r training_data_path
```

3. Import, prepare, and load a few prerequisites, such as `session`, `region`, and `role`, using the **SageMaker Python SDK**:

```
import sagemaker
session = sagemaker.Session()
region = session.boto_region_name
role = sagemaker.get_execution_role()
```

4. Initialize the variable values for `s3_training_data_path` and `s3_output_path` as well:

```
s3_training_data_path = training_data_path
s3_output_path = f"s3://{s3_bucket_name}/{prefix}/output"
```

5. Use the AWS CLI to download the test dataset CSV file from the S3 bucket to the `tmp` directory:

```
!aws s3 cp {s3_training_data_path} tmp/training_data.csv
```

6. Load the training dataset using the `read_csv()` function:

```
import pandas as pd
training_data = pd.read_csv("tmp/training_data.csv")
```

7. Use the `retrieve()` function to get the container URI:

```
from sagemaker.image_uris import retrieve
container = retrieve('xgboost', region, version='1.2-1')
```

8. Initialize the `Estimator` object:

```
from sagemaker.estimator import Estimator
estimator = Estimator(
    container,
    role,
    instance_count=1,
```

```
        instance_type='ml.m5.large',
        sagemaker_session=session
    )
```

9. Specify the hyperparameters using the `set_hyperparameters()` function:

```
estimator.set_hyperparameters(
    objective='binary:logistic',
    max_depth=8,
    eta=0.1,
    min_child_weight=4,
    num_round=500
)
```

10. Use `TrainingInput`:

```
from sagemaker.inputs import TrainingInput
train_input = TrainingInput(
    s3_training_data_path,
    content_type='csv'
)
```

11. Call the `fit()` function to start the training job:

```
%%time
estimator.fit({'train': train_input}, wait='True')
```

This should take about 4 to 7 minutes to complete.

12. Define the `generate_model_name()` function:

```
import random
from string import ascii_uppercase
def generate_model_name():
    chars = random.choices(ascii_uppercase, k=5)
    output = 'model-' + ''.join(chars)
    return output
```

13. Use the `generate_model()` function to generate the model name:

```
model_name = generate_model_name()
```

This should give us a value similar to `'model-RVJML'`.

> **Note**
>
> Why are we generating a random string for the model name? In the next step, we will use the `create_model()` function, which accepts the `name` parameter value. As you have probably guessed already, this `name` parameter value should be unique.

14. Use the `create_model()` function from the `Estimator` object to create the model entity:

```
model = estimator.create_model(name=model_name)
```

At this point, the model has been trained and prepared for the succeeding set of steps.

The next set of steps focus on using the prerequisites from the first half of this recipe to run the post-training bias detection job:

15. Use the `prepare_container_def()` function and store the dictionary with values for `Image`, `Environment`, and `ModelDataUrl` in the `container_def` variable:

```
container_def = model.prepare_container_def()
```

16. Use the `create_model()` function from the session:

```
session.create_model(
    model_name,
    role,
    container_def
)
```

17. Initialize the `SageMakerClarifyProcessor` object:

```
from sagemaker.clarify import SageMakerClarifyProcessor
processor = SageMakerClarifyProcessor(
    role=role,
    instance_count=1,
    instance_type='ml.m5.large',
```

```
        sagemaker_session=session
    )
```

18. Initialize the `DataConfig` object with the prerequisites we prepared in the previous steps:

```
from sagemaker.clarify import DataConfig
data_config = DataConfig(
    s3_data_input_path=s3_training_data_path,
    s3_output_path=s3_output_path,
    label='approved',
    headers=training_data.columns.to_list(),
    dataset_type='text/csv'
)
```

19. Initialize the `ModelConfig` and `ModelPredictedLabelConfig` objects as well:

```
from sagemaker.clarify import ModelConfig
model_config = ModelConfig(
    model_name=model_name,
    instance_type='ml.c5.xlarge',
    instance_count=1,
    accept_type='text/csv'
)
```

20. Initialize the `ModelPredictedLabelConfig` object with a probability threshold of `0.5`:

```
from sagemaker.clarify import ModelPredictedLabelConfig
predictions_config = ModelPredictedLabelConfig(
    probability_threshold=0.5
)
```

The value specified for the `probability_threshold` parameter is used to select the binary label, depending on the predicted value. For example, if the predicted value is `0.4`, then the predicted value will be set to `0` since `0.4` is less than `0.5`.

21. Initialize the `BiasConfig` object:

```
from sagemaker.clarify import BiasConfig
bias_config = BiasConfig(
```

```
label_values_or_threshold=[1],
facet_name='sex'
)
```

Note that this configuration may not make sense at the moment! Do not worry, though – we will discuss these values in detail in the *How it works…* section.

22. Call the `run_post_training_bias()` function and use the configuration objects we initialized in the previous steps as the parameter values:

```
%%time
processor.run_post_training_bias(
    data_config=data_config,
    data_bias_config=bias_config,
    methods=['DPPL', 'RD'],
    model_config=model_config,
    model_predicted_label_config=predictions_config
)
```

Here, we have configured the post-training bias detection job to focus on DPPL and RD. If you are wondering what these are, do not worry – we will explain these in detail in the *How it works…* section.

> **Note**
> This step should take around 10 to 15 minutes to complete. While waiting, feel free to grab a cup of coffee or tea!

23. Store the S3 location of the output files of the processing job in the `output_destination` variable:

```
output = processor.latest_job.outputs[0]
output_destination = output.destination
```

24. Download and inspect the `analysis.json` file that was generated by the **SageMaker Clarify** processing job:

```
!aws s3 cp {output_destination}/ tmp/ --recursive
!cat tmp/analysis.json
```

This should give us a nested structure, similar to what is shown in the following screenshot:

```
{
    "version": "1.0",
    "post_training_bias_metrics": {
        "label": "approved",
        "facets": {
            "sex": [
                {
                    "value_or_threshold": "1",
                    "metrics": [
                        {
                            "name": "DPPL",
                            "description": "Difference in Positive Proportions in Predicted Labels (DPPL)",
                            "value": -0.33541395157418197
                        },
                        {
                            "name": "RD",
                            "description": "Recall Difference (RD)",
                            "value": 0.0
                        }
                    ]
                },
                {
                    "value_or_threshold": "0",
                    "metrics": [
                        {
                            "name": "DPPL",
                            "description": "Difference in Positive Proportions in Predicted Labels (DPPL)",
                            "value": 0.33541395157418197
                        },
                        {
                            "name": "RD",
                            "description": "Recall Difference (RD)",
                            "value": 0.0
                        }
                    ]
                }
            ]
        },
        "label_value_or_threshold": "1"
    }
}
```

Figure 7.11 – Post-training bias metric values for DDPL and RD

Here, we can see that the **Difference in Positive Proportions in Predicted Labels (DDPL)** metric value is -0.33541395157418197, while the **Recall Difference (RD)** metric value is 0.0. If you are wondering what these values mean, do not worry – we will discuss this shortly in the *How it works...* section!

25. Finally, use the %store magic to store the variable value for model_name:

```
%store model_name
```

We will use this value again in the *Enabling ML explainability with SageMaker Clarify* recipe.

Let's see how this works!

How it works...

SageMaker Clarify makes use of **SageMaker Processing** to run a job that assesses data. In this case, we specified that we want to only get the **DPPL** and **RD** metrics from a larger list of post-training bias metrics. In addition to these metrics, there are other bias metrics we can compute using **SageMaker Clarify**. These include **Disparate Impact (DI)**, **Difference in Conditional Acceptance (DCA)**, **Difference in Conditional Rejection (DCR)**, **Difference in Acceptance Rates (DAR)**, **Difference in Rejection Rates (DRR)**, **Accuracy Difference (AD)**, **Conditional Demographic Disparity of Predicted Labels (CDDPL)**, **Treatment Equality (TE)**, and **Flip Test (FT)**. These metric values help quantify and detect model and data bias:

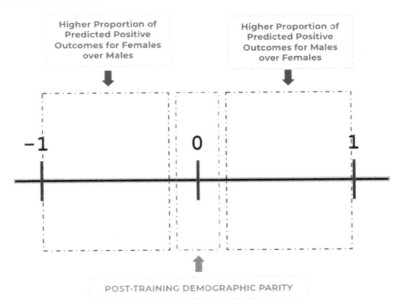

Figure 7.12 – How to interpret the DPPL metric score

So, how do we interpret what is shown in the preceding diagram? A positive **DPPL** value such as 0.335414 means that the sex facet has a relatively higher proportion of predicted possible outcomes for a given numeric value (for example, 1). This means that we are getting a higher approval rate for males than females for scholarship applications.

On the other hand, a value of 0.0 for **RD** means that no bias has been detected against a disfavored group when it comes to correct predictions. Since the formula for **RD** involves the difference in the true positive rates of the two groups involved (for example, TPR(a) - TPR(b)), a value of 0.0 suggests that the model has an equal true positive rate for both groups. This means that the model has an equal chance of getting the prediction correct for positive outcomes for both males and females.

> **Important note**
>
> For more information on post-processing bias metrics for training data, feel free to check out the whitepaper here: `https://pages.awscloud.com/rs/112-TZM-766/images/Amazon.AI.Fairness.and.Explainability.Whitepaper.pdf`.

Enabling ML explainability with SageMaker Clarify

In the previous two recipes, we used **SageMaker Clarify** to detect pre-training and post-training bias. In this recipe, we will take a closer look at ML explainability and how we can use **SageMaker Clarify** to generate an ML explainability report.

We will see the importance of ML explainability as we deal with ethical and legal concerns. For example, customers will want a better idea of how their information is used by a machine learning system to perform recommendations or predictions. In addition to this, ML explainability empowers data scientists and machine learning practitioners to make more accurate and fair models.

> **Note**
>
> It is important to distinguish **model interpretability** from **model explainability**. **Model interpretability** focuses on understanding what a machine learning model is doing internally. On the other hand, **model explainability** involves understanding how a machine learning model performed a prediction using certain feature values in human terms.

With **SageMaker Clarify**, we will be able to automatically compute the SHAP values, which help us determine the importance of each feature concerning the predictions a model makes.

Getting ready

This recipe continues from the *Detecting post-training bias with SageMaker Clarify* recipe.

How to do it...

The steps for this recipe are as follows:

1. Create a new notebook using the `Python 3 (Data Science)` kernel inside the `my-experiments/chapter07` directory and rename it to the name of this recipe (`Enabling ML explainability with SageMaker Clarify`). When prompted for the kernel to use, choose `Python 3 (Data Science)`.

2. Use the `%store` magic to load the variable values for `s3_bucket_name`, `prefix`, `training_data_path`, `test_data_path`, and `model_name`:

   ```
   %store -r s3_bucket_name
   %store -r prefix
   %store -r training_data_path
   %store -r test_data_path
   %store -r model_name
   ```

3. Import, prepare, and load a few prerequisites, such as `session`, `region`, and `role`, from the **SageMaker Python SDK**:

   ```
   import sagemaker
   session = sagemaker.Session()
   region = session.boto_region_name
   role = sagemaker.get_execution_role()
   ```

4. Prepare the S3 target paths as well:

   ```
   s3_training_data_path = training_data_path
   s3_test_data_path = test_data_path
   s3_output_path = f"s3://{s3_bucket_name}/{prefix}/output"
   ```

5. Use the **AWS CLI** to copy the CSV files containing the training and test sets:

   ```
   !aws s3 cp {s3_training_data_path} tmp/training_data.csv
   !aws s3 cp {s3_test_data_path} tmp/test_data.csv
   ```

6. Load the training and test datasets and then generate the `features` CSV file:

   ```
   import pandas as pd
   training_data = pd.read_csv("tmp/training_data.csv")
   test_data = pd.read_csv("tmp/test_data.csv")
   target = test_data['approved']
   ```

```
features = test_data.drop(columns=['approved'])
features.to_csv('tmp/test_features.csv',
                index=False,
                header=False)
```

Note that we have used the test dataset minus the approved column as the source of data for the test_features.csv file.

7. Copy the test_features.csv file to an Amazon S3 bucket:

```
base = f"s3://{s3_bucket_name}/{prefix}/input"
s3_feature_path = f"{base}/test_features.csv"
!aws s3 cp tmp/test_features.csv {s3_feature_path}
```

8. Initialize the ModelConfig object and specify the model_name parameter value. Note that a model with this name must exist already before running the following lines of code. If you have used the create_model() function from the *Detecting post-training bias with SageMaker Clarify* recipe, then we will not have any problems running the next set of instructions. Otherwise, make sure that you locate and get the name of a model from a completed training job that would fit this recipe:

```
from sagemaker.clarify import ModelConfig
model_config = ModelConfig(
    model_name=model_name,
    instance_type='ml.c5.xlarge',
    instance_count=1,
    accept_type='text/csv'
)
```

9. Initialize the SageMakerClarifyProcessor object:

```
from sagemaker.clarify import SageMakerClarifyProcessor
processor = SageMakerClarifyProcessor(
    role=role,
    instance_count=1,
    instance_type='ml.m5.large',
    sagemaker_session=session
)
```

10. Prepare the baseline variable:

```
baseline = features.iloc[0:200].values.tolist()
```

Here, we are using 200 records from the features DataFrame for the baseline.

11. Initialize the SHAPConfig object and pass the baseline as one of the parameter values. Specify the number of samples and the aggregation method for the global SHAP values:

```
from sagemaker.clarify import SHAPConfig
shap_config = SHAPConfig(
    baseline=baseline,
    num_samples=50,
    agg_method='median'
)
```

There are three possible values for the agg_method parameter – mean_abs, median, and mean_sq. This configuration parameter simply tells the training job how it aggregates and adds up the SHAP values.

12. Prepare s3_output_path. Once the processing job has been completed, the output files will be stored in this S3 output path:

```
headers = training_data.columns.to_list()
```

13. Initialize the DataConfig object. Use the variables and objects we prepared and initialized in the previous step as the arguments during this initialization step:

```
from sagemaker.clarify import DataConfig
data_config = DataConfig(
    s3_data_input_path=s3_training_data_path,
    s3_output_path=s3_output_path,
    label='approved',
    headers=headers,
    dataset_type='text/csv'
)
```

14. Call the run_explainability() function and pass the data_config, model_config, and shap_config variable values when calling the function. This step should about 12 to 18 minutes to complete:

```
%%time
processor.run_explainability(
    data_config=data_config,
    model_config=model_config,
```

```
        explainability_config=shap_config
    )
```

This should yield a set of logs, similar to what is shown in the following screenshot:

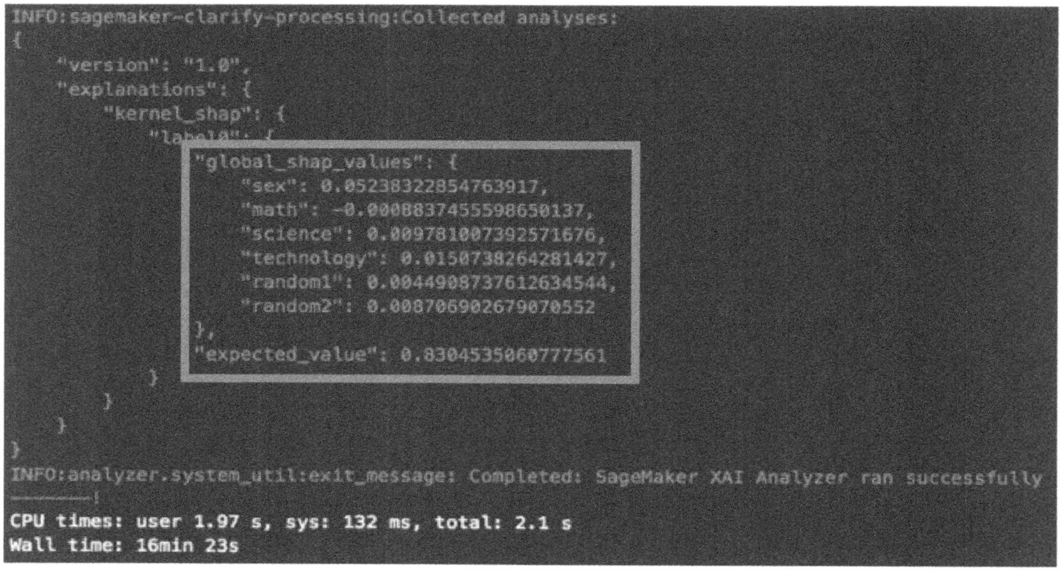

Figure 7.13 – Logs after calling the run_explainability() function

Here, we can see that `sex`, `science`, and `technology` have the highest SHAP values from our list of predictors. Given that SHAP values quantify the contribution of each predictor getting approved or not, we can say that `sex`, `science`, and `technology` are the most important features based on their SHAP values.

Let's see how this works!

How it works...

Sometimes, it is not enough that a model has good predictive performance. In this recipe, we used **SageMaker Clarify** to generate a report that helps explain the model. **SageMaker Clarify** uses **Shapley values** to interpret machine learning models and provide details regarding the importance of each feature in the dataset. The importance of each feature maps to the relative contribution of the features to the target value when making a prediction.

This recipe can be divided into six major parts:

1. Initializing the `SageMakerClarifyProcessor` object
2. Initializing the configuration objects

3. Creating a baseline

4. Initializing the `SHAPConfig` object

5. Calling the `run_explainability()` function

6. Inspecting the results

Once the processing job had been completed toward the end of the *How to do it...* section, we got the global SHAP values, as shown in the preceding screenshot. Given that SHAP values quantify the marginal contribution of each predictor to predicting the target label, this means that the higher the SHAP value of a predictor, the higher its contribution to predicting the target label. That said, we can see that the `math`, `random1`, and `random2` features are not as important as the `sex`, `science`, and `technology` features, as indicated by their scores.

Note

If you can still remember how we generated the synthetic dataset in the *Generating a synthetic dataset and using SageMaker Feature Store for storage and management* recipe, the `random1` and `random2` columns do not really contribute to predicting the target label. This means that if we change the values for `random1` and `random2`, there's a big chance that this set of changes will not significantly impact the value of the `approved` target label, compared to the impact of changing the values for `sex`, `science`, and `technology`.

Deploying an endpoint from a model and enabling data capture with SageMaker Model Monitor

In this recipe, we will deploy the model we trained in the *Detecting post-training bias with SageMaker Clarify* recipe to an inference endpoint. We must be aware that the machine learning process does not end after a model has been deployed to production. We will only know the deployed model's true performance once it is exposed to more data that it has not seen before. That said, we must capture the request and response pairs when the inference endpoint is invoked. This gives us the ability to analyze if there are issues in the deployed model, or if there are issues in the data that is being passed as the payload to the inference endpoint.

The great thing about using **Amazon SageMaker** is that we do not have to build this ourselves, since these challenges and potential issues can already be solved and handled using **SageMaker Model Monitor**. Finally, we will demonstrate how to use the **SageMaker Feature Store** online store when pulling test records for inference after the model has been deployed and data capture has been enabled.

Getting ready

This recipe continues from the *Enabling ML explainability with SageMaker Clarify* recipe.

How to do it...

In the first set of steps, we will focus on deploying the model we created in the *Detecting post-training bias with SageMaker Clarify* recipe. Let's get started:

1. Create a new notebook using the Python 3 (Data Science) kernel inside the my-experiments/chapter07 directory and rename it to the name of this recipe (Deploying an endpoint from a model and enabling data capture with SageMaker Model Monitor). When prompted for the kernel to use, choose Python 3 (Data Science).

2. Import, prepare, and load a few prerequisites, such as session, region, and role, from the **SageMaker Python SDK**:

    ```
    import sagemaker
    session = sagemaker.Session()
    region = session.boto_region_name
    role = sagemaker.get_execution_role()
    ```

3. Use the %store magic to load the variable value for model_name:

    ```
    %store -r model_name
    ```

4. Initialize the Sagemaker boto3 client:

    ```
    import boto3
    client = boto3.client('sagemaker')
    ```

5. Use the describe_model() function to load the details of the model:

    ```
    response = client.describe_model(
        ModelName=model_name
    )
    ```

6. Extract the container image URI from the resulting nested response dictionary values:

```
container = response['PrimaryContainer']['Image']
```

7. Extract the model data URL from the resulting nested response dictionary values:

```
container = response['PrimaryContainer']
model_data = container['ModelDataUrl']
```

8. Initialize the Model object and use the model_name, container, model_data, role, and session variables as the arguments when initializing the Model object:

```
model = sagemaker.model.Model(
    name=model_name,
    image_uri=container,
    model_data=model_data,
    role=role,
    sagemaker_session=session
)
```

> **Note**
>
> We can also add predictor_cls=sagemaker.Predictor when initializing sagemaker.model.Model. This will make the model.deploy() function call in the next step return a Predictor object. This should allow us to skip a step of initializing the Predictor object separately using the model.endpoint_name value.

9. Deploy the model using the deploy() function. Wait a few minutes for the model to be deployed into an inference endpoint:

```
%%time
model.deploy(
    initial_instance_count = 1,
    instance_type = 'ml.m5.large'
)
```

If you are wondering what model we are deploying here, we are deploying the **XGBoost** model we trained in the *Detecting post-training bias with SageMaker Clarify* recipe.

> **Note**
>
> This step should take around 7 to 15 minutes to complete. Feel free to grab a
> cup of coffee or tea while waiting!

10. Initialize the `Predictor` object by specifying the endpoint name from the
    previous step as the parameter value for `endpoint_name`:

```
from sagemaker import Predictor
predictor = Predictor(
    endpoint_name=model.endpoint_name
)
```

The next set of steps focus on loading a record from the online feature store, and then
using the loaded data as the payload to test the inference endpoint that's been deployed:

11. Make sure a few variable values, such as `s3_bucket_name`, `prefix`, and
    `s3_capture_upload_path`, are ready:

```
%store -r s3_bucket_name
%store -r prefix
base = f"s3://{s3_bucket_name}/{prefix}"
s3_capture_upload_path = f"{base}/model-monitor"
```

12. Initialize the `DataCaptureConfig` object and configure it so that it captures
    100% of the request and response pairs during endpoint invocation:

```
from sagemaker.model_monitor import DataCaptureConfig
data_capture_config = DataCaptureConfig(
    enable_capture = True,
    sampling_percentage=100,
    destination_s3_uri=s3_capture_upload_path,
    kms_key_id=None,
    capture_options=["REQUEST", "RESPONSE"],
    csv_content_types=["text/csv"],
    json_content_types=["application/json"]
)
```

13. Use the `update_data_capture_config()` function to enable data capture in our inference endpoint:

```
%%time
predictor.update_data_capture_config(
    data_capture_config=data_capture_config
)
```

This should take approximately 7 to 10 minutes to complete.

Now that we have deployed the model and enabled data capture, we will get a record from the online store and use it as the payload to test the inference endpoint:

14. Initialize the `boto_session` and `runtime` variables:

```
boto_session = boto3.Session(region_name=region)
runtime = boto_session.client(
    service_name='sagemaker-featurestore-runtime',
    region_name=region
)
```

15. Get one of the records inside the online feature store using the `get_record()` function:

```
feature_group_name = 'cookbook-feature-group'
record_response = runtime.get_record(
    FeatureGroupName=feature_group_name,
    RecordIdentifierValueAsString="950"
)
```

Here, we made use of the feature store we created in the *Generating a synthetic dataset and using SageMaker Feature Store for storage and management* recipe at the start of this chapter.

> **Note**
>
> When using **SageMaker Feature Store**, we are charged for writing, reading, and storing the data on the feature store. As part of the free tier, we get 10M write units, 10M read units, and 25 GB of storage for free usage when using **SageMaker Feature Store**. For more information, feel free to check out the pricing page: `https://aws.amazon.com/sagemaker/pricing/`.

16. Inspect the `record_response` structure and internal values:

```
record_response['Record']
```

This should give us a structure similar to what is shown in the following screenshot:

```
[{'FeatureName': 'approved', 'ValueAsString': '1'},
 {'FeatureName': 'sex', 'ValueAsString': '1'},
 {'FeatureName': 'math', 'ValueAsString': '92'},
 {'FeatureName': 'science', 'ValueAsString': '83'},
 {'FeatureName': 'technology', 'ValueAsString': '86'},
 {'FeatureName': 'random1', 'ValueAsString': '96'},
 {'FeatureName': 'random2', 'ValueAsString': '67'},
 {'FeatureName': 'index', 'ValueAsString': '950'},
 {'FeatureName': 'event_time', 'ValueAsString': '1623604510.0'}]
```

Figure 7.14 – Value of record_response['Record']

Here, we can see the record that's been stored in the feature group.

17. Extract the values from the `record_response` nested dictionary and store the feature values inside the `test_record_list` list. Note that we will not include the value of the `approved` variable as we will only be needing the values of the predictor columns – `sex`, `math`, `science`, `technology`, `random1`, and `random2` – when testing the inference endpoint:

```
test_record_list = [
    record_response['Record'][1]['ValueAsString'],
    record_response['Record'][2]['ValueAsString'],
    record_response['Record'][3]['ValueAsString'],
    record_response['Record'][4]['ValueAsString'],
    record_response['Record'][5]['ValueAsString'],
    record_response['Record'][6]['ValueAsString'],
]
test_record_list
```

We should get an output value similar to `['1', '92', '83', '86', '96', '67']`.

18. Once we have the `test_record_list` list ready, we must prepare the `csv_input` string value by joining the elements of `test_record_list`:

```
csv_input = ','.join(test_record_list)
csv_input
```

We should get an output value similar to `'1,92,83,86,96,67'`.

19. Update the `serializer` and `deserializer` attributes of the `Predictor` object:

```
from sagemaker.deserializers import JSONDeserializer
from sagemaker.serializers import CSVSerializer
predictor.serializer = CSVSerializer()
predictor.deserializer = JSONDeserializer()
```

20. Use the `predict()` function and check if the inference endpoint works as expected:

```
predictor.predict(csv_input)
```

21. Use the `%store` magic to store the variable values for `endpoint_name` and `csv_input`:

```
endpoint_name = predictor.endpoint_name
%store endpoint_name
%store csv_input
```

At this point, the request and response pair data should be recorded and stored in the S3 bucket. This is because we enabled data capture with **Model Monitor** before we performed the test prediction.

> **Important note**
> Do not delete this endpoint yet as we will use this in the next recipe.

Now, let's see how this works!

How it works...

In this recipe, we made use of several capabilities of SageMaker, such as **SageMaker Model Monitor** and **SageMaker Feature Store**, to accomplish what we needed to do. With that, we can divide this recipe into three major parts:

1. Deploying an existing model to an inference endpoint using the `Model` object's `deploy()` function
2. Enabling data capture with **SageMaker Model Monitor**
3. Using a record from the online feature store as the payload when using the `predict()` function

In the *Detecting post-training bias with SageMaker Clarify* recipe, we created the model entity using the `create_model()` function. In this recipe, we initialized a `Model` object with the name, container image, and the model data we obtained using the `describe_model()` function. Next, we used the `deploy()` function to deploy the model into an inference endpoint.

After the deployment step, we configured a `DataCaptureConfig` object to capture 100% of the request-response pairs during endpoint invocation. We then used the `update_data_capture_config()` function to apply this configuration to the existing inference endpoint. Toward the latter part of the *How to do it...* section, we used the **SageMaker FeatureStore Runtime Client** with `boto3` to load a single record from the online feature store. Finally, we used this record as the payload when using the `predict()` function.

> **Note**
> Why not use the offline store for this? The offline store is designed to load data for training the model and for batch predictions. Using the online store is more appropriate for this use case as it is designed to support low millisecond latency reads.

As we have enabled data capture in this recipe, we expect that the S3 upload path we specified in the `destination_s3_uri` parameter, when initializing the `DataCaptureConfig` object, will contain the captured request and response pairs, similar to what is shown in the following screenshot:

```
{'captureData': {'endpointInput': {'data': '1,92,83,86,96,67'
                                   'encoding': 'CSV',
                                   'mode': 'INPUT'
                                   'observedContentType': 'text/csv'},
                 'endpointOutput': {'data': '0.9916712045669556',
                                    'encoding': 'CSV',
                                    'mode': 'OUTPUT',
                                    'observedContentType': 'text/csv; '
                                                           'charset=utf-8'}},
 'eventMetadata': {'eventId': '978d0918-6e55-4b9d-91ca-3aac8d39231f',
                   'inferenceTime': '2021-06-14T06:53:23Z'},
 'eventVersion': '0'}
```

Figure 7.15 – Captured data downloaded from the S3 upload path

Here, we can see that the record we used as the payload when calling the `predict()` function had its counterpart `jsonl` file stored in the S3 upload path. That said, the S3 upload path will contain more `jsonl` files as we make more attempts to invoke the endpoint. Of course, collecting the request and response pairs is just the first step, as these can be used to automatically detect different issues and violations using **SageMaker Model Monitor** as well. That said, let's proceed with the next recipe!

Baselining and scheduled monitoring with SageMaker Model Monitor

In the previous recipe, we deployed the model to an inference endpoint and enabled data capture using **SageMaker Model Monitor**. This allows us to collect the request and response pairs when the endpoint is invoked during inference. Note that we are just scratching the surface here in terms of what we can do with **SageMaker Model Monitor**. Using **SageMaker Model Monitor**, we can also automatically monitor and detect the following issues:

- Drift in data quality
- Drift in model quality metric values
- Bias drift during prediction
- Feature attribution drift

This is important as there's a lot of things that can happen after our model gets deployed to production.

In this recipe, we will focus our efforts on detecting **data quality drift**. We will start by preparing a baseline, and then creating a scheduled monitoring job that processes the data captured by **Model Monitor** and outputs summary statistics and violations reports. This will help us debug issues in our data and model when they're used in production environments.

Getting ready

This recipe continues from the *Deploying an endpoint from a model and enabling data capture with SageMaker Model Monitor* recipe.

How to do it...

The first set of steps in this recipe focus on running a **Model Monitor** baselining job to generate a suggested constraints configuration, along with the baseline statistics report. Let's get started:

1. Create a new notebook using the Python 3 (Data Science) kernel inside the my-experiments/chapter07 directory and rename it to the name of this recipe (Baselining and scheduled monitoring with SageMaker Model Monitor).

2. Use the `%store` magic to load the variable values for `s3_bucket_name` and `prefix`. After that, prepare the S3 paths stored in the `baseline_data_uri` and `baseline_results_uri` variables:

```
%store -r s3_bucket_name
%store -r prefix
base = f's3://{s3_bucket_name}/{prefix}'
baseline_data_uri = f'{base}/input/training_data.csv'
baseline_results_uri = f"{base}/model-monitor/baseline-
results"
```

3. Load the contents of the CSV file containing the baseline data:

```
local_file = "tmp/baseline.csv"
!aws s3 cp {baseline_data_uri} {local_file}
import pandas as pd
baseline_df = pd.read_csv(local_file)
```

4. Initialize the `DefaultModelMonitor` object:

```
import sagemaker
role = sagemaker.get_execution_role()
from sagemaker.model_monitor import DefaultModelMonitor
default_monitor = DefaultModelMonitor(
    role=role,
    instance_count=1,
    instance_type='ml.m5.large',
    volume_size_in_gb=20,
    max_runtime_in_seconds=3600,
)
```

5. Run the **Model Monitor** baselining job using the `suggest_baseline()` function. We are using the `baseline_data_uri` and `baseline_results_uri` variables we prepared in the previous steps as the parameter values for `baseline_dataset` and `output_s3_uri`, respectively, when calling the `suggest_baseline()` function:

```
%%time
from sagemaker.model_monitor import dataset_format
dsf = dataset_format.DatasetFormat.csv(header=True)
```

```
default_monitor.suggest_baseline(
    baseline_dataset=baseline_data_uri,
    dataset_format=dsf,
    output_s3_uri=baseline_results_uri,
    wait=True
)
```

When the `suggest_baseline()` function is called, a **SageMaker Processing** job makes use of the `sagemaker-model-monitor-analyzer` container, which creates the baseline and suggests the constraints.

> **Note**
>
> This step should take around 5 to 7 minutes to complete. Feel free to grab a cup of coffee or tea while waiting!

6. Use the `baseline_statistics()` function of the `DefaultModelMonitor` object and the `json_normalize()` function from `pandas` to inspect the baseline statistics generated by the baselining job:

```
baseline_job = default_monitor.latest_baselining_job
stats = baseline_job.baseline_statistics()
schema_dict = stats.body_dict["features"]

import pandas as pd
schema_df = pd.json_normalize(schema_dict)
schema_df.head(5)
```

This should give us a `DataFrame` of values, similar to what is shown in the following screenshot:

	name	inferred_type	numerical_statistics.common.num_present	numerical_statistics.common.num_missing	numerical_statistics.mean
0	approved	Integral	600	0	0.798333
1	sex	Integral	600	0	0.796667
2	math	Integral	600	0	79.303333
3	science	Integral	600	0	79.430000
4	technology	Integral	600	0	80.255000

Figure 7.16 – Value schema_df

Here, we have the schema dictionary and the baseline statistics. We were able to convert the dictionary results that contain the baseline statistics into a `DataFrame` using the `json_normalize()` function.

7. Use the `suggested_constraints()` function to get the suggested constraints. Then, use the `json_normalize()` function from `pandas` to convert the dictionary results into a `DataFrame` as well:

```
constraints = baseline_job.suggested_constraints()
constraints_dict = constraints.body_dict["features"]
constraints_df = pd.json_normalize(constraints_dict)
constraints_df.head(7)
```

This should give us a `DataFrame` of values, similar to what is shown in the following screenshot:

	name	inferred_type	completeness	num_constraints.is_non_negative
0	approved	Integral	1.0	True
1	sex	Integral	1.0	True
2	math	Integral	1.0	True
3	science	Integral	1.0	True
4	technology	Integral	1.0	True
5	random1	Integral	1.0	True
6	random2	Integral	1.0	True

Figure 7.17 – Value of constraints_df

Here, we have a `DataFrame` containing the suggested constraints. We can see that the `approved`, `sex`, `math`, `science`, `technology`, `random1`, and `random2` features have `Integral` are their `inferred_type` values. This means that the baselining job detected that the values that were used from the baseline CSV file contained only integer values for these features. Later, we will see how these suggested constraint values in our baseline will be used to detect if there are issues with future values captured by **Model Monitor**.

The next set of steps focus on scheduling a monitoring job that checks for constraint violations and generates statistics reports:

8. Define the `generate_schedule_name()` function, as shown in the following block of code. This function simply generates a random string for the name of the monitoring schedule. After that, use this function to generate the schedule name:

```
import random
from string import ascii_uppercase
def generate_schedule_name():
    chars = random.choices(ascii_uppercase, k=5)
    output = 'schedule-' + ''.join(chars)
    return output


schedule_name = generate_schedule_name()
```

This should generate a string in a format similar to `'schedule-KYTXY'`.

9. Specify the S3 target report path and store it inside the `s3_report_path` variable:

```
s3_report_path = f'{base}/report-path'
```

10. Load the baseline statistics using the `baseline_statistics()` function and store it in the `baseline_statistics` variable. Load the suggested constraints values in the `constraints` variable as well:

```
baseline_statistics = default_monitor.baseline_
statistics()
constraints = default_monitor.suggested_constraints()
```

11. Prepare the `cron_expression` variable value:

```
from sagemaker.model_monitor import
CronExpressionGenerator
cron_expression = CronExpressionGenerator.hourly()
cron_expression
```

We should get an output equal or similar to `'cron(0 * ? * * *)'`. Here, we are planning to use this value to configure the monitoring job we will schedule later, so that we can perform its scheduled monitoring job on an hourly basis.

12. Load the `Predictor` object by specifying the endpoint name from the *Deploying an endpoint from a model and enabling data capture with SageMaker Model Monitor* recipe:

```
%store -r endpoint_name
from sagemaker import Predictor
predictor = Predictor(endpoint_name=endpoint_name)
```

Note that this endpoint exists already. All we are doing here is attaching a `Predictor` object to the endpoint using the endpoint name.

13. Use the `%store` magic to load the variable value for `csv_input`:

```
%store -r csv_input
csv_input
```

This should give us a value similar to `'1,92,83,86,96,67'`.

14. Perform a test prediction using the `csv_input` variable from the previous step and the `predict()` function of the `Predictor` object:

```
from sagemaker.deserializers import JSONDeserializer
from sagemaker.serializers import CSVSerializer
predictor.serializer = CSVSerializer()
predictor.deserializer = JSONDeserializer()
predictor.predict(csv_input)
```

15. Inspect the `constraints` variable using the `__dict__` attribute:

```
constraints.__dict__
```

This should give us a nested dictionary of values, similar to what is shown in the following screenshot:

```
{'body_dict': {'version': 0.0,
  'features': [{'name': 'approved',
    'inferred_type': 'Integral',
    'completeness': 1.0,
    'num_constraints': {'is_non_negative': True}},
  {'name': 'sex',
    'inferred_type': 'Integral',
    'completeness': 1.0,
    'num_constraints': {'is_non_negative': True}},
  {'name': 'math',
    'inferred_type': 'Integral',
    'completeness': 1.0,
    'num_constraints': {'is_non_negative': True}},
  {'name': 'science',
    'inferred_type': 'Integral',
    'completeness': 1.0,
    'num_constraints': {'is_non_negative': True}},
  {'name': 'technology',
    'inferred_type': 'Integral',
    'completeness': 1.0,
    'num_constraints': {'is_non_negative': True}},
  {'name': 'random1',
    'inferred_type': 'Integral',
    'completeness': 1.0,
    'num_constraints': {'is_non_negative': True}},
```

Figure 7.18 – Suggested constraints stored inside the constraints variable

Here, we have the suggested constraints that were generated by the baselining job. We can see that to change the `inferred_type` value of a certain feature or field, we would have to go through the nested structure of values, as we will see in the next step.

16. Next, we will override the `inferred_type` value from the suggested constraints using the following lines of code. Note that we need to use the `save()` function, similar to what is shown in the following block of code, to make sure this constraints configuration change is applied:

```
constraints.body_dict['features'][0]['inferred_type'] =
'Fractional'
constraints.save()
```

This should return a value similar to `'s3://<s3 bucket name>/chapter07/model-monitor/baseline-results/constraints.json'`.

17. Use the `create_monitoring_schedule()` function:

```
default_monitor.create_monitoring_schedule(
    monitor_schedule_name=schedule_name,
    endpoint_input=predictor.endpoint,
```

```
            output_s3_uri=s3_report_path,
            statistics=baseline_statistics,
            constraints=constraints,
            schedule_cron_expression=cron_expression,
            enable_cloudwatch_metrics=True,
        )
```

Here, we pass the variable values (for example, `baseline_statistics`, `constraints`, and `cron_expression`) we prepared in the previous steps as the parameter values when calling the `create_monitoring_schedule()` function.

18. Use the `sleep()` function to wait a few minutes while the monitoring schedule is being created:

```
from time import sleep
sleep(300)
```

19. Define the `perform_good_input()` and `perform_bad_input()` functions:

```
def perform_good_input():
    predictor.predict(csv_input)
    print("good input")

def perform_bad_input():
    csv_bad_input = '1,92,-83.3,86,-96,67'
    predictor.predict(csv_bad_input)
    print("bad input")
```

In the preceding block of code, we can see that the payload that's used in the `perform_bad_input()` function contains a negative floating-point value for the `science` feature. We expect to see a violation report later for this feature since, in our constraints configuration, we have specified that the `science` feature values our endpoint accepts should be non-negative integer values.

> **Note**
>
> The comma-separated values (`'1,92,-83.3,86,-96,67'`) map to the `sex`, `math`, `science`, `technology`, `random1`, and `random2` features.

20. Next, get the latest constraint violations and monitoring statistics that were collected by **Model Monitor** using the `latest_monitoring_constraint_violations()` and `latest_monitoring_statistics()` functions, respectively:

```
dm = default_monitor
gcv = dm.latest_monitoring_constraint_violations
lms = dm.latest_monitoring_statistics
monitoring_violations = gcv()
monitoring_statistics = lms()
```

Note that at this point, the values of `monitoring_violations` and `monitoring_statistics` will be `None`. This is because they will only have values after the scheduled processing job for collecting the monitoring violations and statistics has been executed and completed, which will happen after an hour or so.

21. Wait until we have data from **Model Monitor** using the following block of code:

```
%%time
from time import sleep
violations = monitoring_violations
while not violations:
    print("No executions yet. [Sleep - 5-min]")
    sleep(300)
    perform_good_input()
    perform_bad_input()
    try:
        violations = gcv()
    except:
        pass

print("Executions found!")
```

Here, we loop forever until we have a violations report available using the `latest_monitoring_constraint_violations()` function. We also call the `perform_good_input()` and `perform_bad_input()` functions inside the loop so that we have data to analyze when the scheduled monitoring job is executed.

> **Important note**
> This step will take about an hour to complete. Feel free to read the *How it works…* section of this recipe or even start the next chapter while waiting!

22. At this point, the previous block of code has finished running and the value of the `violations` variable is no longer `None`:

```
violations = gcv()
```

23. Copy the generated report on the constraints violations from the S3 bucket to the `tmp` directory using the **AWS CLI**:

```
!aws s3 cp {violations.file_s3_uri} tmp/violations.json
```

24. Inspect the constraint violations report:

```
!cat tmp/violations.json
```

This should yield a summary report inside a dictionary, similar to what is shown in the following screenshot:

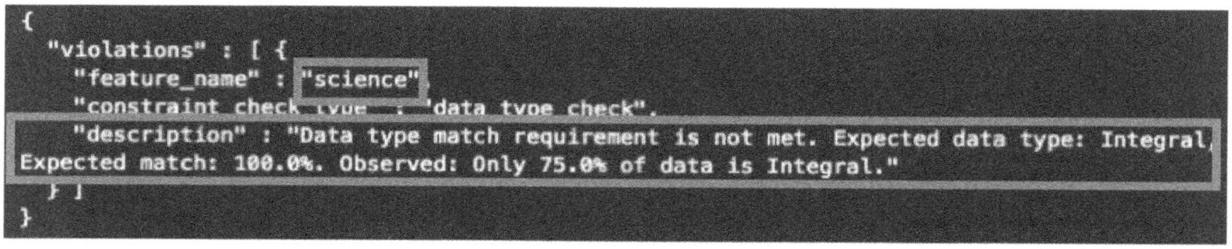

Figure 7.19 – Violations report

Here, we can see the violations report that was generated by **Model Monitor**. It has detected an issue in the `science` field values that were passed as a payload to the endpoint. In this case, the expected data type would be an integer. We passed a float value instead using the `perform_bad_input()` function.

25. Inspect the statistics report using the `latest_monitoring_statistics()` function. Note that the `lms()` function points to the `latest_monitoring_statistics()` function of the `DefaultModelMonitor` object:

```
monitoring_statistics = lms()
monitoring_statistics.__dict__
```

This should yield a nested structure of values, similar to what is shown in the following screenshot:

```
{'body_dict': {'version': 0.0,
  'dataset': {'item_count': 4},
  'features': [{'name': 'approved',
    'inferred_type': 'Fractional',
    'numerical_statistics': {'common': {'num_present': 4, 'num_missing': 0},
      'mean': 0.9229514747858047,
      'sum': 3.691805899143219,
      'std_dev': 0.1190260634633573,
      'min': 0.7167922854423523,
      'max': 0.9916712045669556,
```

Figure 7.20 – Latest monitoring statistics

Here, we can see a trimmed version of the monitoring statistics for the data captured by **Model Monitor**. Using the values from these summary statistics, we will have a better idea of the values in the request-response pairs that have been captured.

26. Finally, let's delete the monitoring schedule and the endpoint:

```
default_monitor.delete_monitoring_schedule()
predictor.delete_endpoint()
```

Do not forget this step as you will be charged for the amount of time the endpoint is running.

Now, let's see how this works!

How it works...

A lot of things can happen after a model gets deployed into production. There is a chance that the data that's sent to our inference endpoint contains data quality issues, similar to what we emulated in this recipe using the perform_bad_input() function. In the *How to do it...* section, we enabled continuous monitoring for the model and the data using **SageMaker Model Monitor** to detect these types of data quality issues. Here are the steps we performed to accomplish this:

1. We enabled data capture in the *Deploying an endpoint from a model and enabling data capture with SageMaker Model Monitor* recipe to collect the request and response pairs when invoking the machine learning model.

2. In this recipe, we generated a suggested baseline using a baselining job to serve as a reference point for metrics and suggested constraints. These constraints are rules that will be used to detect violations when there are data quality issues in future unseen data. We can think of constraints as the boundaries that let us know if the data that was used to invoke our model is still within bounds.

3. We modified the suggested constraints configuration so that **SageMaker Model Monitor** will not report false positives.

4. We configured a monitoring schedule that runs a processing job once every hour. This processing job makes use of the captured data, the baseline statistics, and the constraints configuration to generate statistics and violation reports. Note that we configured this monitoring schedule to automatically emit metrics to **Amazon CloudWatch** after each processing job's execution by specifying `enable_cloudwatch_metrics=True` when calling the `create_monitoring_schedule()` function.

Optionally, we can create an alarm to detect if a specific metric that's been collected in **Amazon CloudWatch** falls below a specific threshold value. This alarm can be then configured to trigger an **AWS Lambda** function, which will perform an automated action such as updating or retraining the model that was used in production. We can also use these alarms to notify specific members of the team to inspect and debug any detected issues in the production systems.

There's more...

How do rule violations work? Rule violations start with a set of rules and constraints that have been configured when the monitoring job is scheduled using `create_monitoring_schedule()`. There are different types of constraints we can specify, and one of them involves the data types constraint, similar to what is shown in the following diagram:

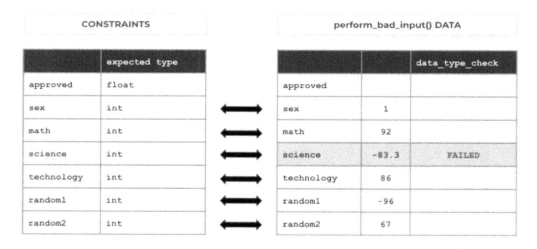

Figure 7.21 – Data type check violation

Here, we have the expected data types for the different features involved in this recipe. This means that if the data type of a feature value does not match what is specified in the constraints configuration, **Model Monitor** will flag that in the violations report after a scheduled job has finished executing.

> **Important note**
> There are other types of violations we can monitor using **SageMaker Model Monitor**. These include the `completeness_check`, `baseline_drift_check`, `missing_column_check`, `extra_column_check`, and `categorical_values_check` violation check types. For more information on this topic, feel free to check out `https://docs.aws.amazon.com/sagemaker/latest/dg/model-monitor-interpreting-violations.html`.

See also

If you are looking for examples of using **SageMaker Model Monitor** on real datasets and more complex examples, feel free to check out some of the notebooks in the `aws/amazon-sagemaker-examples` GitHub repository:

- Detecting model quality drift using **SageMaker Model Monitor**: `https://github.com/aws/amazon-sagemaker-examples/blob/master/sagemaker_model_monitor/model_quality/model_quality_churn_sdk.ipynb`

- Detecting bias drift and feature attribute drift using the bias monitor and explainability monitor: `https://github.com/aws/amazon-sagemaker-examples/blob/master/sagemaker_model_monitor/fairness_and_explainability/SageMaker-Model-Monitor-Fairness-and-Explainability.ipynb`

At this point, we should have a good idea of what we can do with **SageMaker Model Monitor**. Feel free to check out this link for more information about this topic: `https://docs.aws.amazon.com/sagemaker/latest/dg/model-monitor.html`.

8

Solving NLP, Image Classification, and Time-Series Forecasting Problems with Built-in Algorithms

In the previous chapter, we had a closer look at several capabilities of SageMaker, such as **SageMaker Feature Store**, **SageMaker Clarify**, and **SageMaker Model Monitor**. These capabilities help machine learning practitioners handle relevant requirements when working on production-level machine learning experiments and deployments. In this chapter, we will take a look at using the SageMaker built-in algorithms to solve **natural language processing (NLP)**, **image classification**, and **time-series forecasting** problems.

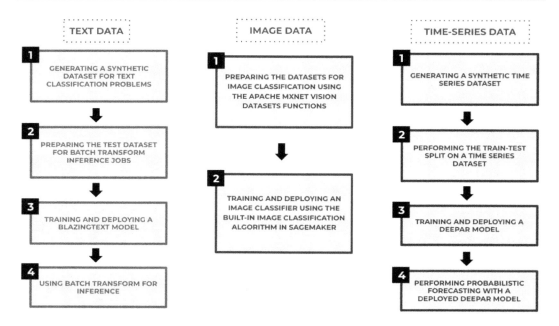

Figure 8.1 – Working with text classification, image classification, and time-series
forecasting problems with built-in algorithms

As in *Figure 8.1*, we will take a look at using **BlazingText** to solve one of the most
common NLP problems—text classification. In addition to this, we will also take a closer
look at using the built-in **Image Classification Algorithm** to solve image classification
problems with the MNIST handwritten digit dataset. We will also have a chance to use
the built-in **DeepAR Forecasting Algorithm** to solve time-series forecasting problems.

That said, we will cover the following recipes in this chapter:

- Generating a synthetic dataset for text classification problems
- Preparing the test dataset for batch transform inference jobs
- Training and deploying a **BlazingText** model
- Using **Batch Transform** for inference
- Preparing the datasets for image classification using the **Apache MXNet** Vision
  Datasets classes

- Training and deploying an image classifier using the built-in **Image Classification Algorithm** in SageMaker

- Generating a synthetic time-series dataset

- Performing the train-test split on a time series dataset

- Training and deploying a **DeepAR** model

- Performing probabilistic forecasting with a deployed **DeepAR** model

After we have completed the recipes in this chapter, we will be able to solve NLP, image classification, and time series forecasting problems and requirements more confidently with the built-in algorithms in SageMaker.

Technical requirements

To execute the recipes in the chapter, make sure you have the following:

- An Amazon S3 bucket

- Permissions to manage the **Amazon SageMaker** and **Amazon S3** resources if using an **AWS IAM** user with a custom URL. If you are using the root account, then you should be able to proceed with the recipes of this chapter. However, it is recommended to be signed in as an AWS IAM user instead of using the root account in most cases. For more information, feel free to take a look at the following guide: `https://docs.aws.amazon.com/IAM/latest/UserGuide/best-practices.html`.

As the recipes in this chapter involve a bit of code, we have made the notebooks available in this repository: `https://github.com/PacktPublishing/Machine-Learning-with-Amazon-SageMaker-Cookbook/tree/master/Chapter08`. Before starting on each of the recipes of this chapter, make sure that the `my-experiments/chapter08` directory is ready. If it has not yet been created, please do so now as this keeps things organized as we go through each of the recipes in this book.

Check out the following link to see the relevant Code in Action video:

`https://bit.ly/3tFF70t`

Generating a synthetic dataset for text classification problems

In this recipe, we will generate a synthetic dataset for a **binary text classification** problem. The dataset to be generated in this recipe has two primary fields: the text field containing a statement in string format and the target label that specifies whether the text is POSITIVE or NEGATIVE.

```
__label__positive Food In The Restaurant donut donut this is
good very delicious very delicious food in the restaurant din
ner time
__label__positive I Like It very good very good spaghetti chi
cken soup impressive spaghetti chicken soup
__label__positive Donut food in the restaurant donut very deli
cious spaghetti chicken soup
__label__positive Very Delicious impressive dinner time i like
it dinner time
__label__negative There Are Better Restaurants Out There dinner
time food in the restaurant i will not recommend this to my
friends there are better restaurants out there this is bad d
onut
__label__negative There Are Better Restaurants Out There this
is bad donut donut spaghetti chicken soup
```

Figure 8.2 – Synthetic dataset for text classification problems

In *Figure 8.2*, we can see that the sentences with the POSITIVE tag have the __label__ positive label while the sentences with the NEGATIVE tag have the __label__ negative label. We will use this dataset to train and deploy a **BlazingText** model in the next recipes to solve a sentiment analysis requirement.

Getting ready

A **SageMaker Studio** notebook running the **Python 3 (Data Science)** kernel is the only prerequisite for this recipe.

How to do it...

The first steps in this recipe focus on generating a list of POSITIVE and NEGATIVE statements and storing them inside a DataFrame:

1. Create a new notebook using the Python 3 (Data Science) kernel inside the my-experiments/chapter08 directory and rename it with the name of this recipe.

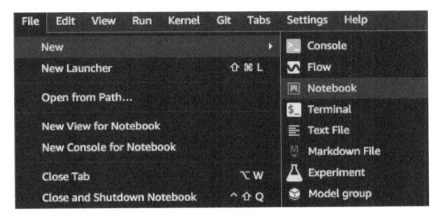

Figure 8.3 – Creating a new notebook

In *Figure 8.3*, we can see how to create a new **Notebook** from the **File** menu. When prompted for which kernel to use, choose **Python 3 (Data Science)**.

2. Install faker using the pip install command:

```
!pip install faker
```

Faker is a Python package that helps provide the utility functions that generate fake data. Later, we will use the sentence() function to generate a fake sentence based on a word list.

3. Initialize faker using the following lines of code:

```
from faker import Faker
faker = Faker()
```

4. Define a list of strings that will be used in a later step to generate the sentences classified as POSITIVE:

```
positive_custom_list = [
    'this is good',
    'i like it',
```

```
        'very delicious',
        'i would recommend this to my friends',
        'food in the restaurant',
        'spaghetti chicken soup',
        'dinner time',
        'tastes good',
        'donut',
        'very good',
        'impressive']
```

Here we can see that this list of strings contains tokens that are generally considered POSITIVE.

5. Define the `generate_positive_sentences()` function. Internally, this function uses the `faker.sentence()` function, which makes use of the `positive_custom_list` variable we defined in the previous step:

```
def generate_positive_sentences():
    return faker.sentence(
        ext_word_list=positive_custom_list
    )
```

6. Next, define a list of strings that will be used in a later step to generate the sentences classified as NEGATIVE:

```
negative_custom_list = [
    'this is bad',
    'i hate it',
    'there are better restaurants out there',
    'i will not recommend this to my friends',
    'food in the restaurant',
    'spaghetti chicken soup',
    'dinner time',
    'tastes bad',
    'donut',
    'very bad',
    'not impressive']
```

Here, we can see that this list of strings contains tokens that are generally considered NEGATIVE.

7. Define the `generate_negative_sentences()` function. Internally, this function uses the `faker.sentence()` function, which uses the `negative_custom_list` list we defined in the previous step:

```
def generate_negative_sentences():
    return faker.sentence(
        ext_word_list=negative_custom_list
    )
```

8. Generate `1000 POSITIVE` sentences using the following block of code and store them in the `positive_sentences` list:

```
positive_sentences = []
for i in range(0, 1000):
    item = generate_positive_sentences()
    item = item.replace(".", "")
    positive_sentences.append(item)
```

9. Inspect the `positive_sentences` variable:

```
positive_sentences
```

This should give us a list of strings similar to what is shown in *Figure 8.4*.

```
['Dinner Time this is good this is good this is good i like it very
delicious food in the restaurant',
 'I Like It food in the restaurant donut',
 'Tastes Good i like it tastes good food in the restaurant i would r
ecommend this to my friends',
 'Very Delicious very good very good spaghetti chicken soup i would
recommend this to my friends i like it',
 'Donut i would recommend this to my friends this is good impressive
spaghetti chicken soup very delicious impressive',
 'Very Delicious dinner time i like it very good i would recommend t
his to my friends food in the restaurant food in the restaurant',
```

Figure 8.4 – Generated list of POSITIVE sentences

In *Figure 8.4*, we have the list of strings generated by the `generate_positive_sentences()` function.

10. Similarly, generate `1000 NEGATIVE` sentences using the following block of code and store them in the `negative_sentences` list variable:

```
negative_sentences = []
for i in range(0, 1000):
```

```
      item = generate_negative_sentences()
      item = item.replace(".", "")
      negative_sentences.append(item)
```

This should give us a list of strings similar to what is shown in *Figure 8.5*.

```
['Very Bad food in the restaurant food in the restaurant very bad th
is is bad spaghetti chicken soup',
 'Spaghetti Chicken Soup i hate it this is bad this is bad tastes ba
d spaghetti chicken soup food in the restaurant spaghetti chicken so
up',
 'Donut food in the restaurant spaghetti chicken soup',
 'Dinner Time not impressive very bad i will not recommend this to m
y friends dinner time donut very bad',
 'This Is Bad very bad very bad not impressive',
 'This Is Bad dinner time not impressive there are better restaurant
s out there this is bad i hate it food in the restaurant',
```

Figure 8.5 – Generated list of NEGATIVE sentences

In *Figure 8.5*, we can see a list of sentences generated using the generate_ negative_sentences() function.

11. Prepare the DataFrame containing the POSITIVE sentences using the following block of code:

```
import pandas as pd
positive_df = pd.DataFrame(
    positive_sentences,
    columns=['text']
)
positive_df.insert(
    0,
    "label",
    "__label__positive"
)
```

This should give us a DataFrame of values similar to what is shown in *Figure 8.6.*

	label	text
0	_label_positive	Dinner Time this is good this is good this is ...
1	_label_positive	I Like It food in the restaurant donut
2	_label_positive	Tastes Good i like it tastes good food in the ...
3	_label_positive	Very Delicious very good very good spaghetti c...
4	_label_positive	Donut I would recommend this to my friends thi...
...	...	...
995	_label_positive	Dinner Time dinner time dinner time donut i li...
996	_label_positive	This Is Good this is good this is good impress...
997	_label_positive	Dinner Time very good spaghetti chicken soup t...
998	_label_positive	Food In The Restaurant dinner time donut spagh...
999	_label_positive	Very Good food in the restaurant impressive donut

1000 rows × 2 columns

Figure 8.6 – DataFrame containing the POSITIVE sentences

Here, we have the `label` column containing the `__label__positive` string.

12. Next, prepare the DataFrame containing the `NEGATIVE` sentences using the following lines of code:

```
negative_df = pd.DataFrame(
    negative_sentences,
    columns=['text']
)
negative_df.insert(
    0,
    "label",
    "__label__negative"
)
```

This should give us a DataFrame of values similar to what is shown in *Figure 8.7*.

	label	text
0	__label__negative	Very Bad food in the restaurant food in the re...
1	__label__negative	Spaghetti Chicken Soup i hate it this is bad t...
2	__label__negative	Donut food in the restaurant spaghetti chicken...
3	__label__negative	Dinner Time not impressive very bad i will not...
4	__label__negative	This Is Bad very bad very bad not impressive
...	...	...
995	__label__negative	Very Bad not impressive dinner time donut tast...
996	__label__negative	I Will Not Recommend This To My Friends not im...
997	__label__negative	Spaghetti Chicken Soup i will not recommend th...
998	__label__negative	Very Bad dinner time this is bad this is bad d...
999	__label__negative	Spaghetti Chicken Soup this is bad donut donut...

1000 rows × 2 columns

Figure 8.7 – DataFrame containing the NEGATIVE sentences

Here, we have the `label` column containing the `__label__negative` string.

13. Merge the two DataFrames using the `concat()` function:

```
all_df = pd.concat(
    [positive_df, negative_df],
    ignore_index=True
)
```

The last set of steps in this recipe focus on the train-test split and uploading the resulting train, validation, and test sets into S3:

14. Perform the train-validation-test split on `all_df`:

```
from sklearn.model_selection import train_test_split
train_val_df, test_df = train_test_split(
    all_df,
    test_size=0.2
)
train_df, val_df = train_test_split(
```

```
        train_val_df,
        test_size=0.25
    )
```

This should give us 600 records for the training set (train_df), 200 records for the validation set (val_df), and 200 records for the test set (test_df).

15. Export the DataFrames into their corresponding CSV files using the to_csv() function:

```
!mkdir tmp
train_df.to_csv(
    "tmp/synthetic.train.txt",
    header=False,
    index=False,
    sep=" ",
    quotechar=" "
)
val_df.to_csv(
    "tmp/synthetic.validation.txt",
    header=False,
    index=False,
    sep=" ",
    quotechar=" "
)
test_df.to_csv(
    "tmp/synthetic.test.txt",
    header=False,
    index=False,
    sep=" ",
    quotechar=" "
)
```

Note that we have set the parameter values for header and index to False.

16. Inspect the contents of the synthetic.train.txt file inside the tmp directory:

```
!head tmp/synthetic.train.txt
```

This should give us lines of text similar to what is shown in *Figure 8.8*.

```
__label__positive Food In The Restaurant donut donut this is
good very delicious very delicious food in the restaurant din
ner time
__label__positive I Like It very good very good spaghetti chi
cken soup impressive spaghetti chicken soup
__label__positive Donut food in the restaurant donut very deli
cious spaghetti chicken soup
__label__positive Very Delicious impressive dinner time i like
it dinner time
__label__negative There Are Better Restaurants Out There dinner
time food in the restaurant i will not recommend this to my
friends there are better restaurants out there this is bad d
onut
__label__negative There Are Better Restaurants Out There this
is bad donut donut spaghetti chicken soup
```

Figure 8.8 – Contents of the synthetic.train.txt file

Here, we have a TXT file containing the labels and the statements.

17. Specify the S3 bucket name and the prefix where the data will be stored. Make sure to replace the value of "<insert bucket name here>" with the name of the bucket we created in the recipe *Preparing the Amazon S3 bucket and the training dataset for the linear regression experiment* in *Chapter 1, Getting Started with Machine Learning Using Amazon SageMaker*:

```
s3_bucket = "<insert bucket name here>"
prefix = "chapter08"
!aws s3 cp tmp/synthetic.train.txt s3://{s3_bucket}/
{prefix}/input/synthetic.train.txt
!aws s3 cp tmp/synthetic.validation.txt s3://{s3_bucket}/
{prefix}/input/synthetic.validation.txt
```

18. Finally, use the %store magic to store the variable values for test_df, s3_bucket, and prefix:

```
%store test_df
%store s3_bucket
%store prefix
```

We will use these variable values in the succeeding recipes.

Now, let's see how this works!

How it works...

In this recipe, we have generated the synthetic dataset that will be used in the next three recipes of this chapter. Compared to the other synthetic datasets generated in this book, we generated a synthetic dataset containing text data in this recipe instead of tabular and numeric data.

The dataset we generated has two primary fields: (1) the text field containing a statement in string format and (2) the target label, which can be `__label__positive` or `__label__negative`. It is important that the labels have the `__label__` prefix as we are planning to train a **BlazingText** model in file mode in the recipe *Training and deploying a BlazingText model*. Note that we are not limited to just having two classes in this dataset. If we wish to build on top of this recipe, one example would be to have three classes instead of two—`__label__positive`, `__label__negative`, and `__label__neutral`. Of course, we will need to make sure that we update the configuration and hyperparameter values accordingly before running the training job using a specific algorithm.

There's more...

We also have the option to store and export the synthetic dataset using the **augmented manifest text format** as this should also work when training the **BlazingText** model. This involves using the JSON Lines format where each line in the file contains a valid JSON value similar to what is shown in the following block of code:

```
{"source":"i will not recommend this", "label":0}
{"source":"i would recommend this", "label":1}
```

One advantage when using the augmented manifest text format is that we can train the model in pipe input mode where the dataset is streamed directly to the training instances. In the recipe *Converting CSV data into protobuf recordIO format* from *Chapter 4, Preparing, Processing, and Analyzing the Data*, we mentioned that we can use **pipe mode** during training when our data is serialized into the `protobuf recordIO` format. In this case, as long as our data is using the augmented manifest text format, we will be able to use **pipe mode** even without having to serialize our data into `protobuf recordIO` format.

Preparing the test dataset for batch transform inference jobs

In this recipe, we will prepare the test dataset that will be used in the recipe *Using batch transform for inference*, which makes use of the **Batch Transform** capability of SageMaker. With **Batch Transform**, we can perform inference on multiple records all at the same time without having a persistent endpoint running.

```
{"source": "There Are Better Restaurants Out There tastes bad
spaghetti chicken soup food in the restaurant"}
{"source": "Not Impressive i will not recommend this to my
friends there are better restaurants out there spaghetti chick
en soup food in the restaurant donut"}
{"source": "I Will Not Recommend This To My Friends spaghett
i chicken soup food in the restaurant"}
{"source": "There Are Better Restaurants Out There there are
better restaurants out there spaghetti chicken soup tastes bad
dinner time tastes bad this is bad"}
{"source": "Spaghetti Chicken Soup this is bad i will not r
ecommend this to my friends i hate it very bad"}
{"source": "Not Impressive spaghetti chicken soup i will not
recommend this to my friends spaghetti chicken soup"}
{"source": "There Are Better Restaurants Out There spaghetti
chicken soup i hate it this is bad i hate it"}
{"source": "Food In The Restaurant tastes bad tastes bad the
re are better restaurants out there i hate it"}
{"source": "This Is Good very good i like it very deliciou
s"}
{"source": "Dinner Time very good very delicious very good v
ery good very good very delicious very good"}
```

Figure 8.9 – Text file containing the test data in JSON lines format

Note that when using **Batch Transform** with a **BlazingText** model, it is important that the input test dataset is in `jsonlines` format. As we have in *Figure 8.9*, each line in the file is a valid JSON value.

Getting ready

Here are the prerequisites for this recipe:

- This recipe continues from *Generating a synthetic dataset for text classification problems*.

- A **SageMaker Studio** notebook running the **Python 3 (Data Science)** kernel.

How to do it...

The steps in this recipe focus on converting the data from the previous recipe into jsonlines format and uploading the resulting file into S3:

1. Create a new notebook using the Python 3 (Data Science) kernel inside the my-experiments/chapter08 directory and rename it to the name of this recipe. When prompted for the kernel to use, choose **Python 3 (Data Science)**.

2. Use the %store magic to load the variable values for test_df, s3_bucket, and prefix:

    ```
    %store -r test_df
    %store -r s3_bucket
    %store -r prefix
    ```

3. Use the drop() function to remove the label column:

    ```
    test_df_without_label = test_df.drop(
        columns="label"
    )
    ```

4. Define the to_jsonlines() function:

    ```
    def to_jsonlines(text):
        return '{"source": "' + text +'"}'
    ```

    > **Note**
    >
    > Take note that in this recipe, we are showing one of the ways to convert what we have to the JSON lines format. It is also possible to use the to_json() function with lines set to True to convert a DataFrame to jsonlines format.

5. Use the apply() function to convert each cell in the text column to jsonlines format. Next, check how the test_df_without_label DataFrame looks after using the apply() function as well:

    ```
    tmp = test_df_without_label['text'].apply(
        to_jsonlines
    )
    test_df_without_label['text'] = tmp
    test_df_without_label
    ```

After running the previous block of code, the string values in the text column in the DataFrame will be converted into a dictionary instead. This should give us a DataFrame of row index and `jsonline` dictionary pairs similar to what is shown in *Figure 8.10*.

Figure 8.10 – DataFrame containing test data without labels

We can see in *Figure 8.10* that the text column of the DataFrame now contains a dictionary with source as the key with the value set to the original text value of the cells in the text column.

6. We then run the following code to store the content of the DataFrame in a file:

```
test_df_without_label.to_csv(
    "tmp/synthetic.test_without_labels.txt",
    header=False,
    index=False,
    sep=" ",
    quotechar=" "
)

!head tmp/synthetic.test_without_labels.txt
```

This should give us the first few lines of the text file, similar to what is shown in *Figure 8.11*.

```
{"source":  "There  Are  Better  Restaurants  Out  There  tastes  bad
spaghetti  chicken  soup  food  in  the  restaurant"}
{"source":  "Not  Impressive  i  will  not  recommend  this  to  my
friends  there  are  better  restaurants  out  there  spaghetti  chick
en  soup  food  in  the  restaurant  donut"}
{"source":  "I  Will  Not  Recommend  This  To  My  Friends  spaghett
i  chicken  soup  food  in  the  restaurant"}
{"source":  "There  Are  Better  Restaurants  Out  There  there  are
better  restaurants  out  there  spaghetti  chicken  soup  tastes  bad
dinner  time  tastes  bad  this  is  bad"}
{"source":  "Spaghetti  Chicken  Soup  this  is  bad  i  will  not  r
ecommend  this  to  my  friends  i  hate  it  very  bad"}
{"source":  "Not  Impressive  spaghetti  chicken  soup  i  will  not
recommend  this  to  my  friends  spaghetti  chicken  soup"}
{"source":  "There  Are  Better  Restaurants  Out  There  spaghetti
chicken  soup  i  hate  it  this  is  bad  i  hate  it"}
{"source":  "Food  In  The  Restaurant  tastes  bad  tastes  bad  the
re  are  better  restaurants  out  there  i  hate  it"}
{"source":  "This  Is  Good  very  good  i  like  it  very  deliciou
s"}
{"source":  "Dinner  Time  very  good  very  delicious  very  good  v
ery  good  very  good  very  delicious  very  good"}
```

Figure 8.11 – Text file containing the test data in jsonlines format

In *Figure 8.11*, we can see that we were able to successfully generate a text file with our data in `jsonlines` format.

7. Now that we have our TXT file ready, we use the **AWS CLI** to upload the generated file to the target S3 location:

```
!aws s3 cp tmp/synthetic.test_without_labels.txt s3://
{s3_bucket}/{prefix}/input/synthetic.test_without_labels.
txt
```

Take note that the file in `jsonlines` format we have prepared in this recipe will be used in the recipe *Using batch transform for inference*.

Now that we are done with the needed preparation work in these last two recipes, we can proceed with the next recipe, where we will use these datasets to train and deploy a **BlazingText** model. In the meantime, let's see how this works!

How it works...

In this recipe, we have prepared the test dataset that we will use in the recipe *Using batch transform for inference*, which involves the **Batch Transform** capability of SageMaker for performing inference without having a persistent real-time endpoint.

Why convert the dataset into jsonlines format? That's because the dataset needs to be in jsonlines format in order to use **Batch Transform** with the **BlazingText** model. As mentioned earlier, each line in the file is a valid JSON value. **Batch Transform** treats each line as one input payload, which means that if we have 1,000 lines in our jsonlines file, we will get 1,000 inference results after the job has been completed. Note that when preparing the jsonlines file, we need to make sure that the dataset we will prepare for the **Batch Transform** job does not contain the label field value. That said, we will only need to provide the values for the source field.

Training and deploying a BlazingText model

In the recipe *Generating a synthetic dataset for text classification problems*, we prepared the dataset we will use to train the **BlazingText** model. In this recipe, we will use the **SageMaker Python SDK** to train and deploy a **BlazingText** model that can be used for sentiment analysis applications.

After we have completed this recipe, we will be able to pass a sentence such as I would recommend this to my friends as the payload to an inference endpoint and get the correct classification, which is POSITIVE.

Getting ready

Here are the prerequisites for this recipe:

- This recipe continues from *Generating a synthetic dataset for text classification problems*.
- A **SageMaker Studio** notebook running the **Python 3 (Data Science)** kernel.

How to do it...

The first steps in this recipe focus on preparing the prerequisites for training and deploying the **BlazingText** model using the **SageMaker Python SDK**:

1. Create a new notebook using the Python 3 (Data Science) kernel inside the my-experiments/chapter08 directory and rename it to the name of this recipe. When prompted for the kernel to use, choose **Python 3 (Data Science)**.

2. Import and initialize a few prerequisites required to use the **SageMaker Python SDK** for training and deployment:

```
import sagemaker
from sagemaker import get_execution_role
import json
import boto3
session = sagemaker.Session()
role = get_execution_role()
region_name = boto3.Session().region_name
```

3. Use the `%store` magic to load the variable values for `s3_bucket` and `prefix`:

```
%store -r s3_bucket
%store -r prefix
s3_train_data = 's3://{}/{}/input/{}'.format(
    s3_bucket,
    prefix,
    "synthetic.train.txt"
)
s3_validation_data = 's3://{}/{}/input/{}'.format(
    s3_bucket,
    prefix,
    "synthetic.validation.txt"
)
s3_output_location = 's3://{}/{}/output'.format(
    s3_bucket,
    prefix
)
```

4. Use the `retrieve()` function to get the ECR Image URI of the built-in algorithm **BlazingText**:

```
from sagemaker.image_uris import retrieve
container = retrieve(
    "blazingtext",
    region_name,
    "1"
)
```

The last set of steps in this recipe focus on using the prerequisites prepared in the previous set of steps in training and deploying the **BlazingText** model:

5. Initialize the `Estimator` object and use the `container` variable from the previous step as the first argument during initialization:

```
estimator = sagemaker.estimator.Estimator(
    container,
    role,
    instance_count=1,
    instance_type='ml.c4.xlarge',
    input_mode= 'File',
    output_path=s3_output_location,
    sagemaker_session=session
)
```

6. Specify a few hyperparameters using the `set_hyperparameters()` function:

```
estimator.set_hyperparameters(
    mode="supervised",
    min_count=2
)
```

7. Next, we prepare the input data channels using the following block of code:

```
from sagemaker.inputs import TrainingInput

train_data = TrainingInput(
    s3_train_data,
    distribution='FullyReplicated',
    content_type='text/plain',
    s3_data_type='S3Prefix'
)

validation_data = TrainingInput(
    s3_validation_data,
    distribution='FullyReplicated',
    content_type='text/plain',
    s3_data_type='S3Prefix'
```

```
)
data_channels = {
    'train': train_data,
    'validation': validation_data
}
```

8. With everything ready, we use the `fit()` function to start the training job. Wait for about 5-10 minutes for the training job to complete:

```
%%time
estimator.fit(
    inputs=data_channels,
    logs=True
)
```

This should yield a set of logs similar to what is shown in *Figure 8.12*.

```
[06/01/2021 10:11:25 WARNING 140448806991232] Loggers have already been setup.
[06/01/2021 10:11:25 WARNING 140448806991232] Loggers have already been setup.
[06/01/2021 10:11:25 INFO 140448806991232] nvidia-smi took: 0.025360822677612305 secs to identify 0 g
pus
[06/01/2021 10:11:25 INFO 140448806991232] Running single machine CPU BlazingText training using supe
rvised mode.
Number of CPU sockets found in instance is  1
[06/01/2021 10:11:25 INFO 140448806991232] Processing /opt/ml/input/data/train/synthetic.train.txt .
File size: 0.1503753662109375 MB
[06/01/2021 10:11:25 INFO 140448806991232] Processing /opt/ml/input/data/validation/synthetic.validat
ion.txt . File size: 0.049597740173339844 MB
Read 0M words
Number of words:  69
##### Alpha: -0.0001  Progress: 100.24%  Million Words/sec: 1.13 #####
##### Alpha: 0.0000  Progress: 100.00%  Million Words/sec: 1.13 #####
Training finished.
Average throughput in Million words/sec: 1.13
Total training time in seconds: 0.10

#train_accuracy: 0.9908
Number of train examples: 1200

#validation_accuracy: 0.9925
Number of validation examples: 400

2021-06-01 10:11:57 Uploading - Uploading generated training model
2021-06-01 10:13:20 Completed - Training job completed
Training seconds: 148
Billable seconds: 148
CPU times: user 597 ms, sys: 40.2 ms, total: 638 ms
Wall time: 5min 13s
```

Figure 8.12 – Training job logs

In *Figure 8.12*, we can see that our `validation_accuracy` value is `99.25%`! Of course, we are using a simplified example with a synthetic dataset, but this is a good start for us.

9. Use the `deploy()` function to deploy our `BlazingText` model:

```
endpoint = estimator.deploy(
    initial_instance_count = 1,
    instance_type = 'ml.r5.large'
)
```

This step may take 5 to 10 minutes to complete.

> **Important note**
>
> Running the `deploy()` function would launch an instance that would continue running until the delete resource operation is performed. While the instance is running, you will be charged for the amount of time it is running. Make sure to delete the inference endpoint after completing this recipe.

10. Next, prepare the sentences we will include in our payload to test our deployed model:

```
sentences = [
    "that is bad",
    "the apple tastes good",
    "i would recommend it to my friends"
]
payload = {"instances" : sentences}
```

11. After that, we use the `predict()` function to test our deployed model:

```
from sagemaker.serializers import JSONSerializer

endpoint.serializer = JSONSerializer()
response = endpoint.predict(payload)
predictions = json.loads(response)
print(json.dumps(predictions, indent=2))
```

Running the previous block of code would yield a set of results similar to the one shown in the following figure.

```
[
    {
        "label": [
            "__label__negative"
        ],
        "prob": [
            0.8144211173057556
        ]
    },
    {
        "label": [
            "__label__positive"
        ],
        "prob": [
            0.7917395234107971
        ]
    },
    {
        "label": [
            "__label__positive"
        ],
        "prob": [
            0.6492884159088135
        ]
    }
]
```

Figure 8.13 – Prediction results

In *Figure 8.13*, we got one label for each of the sentences we included in our payload. Our first sentence, that is bad, was tagged as NEGATIVE with an 81% probability score. Our second and third sentences, the apple tastes good and i would recommend it to my friends, were tagged as POSITIVE with 79% and 65% probability scores respectively. I think our deployed model is working just fine!

12. Use the %store magic to store the training_job_name variable value:

```
tn = estimator.latest_training_job.name
training_job_name = tn
%store training_job_name
```

Now that we have completed the steps in this recipe, feel free to play around with the deployed model. Once you are done testing different sentences, do not forget to delete the endpoint by using endpoint.delete_endpoint().

Now let's see how this works!

How it works...

Before discussing the details and hyperparameters when using **BlazingText**, it is important to note that we have two modes when using this algorithm:

- Unsupervised learning with the **Word2vec** algorithm
- Supervised learning for text classification problems

In this recipe, we set the mode to `supervised` using the `set_hyperparameters()` function since we are solving a **text classification** problem. If we were to use the `Word2vec` algorithm to map words to distributed vectors and learn word associations from text data such as `EAT - SPAGHETTI` and `DRINK - JUICE`, we could set the value of `mode` to `batch_skipgram`, `skipgram`, or `cbow`. If we will work with a single GPU instance, we can set the value of `mode` to `cbow` or `skipgram`. If we want to get faster training times with distributed processing across multiple CPU instances, you can specify `batch_skipgram` for the `mode` value.

> **Note**
>
> There are also other hyperparameters we can specify and configure, such as `early_stopping`, `learning_rate`, `epochs`, and `word_ngrams`. We will not discuss these in detail in this book so feel free to check the following link: `https://docs.aws.amazon.com/sagemaker/latest/dg/blazingtext_hyperparameters.html`.

See more

If you are looking for examples of using the built-in **BlazingText algorithm** on real datasets and more complex examples, feel free to check some of the notebooks in the `aws/amazon-sagemaker-examples` GitHub repository:

- Generate Word2Vec embeddings using **BlazingText**—`https://github.com/aws/amazon-sagemaker-examples/tree/master/introduction_to_amazon_algorithms/blazingtext_word2vec_text8`
- Create an inference pipeline using **SparkML** and **BlazingText**—`https://github.com/aws/amazon-sagemaker-examples/tree/master/advanced_functionality/inference_pipeline_sparkml_blazingtext_dbpedia`

As we will not be able to dive deep into the different features of the **BlazingText algorithm** in this book, feel free to check this link for more information: `https://docs.aws.amazon.com/sagemaker/latest/dg/blazingtext.html`.

Using Batch Transform for inference

In the previous recipe, we trained and deployed a **BlazingText** model that accepts a string statement and returns whether the statement is POSITIVE or NEGATIVE. In this recipe, we will use this model along with the **Batch Transform** capability of **SageMaker** to perform text classification on the entire test dataset all at the same time without having a persistent inference endpoint.

Getting ready

Here are the prerequisites for this recipe:

- This recipe continues from *Training and deploying a BlazingText model*.

- A **SageMaker Studio** notebook running the **Python 3 (Data Science)** kernel.

How to do it...

The steps in this recipe focus on using the prerequisites we prepared in the previous recipes to run a Batch Transform job using the **SageMaker Python SDK**:

1. Create a new notebook using the Python 3 (Data Science) kernel inside the my-experiments/chapter08 directory and rename it to the name of this recipe. When prompted for the kernel to use, choose **Python 3 (Data Science)**.

2. Use the %store magic to load the variable values for s3_bucket, prefix, and training_job_name. At the same time, initialize and set the values of a few prerequisites of the batch transform job. We set the value of the batch_output variable as well. This variable will point to the location where the batch transform job output artifacts will be stored:

```
%store -r s3_bucket
%store -r prefix
%store -r training_job_name
path = 's3://{}/{}/input/{}'.format(
    s3_bucket,
    prefix,
    "synthetic.test_without_labels.txt"
```

```
)
s3_test_without_labels_data = path

path = 's3://{}/{}/batch-prediction'.format(
    s3_bucket,
    prefix
)
batch_output = path
```

Note that we have loaded the variable values for s3_bucket and prefix from the recipe *Generating a synthetic dataset for text classification problems*. That said, the variable prefix should have a value of chapter08 after we have used the %store magic to load its value.

3. We use the transformer() function of the Estimator object to get the transformer object, which we will use in the next step:

```
from sagemaker.estimator import Estimator
estimator = Estimator.attach(training_job_name)
transformer = estimator.transformer(
    instance_count=1,
    instance_type='ml.m4.xlarge',
    output_path=batch_output
)
```

Here, we also used the Estimator.attach() function to load the Estimator object from the name of the training job we ran in the recipe *Training and deploying a BlazingText model*.

4. Use the transform() function to start the batch transform job. We then use the wait() function to wait for the job to complete before proceeding to the next step:

```
transformer.transform(
    data=s3_test_without_labels_data,
    data_type='S3Prefix',
    content_type='application/jsonlines',
    split_type='Line'
)
transformer.wait()
```

This should yield a set of logs similar to what is shown in *Figure 8.14*.

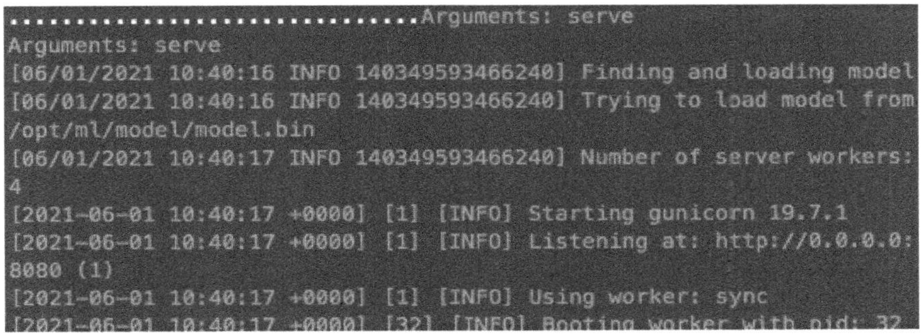

```
····························Arguments: serve
Arguments: serve
[06/01/2021 10:40:16 INFO 140349593466240] Finding and loading model
[06/01/2021 10:40:16 INFO 140349593466240] Trying to load model from
/opt/ml/model/model.bin
[06/01/2021 10:40:17 INFO 140349593466240] Number of server workers:
4
[2021-06-01 10:40:17 +0000] [1] [INFO] Starting gunicorn 19.7.1
[2021-06-01 10:40:17 +0000] [1] [INFO] Listening at: http://0.0.0.0:
8080 (1)
[2021-06-01 10:40:17 +0000] [1] [INFO] Using worker: sync
[2021-06-01 10:40:17 +0000] [32] [INFO] Booting worker with pid: 32
```

Figure 8.14 – Batch transform job logs

Remember in the previous recipe that we deleted the endpoint for real-time predictions? Here, we can see that the Batch Transform job involves running inference without a persistent endpoint.

> **Note**
>
> This step may take around 4 to 8 minutes to complete. While waiting, feel free to grab a cup of coffee or tea!

5. Use the **AWS S3 CLI** to copy the output of the batch transform job to the tmp directory in the same directory as the Jupyter notebook with the code blocks and scripts in this recipe:

```
!aws s3 cp {batch_output} ./tmp --recursive
```

6. Use the head bash command to check a few values:

```
!head tmp/synthetic.test_without_labels.txt.out
```

This should give us lines of text similar to what is shown in *Figure 8.15*.

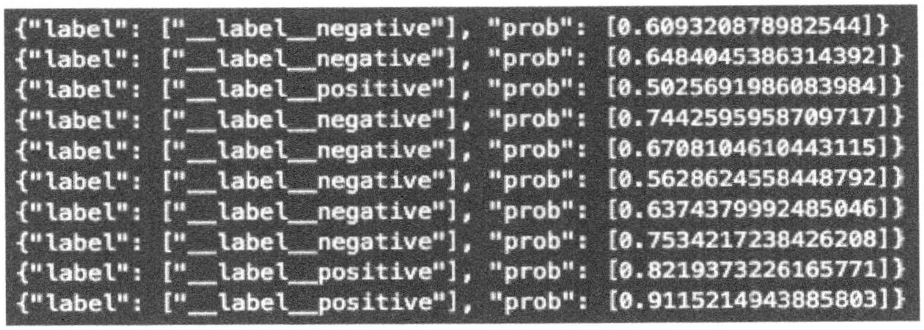

```
{"label": ["__label__negative"], "prob": [0.609320878982544]}
{"label": ["__label__negative"], "prob": [0.6484045386314392]}
{"label": ["__label__positive"], "prob": [0.5025691986083984]}
{"label": ["__label__negative"], "prob": [0.7442595958709717]}
{"label": ["__label__negative"], "prob": [0.6708104610443115]}
{"label": ["__label__negative"], "prob": [0.5628624558448792]}
{"label": ["__label__negative"], "prob": [0.6374379992485046]}
{"label": ["__label__negative"], "prob": [0.7534217238426208]}
{"label": ["__label__positive"], "prob": [0.8219373226165771]}
{"label": ["__label__positive"], "prob": [0.9115214943885803]}
```

Figure 8.15 – Batch transform results

In *Figure 8.15*, we can see that we have multiple results, with each line containing a label and a probability score between `0.0` and `1.0`. The closer the value of `prob` to `1.0`, the higher the probability score is for a certain prediction.

7. We check their corresponding input *jsonline* values as well:

```
!head tmp/synthetic.test_without_labels.txt
```

This should give us lines of text similar to what is shown in *Figure 8.16*.

```
{"source": "There Are Better Restaurants Out There tastes bad
spaghetti chicken soup food in the restaurant"}
{"source": "Not Impressive i will not recommend this to my
friends there are better restaurants out there spaghetti chick
en soup food in the restaurant donut"}
{"source": "I Will Not Recommend This To My Friends spaghett
i chicken soup food in the restaurant"}
{"source": "There Are Better Restaurants Out There there are
better restaurants out there spaghetti chicken soup tastes bad
dinner time tastes bad this is bad"}
{"source": "Spaghetti Chicken Soup this is bad i will not r
ecommend this to my friends i hate it very bad"}
{"source": "Not Impressive spaghetti chicken soup i will not
recommend this to my friends spaghetti chicken soup"}
{"source": "There Are Better Restaurants Out There spaghetti
chicken soup i hate it this is bad i hate it"}
{"source": "Food In The Restaurant tastes bad tastes bad the
re are better restaurants out there i hate it"}
{"source": "This Is Good very good i like it very deliciou
s"}
{"source": "Dinner Time very good very delicious very good v
ery good very good very delicious very good"}
```

Figure 8.16 – Test data without labels in jsonlines format

In *Figure 8.16*, we can see that the first eight sentences should be tagged with the `NEGATIVE` class while the next two sentences should be tagged with the `POSITIVE` class. Comparing it with the results of the batch transform job in *Figure 8.15*, we can see that our model got 9 out of 10 predictions correct.

Now let's see how this works!

How it works...

In this recipe, we used the **Batch Transform** capability of SageMaker to perform predictions and get inferences using a number of test records as input.

When do we use **Batch Transform**? We use this capability when we do not need a real-time inference endpoint and when we need to get inferences by batch. This means that we can pass 1,000 test records as "payload" and get 1,000 inference values as the return output of the batch transform job. With **Batch Transform**, we will not have to worry about resource management as we do not have to delete the endpoint manually after performing inferences compared to having a real-time endpoint running 24/7.

Note that we can significantly increase the number of sentences to classify with our batch transform job. Given that **Batch Transform** makes it easy for us to easily handle these types of scenarios and requirements, we can just change the `instance_type` parameter value from `ml.m4.xlarge` to an **ML** (**machine learning**) instance that has more memory or computing power. Again, there is nothing to worry about as the instance used to process this batch classification work is automatically deleted after the job has been completed.

See also

If you are looking for examples of using **Batch Transform** on real datasets and more complex examples, feel free to check some of the notebooks focusing on this topic in the GitHub repository here: `https://github.com/aws/amazon-sagemaker-examples/tree/master/sagemaker_batch_transform`.

Preparing the datasets for image classification using the Apache MXNet Vision Datasets classes

In this recipe, we will set up the file and directory structure needed for the image classification experiments in this chapter. We will create five directories inside the `tmp` directory—`train`, `validation`, `train_1st`, `validation_1st`, and `test`. After that, we will use the **Apache MXNet Vision Datasets classes** to load the datasets required to train and test the image classification models in this chapter. We will perform the train-test split, store the loaded data as image files, and generate the `.1st` files that will be used for the training job.

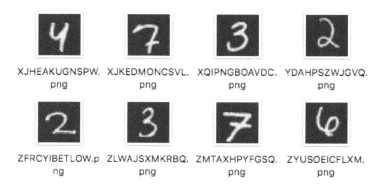

XJHEAKUGNSPW.
png

XJKEDMONCSVL.
png

XQIPNGBOAVDC.
png

YDAHPSZWJGVQ.
png

ZFRCYIBETLOW.p
ng

ZLWAJSXMKRBQ.
png

ZMTAXHPYFGSQ.
png

ZYUSOEICFLXM.
png

Figure 8.17 – MNIST dataset

We have in *Figure 8.17* a few sample image files that will be prepared in this recipe. In the recipe *Training and deploying an image classifier using the built-in image classification algorithm in SageMaker*, we will use these image files to train an image classifier model that can recognize the labels (digits) of the provided images.

Getting ready

A **SageMaker Studio** notebook running the **Python 3 (MXNet 1.8 Python 3.7 CPU Optimized)** kernel is the only prerequisite for this recipe.

> **Important note**
>
> Make sure to choose the **CPU Optimized** option instead of the **GPU Optimized** option as the **GPU Optimized** option will launch an instance with a default type of `ml.g4dn.xlarge`. On the other hand, the **CPU Optimized** option will launch an instance with a default type of `ml.t3.medium`. For more information, feel free to check `https://docs.aws.amazon.com/sagemaker/latest/dg/notebooks-usage-metering.html`.

How to do it...

The first steps in this recipe focus on preparing the file and directory structure where we will store the generated images and files:

1. Create a new notebook using the `Python 3 (MXNet 1.8 Python 3.7 CPU Optimized)` kernel inside the `my-experiments/chapter08` directory and rename it to the name of this recipe. When prompted for the kernel to use, choose **Python 3 (Data Science)**.

    > **Note**
    >
    > Feel free to clean the `tmp` directory (if it exists) before running the next set of steps.

2. Using the `mkdir` command, create the directories where we will store the training and validation dataset images:

    ```
    %%bash
    mkdir -p tmp/train/0 tmp/train/1 tmp/train/2 tmp/train/3
    tmp/train/4
    mkdir -p tmp/train/5 tmp/train/6 tmp/train/7 tmp/train/8
    tmp/train/9
    mkdir -p tmp/validation/0 tmp/validation/1 tmp/
    validation/2 tmp/validation/3 tmp/validation/4
    mkdir -p tmp/validation/5 tmp/validation/6 tmp/
    validation/7 tmp/validation/8 tmp/validation/9
    ```

3. Create the directory that will contain the training data .1st file:

```
%%bash
mkdir -p tmp/train_lst
mkdir -p tmp/validation_lst
mkdir -p tmp/test
```

The next set of steps in this recipe focus on preparing the datasets for image classification:

4. Import mxnet and set the seed value to any number using the mx.random.
    seed() function:

```
import mxnet as mx
mx.random.seed(21)
```

5. Define the transform_fxn() function and use this function as the value for the
    transform parameter when using the **Apache MXNet Vision Datasets** MNIST
    class:

```
def transform_fxn(data, label):
    data = data.astype('float32')
    data = data / 255
    return data, label


ds = mx.gluon.data.vision.datasets.MNIST(
    train=True,
    transform=transform_fxn
)
training_and_validation_dataset = ds

ds = mx.gluon.data.vision.datasets.MNIST(
    train=False,
    transform=transform_fxn
)
test_dataset = ds
```

6. Define the get_training_row_indexes() function:

```
def get_training_row_indexes(row_count,
                             percent=0.5,
                             ratio=0.8):
```

```
training_index_start = 0
end = int(row_count * ratio * percent)
training_index_end = end

print("Range Index Start:",
        training_index_start)
print("Range Index End:",
        training_index_end)

output = list(range(training_index_start,
                        training_index_end))

print("Output Length:", len(output))
print("Last Index:", output[-1])

return output
```

As its name suggests, this function returns a list of indexes that will map to the corresponding images that will be part of the training dataset. With this function, we are expecting the start and end index values to change depending on the parameter values for row_count, percent, and ratio.

7. Define the get_validation_row_indexes() function:

```
def get_validation_row_indexes(row_count,
                                    percent=0.5,
                                    ratio=0.8):
    start = int(row_count * ratio)
    validation_index_start = start

    count = int((1 - ratio)*row_count*percent) + 1
    element_count = count
    validation_index_end = validation_index_start +
element_count

    print("Range Index Start:",
            validation_index_start)
    print("Element Count:",
```

```
                    element_count)
        print("Range Index End:",
              validation_index_end)


        output = list(range(validation_index_start,
                            validation_index_end))


        print("Output Length:", len(output))
        print("Last Index:", output[-1])


        return output
```

Similar to the `get_training_row_indexes()`, this function returns a list of indexes that will map to the corresponding images that will be part of the validation dataset. With this function, we are expecting the start and end index values to change as well depending on the parameter values for `row_count`, `percent`, and `ratio`. As you might expect, the index values generated by `get_training_row_indexes()` and `get_validation_row_indexes()` do not overlap.

8. Define the `get_test_row_indexes()` function:

```
   def get_test_row_indexes(row_count,
                                percent=0.5):
        test_index_start = 0
        test_index_end = int(row_count * percent)

        print("Range Index Start:",
              test_index_start)
        print("Range Index End:",
              test_index_end)


        output = list(range(test_index_start,
                            test_index_end))


        print("Output Length:", len(output))
        print("Last Index:", output[-1])


        return output
```

9. Define the `generate_random_string()` function, which will be used to generate the filenames of the image files in a later step:

```
import string
import random

def generate_random_string():
    return ''.join(
        random.sample(
            string.ascii_uppercase,12)
    )
```

When used, this function should generate a random string similar to `'FCTASXQNPOVY'`. Of course, we will get a different set of values every time we use this function. We will use this later to assign a random string filename to each image in our dataset.

10. Define the `save_image()` function:

```
import matplotlib
import matplotlib.pyplot
def save_image(image_data, filename):
    matplotlib.pyplot.imsave(
        f"tmp/{filename}",
        image_data[:,:,0].asnumpy())
```

This function accepts the image data and filename as the parameters and uses the `matplotlib.pyplot.imsave()` function to save the image with the specified filename.

11. Define the `generate_image_files_and_lst_dict()` function:

```
def generate_image_files_and_lst_dict(
    dataset,
    indexes,
    tag
):
    list_of_lst_dicts = []

    for index in indexes:
        image_label_pair = dataset[index]
```

```
        image_data = image_label_pair[0]
        label = image_label_pair[1]
        random_string = generate_random_string()

        if tag == "test":
            rp = f"{random_string}.png"
            relative_path = rp
            filename = f"{tag}/{relative_path}"
        else:
            rp = f"{label}/{random_string}.png"
            relative_path = rp
            filename = f"{tag}/{relative_path}"

        save_image(
            image_data,
            filename=filename
        )

        lst_dict = {
            'relative_path': relative_path,
            'class': label
        }
        list_of_lst_dicts.append(lst_dict)

    return list_of_lst_dicts
```

This function performs the following:

- Accepts the dataset, indexes, and test tag
- Loops through each of the index values in the list of indexes specified
- Saves the corresponding images from the dataset based on the index values

The last set of steps in this recipe focus on using the prerequisites and defined functions prepared in the previous steps to generate the image files:

12. Use the functions prepared in the previous steps to generate the training image data and the dictionary that will contain the `train.lst` file data:

```
train_dataset_length = len(
    training_and_validation_dataset
)
train_indexes = get_training_row_indexes(
    row_count=train_dataset_length,
    percent=0.01)

t = generate_image_files_and_lst_dict(
    dataset=training_and_validation_dataset,
    indexes=train_indexes,
    tag = "train"
)
train_lst_dict = t
```

13. Inspect the `train_lst_dict` variable:

```
train_lst_dict
```

This should give us a nested structure of values similar to what is shown in *Figure 8.18*.

```
[{'relative_path': '5/RIDQNTFUPJME.png', 'class': 5},
 {'relative_path': '0/RSXIPQJWAKFU.png', 'class': 0},
 {'relative_path': '4/RQSGDZIMUAPC.png', 'class': 4},
 {'relative_path': '1/QGSLNAWKDOHU.png', 'class': 1},
 {'relative_path': '9/JYBUESOIAVDZ.png', 'class': 9},
 {'relative_path': '2/TUPYWXORSIFH.png', 'class': 2},
 {'relative_path': '1/VBGPEZXTQFDS.png', 'class': 1},
 {'relative_path': '3/PIOYMDBWFELC.png', 'class': 3},
 {'relative_path': '1/PIYOTQVWKSXM.png', 'class': 1},
 {'relative_path': '4/XFSIWUCMTNPZ.png', 'class': 4},
 {'relative_path': '3/LZXNFDCRYEQA.png', 'class': 3},
```

Figure 8.18 – Image and label pairs

In *Figure 8.18*, we have a list of dictionaries containing pairs of image paths and classes. If you are familiar with the **MNIST** dataset, you are probably aware that this dataset involves images of numbers with their corresponding numerical labels.

14. In a similar way, generate the validation image data and the dictionary that will contain the validation `.lst` file data:

```
train_dataset_length = len(
    training_and_validation_dataset
)
validation_indexes = get_validation_row_indexes(
    row_count=train_dataset_length,
    percent=0.01)

v = generate_image_files_and_lst_dict(
    dataset=training_and_validation_dataset,
    indexes=validation_indexes,
    tag = "validation"
)
validation_lst_dict = v
```

15. Finally, generate the test image data and the dictionary that will contain the validation `.lst` file data:

```
test_dataset_length = len(test_dataset)
test_indexes = get_test_row_indexes(
    row_count=test_dataset_length,
    percent=0.01)

test_lst_dict = generate_image_files_and_lst_dict(
    dataset=test_dataset,
    indexes=test_indexes,
    tag = "test"
)
```

16. Define the `save_lsts_to_file()` function:

```
def save_lsts_to_file(values, filename):
    with open(filename, 'w') as output:
        for index, row in enumerate(
            values,
            start=1
        ):
```

```
            relative_path = row['relative_path']
            cls = row['class']
            t = f"{index}\t{cls}\t{relative_path}\n"
            output.write(t)
```

17. Use the `save_lsts_to_file()` function to generate the `train.lst` and `validation.lst` files:

```
save_lsts_to_file(
    train_lst_dict,
    filename="tmp/train_lst/train.lst"
)
save_lsts_to_file(
    validation_lst_dict,
    filename="tmp/validation_lst/validation.lst"
)
```

18. Inspect the structure and content of the `train.lst` file:

```
%%bash
head tmp/train_lst/train.lst
```

This should give us a list of label and filename pairs similar to what is shown in *Figure 8.19*.

```
1       5       5/RIDQNTFUPJME.png
2       0       0/RSXIPQJWAKFU.png
3       4       4/RQSGDZIMUAPC.png
4       1       1/QGSLNAWKDOHU.png
5       9       9/JYBUESOIAVDZ.png
6       2       2/TUPYWXORSIFH.png
7       1       1/VBGPEZXTQFDS.png
8       3       3/PIOYMDBWFELC.png
9       1       1/PIYOTQVWKSXM.png
10      4       4/XFSIWUCMTNPZ.png
```

Figure 8.19 – List of test image files

This file should contain around 480 label and filename pairs.

19. Specify the S3 bucket name and the prefix where the data will be stored. Make sure to replace the value of "<insert s3 bucket name here>" with the name of the bucket we created in the recipe *Preparing the Amazon S3 bucket and the training dataset for the linear regression experiment* from *Chapter 1, Getting Started with Machine Learning Using Amazon SageMaker*:

```
s3_bucket = "<insert s3 bucket name here>"
prefix = "image-experiments"
!aws s3 cp tmp/. s3://{s3_bucket}/{prefix}/ --recursive
```

20. Finally, use the %store magic to store the variable values for s3_bucket and prefix:

```
%store s3_bucket
%store prefix
```

We will use these variable values in the succeeding recipes.

Let's see how this works!

How it works...

In this recipe, we performed the required steps to prepare the training, validation, and test datasets before we proceed with the training step. In this recipe, we saved and generated the image files into their respective directories using the **Apache MXNet Vision Datasets classes**. We specifically used the mx.gluon.data.vision.datasets.MNIST class to generate the image dataset, similar to what is shown in *Figure 8.20*.

Figure 8.20 – MNIST dataset

Here, we have 1 class for each digit from 0 to 9. This gives us a total of 10 classes similar to what is shown in *Figure 8.20*. The objective of the image classifier model we will train with this dataset will be to correctly identify which digit is mapped to a given input image from the test dataset.

See also

There are other pre-defined datasets in the **Apache MXNet Vision Datasets**. These include the **Fashion MNIST**, **CIFAR10**, and **CIFAR100** datasets. Feel free to check the other pre-defined datasets we can load and generate here: https://mxnet.apache. org/versions/1.7.0/api/python/docs/api/gluon/data/vision/ datasets/index.html.

Training and deploying an image classifier using the built-in Image Classification Algorithm in SageMaker

In the previous recipe, we prepared the image files and a few other prerequisites using the **Apache MXNet Vision Dataset** classes. In this recipe, we will use the **SageMaker Python SDK** and the built-in **Image Classification Algorithm** to train a model using these image files and prerequisites. The image classifier trained and deployed in this recipe will be used to classify the images in the test dataset.

Getting ready

Here are the prerequisites for this recipe:

- This recipe continues from *Preparing the datasets for image classification using the Apache MXNet Vision Datasets classes*.

- A **SageMaker Studio** notebook running the **Python 3 (Data Science)** kernel.

How to do it...

The first set of steps in this recipe focus on preparing the prerequisites of the training and deployment steps:

1. Create a new notebook using the `Python 3 (Data Science)` kernel inside the `my-experiments/chapter08` directory and rename it to the name of this recipe. When prompted for the kernel to use, choose **Python 3 (Data Science)**.

2. Import and initialize a few prerequisites for the training job:

```
import sagemaker
from sagemaker import get_execution_role
import json
import boto3
session = sagemaker.Session()
role = get_execution_role()
region_name = boto3.Session().region_name
```

3. Use the `%store` magic to load the variable values for `s3_bucket` and `prefix`:

```
%store -r s3_bucket
%store -r prefix
```

4. Initialize the location of the S3 training and validation data as well as the location of the `.lst` files generated for each of the datasets. Set the `s3_output_location` value as well:

```
s3_train_data = 's3://{}/{}/{}'.format(
    s3_bucket,
    prefix,
    "train"
)
s3_validation_data = 's3://{}/{}/{}'.format(
    s3_bucket,
    prefix,
    "validation"
)
s3_train_lst_path = 's3://{}/{}/{}'.format(
    s3_bucket,
    prefix,
```

```
    "train_lst"
)
s3_validation_lst_path = 's3://{}/{}/{}'.format(
    s3_bucket,
    prefix,
    "validation_lst"
)
s3_output_location = 's3://{}/{}/output'.format(
    s3_bucket,
    prefix
)
```

5. Use the `retrieve()` function to get the container image URI of the **Image Classification Algorithm**:

```
from sagemaker.image_uris import retrieve
container = retrieve(
    "image-classification",
    region_name,
    "1"
)
container
```

This should give us a string value for the `container` variable similar to `'811284229777.dkr.ecr.us-east-1.amazonaws.com/image-classification:1'`.

The next set of steps focus on using the prerequisites from the previous set of steps for training and deploying the image classification model:

6. Initialize the `Estimator` object:

```
estimator = sagemaker.estimator.Estimator(
    container,
    role,
    instance_count=1,
    instance_type='ml.p2.xlarge',
    output_path=s3_output_location,
    sagemaker_session=session
)
```

Here, we make use of P2 instances that provide GPU-based parallel compute capabilities.

7. Use the `set_hyperparameters()` function to specify the hyperparameters of the training job:

```
estimator.set_hyperparameters(
    num_layers=18,
    image_shape = "1,28,28",
    num_classes=10,
    num_training_samples=600,
    mini_batch_size=20,
    epochs=5,
    learning_rate=0.01,
    top_k=2,
    precision_dtype='float32'
)
```

> **Note**
>
> If you have no idea what these hyperparameter values mean, do not worry as we will take a closer look at these values in the *How it works* section!

8. Prepare the training input channels for the actual image files:

```
from sagemaker.inputs import TrainingInput
train = TrainingInput(
    s3_train_data,
    distribution='FullyReplicated',
    content_type='application/x-image',
    s3_data_type='S3Prefix'
)
validation = TrainingInput(
    s3_validation_data,
    distribution='FullyReplicated',
    content_type='application/x-image',
    s3_data_type='S3Prefix'
)
```

9. Prepare the training input channels for the .1st files:

```
content_type = 'application/x-image'
train_1st = TrainingInput(
    s3_train_1st_path,
    distribution='FullyReplicated',
    content_type=content_type,
    s3_data_type='S3Prefix'
)
validation_1st = TrainingInput(
    s3_validation_1st_path,
    distribution='FullyReplicated',
    content_type=content_type,
    s3_data_type='S3Prefix'
)
```

10. Use the fit() function to start the training job with the data channels from the previous steps as the input values:

```
%%time
data_channels = {
    'train': train,
    'validation': validation,
    'train_1st': train_1st,
    'validation_1st': validation_1st
}
estimator.fit(inputs=data_channels, logs=True)
```

This should yield a set of logs similar to what is shown in *Figure 8.21*.

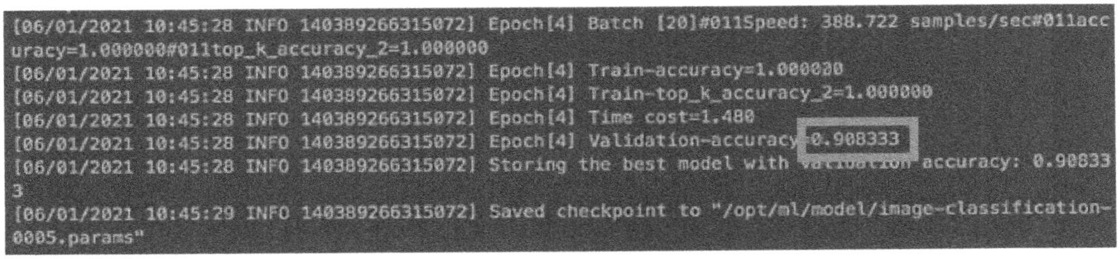

```
[06/01/2021 10:45:28 INFO 140389266315072] Epoch[4] Batch [20]#011Speed: 388.722 samples/sec#011acc
uracy=1.000000#011top_k_accuracy_2=1.000000
[06/01/2021 10:45:28 INFO 140389266315072] Epoch[4] Train-accuracy=1.000000
[06/01/2021 10:45:28 INFO 140389266315072] Epoch[4] Train-top_k_accuracy_2=1.000000
[06/01/2021 10:45:28 INFO 140389266315072] Epoch[4] Time cost=1.480
[06/01/2021 10:45:28 INFO 140389266315072] Epoch[4] Validation-accuracy=0.908333
[06/01/2021 10:45:28 INFO 140389266315072] Storing the best model with validation accuracy: 0.90833
3
[06/01/2021 10:45:29 INFO 140389266315072] Saved checkpoint to "/opt/ml/model/image-classification-
0005.params"
```

Figure 8.21 – Training job logs

Here, we can see that the validation accuracy is at `90.83%`. As we are dealing with a simplified example in this recipe, this should be a good start for us.

> **Note**
>
> This step may take around 5 to 10 minutes to complete. While waiting, feel free to take a quick break and grab a cup of coffee or tea!

11. Use the `deploy()` function to deploy the model to an inference endpoint:

```
endpoint = estimator.deploy(
    initial_instance_count = 1,
    instance_type = 'ml.m4.xlarge'
)
```

This step should take about 5 to 10 minutes to complete.

> **Important note**
>
> Running the `deploy()` function would launch an instance that would continue running until the delete resource operation is performed. While the instance is running, you will be charged for the amount of time it is running. Make sure to delete the inference endpoint after completing this recipe.

Now that we have the endpoint deployed in the previous step, the last set of steps in this recipe focus on testing this endpoint using the data from the test set:

12. Update the `serializer` property of the endpoint using the `IdentitySerializer` from the **SageMaker Python SDK**:

```
from sagemaker.serializers import IdentitySerializer
endpoint.serializer = IdentitySerializer(
    content_type="application/x-image"
)
```

13. Define the `get_class_from_results()` function:

```
import json


def get_class_from_results(results):
    results_prob_list = json.loads(results)
    best_index = results_prob_list.index(
        max(results_prob_list)
```

```
    )

    return {
        0: "ZERO",
        1: "ONE",
        2: "TWO",
        3: "THREE",
        4: "FOUR",
        5: "FIVE",
        6: "SIX",
        7: "SEVEN",
        8: "EIGHT",
        9: "NINE"
    }[best_index]
```

14. Define the `predict()` function, which displays the image specified as the payload along with the predicted label:

```
from IPython.display import Image, display

def predict(filename, endpoint=endpoint):
    byte_array_input = None

    with open(filename, 'rb') as image:
        f = image.read()
        byte_array_input = bytearray(f)

    display(Image(filename))

    results = endpoint.predict(byte_array_input)
    return get_class_from_results(results)
```

15. Use the `predict()` function on each of the files inside the `tmp/test` directory:

```
results = !ls -1 tmp/test
for filename in results:
    print(predict(f"tmp/test/{filename}"))
```

This should give us a list of images with their corresponding predicted labels, similar to what is shown in *Figure 8.22*.

Figure 8.22 – List of images with their corresponding predicted labels

In *Figure 8.22*, we can see that our model was able to get correct predictions based on the images used from the test dataset. It is important to note also that it makes mistakes from time to time as we can see that the fourth image got a predicted label of SEVEN instead of a predicted label of FIVE.

16. Delete the endpoint using the delete_endpoint() function:

```
endpoint.delete_endpoint()
```

At this point, we should be comfortable using the SageMaker built-in **Image Classification Algorithm**.

Now, let's see how this works!

How it works...

In this recipe, we used the **SageMaker Python SDK** to train and deploy our image classifier model. As we have already performed a couple of training jobs and deployments using other built-in algorithms, it is safe to say that this recipe has the same pattern as the other similar recipes on training and deployment. That said, let's focus our discussion on the hyperparameters used when training the model.

Let's start with the easy ones. The `num_classes` hyperparameter defines the number of output classes in our multi-label classification problem. In this recipe, we set the value to `10` as we have 10 classes to group our MNIST dataset (for example, the digits 0 to 9). The `num_training_samples` hyperparameter value should be equivalent to the number of training samples or records in the dataset (for example, `600`).

The built-in **Image Classification Algorithm** makes use of a **convolutional neural network** (**CNN**) to perform multi-label classification. This means that its configurable hyperparameters will revolve around what we can configure with the neural network structure. These include the `num_layers` hyperparameter, which determines the number of layers for the network. These hyperparameters also include `image_shape`, which should be equal to the dimensions of the image (for example, `28 x 28`). This will define the size of the input layer of the network. There are other hyperparameters we will not discuss in this section, so feel free to check `https://docs.aws.amazon.com/sagemaker/latest/dg/IC-Hyperparameter.html`.

There's more...

It is important to note that we also have the option to use **transfer learning** to fine-tune a pre-trained model instead of training the image classifier from scratch. **Transfer learning** involves using a pre-trained model as the starting point when producing a new model. This allows us to use significantly fewer images when producing a high-quality model. The steps are mostly similar, and we simply need to specify the hyperparameter value of `use_pretrained_model` to 1.

We can also use **incremental training** to start another training job using a generated model from a previous training job and produce a more accurate model. This will save machine learning practitioners time when working with a similar dataset:

```
input_data = {
    "train": train_data,
    "validation": validation_data,
    "model": model_data
}
estimator.fit(inputs=input_data)
```

Here, we simply pass the S3 path where the model data of the previous training job is stored as one of the input channels when using the `fit()` function.

See also

If you are looking for examples of using the built-in **Image Classification Algorithm** on real datasets and more complex examples, feel free to check some of the notebooks in the `aws/amazon-sagemaker-examples` GitHub repository:

- Transfer learning with the **Image Classification algorithm**—`https://github.com/aws/amazon-sagemaker-examples/blob/master/introduction_to_amazon_algorithms/imageclassification_caltech/Image-classification-transfer-learning-highlevel.ipynb`

- Incremental training with the **Image Classification algorithm**—`https://github.com/aws/amazon-sagemaker-examples/blob/master/introduction_to_amazon_algorithms/imageclassification_caltech/Image-classification-incremental-training-highlevel.ipynb`

As we will not be able to dive deep into the different features of the built-in **Image Classification algorithm** in this book, feel free to check this link for more information: `https://docs.aws.amazon.com/sagemaker/latest/dg/image-classification.html`.

Generating a synthetic time series dataset

In the previous recipes of this chapter, we trained and deployed models that deal with text classification and image classification requirements. In this recipe, we will generate a synthetic time series dataset similar to what is shown in *Figure 8.23*. This dataset will then be used later for training the **DeepAR** model in the recipe *Training and deploying a DeepAR model*.

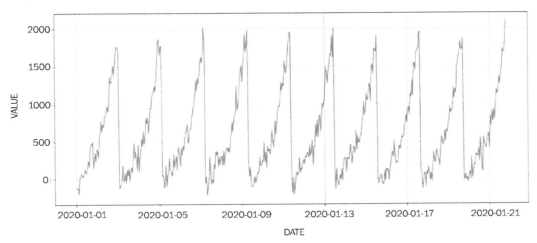

Figure 8.23 – Time series plot

We can see that seasonal variations or **seasonality** are present in this time series dataset. At the same time, we can see that there is a bit of noise added to make the dataset a bit more realistic and enhance the robustness of trained machine learning models.

Getting ready

A **SageMaker Studio** notebook running the **Python 3 (Data Science)** kernel is the only prerequisite for this recipe.

How to do it...

The steps in this recipe focus on generating and plotting the synthetic time series dataset:

1. Create a new notebook using the `Python 3 (Data Science)` kernel inside the `my-experiments/chapter08` directory and rename it to the name of this recipe. When prompted for the kernel to use, choose **Python 3 (Data Science)**.

2. Import a few prerequisites as follows:

```
import numpy as np
import matplotlib.pyplot as plt
import pandas as pd
%matplotlib inline
```

3.   Define the `generate_time_series()` function:

```
def generate_time_series(
    t0="2020-01-01 00:00:00"
):
    time = np.arange(50)
    values = np.where(time < 20, time**2,
                      (time-5)**2)

    base = []
    for iteration in range(10):
        for y in range(50):
            base.append(values[y])

    base += np.random.randn(500)*100

    freq = "H"
    data_length = len(base)
    index = pd.date_range(start=t0,
                          freq=freq,
                          periods=data_length)
    ts = pd.Series(data=base, index=index)

    return {
        "freq": freq,
        "t0": t0,
        "length": len(ts),
        "data": ts
    }
```

This function accepts an optional `t0` parameter value that lets us configure the start date and time value of the time series dataset.

4. Use the `generate_time_series()` function to generate the synthetic time series dataset:

```
time_series_data = generate_time_series()
time_series_data
```

This should give us a dictionary of values similar to what is shown in *Figure 8.24*.

```
{'freq': 'H',
 't0': '2020-01-01 00:00:00',
 'length': 500,
 'data': 2020-01-01 00:00:00     -118.520821
2020-01-01 01:00:00    -121.417558
2020-01-01 02:00:00    -113.624594
2020-01-01 03:00:00    -197.074225
2020-01-01 04:00:00      11.191463
                         ...
2020-01-21 15:00:00    1732.537542
2020-01-21 16:00:00    1675.656558
2020-01-21 17:00:00    1684.402002
2020-01-21 18:00:00    1844.359636
2020-01-21 19:00:00    2092.252685
Freq: H, Length: 500, dtype: float64}
```

Figure 8.24 – Time series data and properties

Here, we can see that we have generated a dictionary of values with the keys `freq`, `t0`, `length`, and `data`.

5. Visualize how the time series dataset looks using `matplotlib`:

```
data = time_series_data["data"]
time = data.index
values = data
plt.figure(figsize=(14,6))
plt.plot(time, values)
plt.grid(True)
plt.xlabel("DATE")
plt.ylabel("VALUE")
```

This should render a chart similar to what is shown in *Figure 8.25*.

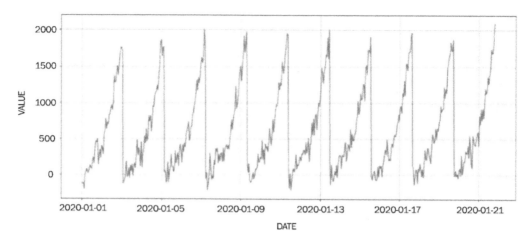

Figure 8.25 – Time series plot

In *Figure 8.25*, we can see the time series with a bit of noise we added while generating the dataset of values.

6. Create a temporary directory using the mkdir command:

```
!mkdir -p tmp
```

7. Define the save_data_to_json() function, which serializes our time-series data by storing the data points along with the frequency, start time, and length in a file:

```python
import json


def save_data_to_json(time_series_data,
                      filename):
    tmp = {}
    tmp["freq"] = time_series_data["freq"]
    tmp["t0"] = time_series_data["t0"]
    tmp["length"] = time_series_data["length"]
    tmp["data"] = list(time_series_data["data"])

    with open(filename, 'w') as file:
        json.dump(tmp, file)
```

8. Use the `save_data_to_json()` function to store the time series dataset in a JSON file inside the `tmp` directory:

```
save_data_to_json(time_series_data,
                  "tmp/all.json")
```

9. Use the `head` command to check the content of the `tmp/all.json` file:

```
!head tmp/all.json
```

This should give us a structure of values similar to what is shown in *Figure 8.26*.

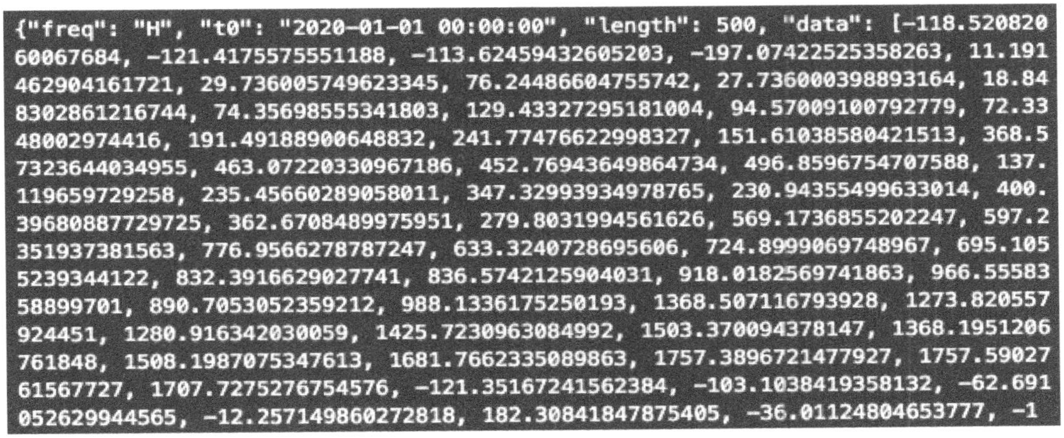

{"freq": "H", "t0": "2020-01-01 00:00:00", "length": 500, "data": [-118.520820
60067684, -121.4175575551188, -113.62459432605203, -197.07422525358263, 11.191
462904161721, 29.736005749623345, 76.24486604755742, 27.736000398893164, 18.84
8302861216744, 74.35698555341803, 129.43327295181004, 94.57009100792779, 72.33
48002974416, 191.49188900648832, 241.77476622998327, 151.61038580421513, 368.5
7323644034955, 463.07220330967186, 452.76943649864734, 496.8596754707588, 137.
119659729258, 235.45660289058011, 347.32993934978765, 230.94355499633014, 400.
39680887729725, 362.6708489975951, 279.8031994561626, 569.1736855202247, 597.2
351937381563, 776.9566278787247, 633.3240728695606, 724.8999069748967, 695.105
5239344122, 832.3916629027741, 836.5742125904031, 918.0182569741863, 966.55583
58899701, 890.7053052359212, 988.1336175250193, 1368.507116793928, 1273.820557
924451, 1280.916342030059, 1425.7230963084992, 1503.370094378147, 1368.1951206
761848, 1508.1987075347613, 1681.7662335089863, 1757.3896721477927, 1757.59027
61567727, 1707.7275276754576, -121.35167241562384, -103.1038419358132, -62.691
052629944565, -12.257149860272818, 182.30841847875405, -36.01124804653777, -1

Figure 8.26 – Contents of the all.json file

At this point, our synthetic dataset is ready. In the next recipe, we will perform the train-test split on this synthetic dataset.

For now, let's see how this works!

How it works...

In this recipe, we have generated a synthetic time series dataset with the help of the `generate_time_series()` function we defined in this recipe. Time series data involves a sequence of values indexed with timestamps. A time series dataset is composed of the following parts: the value and the date and time index. If you have worked with sales and stock price data represented in time series format, then that would be an example of time series data.

While preparing the synthetic dataset, we added **seasonality** and **noise** within the `generate_time_series()` function to make it a bit more realistic. The presence of noise in the dataset enhances the robustness of trained machine learning models. At the same time, it would be rare to see real time series data without noise.

In the next couple of recipes, we will process this synthetic dataset and use this to train our **DeepAR** model. Since we are going to work with a simplified example of using the **DeepAR Forecasting algorithm** in this book, we will not have multiple time series datasets for our training dataset.

> **Important note**
> Note that **DeepAR** models start to outperform other models when trained with multiple time series data. That said, you may wish to extend this recipe later on to generate multiple time series data that will be used to train a **DeepAR** model.

Performing the train-test split on a time series dataset

In the previous recipe, we generated a synthetic time-series dataset that we will use to train a **DeepAR** model in the next two recipes. Before we proceed with the actual training of the model, Before we proceed with the actual training of the model, we need to properly split the data first into the train and test sets. That is what we will do in this recipe!

When performing the train-test split with a time series dataset, it is important to note that we do not perform random splitting of the data as this would not preserve the temporal order of the observations.

Getting ready

Here are the prerequisites of this recipe:

- This recipe continues from *Generating a synthetic time series dataset*.

- A **SageMaker Studio** notebook running the **Python 3 (Data Science)** kernel.

How to do it...

1. Create a new notebook using the `Python 3 (Data Science)` kernel inside the `my-experiments/chapter08` directory and rename it to the name of this recipe. When prompted for the kernel to use, choose **Python 3 (Data Science)**.

2. Import a few prerequisites as follows:

```
import json
import numpy as np
import matplotlib.pyplot as plt
import pandas as pd
%matplotlib inline
```

3. Define the `load_data_from_json()` function. This will load the time series dataset previously stored using `save_data_to_json()`, defined in the recipe *Generating a synthetic time series dataset*:

```
def load_data_from_json(filename):
    tmp = {}
    with open(filename) as file:
        tmp = json.load(file)

    index = pd.date_range(
        start=tmp["t0"],
        freq=tmp["freq"],
        periods=tmp["length"])
    tmp["data"] = pd.Series(
        data=tmp["data"],
        index=index)

    return tmp
```

4. Use the `load_data_from_json()` function to load the time series dataset generated in the recipe *Generating a synthetic time series dataset*:

```
time_series_data = load_data_from_json(
    "tmp/all.json"
)
```

5. Define the `train_test_split()` function:

```
def train_test_split(data, ratio=0.9):
    train_length = int(len(data) * ratio)
    pl = int(len(data)) - train_length
    prediction_length = pl
    training_dataset = data[:-prediction_length]
    target_dataset = data[train_length-1:]
    test_dataset = data

    return {
        "prediction_length": prediction_length,
        "training_dataset": training_dataset,
        "target_dataset": target_dataset,
        "test_dataset": test_dataset
    }
```

6. Use the `train_test_split()` function:

```
results = train_test_split(
    time_series_data["data"]
)
print(results["prediction_length"])
```

This should give us 50 as the `prediction_length` value.

7. Use `matplotlib` to prepare a plot:

```
training_dataset = results["training_dataset"]
target_dataset = results["target_dataset"]

plt.figure(figsize=(14,6))
plt.plot(training_dataset.index,
         training_dataset, label="training")
plt.plot(target_dataset.index,
         target_dataset,
         label="target")
plt.grid(True)
plt.xlabel("DATE")
plt.ylabel("VALUE")
```

```
plt.legend()
plt.show()
```

This should render a chart similar to what is shown in *Figure 8.27*.

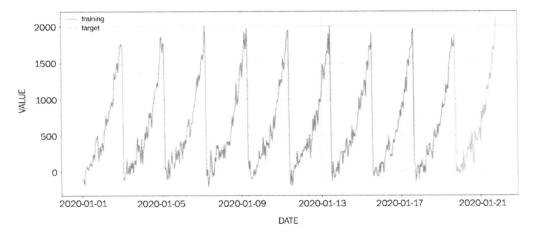

Figure 8.27 – Time series plot

We have in *Figure 8.27* the training and target datasets.

8. Define the series_to_object() function:

```
def series_to_object(data):
    return {"start": str(data.index[0]),
            "target": list(data)}
```

9. Define the series_to_jsonline() function:

```
def series_to_jsonline(data):
    return json.dumps(series_to_object(data))
```

10. Define the save_data_to_jsonlines() function:

```
def save_data_to_jsonlines(data, filename):
    with open(filename, 'wb') as file:
        t = series_to_jsonline(data)
        t = t.encode("utf-8")
        file.write(t)
        file.write("\n".encode("utf-8"))
```

11. Use the `save_data_to_jsonlines()` function for the training and test datasets:

```
save_data_to_jsonlines(
    results["training_dataset"],
    "tmp/training.jsonl"
)
save_data_to_jsonlines(
    results["test_dataset"],
    "tmp/test.jsonl"
)
```

This should generate the jsonl files inside the tmp directory.

12. Specify the S3 bucket name and the prefix where the data will be stored. Make sure to replace the value of `"<insert s3 bucket name here>"` with the name of the bucket we created in the recipe *Preparing the Amazon S3 bucket and the training dataset for the linear regression experiment* from *Chapter 1, Getting Started with Machine Learning Using Amazon SageMaker*:

```
s3_bucket = '<insert s3 bucket name here>'
prefix = 'chapter08'
```

13. Upload the `training.jsonl` and `test.jsonl` files to the Amazon S3 bucket using the following command:

```
!aws s3 cp tmp/training.jsonl s3://{s3_bucket}/{prefix}/
input/training.jsonl
!aws s3 cp tmp/test.jsonl s3://{s3_bucket}/{prefix}/
input/test.jsonl
```

14. Store the prediction length using the `%store` magic:

```
prediction_length = results["prediction_length"]
%store prediction_length
```

15. Store the frequency using the `%store` magic:

```
freq = time_series_data["freq"]
%store freq
```

16. In a similar fashion, use the %store magic to save the training dataset time series:

```
training_dataset = results["training_dataset"]
%store training_dataset
training_dataset
```

This should give us a set of values similar to what is shown in *Figure 8.27*.

```
2020-01-01 00:00:00    -118.520821
2020-01-01 01:00:00    -121.417558
2020-01-01 02:00:00    -113.624594
2020-01-01 03:00:00    -197.074225
2020-01-01 04:00:00      11.191463
                          ...
2020-01-19 13:00:00    1675.098326
2020-01-19 14:00:00    1830.164759
2020-01-19 15:00:00    1717.529411
2020-01-19 16:00:00    1714.456715
2020-01-19 17:00:00    1868.780563
Freq: H, Length: 450, dtype: float64
```

Figure 8.27 – Training dataset

In *Figure 8.27*, we have the timestamp and value pairs in the training dataset.

17. Finally, store the target dataset time series values as well:

```
target_dataset = results["target_dataset"]
%store target_dataset
target_dataset
```

This should give us a set of values similar to what is shown in *Figure 8.28*.

```
2020-01-19 17:00:00    1868.780563
2020-01-19 18:00:00       7.224860
2020-01-19 19:00:00     -17.103169
2020-01-19 20:00:00      61.009508
2020-01-19 21:00:00     -17.969486
2020-01-19 22:00:00      11.795133
2020-01-19 23:00:00     -44.306254
2020-01-20 00:00:00      69.532034
2020-01-20 01:00:00      95.094393
```

Figure 8.28 – Target dataset

We have in *Figure 8.28* the timestamp and value pairs in the target dataset.

18. Finally, use the `%store` magic to store the variable values for `s3_bucket` and `prefix`:

```
%store s3_bucket
%store prefix
```

At this point, we should be ready to train and deploy our **DeepAR** model.

In the meantime, let's see how this works first!

How it works...

To test and estimate the performance of a machine learning model using data it has not seen before, we split a given dataset and use only a certain portion (for example, a training dataset) to train the model and the other portion to evaluate the model produced.

There are different ways to split the data but two of the most common approaches involve **random splitting** and **sequential splitting** of the dataset. As the name suggests, **random splitting** involves randomly selecting a certain percentage of the dataset and using those selected records for the test dataset. Random splitting should do the trick when we are not required to preserve the order of the input data. On the other hand, **sequential splitting** involves dividing the dataset into the training and test sets while preserving the order of the records. This works well for time-series data where random splitting will not work as that will make the randomly selected records lose their context and meaning. That said, the most appropriate way to divide the time-series dataset is through sequential splitting—which is what we performed in this recipe.

Training and deploying a DeepAR model

The goal of forecasting models is to predict future data points based on previous records. There are different forecasting algorithms available, including ARIMA and ETS. One algorithm making use of **recurrent neural networks** (**RNNs**) to forecast time series data is **DeepAR**. In this recipe, we will train and deploy a **DeepAR** model using the **SageMaker Python SDK**. To help us get started with using the built-in **DeepAR** forecasting algorithm, we will only work with a single time series dataset when training the model.

Getting ready

Here are the prerequisites of this recipe:

- This recipe continues from *Performing the train-test split on a time series dataset.*

- A **SageMaker Studio** Notebook running the **Python 3 (Data Science)** kernel.

How to do it...

The first few steps in this recipe focus on preparing the prerequisites for training the **DeepAR** model:

1. Create a new notebook using the `Python 3 (Data Science)` kernel inside the `my-experiments/chapter08` directory and rename it with the name of this recipe. When prompted for the kernel to use, choose **Python 3 (Data Science)**.

2. Import the following prerequisites:

```
import sagemaker
import boto3
from sagemaker import get_execution_role
role = get_execution_role()
session = sagemaker.Session()
region_name = boto3.Session().region_name
```

3. Specify the values for `s3_bucket` and `prefix`:

```
%store -r s3_bucket
%store -r prefix
```

4. Prepare the variable values for `training_s3_input_location`, `test_s3_input_location`, and `training_s3_output_location`:

```
training_s3_input_location = f"s3://{s3_bucket}/{prefix}/input/training.jsonl"
test_s3_input_location = f"s3://{s3_bucket}/{prefix}/input/test.jsonl"
training_s3_output_location = f"s3://{s3_bucket}/{prefix}/output/"
```

5. Use `TrainingInput` and specify the `content_type` as `json`:

```
from sagemaker.inputs import TrainingInput
train = TrainingInput(
    training_s3_input_location,
    content_type="json"
)
test = TrainingInput(
    test_s3_input_location,
    content_type="json"
)
```

6. Read the `prediction_length` value using the `%store` magic. Note that we will use the `prediction_length` value to control how far into the future we will make predictions using the trained model when we specify this as one of the hyperparameter values during the training step:

```
%store -r prediction_length
```

7. Read the `freq` value using the `%store` magic:

```
%store -r freq
```

Note that we will pass this as the hyperparameter value for `time_freq` during the training step, which corresponds to the granularity of the time series data.

8. Set the `context_length` value to be equal to the `prediction_length`:

```
context_length = prediction_length
```

9. Use the `retrieve()` function to get the ECR image URI for the built-in **DeepAR** algorithm:

```
from sagemaker.image_uris import retrieve
container = retrieve(
    "forecasting-deepar",
    region_name,
    "1"
)
container
```

We should get a value similar to `'522234722520.dkr.ecr.us-east-1. amazonaws.com/forecasting-deepar:1'`.

The next set of steps focus on training and deploying our **DeepAR** model using the prerequisites prepared in the previous steps:

10. Initialize the `Estimator` object using the following block of code:

```
estimator = sagemaker.estimator.Estimator(
    container,
    role,
    instance_count=1,
    instance_type='ml.c4.2xlarge',
    output_path=training_s3_output_location,
    sagemaker_session=session
)
```

11. Use the `set_hyperparameters()` function:

```
estimator.set_hyperparameters(
    time_freq=freq,
    context_length=str(context_length),
    prediction_length=str(prediction_length),
    num_cells=40,
    num_layers=3,
    likelihood="gaussian",
    epochs=20,
    mini_batch_size=32,
    learning_rate=0.001,
    dropout_rate=0.05,
    early_stopping_patience=10
)
```

12. Use the `fit()` function to start the training job:

```
%%time
data_channels = {"train": train, "test": test}
estimator.fit(inputs=data_channels)
```

This should yield a set of logs similar to what is shown in *Figure 8.29*.

```
2021-06-01 11:05:17 Starting - Starting the training job...
2021-06-01 11:05:18 Starting - Launching requested ML instancesProfilerReport-
1622545517: InProgress
......
2021-06-01 11:06:44 Starting - Preparing the instances for training.........
2021-06-01 11:08:14 Downloading - Downloading input data
2021-06-01 11:08:14 Training - Downloading the training image...
2021-06-01 11:08:35 Training - Training image download completed. Training in
progress.Arguments: train
[06/01/2021 11:08:39 INFO 139987855828352] Reading default configuration from
/opt/amazon/lib/python3.6/site-packages/algorithm/resources/default-input.jso
n: {'_kvstore': 'auto', '_num_gpus': 'auto', '_num_kv_servers': 'auto', '_tuni
ng_objective_metric': '', 'cardinality': 'auto', 'dropout_rate': '0.10', 'earl
y_stopping_patience': '', 'embedding_dimension': '10', 'learning_rate': '0.00
1', 'likelihood': 'student-t', 'mini_batch_size': '128', 'num_cells': '40', 'n
um_dynamic_feat': 'auto', 'num_eval_samples': '100', 'num_layers': '2', 'test_
quantiles': '[0.1, 0.2, 0.3, 0.4, 0.5, 0.6, 0.7, 0.8, 0.9]'}
```

Figure 8.29 – Training job logs

Note that this step may take about 4 to 8 minutes to complete.

13. Use the `deploy()` function to deploy the DeepAR model:

```
predictor = estimator.deploy(
    initial_instance_count=1,
    instance_type="ml.m4.xlarge"
)
```

This step should take about 5 to 10 minutes to complete.

> **Important note**
>
> Running the `deploy()` function would launch an instance that would continue running until the delete resource operation is performed. While the instance is running, you will be charged for the amount of time it is running. Make sure to delete the inference endpoint after completing the recipe *Performing probabilistic forecasting with a deployed DeepAR model*.

14. Use the `%store` magic to save the value of the endpoint name:

```
endpoint_name = predictor.endpoint_name
%store endpoint_name
```

We will use this variable value in the next recipe.

Now, let's see how this works!

How it works...

In this recipe, we used the **SageMaker Python SDK** to train and deploy our **DeepAR** model. As we have already performed a couple of training jobs and deployments using other built-in algorithms, it is safe to say that about 80% of this recipe works pretty much the same way as with the other similar recipes. That said, let's focus our discussion on the hyperparameters used so that we will have a better understanding of how this built-in algorithm works.

We have specified a couple of hyperparameters, such as `context_length` and `prediction_length`. The hyperparameter `context_length` is generally close to the value of `prediction_length`. The `context_length` hyperparameter refers to the number of records the model gets to see before making a prediction. On the other hand, the hyperparameter `prediction_length` refers to the number of records we want to forecast. This means that the higher the `prediction_length` value, the longer the forecast output data will be.

The **DeepAR** forecasting algorithm makes use **RNNs** to forecast time-series values, which means that the configurable hyperparameters will revolve around what we can configure with the neural network structure. The `num_layers` hyperparameter, for example, determines the number of hidden layers in the RNN. Generally, the more hidden layers we use, the greater the model can detect complex features.

> **Tip**
> Of course, if we were to use an excessive amount of layers in our neural network, this could result in overfitting.

Next would be the `dropout_rate` hyperparameter. This value impacts a model's susceptibility to overfitting. That said, we can prevent overfitting by experimenting with different values for `dropout_rate`. There are other hyperparameter values that we will not discuss in this section so feel free to check `https://docs.aws.amazon.com/sagemaker/latest/dg/deepar_hyperparameters.html` for more information.

Performing probabilistic forecasting with a deployed DeepAR model

In the previous recipe, we trained a **DeepAR** model using the synthetic time-series dataset generated. After the training step, we also deployed this model to a real-time inference endpoint. In this recipe, we will use the deployed **DeepAR** model we deployed in the previous recipe for inference. We will also take a look at how to use one of the advantages of using **DeepAR** models—the ability to estimate the probability distribution of the future state of a time series dataset.

Getting ready

Here are the prerequisites of this recipe:

- This recipe continues from *Training and deploying a DeepAR model*.
- A **SageMaker Studio** notebook running the **Python 3 (Data Science)** kernel.

How to do it...

The steps in this recipe focus on using the endpoint from the previous recipe to perform probabilistic forecasting:

1. Create a new notebook using the Python 3 (Data Science) kernel inside the my-experiments/chapter08 directory and rename it to the name of this recipe. When prompted for the kernel to use, choose **Python 3 (Data Science)**.

2. Use the %store magic to read the value for endpoint_name, training_dataset, target_dataset, freq, and prediction_length:

```
%store -r endpoint_name
%store -r training_dataset
%store -r target_dataset
%store -r freq
%store -r prediction_length
```

3. Import and prepare a few prerequisites such as the role and the session:

```
import sagemaker
import boto3
from sagemaker import get_execution_role
role = get_execution_role()
session = sagemaker.Session()
```

4. Initialize the `Predictor` object with the endpoint name and the SageMaker session:

```
Predictor = sagemaker.predictor.Predictor
predictor = Predictor(
    endpoint_name=endpoint_name,
    sagemaker_session=session
)
```

5. Specify the `serializer` of the predictor with the `JSONSerializer`:

```
predictor.serializer = sagemaker.serializers.
JSONSerializer()
```

6. Prepare the quantiles and configuration parameters:

```
quantiles=["0.1", "0.5", "0.9"]
configuration = {
    "num_samples": 100,
    "output_types": ["quantiles"],
    "quantiles": quantiles,
}
```

In a later step, we will use the 0.5 quantile values returned by the `predict()` function for the **deterministic point forecast**. For the 0.1 and 0.9 quantile values, we will use these values as the lower and upper bounds of possible values respectively for the **probabilistic forecast**.

7. Define the `series_to_object()` function:

```
def series_to_object(data):
    return {
        "start": str(data.index[0]),
        "target": list(data)
    }
```

8. Use the `series_to_object()` function to prepare the `http_request_data` payload to the `predict()` function:

```
instances = [series_to_object(training_dataset)]
http_request_data = {
    "instances": instances,
    "configuration": configuration
}
http_request_data
```

This should give us a dictionary of values similar to what is shown in *Figure 8.30*.

{'instances': [{'start': '2020—01—01 00:00:00',
    'target': [-118.52082060067684,
    -121.4175575551188,
    -113.62459432605203,
    -197.07422525358263,
    11.191462904161721,
    29.736005749623345,
    76.24486604755742,
    27.736000398893164,
    18.848302861216744,
    74.35698555341803,
    129.43327295181004,
    94.57009100792779,
    72.3348002974416,
    191.49188900648832,
    241.77476622998327,

Figure 8.30 – Payload to the predict() function

We will use this dictionary of values in the next step.

9. Use the `predict()` function to perform the prediction:

```
response = predictor.predict(http_request_data)
```

10. Use the `loads()` function from the `json` library to convert the string response to a dictionary:

```
import json
response_data = json.loads(response)
response_data
```

This should give us a nested structure of values similar to what is shown in *Figure 8.31*.

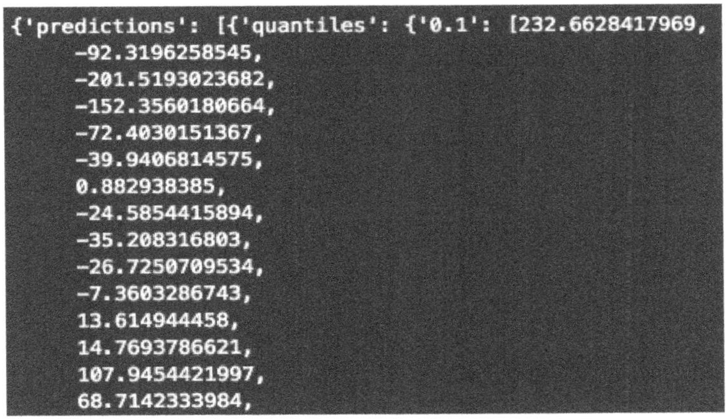

```
{'predictions': [{'quantiles': {'0.1': [232.6628417969,
        -92.3196258545,
        -201.5193023682,
        -152.3560180664,
        -72.4030151367,
        -39.9406814575,
        0.882938385,
        -24.5854415894,
        -35.208316803,
        -26.7250709534,
        -7.3603286743,
        13.614944458,
        14.7693786621,
        107.9454421997,
        68.7142333984,
```

Figure 8.31 – Prediction results

Here, we have the prediction values returned by the inference endpoint. As we have only passed a single time series dataset as the payload to the `predict()` function, we also expect to get a single set of values.

11. Store the results in the `single_result` variable:

```
single_result = response_data['predictions'][0]
```

12. Prepare the `prediction_time` variable, which contains the start time of the predicted time series values:

```
import pandas as pd
prediction_time = training_dataset.index[-1] +
pd.Timedelta(1, unit=freq)
```

13. Set up the `prediction_index` variable containing a list of date and time values:

```
prediction_index = pd.date_range(
    start=prediction_time,
    freq=freq,
    periods=prediction_length
)
```

14. Next, prepare the output variable containing the prediction results:

```
output = pd.DataFrame(
    data=single_result['quantiles'],
    index=prediction_index
)
output
```

This should give us a DataFrame of values similar to what is shown in *Figure 8.32*.

	0.1	0.5	0.9
2020-01-19 18:00:00	232.662842	682.048706	1095.776245
2020-01-19 19:00:00	-92.319626	156.423843	375.291443
2020-01-19 20:00:00	-201.519302	-28.107492	165.889114
2020-01-19 21:00:00	-152.356018	-29.349388	94.649643
2020-01-19 22:00:00	-72.403015	9.842265	109.686554
2020-01-19 23:00:00	-39.940681	28.484755	110.671051
2020-01-20 00:00:00	0.882938	80.206360	159.140900
2020-01-20 01:00:00	-24.585442	55.242252	127.168518
2020-01-20 02:00:00	-35.208317	52.206253	128.978928

Figure 8.32 – DataFrame containing the prediction results with quantile values

Here, we can see the prediction results with the quantile values. As **DeepAR** returns probabilistic forecasts using quantile values, the quantile values dictate the chance of observing a certain percentage of values lower and higher than said quantile values. For example, we have a 50% chance of observing a value lower than the 0.5 quantile and a 50% chance of observing a value higher than the 0.5 quantile. This makes it a good choice for **deterministic point forecasts**. On the other hand, we have a 10% chance of observing a value lower than the 0.1 quantile and a 90% chance of observing a value lower than the 0.9 quantile. That said, we will use the 0.9 and 0.1 quantile values for the upper bound and lower bound values of the **probabilistic forecast**.

> **Note**
> We will discuss what **probabilistic forecasting** means in the *How it works...* section.

15. Visualize the predicted values with the training dataset:

```python
import matplotlib.pyplot as plt
%matplotlib inline

plt.figure(figsize=(14,6))
plt.plot(target_dataset.index,
         target_dataset,
         label="target")
plt.plot(training_dataset.index,
         training_dataset,
         label="training")
plt.grid(True)
plt.xlabel("DATE")
plt.ylabel("VALUE")

p10 = output["0.1"]
p90 = output["0.9"]
plt.fill_between(
    p10.index,
    p10,
    p90,
    color="y",
    alpha=0.5,
    label="80% confidence interval"
)

plt.plot(output["0.5"].index,
         output["0.5"],
         label="prediction median")
plt.legend()
plt.show()
```

This should render a chart similar to what is shown in *Figure 8.33*.

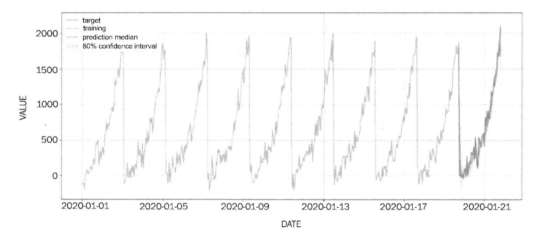

Figure 8.33 – Time series chart with predicted values

We can see in *Figure 8.33* the predicted values along with the confidence interval towards the right side of the chart. Here, we can see that the known target values from our synthetic dataset more or less fall within the `80%` confidence interval predicted by the **DeepAR** model. Note that this `80%` confidence interval was obtained using the upper bound and lower bound values with the predicted `0.9` and `0.1` quantile values respectively.

16. Finally, let's not forget to delete the endpoint using the `delete_endpoint()` function:

```
predictor.delete_endpoint()
```

At this point, we should be comfortable using the SageMaker built-in **DeepAR forecasting algorithm**.

Now, let's see how this works!

How it works...

In this recipe, we successfully used the deployed **DeepAR** model to perform inference as seen in the plot. One of the advantages when using **DeepAR** is that we are able to perform probabilistic forecasting when performing predictions. Probabilistic forecasting involves making use of a range of possible outcomes instead of simply just providing the average or median value. This allows for better decisions due to the availability of additional information—the predicted probability of occurrence of a certain event or value.

How does this work? As shown in *Figure 8.34*, we simply use the 0.5 quantile values returned by the predict() function for the **deterministic point forecast**. We expect a 50% chance of observing a value lower than the 0.5 quantile values and we also expect a 50% chance of observing a value higher than the 0.5 quantile values—making it a good choice for deterministic point forecast values.

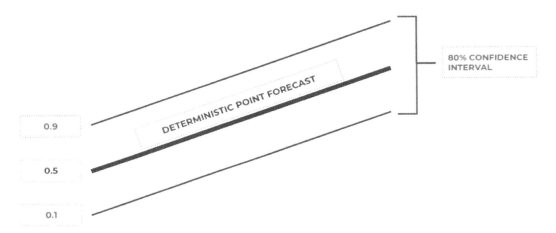

Figure 8.34 – Deterministic and probabilistic forecasts

On the other hand, for the **probabilistic forecast**, we first obtain the 0.1 and 0.9 quantile values and then use these values as the lower and upper bounds of possible values respectively. Of course, if we use a smaller range, the difference of the quantile values will be smaller and the window will be smaller as well. This is useful when we want probabilistic forecast values closer to the deterministic point forecast value.

See also

If you are looking for examples of using the **DeepAR forecasting algorithm** on real datasets and more complex examples, feel free to check some of the notebooks in the **aws/amazon-sagemaker-examples** GitHub repository:

- Predicting driving speed violations using a **DeepAR** model—https://
  github.com/aws/amazon-sagemaker-examples/blob/master/
  introduction_to_applying_machine_learning/deepar_chicago_
  traffic_violations/deepar_chicago_traffic_violations.ipynb

- Using the **DeepAR forecasting algorithm** on the electricity dataset—https://
  github.com/aws/amazon-sagemaker-examples/blob/master/
  introduction_to_amazon_algorithms/deepar_electricity/
  DeepAR-Electricity.ipynb

As we will not be able to dive deep into the different features of the **DeepAR forecasting algorithm** in this book, feel free to check this link for more information: `https://docs.aws.amazon.com/sagemaker/latest/dg/deepar.html`.

9
Managing Machine Learning Workflows and Deployments

In the previous chapters, we focused on relatively straightforward machine learning model deployments with SageMaker; that is, using the `deploy()` function to deploy a single model to an inference endpoint. In simple experiments and deployments, this would do the trick. However, when dealing with requirements that involve a more complex setup, we need to have a few more tricks up our sleeves.

In this chapter, we will work with a relatively more complex set of deployment solutions for real-time endpoint deployments and automated workflows. As shown in the following diagram, this chapter has three primary focus areas – deep learning model deployment for **Hugging Face** models, **multi-model endpoint** deployments, and ML workflows:

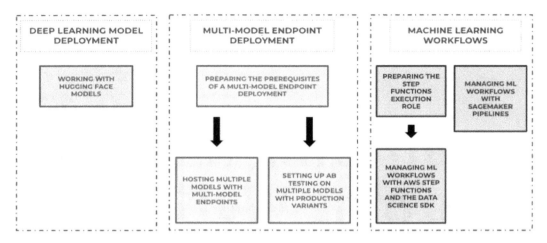

Figure 9.1 – How the recipes in this chapter are divided

The first focus area involves fine-tuning and deploying state-of-the-art NLP models in SageMaker. We will fine-tune a **Transformer** model in **PyTorch** using **Hugging Face Transformers**. Back in 2017, **Transformers** took the world by storm as these allowed training to be performed on massive amounts of data due to its architecture and approach. This allowed these models to perform significantly better compared to RNNs and CNNs. The second focus area involves a couple of relevant variations when deploying multiple models within a single endpoint. These include the different ways to perform **multi-model endpoint** deployments, as well as setting up **A/B testing** to help us identify which model has better performance in production before updating the models. The third focus area in this chapter involves working with **Step Functions** and **SageMaker Pipelines** to build workflows that help orchestrate and automate machine learning processes and tasks. With these workflow management solutions, we will be able to automate and orchestrate tasks such as data preparation, training, evaluation, and deployment without having to worry about server administration and management.

That said, we will cover the following recipes in this chapter:

- Working with **Hugging Face** models

- Preparing the prerequisites of a **multi-model endpoint** deployment

- Hosting multiple models with **multi-model endpoints**

- Setting up **A/B testing** on multiple models with production variants

- Preparing the **Step Functions** execution role

- Managing ML workflows with **AWS Step Functions** and the **Data Science SDK**

- Managing ML workflows with **SageMaker Pipelines**

Once we have completed the recipes in this chapter, we will have a better understanding of some of the relevant and useful solutions available when performing model deployments and automating machine learning pipelines.

Technical requirements

To execute the recipes in the chapter, make sure you have the following:

- An Amazon S3 bucket.

- Permission to manage the **Amazon SageMaker** and **Amazon S3** resources if you're using an **AWS IAM** user with a custom URL. If you are using the root account, then you should be able to proceed with the recipes in this chapter. However, it is recommended that you're signed in as an AWS IAM user instead of using the root account in most cases. For more information, feel free to take a look at the following guide: `https://docs.aws.amazon.com/IAM/latest/UserGuide/best-practices.html`.

As the recipes in this chapter involve a bit of code, we have made the notebooks available in this book's GitHub repository: `https://github.com/PacktPublishing/Machine-Learning-with-Amazon-SageMaker-Cookbook/tree/master/Chapter09`.

Before starting on each of the recipes in this chapter, make sure that the `my-experiments/chapter09` directory is ready. If it has not been created yet, please do so now as this helps keep things organized as we go through each of the recipes in this book.

Check out the following link to see the relevant Code in Action video:

`https://bit.ly/3yViSEY`

Working with Hugging Face models

The **Hugging Face Transformers** library has brought together multiple pre-trained transformer models to solve relevant NLP tasks such as text classification, text generation, and information extraction. These models include **BERT**, **RoBERTa**, **GPT**, **GPT-2**, and other state-of-the-art transformer models. What is the advantage of using pre-trained models? With pre-trained models and **transfer learning**, we can come up with accurate models in a shorter period of time. That's because we can start with a good set of weights from models that have been trained to solve a similar set of problems.

In this recipe, we will start with the pre-trained **DistilBERT** model and use the `HuggingFace` estimator class from the **SageMaker Python SDK**, along with a custom Python script file, to fine-tune our **DistilBERT** model. We will use the synthetic text data from the *Generating a synthetic dataset for text classification problems* recipe of *Chapter 8, Solving NLP, Image Classification, and Time-Series Forecasting Problems with Built-In Algorithms*, to fine-tune the **DistilBERT** model and make it more suited to our data. In addition to this, we will demonstrate how to use the `PyTorchModel` class to deploy the fine-tuned model.

> **Note**
>
> If you are wondering what **DistilBERT** is, it is a "distilled version of **BERT**." A few years ago, the **Bidirectional Encoder Representations from Transformers (BERT)** framework achieved state-of-the-art results in sentiment analysis, sentence classification, semantic role labeling, and other relevant NLP tasks. Through the use of **knowledge distillation**, the researchers at **Hugging Face** were able to prepare a smaller and faster model called **DistilBERT** without taking a huge performance hit. For more information on this topic, feel free to check out `https://huggingface.co/transformers/model_doc/distilbert.html`.

We will use this model to solve a text classification problem, similar to what we worked on using a **BlazingText** model in the *Training and deploying a BlazingText model* recipe of *Chapter 8, Solving NLP, Image Classification, and Time-Series Forecasting Problems with Built-In Algorithms*.

Getting ready

For this recipe, you will need a **SageMaker Studio** notebook running the **Python 3 (Data Science)** kernel.

How to do it...

The first set of steps in this recipe focus on preparing the prerequisites for the training job. Let's get started:

1. Create a new notebook using the **Python 3 (Data Science)** kernel inside the `my-experiments/chapter09` directory and rename it to the name of this recipe (`Working with Hugging Face models`):

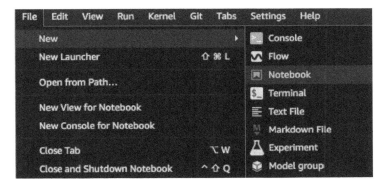

Figure 9.2 – Creating a new notebook

In *Figure 9.2*, we can see how to create a new **notebook** from the **File** menu. When prompted for the kernel to use, choose `Python 3 (Data Science)`.

2. Create the `scripts` directory using the `mkdir` command if it does not exist yet:

```
!mkdir -p scripts
```

3. Prepare the `path` variable so that it points to where the `entry_point` script files, along with a few prerequisite files, are stored:

```
g = "raw.githubusercontent.com"
p = "PacktPublishing"
a = "Machine-Learning-with-Amazon-SageMaker-Cookbook"
mc = "master/Chapter09"
path = f"https://{g}/{p}/{a}/{mc}/scripts"
```

4. Download the `setup.py`, `train.py`, `inference.py`, and `requirements.txt` files to the `scripts` directory using wget:

```
!wget -P scripts {path}/setup.py
!wget -P scripts {path}/train.py
!wget -P scripts {path}/inference.py
!wget -P scripts {path}/requirements.txt
```

5. Locate the `scripts` directory in the **File Browser**:

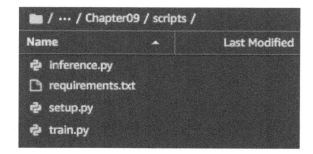

Figure 9.3 – Downloaded files inside the scripts directory

Here, we can see the `scripts` directory, which contains the `requirements.txt`, `setup.py`, `train.py`, and `inference.py` files.

6. Inspect the `train.py` file inside the `scripts` directory:

```
≡ train.py                    ×
 1  import random
 2  import logging
 3  import sys
 4  import argparse
 5  import os
 6  import torch
 7
 8  from sklearn.metrics import accuracy_score, precision_recall_fscore_support
 9  from datasets import load_from_disk
10  from transformers import DistilBertTokenizerFast
11  from transformers import DistilBertForSequenceClassification
12  from transformers import Trainer, TrainingArguments
13
14
15  tokenizer = DistilBertTokenizerFast.from_pretrained('distilbert-base-uncased')
16
17
18  class CustomDataset(torch.utils.data.Dataset):
19      def __init__(self, encodings, labels):
20          self.encodings = encodings
21          self.labels = labels
22
23      def __getitem__(self, idx):
24          item = {key: torch.tensor(val[idx]) for key, val in self.encodings.items()}
25          item['labels'] = torch.tensor(self.labels[idx])
26          return item
```

Figure 9.4 – train.py

Here, we have an `entry_point` script file called `train.py`. It contains Python code that processes the arguments, loads the training and validation data, tokenizes and processes the data, loads the pre-trained `DistilBert` model, fine-tunes the model using the processed training and validation data, and saves the model.

> **Note**
>
> The structure, flow, and objective of the `train.py` file's `entry_point` script file in this recipe should be similar to what we had in the *Preparing the entrypoint PyTorch training script* recipe of *Chapter 3, Using Machine Learning and Deep Learning Frameworks with Amazon SageMaker*. Of course, some of the major differences between these script files include the libraries and the data that's used to train the model during the training step.

7. Quickly inspect the `inference.py` file inside the `scripts` directory:

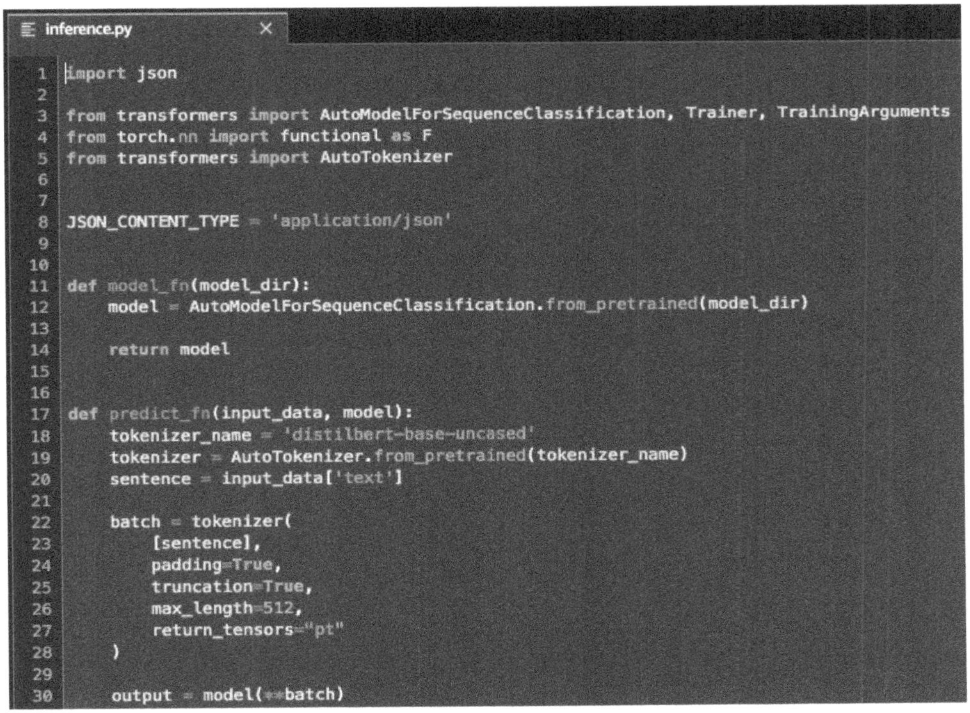

```python
import json

from transformers import AutoModelForSequenceClassification, Trainer, TrainingArguments
from torch.nn import functional as F
from transformers import AutoTokenizer

JSON_CONTENT_TYPE = 'application/json'


def model_fn(model_dir):
    model = AutoModelForSequenceClassification.from_pretrained(model_dir)

    return model


def predict_fn(input_data, model):
    tokenizer_name = 'distilbert-base-uncased'
    tokenizer = AutoTokenizer.from_pretrained(tokenizer_name)
    sentence = input_data['text']

    batch = tokenizer(
        [sentence],
        padding=True,
        truncation=True,
        max_length=512,
        return_tensors="pt"
    )

    output = model(**batch)
```

Figure 9.5 – The inference.py file

Here, we have an `entry_point` script file called `inference.py`. It contains four functions called `model_fn`, `predict_fn`, `input_fn`, and `output_fn`.

8. Next, inspect the `requirements.txt` file as well:

This file contains a single line – `transformers==4.4.2`. Note that the `requirements.txt` file needs to be in the `scripts` directory so that the package/s inside the `requirements.txt` file will be installed automatically before the `entry_point` script file is run.

9. Back in the notebook, create the tmp directory using the mkdir command if it does not exist yet:

```
!mkdir -p tmp
```

10. Prepare the path variable so that it points to where the training and validation datasets are stored:

```
g = "raw.githubusercontent.com"
p = "PacktPublishing"
a = "Machine-Learning-with-Amazon-SageMaker-Cookbook"
mc = "master/Chapter09"
path = f"https://{g}/{p}/{a}/{mc}/files"
```

11. Use the wget command to download the files containing the training and validation data from this book's GitHub repository to the tmp directory:

```
!wget -P tmp {path}/synthetic.train.txt
!wget -P tmp {path}/synthetic.validation.txt
```

These files should contain label and text pairs, similar to what is shown in the following block of code:

```
__label__negative Donut this is bad not impressive
this is bad spaghetti chicken soup
__label__positive Spaghetti Chicken Soup tastes good
donut this is good this is good this is good
tastes good
```

12. Specify the S3 bucket name and the prefix where the data will be stored. Make sure that you replace the value of <insert s3 bucket here> with the name of the bucket we created in the *Preparing the Amazon S3 bucket and the training dataset for the linear regression experiment* recipe of *Chapter 1, Getting Started with Machine Learning Using Amazon SageMaker*:

```
s3_bucket = "<insert s3 bucket here>"
prefix = "chapter09"
```

13. Initialize the variables that point to where the training and validation data should be stored in **Amazon S3**:

```
s3_train_data = 's3://{}/{}/input/{}'.format(
    s3_bucket,
    prefix,
```

```
    "synthetic.train.txt"
)
s3_validation_data = 's3://{}/{}/input/{}'.format(
    s3_bucket,
    prefix,
    "synthetic.validation.txt"
)
```

14. Use the **AWS CLI** to upload the TXT files containing the training and validation data:

```
!aws s3 cp tmp/synthetic.train.txt {s3_train_data}
!aws s3 cp tmp/synthetic.validation.txt {s3_validation_
data}
```

15. Prepare and load the role ARN using the `get_execution_role()` function from the **SageMaker Python SDK**:

```
import sagemaker
role = sagemaker.get_execution_role()
```

Now that we have the prerequisites ready, we can focus on the training job.

16. Initialize the `HuggingFace` estimator object. Here, we will specify `distilbert-base-uncased` for the `model_name` value:

```
from sagemaker.huggingface import HuggingFace
hyperparameters = {
    'epochs': 1,
    'train_batch_size': 32,
    'model_name':'distilbert-base-uncased'
}
estimator = HuggingFace(
    entry_point='train.py',
    source_dir='./scripts',
    instance_type='ml.p3.2xlarge',
    instance_count=1,
    role=role,
    transformers_version='4.4',
    pytorch_version='1.6',
    py_version='py36',
```

```
        hyperparameters=hyperparameters
    )
```

17. Prepare the `data_channels` dictionary, which contains the `TrainingInput` objects pointing to the training and validation dataset files stored in **Amazon S3**:

```
from sagemaker.inputs import TrainingInput

train_data = TrainingInput(s3_train_data)
validation_data = TrainingInput(s3_validation_data)

data_channels = {
    'train': train_data,
    'valid': validation_data
}
```

18. Use the `fit()` function to start the training job:

```
%%time
estimator.fit(data_channels)
```

This should yield a set of logs, similar to what is shown in the following screenshot:

```
2021-06-08 02:29:10 Starting - Starting the training job...
2021-06-08 02:29:13 Starting - Launching requested ML instancesProfilerReport-1
623119350: InProgress
......
2021-06-08 02:30:26 Starting - Preparing the instances for training.........
2021-06-08 02:32:07 Downloading - Downloading input data...
2021-06-08 02:32:27 Training - Downloading the training image.................b
ash: cannot set terminal process group (-1): Inappropriate ioctl for device
bash: no job control in this shell
2021-06-08 02:35:19,571 sagemaker-training-toolkit INFO     Imported framework
sagemaker_pytorch_container.training
2021-06-08 02:35:19,594 sagemaker_pytorch_container.training INFO     Block unt
il all host DNS lookups succeed.
2021-06-08 02:35:21,012 sagemaker_pytorch_container.training INFO     Invoking
user training script.
2021-06-08 02:35:21,329 sagemaker-training-toolkit INFO     Installing module w
ith the following command:
/opt/conda/bin/python3.6 -m pip install . -r requirements.txt
Processing /opt/ml/code
```

Figure 9.6 – Training job logs

Here, we have the logs that are generated after calling the `fit()` function.

> **Note**
>
> This should take about 7 to 10 minutes to complete. Feel free to grab a cup of coffee or tea while waiting!

The last set of steps focus on deploying the model and testing the inference endpoint:

19. Next, initialize the `PyTorchModel` object and specify `inference.py` as the value of the `entry_point` parameter:

```
from sagemaker.pytorch.model import PyTorchModel
model_data = estimator.model_data

model = PyTorchModel(
    model_data=model_data,
    role=role,
    source_dir="scripts",
    entry_point='inference.py',
    framework_version='1.6.0',
    py_version="py3"
)
```

20. Deploy the model using the `deploy()` function:

```
%%time
predictor = model.deploy(
    instance_type='ml.m5.xlarge',
    initial_instance_count=1
)
```

> **Note**
>
> This step should take 7 to 10 minutes to complete. Feel free to grab a quick bite while waiting!

21. Update `serializer` and `deserializer` of the predictor:

```
from sagemaker.serializers import JSONSerializer
from sagemaker.deserializers import JSONDeserializer
predictor.serializer = JSONSerializer()
predictor.deserializer = JSONDeserializer()
```

22. Use the `predict()` function to test a NEGATIVE scenario:

```
test_data = {
    "text": "This tastes bad. I hate this place."
}
predictor.predict(test_data)
```

This should return a NEGATIVE string value.

23. Use the `predict()` function again to test a POSITIVE scenario:

```
test_data = {
    "text": "Very delicious. I would recommend this to my friends"
}
predictor.predict(test_data)
```

This should return a POSITIVE string value.

24. Delete the endpoint afterward using the `delete_endpoint()` function:

```
predictor.delete_endpoint()
```

Do not forget this step as you will be charged for the amount of time the endpoint is running.

Now, let's see how this works!

How it works...

In this recipe, we used the `HuggingFace` estimator class from the **SageMaker Python SDK** to fine-tune our **DistilBERT** model to solve a text classification problem.

> **Note**
>
> Note that we are not limited to just using **DistilBERT** here. We can make use of the same approach and solutions we have used in this recipe to fine-tune and deploy models such as **BERT**, **RoBERTa**, **GPT-2**, and other Transformers-based NLP models here: `https://huggingface.co/transformers/pretrained_models.html`.

When dealing with Hugging Face model deployments, it is important to note that the `HuggingFace` estimator's `deploy()` function is not supported at the time of writing. This means that we had to perform a workaround and use the `PyTorchModel` class from the **SageMaker Python SDK**, along with a custom `entry_point` script, to help us deploy our model into an inference endpoint. We can always fall back into using the other existing features of Amazon SageMaker and the **SageMaker Python SDK**, along with our customization skills.

> **Note**
>
> Feel free to check out `https://aws.amazon.com/blogs/machine-learning/announcing-managed-inference-for-hugging-face-models-in-amazon-sagemaker/` for more information about this topic.

When using the framework estimators from the **SageMaker Python SDK**, such as the `TensorFlow`, `HuggingFace`, and `PyTorch` estimator classes, we need to make sure that the following files and scripts are ready before the training and deployment steps begin: the training `entry_point` script file, the inference `entry_point` script file, and other files such as the `requirements.txt` file, which help with installing the necessary libraries before the `entry_point` script files are executed. Inside the training `entry_point` script file, we used the following lines of code to fine-tune and save the **DistilBERT** model:

```
trainer = Trainer(model=model,...)
eval_result = trainer.evaluate(...)
trainer.save_model(model_dir)
```

Once the training job had finished running, the model artifacts were uploaded to the S3 bucket inside a `model.tar.gz` file and then downloaded to the inference container when the inference endpoint was deployed. Inside the inference `entry_point` script file, we loaded the model and the tokenizer using the `AutoModelForSequenceClassification.from_pretrained()` and `AutoTokenizer.from_pretrained()` functions, respectively.

Feel free to check out the *Training and deploying a PyTorch model with Amazon SageMaker Local Mode* recipe of *Chapter 3, Using Machine Learning and Deep Learning Frameworks with Amazon SageMaker*, for additional information on how to prepare custom `entry_point` scripts.

There's more...

We are just scratching the surface of what we can do with the `HuggingFace` estimator class from the **SageMaker Python SDK**. As we are dealing with relatively larger ML instances due to the model being fine-tuned, we may need to use different solutions to reduce costs and speed up the training time.

One of the ways to significantly reduce costs when using SageMaker is through the use of **Spot Instances** when training machine learning models. With **Spot Instances**, we can reduce costs by around 70% to 90% compared to on-demand instances by taking into account spot instance interruptions. With SageMaker, we can make use of its **Managed Spot Training** capability, which allows us to utilize Spot Instances without having to worry about the details. Of course, a couple of tweaks are needed to get this to work:

- We must add support for saving and resuming `using` checkpoints during interruptions in the train `entry_point` script file.

- We must configure the estimator so that it can use **Managed Spot Training** (for example, `train_use_spot_instances=True`).

We can also utilize distributed training through **data parallelism** and **model parallelism**. With **data parallelism**, the training set is divided into mini-batches. These mini-batches are then used to train the model across several GPU instances. This helps reduce training time, especially when we're dealing with larger datasets. On the other hand, we can automatically split large deep learning models across multiple GPUs and instances with **model parallelism**. We will not dive deep into the details in this book, so feel free to check out `https://sagemaker-examples.readthedocs.io/en/latest/training/distributed_training/index.html`.

See also

If you are looking for examples of using the `HuggingFace` estimator from the **SageMaker Python SDK** on real datasets and more complex examples, feel free to check out some of the notebooks that focus on this topic here: `https://huggingface.co/transformers/sagemaker.html`.

Preparing the prerequisites of a multi-model endpoint deployment

In this recipe, we will prepare some of the prerequisites of a multi-model endpoint deployment, including pre-trained model files and the S3 paths where the pre-trained model files will be uploaded to. These prerequisites will be used in the *Hosting multiple models with multi-model endpoints* and *Setting up A/B testing on multiple models with production variants* recipes.

Getting ready

For this recipe, you will need a **SageMaker Studio** notebook running the **Python 3 (Data Science)** kernel.

How to do it...

The steps in this recipe focus on downloading the pre-trained model files from this book's GitHub repository and uploading them to the S3 bucket. Let's get started:

1. Create a new notebook using the `Python 3 (Data Science)` kernel inside the `my-experiments/chapter09` directory and rename it to the name of this recipe (`Preparing the prerequisites of a multi-model endpoint deployment`).

2. Prepare the `path` variable so that it points to where the pre-trained model files are stored:

   ```
   path = "https://github.com/PacktPublishing/" + \
       "Machine-Learning-with-Amazon-SageMaker-Cookbook/raw/
   master/" + \
       "Chapter09/files/"
   ```

3. Use the `wget` command to download the pre-trained model files to the `tmp` directory:

   ```
   pretrained_model_a = path + "model.a.tar.gz"
   pretrained_model_b = path + "model.b.tar.gz"
   !wget -O tmp/model.a.tar.gz {pretrained_model_a}
   !wget -O tmp/model.b.tar.gz {pretrained_model_b}
   ```

> **Note**
>
> If you are wondering where we got these pre-trained model files, we simply reused two of the **XGBoost** models we trained in *Chapter 5, Effectively Managing Machine Learning Experiments*. We just decided to download these models in this recipe so that we don't have to train these models again.

4. Specify the S3 bucket's name and the prefix where the data will be stored. Make sure that you replace the value of `<insert s3 bucket name here>` with the name of the bucket we have created in the *Preparing the Amazon S3 bucket and the training dataset for the linear regression experiment* recipe of *Chapter 1, Getting Started with Machine Learning Using Amazon SageMaker*:

```
s3_bucket = "<insert s3 bucket name here>"
prefix = "chapter09"
```

5. Upload the pre-trained model files to **Amazon S3**:

```
m_a = "files/model.a.tar.gz"
m_b = "files/model.b.tar.gz"
a = f"s3://{s3_bucket}/{prefix}/{m_a}"
b = f"s3://{s3_bucket}/{prefix}/{m_b}"
!aws s3 cp tmp/model.a.tar.gz {a}
!aws s3 cp tmp/model.b.tar.gz {b}
```

6. Use the `%store` magic to save the variable values for `model_a_s3_path` and `model_b_s3_path`:

```
model_a_s3_path = a
model_b_s3_path = b
%store model_a_s3_path
%store model_b_s3_path
```

7. Use the `%store` magic to save the variable values for `s3_bucket` and `prefix`:

```
%store s3_bucket
%store prefix
```

We will use these stored variable values in the succeeding recipes.

Now, let's see how this works!

How it works...

In this recipe, we prepared the prerequisites of the multi-model endpoint. This includes the model files, along with the paths where the files are stored in S3. When working with multi-model endpoints, note that we need to specify the model artifacts and the container images that will be used in the inference endpoint.

That said, we simply downloaded pre-trained models from this book's GitHub repository and then uploaded these `model.tar.gz` files into the S3 bucket. Since these pre-trained model artifacts were prepared ahead of time, we also know how they were prepared and which built-in algorithms and container images (for example, **XGBoost**) were used to train the models. In the succeeding recipes, we will use the the `sagemaker.image_uris.retrieve()` function to get the container image URI for **XGBoost**. This will help us complete the prerequisites for the multi-model endpoint deployment.

Hosting multiple models with multi-model endpoints

In the previous recipe, we prepared a few prerequisites for a multi-model endpoint deployment; that is, the pre-trained model files and the paths where the pre-trained model files will be uploaded to in S3.

In this recipe, we will deploy multiple models within a single endpoint using the multi-model endpoint support of SageMaker. With **multi-model endpoints**, we can reduce costs as we can host multiple models inside a single endpoint, compared to having one dedicated endpoint for each model. This approach also works well in staging or test environments, where occasional cold-start delays can be tolerated for infrequently used models.

> **Note**
>
> If you are wondering where we got these pre-trained models, we simply reused two of the **XGBoost** models we trained in *Chapter 5, Effectively Managing Machine Learning Experiments*. These models simply accept numerical values for the a and b features and return the predicted `label` value. The predicted `label` value will be a number between `0.0` and `1.0`; we can simply map the resulting value to `1` if the predicted value is greater than a specified threshold, such as `0.8`.

Getting ready

Here are the prerequisites for this recipe:

- This recipe continues from the *Preparing the prerequisites of a multi-model endpoint deployment* recipe.

- You will need a **SageMaker Studio** notebook running the **Python 3 (Data Science)** kernel.

How to do it...

The first set of steps in this recipe focus on preparing a few prerequisites and parameter values for the multi-model endpoint deployment. Let's get started:

1. Create a new notebook using the `Python 3 (Data Science)` kernel inside the `my-experiments/chapter09` directory and rename it to the name of this recipe (`Hosting multiple models with multi-model endpoints`).

2. Use the `%store` magic to load the variable values for `model_a_s3_path` and `model_b_s3_path`:

```
%store -r model_a_s3_path
%store -r model_b_s3_path
```

3. Use the `%store` magic to load the variable values for `s3_bucket` and `prefix`:

```
%store -r s3_bucket
%store -r prefix
```

4. Import and prepare a few prerequisites, such as `session` and `role`, from the **SageMaker Python SDK**:

```
import sagemaker
from sagemaker import get_execution_role
session = sagemaker.Session()
role = get_execution_role()
```

5. Use the `retrieve()` function to get the ECR image URI of the **XGBoost** built-in algorithm container image:

```
from sagemaker.image_uris import retrieve
image_uri = retrieve(
    "xgboost",
    region="us-east-1",
```

```
    version="0.90-2"
)
image_uri
```

> **Tip**
> Feel free to change the `region` value if you are operating in another region.

6. Prepare the `models_path` variable where the model data for the multi-model endpoint will be stored:

    ```
    models_path = f"s3://{s3_bucket}/model-artifacts/"
    ```

    Now that the prerequisites are ready, we will focus on initializing the `MultiDataModel` object and proceeding with the other steps needed to deploy the **multi-model endpoint**.

7. Initialize the `MultiDataModel` object:

    ```
    from sagemaker.multidatamodel import MultiDataModel
    multi_model = MultiDataModel(
        name="chapter09-multi",
        model_data_prefix=models_path,
        image_uri=image_uri,
        role=role
    )
    ```

8. Use the `add_model()` function to transfer and store the model data from the original S3 locations to the location specified in the `models_path` variable:

    ```
    multi_model.add_model(model_a_s3_path)
    multi_model.add_model(model_b_s3_path)
    ```

9. Use the `deploy()` function to start the multi-model endpoint deployment:

    ```
    %%time
    endpoint_name = "chapter09-mma"
    multi_model.deploy(
        initial_instance_count=1,
        instance_type='ml.t2.medium',
        endpoint_name=endpoint_name
    )
    ```

> **Note**
>
> This step should take approximately 7 to 15 minutes to complete. Feel free to grab a cup of coffee or tea while waiting!

10. Initialize the `Predictor` object and update its `serializer` and `deserializer`:

```
from sagemaker.predictor import Predictor
from sagemaker.serializers import CSVSerializer
from sagemaker.deserializers import JSONDeserializer

predictor = Predictor(
    endpoint_name=endpoint_name
)

predictor.serializer = CSVSerializer()
predictor.deserializer = JSONDeserializer()
```

11. Store the model identifiers inside the a and b variables:

```
a, b = list(multi_model.list_models())
```

This should give us `chapter09/files/model.a.tar.gz` and `chapter09/files/model.b.tar.gz` for the a and b variables, respectively.

12. Use the `predict()` function and specify which model to use by passing a as the value for the `target_model` parameter:

```
predictor.predict(data="10,-5", target_model=a)
```

This should return a value similar to `[0.895996630191803]`.

13. Similarly, use the `predict()` function and specify which model to use by passing b as the value for the `target_model` parameter:

```
predictor.predict(data="10,-5", target_model=b)
```

This should return a value similar to `[0.8308258652687073]`.

14. Finally, delete the endpoint using the delete_endpoint() function:

```
predictor.delete_endpoint()
```

Do not forget this step as you will be charged for the amount of time the endpoint is running.

Now, let's see how this works!

How it works...

In this recipe, we used the MultiDataModel class from the **SageMaker Python SDK** to deploy a multi-model endpoint. With multi-model endpoints, our model files are stored in an S3 bucket. These are then loaded and used for inference when the endpoint is invoked:

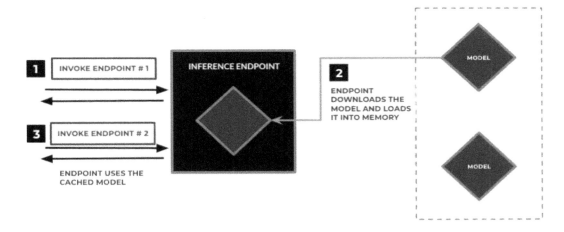

Figure 9.7 – Multi-model endpoint

In the preceding diagram, we can see that a model is dynamically loaded when an invocation request is performed on the inference endpoint. When a model has not been loaded into a container's memory inside the ML instance yet, the model is downloaded from the S3 bucket and is loaded into memory. This means that succeeding invocations will be faster as the model has already been loaded into memory. For more information, feel free to check out https://docs.aws.amazon.com/sagemaker/latest/dg/multi-model-endpoints.html.

Setting up A/B testing on multiple models with production variants

When dealing with production deployments, note that multiple models may be deployed and tested at the same time. This helps data scientists and machine learning engineers compare the performance of models when dealing with data that these models have not seen before. One of the standard ways to manage and test multiple models in production is through the use of **A/B testing** in inference endpoints. What's **A/B testing**? It is an experiment that involves randomly selecting a model from a list of deployed models to perform predictions. It helps identify the better (or best) performing model in production before completely replacing a deployed model.

In this recipe, we will deploy two pre-trained **XGBoost** models within a single endpoint using the multi-model endpoint support of SageMaker. We will configure and set up this endpoint to allow **A/B testing** of the pre-trained models that have been deployed in this endpoint.

> **Note**
>
> If you are wondering where we got these pre-trained models, we simply reused two of the **XGBoost** models we trained in *Chapter 5, Effectively Managing Machine Learning Experiments*. These models simply accept numerical values for the a and b features and return the predicted `label` value. The predicted `label` value will be a number between `0.0` and `1.0`; we can simply map the resulting value to `1` if the predicted value is greater than a specified threshold, such as `0.8`.

Once this **A/B testing** setup has been configured and deployed, traffic should be distributed randomly between the two models.

Getting ready

Here are the prerequisites for this recipe:

- This recipe continues from the *Preparing the prerequisites of a multi-model endpoint deployment* recipe.

- You will need a **SageMaker Studio** notebook running the **Python 3 (Data Science)** kernel.

How to do it...

The steps in this recipe focus on setting up A/B testing. Let's get started:

1. Create a new notebook using the `Python 3 (Data Science)` kernel inside the `my-experiments/chapter09` directory and rename it to the name of this recipe (`Setting up A/B testing on multiple models with production variants`).

2. Use the `%store` magic to load the variable values for `model_a_s3_path` and `model_b_s3_path`:

   ```
   %store -r model_a_s3_path
   %store -r model_b_s3_path
   ```

3. Use the `%store` magic to load the variable values for `s3_bucket` and `prefix`:

   ```
   %store -r s3_bucket
   %store -r prefix
   ```

4. Import and prepare a few prerequisites, such as `session` and `role`, from the **SageMaker Python SDK**:

   ```
   import sagemaker
   from sagemaker import get_execution_role
   session = sagemaker.Session()
   role = get_execution_role()
   ```

5. Use the `retrieve()` function to get the ECR image URI of the **XGBoost** built-in algorithm container image:

   ```
   from sagemaker.image_uris import retrieve
   image_uri = retrieve(
       "xgboost",
       region="us-east-1",
       version="0.90-2"
   )
   image_uri
   ```

   This should give us a value equal or similar to `683313688378.dkr.ecr.us-east-1.amazonaws.com/sagemaker-xgboost:0.90-2-cpu-py3` for `image_uri`.

6. Prepare the dictionary values for the `container1` and `container2` variables:

```
image_uri_a = image_uri
image_uri_b = image_uri
container1 = {
    'Image': image_uri_a,
    'ContainerHostname': 'containerA',
    'ModelDataUrl': model_a_s3_path
}
container2 = {
    'Image': image_uri_b,
    'ContainerHostname': 'containerB',
    'ModelDataUrl': model_b_s3_path
}
```

7. Prepare the low-level `boto3` client for the SageMaker service:

```
import boto3
sm_client = boto3.Session().client('sagemaker')
```

8. Prepare the variable values for `model_name_a`, `model_name_b`, `endpoint_config_name`, and `endpoint_name`:

```
model_name_a = "ab-model-a"
model_name_b = "ab-model-b"
endpoint_config_name = 'ab-endpoint-config'
endpoint_name = 'ab-endpoint'
```

9. Optionally, use the `delete_model()` function to delete any existing models to make sure that the next step succeeds without issues:

```
try:
    sm_client.delete_model(ModelName=model_name_a)
    sm_client.delete_model(ModelName=model_name_b)
except:
    pass
```

10. Use the `create_model()` function to create the two models in SageMaker. For each of these models, we specify the model's name, role, and the container image URI when calling the `create_model()` function:

```
response = sm_client.create_model(
    ModelName         = model_name_a,
    ExecutionRoleArn  = role,
    Containers        = [container1])
print(response)
response = sm_client.create_model(
    ModelName         = model_name_b,
    ExecutionRoleArn  = role,
    Containers        = [container2])
print(response)
```

This should yield a set of logs, similar to what is shown in the following screenshot:

{'ModelArn': 'arn:aws:sagemaker:us-east-1:   :model/ab-model-a', 'ResponseMetadata': {'RequestId': '7a0fbb6d-10c7-4d5d-9f4c-830bb9bcff92', 'HTTPStatusCode': 200 'HTTPHeaders': {'x-amzn-requestid': '7a0fbb6d-10c7-4d5d-9f4c-830bb9bcff92', 'content-type': 'application/x-amz-json-1.1', 'content-length': '72', 'date': 'Sun, 08 Aug 2021 16:56:29 GMT'}, 'RetryAttempts': 0}}
{'ModelArn': 'arn:aws:sagemaker:us-east-1:   :model/ab-model-b', 'ResponseMetadata': {'RequestId': '803e4363-3233-4121-bdbf-d48e95ca869d', 'HTTPStatusCode': 200 'HTTPHeaders': {'x-amzn-requestid': '803e4363-3233-4121-bdbf-d48e95ca869d', 'content-type': 'application/x-amz-json-1.1', 'content-length': '72', 'date': 'Sun, 08 Aug 2021 16:56:29 GMT'}, 'RetryAttempts': 1}}

Figure 9.8 – Response values after using create_model()

Here, we can see that our `create_model()` call has succeeded as the `HTTPStatusCode` value returned `200`.

11. Prepare the corresponding production variants of each of the models that were created in the previous step. Here, we will also specify a 50/50 split for the portion of the traffic that's expected to be received by each of the models:

```
from sagemaker.session import production_variant
variant1 = production_variant(
    model_name=model_name_a,
    instance_type="ml.t2.medium",
    initial_instance_count=1,
    variant_name='VariantA',
    initial_weight=0.5
)
variant2 = production_variant(
    model_name=model_name_b,
    instance_type="ml.t2.medium",
```

```
        initial_instance_count=1,
        variant_name='VariantB',
        initial_weight=0.5
)
```

12. Use the `endpoint_from_production_variants()` function to start the deployment:

```
session.endpoint_from_production_variants(
        name=endpoint_name,
        production_variants=[variant1, variant2]
)
```

> **Note**
>
> This should take approximately 10 minutes to complete. Feel free to grab a cup of coffee or tea!

13. Prepare the `boto3` client for SageMaker Runtime:

```
runtime_sm_client = boto3.client('sagemaker-runtime')
```

> **Note**
>
> What is the difference between `boto3.client('sagemaker')` and `boto3.client('sagemaker-runtime')`? The **SageMaker Runtime** client focuses solely on invoking a SageMaker inference endpoint using the `invoke_endpoint()` function. On the other hand, the **SageMaker** client provides the complete set of functions that map to the API actions that can be performed with the SageMaker service. This includes the `create_auto_ml_job()`, `create_training_job()`, `create_endpoint()`, and `create_transform_job()` functions, along with all the functions in this list: `https://boto3.amazonaws.com/v1/documentation/api/latest/reference/services/sagemaker.html`.

14. Test the A/B testing setup by using the `invoke_endpoint()` function without specifying the target variant:

```python
from time import sleep
body = "10,-5"
def test_ab_testing_setup():
    response = runtime_sm_client.invoke_endpoint(
        EndpointName=endpoint_name,
        ContentType='text/csv',
        Body=body
    )
    variant = response['InvokedProductionVariant']
    b = response['Body'].read()
    prediction = b.decode("utf-8")
    print(variant + " - "+ prediction)


for _ in range(0,10):
    test_ab_testing_setup()
    sleep(1)
```

This should yield log messages similar to what is shown in the following screenshot:

```
VariantA — 0.895996630191803
VariantA — 0.895996630191803
VariantA — 0.895996630191803
VariantA — 0.895996630191803
VariantB — 0.8308258652687073
VariantB — 0.8308258652687073
VariantB — 0.8308258652687073
VariantA — 0.895996630191803
VariantA — 0.895996630191803
VariantA — 0.895996630191803
```

Figure 9.9 – Verifying the A/B testing setup of the inference endpoint

Here, we can see that our **A/B testing** setup works just fine. When the `invoke_endpoint()` function is called, sometimes, `VariantA` is invoked; other times, `VariantB` will be invoked.

> **Note**
>
> In the meantime, do not worry about what these prediction values mean; our goal here is to trigger a random model from a list of deployed models with the **A/B testing** setup.

15. Finally, delete the endpoint by using the `delete_endpoint()` function:

```
sm_client.delete_endpoint(
    EndpointName=endpoint_name
)
```

Do not forget this step as you will be charged for the amount of time the endpoint is running.

Now, let's see how this works!

How it works...

In this recipe, we set up a multi-model endpoint that supports A/B testing of models. Here, we configured our inference endpoint to pass 50% of the traffic to the first model and the other 50% of the traffic to the second model. The traffic distribution was configured when we prepared the production variants, as shown in the following block of code:

```
variant1 = production_variant(
    model_name=model_name_a,
    variant_name='VariantA',
    initial_weight=0.5,
    ...
)
variant2 = production_variant(
    model_name=model_name_b,
    variant_name='VariantB',
    initial_weight=0.5,
    ...
)
```

If you are wondering whether we can change the traffic distribution so that we have the first model receive approximately 80% of the traffic, while the second model receives about 20% of the traffic, then the answer would be *yes*. It is as simple as updating the `initial_weight` values to `0.8` and `0.2` for production variants A and B, respectively.

If you are also wondering whether we can perform A/B testing on models from different model families, then the answer to that would be *yes* as well. We will just need to register models before using these models to prepare the production variants, as shown in the following lines of code:

```
sm_client.create_model(Containers=[container1], ...)
sm_client.create_model(Containers=[container2], ...)
```

All we need to do is specify different container images, depending on the model families of the pre-trained models.

> **Note**
>
> You may also be wondering why the `Containers` parameter in the `create_model()` function accepts a list instead of a single string value. This is because we can set up an inference pipeline that involves a linear sequence of containers to process a single endpoint invocation request. For more information, feel free to check out `https://docs.aws.amazon.com/sagemaker/latest/dg/inference-pipelines.html`.

There's more...

Even with the A/B testing configuration and setup in place, we can also invoke a specific variant directly. We can do this by using the `invoke_endpoint()` function and specifying the name of the target variant we want to invoke as the value of the `TargetVariant` parameter. Here is a sample block of code that performs this action:

```
response = runtime_sm_client.invoke_endpoint(
    EndpointName=endpoint_name,
    ContentType='text/csv',
    TargetVariant='VariantB',
    Body=body
)
```

That said, we can evaluate the performance of each variant with the following set of steps:

1. Use the data from a test dataset as payload when invoking each of the variants directly using the `invoke_endpoint()` function. This should give us two different sets of predicted values as we are working with two variants in this recipe.

2. Using the known target values from the test dataset and the predicted values from the endpoint invocations, we can evaluate the performance of each deployed model and compute the **RMSE, MSE,** and **MAE** for the regression models and **AUC, accuracy, precision,** and **recall** for the classifier models.

3. Once we have computed the metric values of each model, we can identify which model is "better."

After that, we can update the configuration of the traffic distribution for each of the variants we've deployed. We can perform this configuration change by using the `update_endpoint_weights_and_capacities()` function with the SageMaker `boto3` client. We will not dive deep into the details in this book, so feel free to check out `https://docs.aws.amazon.com/sagemaker/latest/dg/model-ab-testing.html` for more information.

Preparing the Step Functions execution role

In this recipe, we will create an IAM execution role that will allow us to create and execute **AWS Step Functions** workflows. First, we will create an execution role with Step Functions as the AWS service trusted entity type. Then, we will add an **inline policy** to this role with the necessary permissions that will be assigned to this execution role.

> **Important Note**
>
> What is an IAM role? An IAM role is an IAM identity that's used to delegate access to entities and resources. This role can be assumed by a resource to gain the permissions needed to perform a specific task. In our case, we will create a role with the permissions to create and execute Step Functions workflows. This role will be used in the *Managing ML workflows with AWS Step Functions and the Data Science SDK* recipe. For more information, feel free to check out `https://docs.aws.amazon.com/IAM/latest/UserGuide/id_roles_terms-and-concepts.html`.

Getting ready

For this recipe, you will need permission to create and manage **AWS IAM** roles, policies, and other resources.

How to do it...

The steps in this recipe focus on preparing the execution role using the AWS console. Let's get started:

1. Navigate to the IAM console using the search bar, similar to what is shown in the following screenshot:

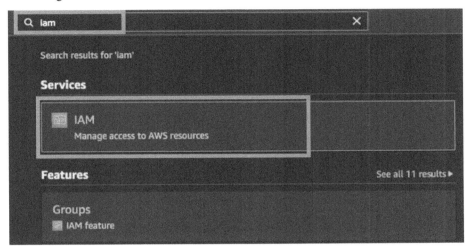

Figure 9.10 – Navigating to the IAM console

As we can see, using `iam` as the keyword in the search bar should allow us to easily search for the IAM service and jump straight to the IAM console.

2. In the navigation pane, locate and click **Roles** under **Access management**:

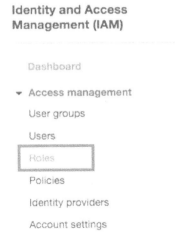

Figure 9.11 – Navigating to the Roles page

As we can see, we can find the link to the **Roles** page between **Users** and **Policies** under **Access management**.

3. Click the **Create role** button, as shown in the following screenshot:

Figure 9.12 – Create role button

We should see the **Create role** button, similar to what is shown in the preceding screenshot, at the top of the page, near the navigation pane.

4. Choose **AWS service** under **Select type of trusted entity**:

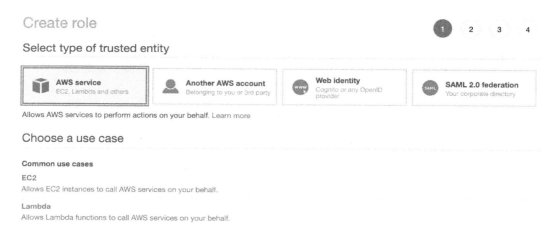

Figure 9.13 – Choosing the AWS service trusted entity type

Here, we have selected **AWS service** from the list of trusted entity types.

5.   Under **Choose a use case**, select **Step Functions**:

Or select a service to view its use cases

API Gateway	CodeBuild	EMR Containers	IoT SiteWise	RDS
AWS Backup	CodeDeploy	ElastiCache	IoT Things Graph	Redshift
AWS Chatbot	CodeGuru	Elastic Beanstalk	KMS	Rekognition
AWS Marketplace	CodeStar Notifications	Elastic Container Registry	Kinesis	RoboMaker
AWS Support	Comprehend	Elastic Container Service	Lake Formation	S3
Amplify	Config	Elastic Transcoder	Lambda	SMS
AppStream 2.0	Connect	ElasticLoadBalancing	Lex	SNS
AppSync	DMS	EventBridge	License Manager	SWF
Application Auto Scaling	Data Lifecycle Manager	Forecast	MQ	SageMaker
Application Discovery Service	Data Pipeline	GameLift	Machine Learning	Security Hub
	DataBrew	Global Accelerator	Macie	Service Catalog
Batch	DataSync	Glue	Managed Blockchain	**Step Functions**
Braket	DeepLens	Greengrass	MediaConvert	Storage Gateway
Budgets	Directory Service	GuardDuty	Migration Hub	Systems Manager
Certificate Manager	DynamoDB	Health Organizational View	Network Firewall	Textract

Figure 9.14 – Choosing a use case

As we can see, we have selected **Step Functions** (bottom right). It should be relatively easy to find as the services are sorted alphabetically.

6.   Click the **Next: Permissions** button:

Select your use case

Step Functions
Allows Step Functions to access AWS resources on your behalf.

* Required   Cancel   **Next: Permissions**

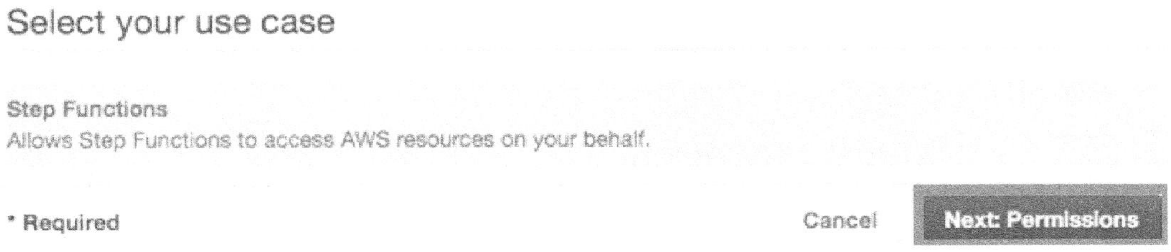

Figure 9.15 – Locating the Next: Permissions button

Note that we may need to scroll down to the bottom of the page to see the **Next: Permissions** button shown in the preceding screenshot.

7. Click the **Next: Tags** button at the bottom of the page, as shown in the following screenshot:

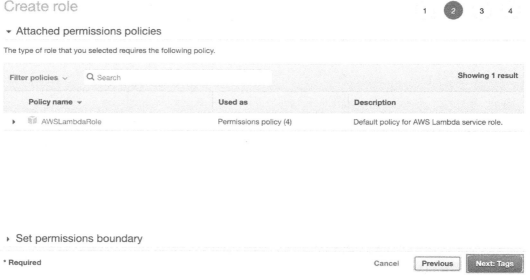

Figure 9.16 – Locating the Next: Tags button (bottom right)

Here, we can see a single attached permissions policy. Note that this permission policy, which is attached by default, is not enough to create and execute **Step Functions** workflows. Later in this recipe, we will attach a custom inline policy to this role, which will help us perform the needed steps in *Managing ML workflows with AWS Step Functions and the Data Science SDK*.

8. On the **Add tags (optional)** page, click the **Next: Review** button.

9. On the **Create Role — Review** page, specify the **Role name** value (for example, `sf-execution-role`). Click the **Create role** button afterward:

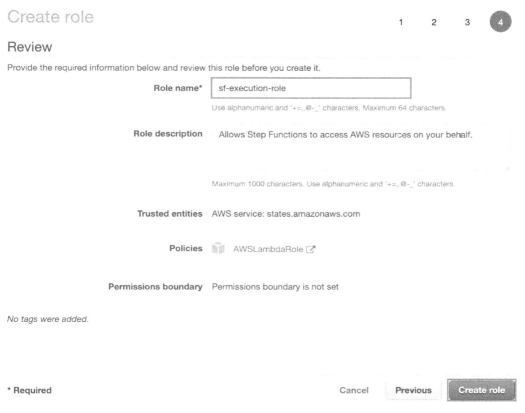

Figure 9.17 – Review page

As we can see, we can specify the role name and description on this page.

10. A success notification will appear, similar to what is shown in the following screenshot. Click the link (for example, **sf-execution-role** in this case) to view and modify the details and configure the role:

Figure 9.18 – Role successfully created success notification

Optionally, navigate to the specific role page by searching for the role using the role's name.

The next set of steps focus on adding an inline policy with additional permissions to the role we created in the first half of this recipe.

11. Inside the **Permissions** tab, click **Add inline policy**:

Figure 9.19 – Locating the Add inline policy button

Here, we can see the **Add inline policy** button at the top right-hand corner of the **Permissions** tab.

12. Navigate to the **JSON** tab, as shown in the following screenshot:

Figure 9.20 – Create policy JSON tab

Here, we can see that we can edit the JSON policy directly without having to use the **Visual editor** tab.

13. Open a new browser tab. Navigate to the **Machine-Learning-with-Amazon-SageMaker-Cookbook** GitHub repository and copy the JSON value from the **05 – Preparing the Step Functions Execution Role.ipynb** notebook:

Figure 9.21 – JSON policy from this book's GitHub repository

You may decide to copy the JSON policy from the rendered **05 – Preparing the Step Functions Execution Role.ipynb** notebook, as shown in the preceding screenshot, or use the policy JSON from the `files/role.json` file inside the `Chapter09` directory instead.

Tip

You can find the JSON policy here: `https://github.com/PacktPublishing/Machine-Learning-with-Amazon-SageMaker-Cookbook/tree/master/Chapter09`. Note that this policy needs to be reviewed and updated with a more secure configuration before it can be used in production environments.

---

14. Navigate back to the AWS console browser tab. Paste the JSON value we copied in the previous step into the text area shown in the following screenshot:

A policy defines the AWS permissions that you can assign to a user, group, or role. You can create and edit a policy in the visual editor and using JSON. Learn more

| Visual editor | **JSON** | Import managed policy |

```
33        Effect : Allow ,
34        "Action": [
35            "events:PutTargets",
36            "events:PutRule",
37            "events:DescribeRule"
38        ],
39        "Resource": [
40            "arn:aws:events:*:*:rule/StepFunctionsGetEventsForSageMakerTrainingJobsRule",
41            "arn:aws:events:*:*:rule/StepFunctionsGetEventsForSageMakerTransformJobsRule",
42            "arn:aws:events:*:*:rule/StepFunctionsGetEventsForSageMakerTuningJobsRule",
43            "arn:aws:events:*:*:rule/StepFunctionsGetEventsForECSTaskRule",
44            "arn:aws:events:*:*:rule/StepFunctionsGetEventsForBatchJobsRule"
45        ]
46    }
47  ]
48 }
```

Security: 0   Errors: 0   Warnings: 0   Suggestions: 0

Character count: 2,340 of 10,240   Cancel   **Review policy**
The current character count includes character for all in the policies in the role: sf-execution-role.

Figure 9.22 – Create policy JSON tab with the JSON policy value in the text area

Here, we can see that we can specify a certain level of granularity when using this option over **Visual editor**. Once you have updated the JSON policy value in the text area, click the **Review policy** button.

15. On the **Review policy** page, specify the name of the policy (for example, sf-policy). Click the **Create policy** button afterward.

16. Copy the **Role ARN** value, as shown in the following screenshot:

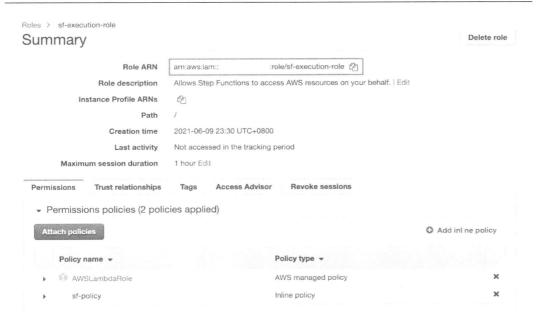

Figure 9.23 – Copying the role ARN from the Summary page

Feel free to copy this value to a text file. It should have a format similar to `arn:aws:iam::01234567890:role/sf-execution-role`. We will use this ARN value in the *Managing ML workflows with AWS Step Functions and the Data Science SDK* recipe.

Now, let's see how this works!

How it works...

In this recipe, we prepared an **IAM role** that will be used when we initialize the `Workflow` object in the next recipe. If this role had not been configured properly, we would not be able to create and execute **Step Functions** workflows using the **Step Functions Data Science SDK** in the *Managing ML workflows with AWS Step Functions and the Data Science SDK* recipe.

Once we had created the IAM role, we created and attached an inline policy that allows the entity assuming the role to create, execute, and manage the Step Functions workflows. What's an **inline policy** anyway? An inline policy is a policy that is associated with a role, user, or group and is useful when preventing the permissions of a policy from accidentally being assigned to another identity. For more information on this topic, feel free to check out the differences between managed policies and inline policies here: `https://docs.aws.amazon.com/IAM/latest/UserGuide/access_policies_managed-vs-inline.html`.

Managing ML workflows with AWS Step Functions and the Data Science SDK

AWS Step Functions is a serverless orchestration service that helps integrate and sequence tasks using multiple AWS services. With this service, we just need to focus on configuring the workflows and worry less about the operational overhead of managing distributed and complex applications.

In this recipe, we will use the **Data Science SDK** to create and manage automated ML workflows with **AWS Step Functions**. We will build on top of the recipes from *Chapter 1, Getting Started with Machine Learning Using Amazon SageMaker*, where we trained and deployed a **linear learner** model to solve a regression problem. Once we have completed the steps in this recipe, we will be able to execute an end-to-end automated workflow using **Step Functions** state machines, without having to run scripts manually inside Jupyter notebooks.

Getting ready

Here are the prerequisites for this recipe:

- You will need a **SageMaker Studio** notebook running the **Python 3 (Data Science)** kernel.

- Make sure that the execution role associated with **SageMaker Studio** has the necessary permissions to work with **AWS Step Functions** resources. You may need to attach the `AWSStepFunctionsFullAccess` policy to the said execution role.

How to do it...

The first set of steps in this recipe focus on getting the prerequisites to prepare the **Step Functions** workflow ready. Let's get started:

1. Create a new notebook using the `Python 3 (Data Science)` kernel inside the `my-experiments/chapter09` directory and rename it to the name of this recipe (`Managing ML workflows with AWS Step Functions and the Data Science SDK`).

2. Create the `tmp` directory using the `mkdir` command if it does not exist yet:

```
!mkdir -p tmp
```

3. Prepare the `path` variable so that it points to the location of the CSV file to be used in this recipe:

```
g = "raw.githubusercontent.com"
p = "PacktPublishing"
a = "Machine-Learning-with-Amazon-SageMaker-Cookbook"
mc = "master/Chapter01"
path = f"https://{g}/{p}/{a}/{mc}/files"
```

4. Use the `wget` command to download the CSV file from this book's GitHub repository to the `tmp` directory:

```
fname = "management_experience_and_salary.csv"
!wget -P tmp {path}/{fname}
```

Note that this is the same CSV file we used in *Chapter 1, Getting Started with Machine Learning Using Amazon SageMaker*.

5. Use the `read_csv()` function from `pandas` to load the contents of the CSV file into a DataFrame:

```
import pandas as pd
filename = f"tmp/{fname}"
df_all_data = pd.read_csv(filename)
```

6. Perform the train-test split using the `train_test_split()` function from `sklearn`:

```
from sklearn.model_selection import train_test_split
dad = df_all_data
X = dad['management_experience_months'].values
y = dad['monthly_salary'].values
X_train, X_test, y_train, y_test = train_test_split(
    X, y,
    test_size=0.3, random_state=0
)
```

7. Prepare the DataFrame containing the training data. Note that the first column contains the target variable values (for example, `monthly_salary`):

```
import pandas as pd
df_training_data = pd.DataFrame({
```

```
    'monthly_salary': y_train,
    'management_experience_months': X_train
})
```

8. Use the `to_csv()` function to store the contents of the `df_training_data` DataFrame in a CSV file:

```
df_training_data.to_csv(
    'tmp/training_data.csv',
    header=False, index=False
)
```

9. Specify the S3 bucket name and the prefix where the data will be stored. Make sure to replace the value of `<insert s3 bucket name here>` with the name of the bucket we created in the *Preparing the Amazon S3 bucket and the training dataset for the linear regression experiment* recipe of *Chapter 1, Getting Started with Machine Learning Using Amazon SageMaker*:

```
s3_bucket = '<insert s3 bucket name here>'
prefix = 'chapter09'
```

10. Use the **AWS CLI** to upload the `training_data.csv` file from the `tmp` directory to the S3 bucket:

```
tn = "training_data.csv"
source = f"tmp/{tn}"
dest = f"s3://{s3_bucket}/{prefix}/input/{tn}"
!aws s3 cp {source} {dest}
```

11. Import and prepare a few prerequisites, such as `session` and `role`, from the **SageMaker Python SDK**:

```
import sagemaker
import boto3
from sagemaker import get_execution_role
role = get_execution_role()
session = sagemaker.Session()
region_name = boto3.Session().region_name
```

12. Prepare the variable values for `training_s3_input_location` and `training_s3_output_location`:

```
training_s3_input_location = f"s3://{s3_bucket}/{prefix}/
input/training_data.csv"
training_s3_output_location = f"s3://{s3_bucket}/
{prefix}/output/"
```

13. Prepare the `TrainingInput` object for the training data:

```
from sagemaker.inputs import TrainingInput
train = TrainingInput(
    training_s3_input_location,
    content_type="text/csv"
)
```

14. Use the `retrieve()` function to get the container image URI for **linear learner**:

```
from sagemaker.image_uris import retrieve
container = retrieve(
    "linear-learner",
    region_name, "1"
)
container
```

The `container` variable should have a value equal or similar to `382416733822.dkr.ecr.us-east-1.amazonaws.com/linear-learner:1`.

The next set of steps in this recipe focus on preparing and executing the **Step Functions** workflow:

15. Initialize the `Estimator` object:

```
estimator = sagemaker.estimator.Estimator(
    container,
    role,
    instance_count=1,
    instance_type='ml.m5.xlarge',
    output_path=training_s3_output_location,
    sagemaker_session=session
)
```

16. Configure the hyperparameter values using the `set_hyperparameters()` function:

```
estimator.set_hyperparameters(
    predictor_type='regressor',
    mini_batch_size=4
)
```

17. Use `pip` to install the **AWS Step Functions Data Science SDK**:

```
!pip -q install --upgrade stepfunctions
```

18. Set the variable value of `execution_role` to the ARN of the IAM role ARN we prepared in the *Preparing the Step Functions execution role* recipe:

```
execution_role = '<insert role arn here>'
```

The `execution_role` variable value should be in a format similar to `arn:aws:iam::01234567890:role/sf-execution-role`.

> **Important Note**
>
> Before proceeding with the next set of steps, make sure that the execution role associated with **SageMaker Studio** has the necessary permissions to work with and create **AWS Step Functions** resources. You may attach the `AWSStepFunctionsFullAccess` policy to said execution role.

19. Initialize the `ExecutionInput` object:

```
from stepfunctions.inputs import ExecutionInput
execution_input = ExecutionInput(
    schema={
        'ModelName': str,
        'EndpointName': str,
        'JobName': str
    }
)

ei = execution_input
```

20. Initialize the `TrainingStep` object:

```python
from stepfunctions.steps import TrainingStep
training_step = TrainingStep(
    'Training Step',
    estimator=estimator,
    data={
        'train': train
    },
    job_name=ei['JobName']
)
```

21. Initialize the `ModelStep` object:

```python
from stepfunctions.steps import ModelStep
model_step = ModelStep(
    'Model Step',
    model=training_step.get_expected_model(),
    model_name=ei['ModelName']
)
```

22. Next, initialize the `EndpointConfigStep` object:

```python
from stepfunctions.steps import EndpointConfigStep
endpoint_config_step = EndpointConfigStep(
    "Create Endpoint Configuration",
    endpoint_config_name=ei['ModelName'],
    model_name=ei['ModelName'],
    initial_instance_count=1,
    instance_type='ml.m5.xlarge'
)
```

23. Initialize the `EndpointStep` object:

```python
from stepfunctions.steps import EndpointStep
endpoint_step = EndpointStep(
    "Deploy Endpoint",
    endpoint_name=ei['EndpointName'],
    endpoint_config_name=ei['ModelName']
)
```

> **Note**
>
> If you are wondering what `TrainingStep`, `ModelStep`, `EndpointConfigStep`, and `EndpointStep` are for, do not worry – we will discuss these in the *How it works...* section.

24. Initialize the `Chain` object. This will be used to connect the `TrainingStep`, `ModelStep`, `EndpointConfigStep`, and `EndpointStep` objects we prepared in the previous set of steps:

```
from stepfunctions.steps import Chain
workflow_definition = Chain([
    training_step,
    model_step,
    endpoint_config_step,
    endpoint_step
])
```

25. Define the `generate_random_string()` function:

```
import uuid

def generate_random_string():
    return uuid.uuid4().hex

grs = generate_random_string
```

26. Initialize the `Workflow` object:

```
from stepfunctions.workflow import Workflow
workflow = Workflow(
    name='{}-{}'.format('Workflow', grs()),
    definition=workflow_definition,
    role=execution_role,
    execution_input=execution_input
)
```

27. Create the workflow using the `create()` function:

```
workflow.create()
```

> **Important Note**
>
> If you encounter `AccessDeniedException` when calling
> the `create()` function, make sure that the execution role
> associated with **SageMaker Studio** has the necessary permissions
> to create **AWS Step Functions** resources. You may attach the
> `AWSStepFunctionsFullAccess` policy to said execution role to solve
> this issue.

28. Use the `execute()` function to execute the workflow:

```
execution = workflow.execute(
    inputs={
        'JobName': 'll-{}'.format(grs()),
        'ModelName': 'll-{}'.format(grs()),
        'EndpointName': 'll-{}'.format(grs())
    }
)
```

> **Tip**
>
> To inspect and debug the configuration of the execution workflow, you can run
> `print(workflow.definition.to_json(pretty=True))`.

29. Navigate back to the Studio notebook and use the `list_events()` function to get
    more details on the execution:

```
import pandas as pd
events = execution.list_events()
pd.json_normalize(events)
```

This should give us a DataFrame of values, similar to what is shown in the following
screenshot:

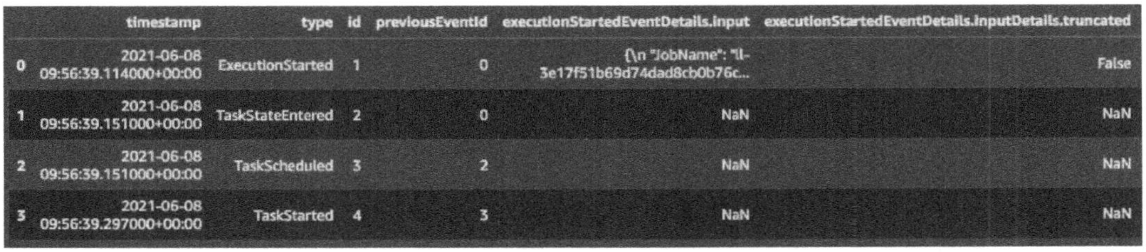

	timestamp	type	id	previousEventId	executionStartedEventDetails.input	executionStartedEventDetails.inputDetails.truncated
0	2021-06-08 09:56:39.114000+00:00	ExecutionStarted	1	0	{\n "JobName": "ll-3e17f51b69d74dad8cb0b76c...	False
1	2021-06-08 09:56:39.151000+00:00	TaskStateEntered	2	0	NaN	NaN
2	2021-06-08 09:56:39.151000+00:00	TaskScheduled	3	2	NaN	NaN
3	2021-06-08 09:56:39.297000+00:00	TaskStarted	4	3	NaN	NaN

Figure 9.24 – Execution event details

Here, we have a better idea of the progress of each step in the Step Functions workflow. At this point, all we need to do is wait for the workflow's execution to complete. If we need to execute this workflow again, all we need to do is attach it to an existing state machine using `Workflow.attach()` and use the `execute()` function.

> **Important Note**
>
> Do not forget to delete the inference endpoint that was created in this recipe as the workflow automatically trains and deploys a model.

Now, let's see how this works!

How it works...

Automating machine learning workflows allows data science and MLOps teams to significantly speed up the process of training and deploying machine learning models. In this recipe, we used the **Data Science SDK** to automatically generate a Step Functions workflow state machine using Python code. With this approach, it will be easy to convert existing notebooks using the **SageMaker Python SDK** that focus on training and deploying models to automated workflows using **Step Functions** and the **Data Science SDK**. There are several classes in the **AWS Step Functions Data Science SDK** that map to the corresponding step or task in a machine learning workflow:

- `TrainingStep`: This focuses on executing a training job to build a machine learning model.
- `ModelStep`: This focuses on creating or registering a model in SageMaker.
- `TransformStep`: This focuses on executing a SageMaker **Batch Transform** job, which uses a trained model to get inferences without having a real-time inference endpoint.
- `EndpointConfigStep`: This focuses on creating an endpoint configuration.
- `EndpointStep`: This focuses on creating or updating an inference endpoint.
- `TuningStep`: This focuses on executing an **Automatic Model Tuning** job.
- `ProcessingStep`: This focuses on executing a **SageMaker Processing** job.

If we were to navigate to the Step Functions console, we would see a visual workflow diagram of the state machine's execution, similar to what is shown in the following diagram. Here, we can see that we have automated the steps we manually ran in the previous chapters of this book:

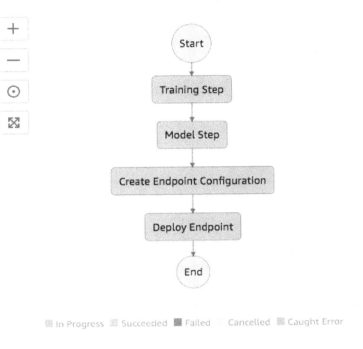

Figure 9.25 – Step Functions visual workflow

In Figure 9.25, we can see that all the steps have been completed successfully. If you are wondering whether we can design a more complex Step Functions workflow using different AWS services, along with conditional branching logic, then the answer to that would be *yes*. Note that we can also create parallel branches of execution in the workflow, as well as handling error conditions when there are runtime errors. We will not dive deep into the details of using **Step Functions** in this book, so feel free to check out `https://docs.aws.amazon.com/step-functions/latest/dg/how-step-functions-works.html` for more information.

See also

If you are looking for examples of using **Step Functions** and the **Data Science SDK** on real datasets and more complex examples, feel free to check out some of the examples focusing on this topic here: `https://sagemaker-examples.readthedocs.io/en/latest/step-functions-data-science-sdk/index.html`.

Managing ML workflows with SageMaker Pipelines

SageMaker Pipelines is a purpose-built CI/CD and orchestration service that helps automate, manage, and reuse machine learning workflows. It has tight integration with the different features and capabilities of SageMaker, which makes it easy for data scientists and machine learning engineers to use it for **MLOps** requirements with the SageMaker service.

In this recipe, we will use **SageMaker Pipelines** to create and manage automated ML workflows. We will work with a simplified example involving a sequential workflow of a processing step, followed by a training step. The processing step makes use of **SageMaker Processing** to perform the train-test split, while the training step focuses on training a **linear learner** model using the training data that's been prepared by the processing step. Once we have completed the steps in this recipe, we will be able to execute an end-to-end automated pipeline using **SageMaker Pipelines**, without having to run scripts manually inside Jupyter notebooks.

Getting ready

For this recipe, you will need a **SageMaker Studio** notebook running the **Python 3 (Data Science)** kernel.

How to do it...

This recipe focuses on creating and executing a basic workflow using **SageMaker Pipelines**. Let's get started:

1. Create a new notebook using the `Python 3 (Data Science)` kernel inside the `my-experiments/chapter09` directory and rename it to the name of this recipe (`Managing ML workflows with SageMaker Pipelines`).

2. Create the `tmp` directory using the `mkdir` command if it does not exist yet:

   ```
   !mkdir -p tmp
   ```

3. Prepare the `path` variable so that it points to the location of the CSV file to be used in this recipe:

   ```
   g = "raw.githubusercontent.com"
   p = "PacktPublishing"
   a = "Machine-Learning-with-Amazon-SageMaker-Cookbook"
   ```

```
mc = "master/Chapter01"
path = f"https://{g}/{p}/{a}/{mc}/files"
```

4. Download the CSV file to the `tmp` directory using the `wget` command:

```
csv = "management_experience_and_salary.csv"
!wget -P tmp {path}/{csv}
```

5. Specify the S3 bucket's name and the prefix where the data will be stored. Make sure to replace the value of `<insert s3 bucket name here>` with the name of the bucket we created in the *Preparing the Amazon S3 bucket and the training dataset for the linear regression experiment* recipe of *Chapter 1, Getting Started with Machine Learning Using Amazon SageMaker*:

```
s3_bucket = '<insert s3 bucket name here>'
prefix = 'chapter09'
input_data_uri = f"s3://{s3_bucket}/{prefix}/input/{csv}"
```

6. Use the **AWS CLI** to upload the CSV file from the `tmp` directory to the S3 bucket:

```
!aws s3 cp tmp/{csv} {input_data_uri}
```

7. Prepare the `ParameterString` objects that correspond to the workflow parameter values for processing the instance type, training instance type, and input data's S3 location:

```
from sagemaker.workflow.parameters import (
    ParameterInteger,
    ParameterString,
)
processing_instance_type = ParameterString(
    name="ProcessingInstanceType",
    default_value="ml.m5.xlarge"
)
training_instance_type = ParameterString(
    name="TrainingInstanceType",
    default_value="ml.m5.xlarge"
)
input_data = ParameterString(
    name="InputData",
```

```
        default_value=input_data_uri,
)
```

8. Import and prepare a few prerequisites, such as `role`, from the **SageMaker Python SDK**:

```
from sagemaker import get_execution_role
role = get_execution_role()
```

9. Prepare the path variable so that it points to where the `preprocessing.py` script file is located in this book's GitHub repository:

```
g = "raw.githubusercontent.com"
p = "PacktPublishing"
a = "Machine-Learning-with-Amazon-SageMaker-Cookbook"
mc = "master/Chapter09"
path = f"https://{g}/{p}/{a}/{mc}/scripts"
```

10. Download the `preprocessing.py` script file from this book's GitHub repository to the `tmp` directory:

```
!wget -P tmp {path}/preprocessing.py
```

This should download a file, similar to what is shown in the following screenshot, inside the `tmp` directory. We can check the contents of `tmp/preprocessing.py` using the **File Browser**:

```
 1  import pandas as pd
 2  from sklearn.model_selection import train_test_split
 3
 4  if __name__ == "__main__":
 5      base_dir = "/opt/ml/processing"
 6
 7      csv = "management_experience_and_salary.csv"
 8      filename = f"{base_dir}/input/{csv}"
 9      df_all_data = pd.read_csv(filename)
10
11      X = df_all_data['management_experience_months'].values
12      y = df_all_data['monthly_salary'].values
13
14      X_train, X_test, y_train, y_test = train_test_split(
15          X, y,
16          test_size=0.3, random_state=0
17      )
```

Figure 9.26 – preprocessing.py

The preprocessing.py script file focuses on performing the train-test split and preparing the training data CSV file so that it's ready for a training job.

11. Initialize the SKLearnProcessor object:

```
from sagemaker.sklearn.processing import SKLearnProcessor

framework_version = "0.23-1"
sklearn_processor = SKLearnProcessor(
    framework_version=framework_version,
    instance_type=processing_instance_type,
    instance_count=1,
    role=role,
)
```

12. Initialize the ProcessingStep object:

```
from sagemaker.processing import ProcessingInput,
ProcessingOutput
from sagemaker.workflow.steps import ProcessingStep

step_process = ProcessingStep(
    name="ProcessingStep",
    processor=sklearn_processor,
    inputs=[
        ProcessingInput(
            source=input_data,
            destination="/opt/ml/processing/input"
        ),
    ],
    outputs=[
        ProcessingOutput(
            output_name="output",
            source="/opt/ml/processing/output"
        ),
    ],
    code="tmp/preprocessing.py",
)
```

13. Import and prepare a few prerequisites, such as `session` and `region_name`, from the **SageMaker Python SDK**:

```
import sagemaker
import boto3


session = sagemaker.Session()
region_name = boto3.Session().region_name
```

14. Initialize the `Estimator` object and specify the hyperparameters using the `set_hyperparameters()` function:

```
from sagemaker.image_uris import retrieve
model_path = f"s3://{s3_bucket}/{prefix}/model"
container = retrieve(
    "linear-learner",
    region_name, "1"
)
estimator = sagemaker.estimator.Estimator(
    container,
    role,
    instance_count=1,
    instance_type='ml.m5.xlarge',
    output_path=model_path,
    sagemaker_session=session
)
estimator.set_hyperparameters(
    predictor_type='regressor',
    mini_batch_size=4
)
```

15. Initialize the `TrainingStep` object:

```
from sagemaker.inputs import TrainingInput
from sagemaker.workflow.steps import TrainingStep
s3_input_data = step_process.properties.
ProcessingOutputConfig.Outputs["output"].S3Output.S3Uri
step_train = TrainingStep(
    name="TrainStep",
```

```
        estimator=estimator,
        inputs={
            "train": TrainingInput(
                s3_data=s3_input_data,
                content_type="text/csv",
            )
        },
    )
```

16. Initialize the `Pipeline` object:

```
from sagemaker.workflow.pipeline import Pipeline
pipeline = Pipeline(
    name="Pipeline",
    parameters=[
        processing_instance_type,
        training_instance_type,
        input_data,
    ],
    steps=[step_process, step_train],
)
```

17. Use the `upsert()` function to submit the pipeline definition we prepared in the previous steps to the **SageMaker Pipelines** service. This will create the pipeline if it does not exist yet:

```
pipeline.upsert(role_arn=role)
```

18. Start the pipeline's execution using the `start()` function:

```
execution = pipeline.start()
```

> **Note**
>
> Note that the `start()` function does not wait for the pipeline to finish executing. You might be surprised that running the preceding line of code will take about a second or less to complete! Behind the scenes, the pipeline is still running and it will take about 8 to 15 minutes for the entire process to complete.

19. Use the `describe()` function to inspect the pipeline's execution:

```
execution.describe()
```

This should give us a nested dictionary of values, similar to what is shown in the following screenshot:

```
[18]: {'PipelineArn': 'arn:aws:sagemaker:us-east-1:        :pipeline/pipeline',
       'PipelineExecutionArn': 'arn:aws:sagemaker:us-east-1:        :pipeline/pipeline/execution/1tb3dly187ct',
       'PipelineExecutionDisplayName': 'execution-1628443739616',
       'PipelineExecutionStatus': 'Executing',
       'PipelineExperimentConfig': {'ExperimentName': 'pipeline',
       'TrialName': '1tb3dly187ct'},
       'CreationTime': datetime.datetime(2021, 8, 8, 17, 28, 59, 541000, tzinfo=tzlocal()),
       'LastModifiedTime': datetime.datetime(2021, 8, 8, 17, 28, 59, 541000, tzinfo=tzlocal()),
       'CreatedBy': {'UserProfileArn': 'arn:aws:sagemaker:us-east-1:        :user-profile/d-rgvubtsq1vug/arvs',
       'UserProfileName': 'arvs',
       'DomainId': 'd-rgvubtsq1vug'},
       'LastModifiedBy': {'UserProfileArn': 'arn:aws:sagemaker:us-east-1:        :user-profile/d-rgvubtsq1vug/arvs',
       'UserProfileName': 'arvs',
       'DomainId': 'd-rgvubtsq1vug'},
       'ResponseMetadata': {'RequestId': '76ce5454-68cd-4321-ad4b-5045eae6b650',
       'HTTPStatusCode': 200,
       'HTTPHeaders': {'x-amzn-requestid': '76ce5454-68cd-4321-ad4b-5045eae6b650',
       'content-type': 'application/x-amz-json-1.1',
       'content-length': '755',
       'date': 'Sun, 08 Aug 2021 17:28:59 GMT'},
       'RetryAttempts': 0}}
```

Figure 9.27 – Results of execution.describe()

Here, we can see that the pipeline's execution status is still `'Executing'`.

20. Call the `wait()` function to wait for the execution to complete:

```
execution.wait()
```

> **Note**
>
> This step should take about 8 to 15 minutes to complete. Feel free to grab a cup of coffee or tea while waiting!

The next few steps focus on inspecting the pipeline we created in the previous steps:

21. Use the `list_steps()` function to inspect the progress of the steps in our execution workflow:

```
execution.list_steps()
```

This should give us a list of dictionaries, similar to what is shown in the following screenshot:

```
[20]: [{'StepName': 'TrainStep',
        'StartTime': datetime.datetime(2021, 8, 8, 17, 33, 19, 76000, tzinfo=tzlocal()),
        'EndTime': datetime.datetime(2021, 8, 8, 17, 37, 1, 580000, tzinfo=tzlocal()),
        'StepStatus': 'Succeeded',
        'Metadata': {'TrainingJob': {'Arn': 'arn:aws:sagemaker:us-east-1:          :training-job/pipelines-1tb3dly187ct-trainstep-
ojcb8cmvwu'}}},
       {'StepName': 'ProcessingStep',
        'StartTime': datetime.datetime(2021, 8, 8, 17, 29, 0, 856000, tzinfo=tzlocal()),
        'EndTime': datetime.datetime(2021, 8, 8, 17, 33, 18, 800000, tzinfo=tzlocal()),
        'StepStatus': 'Succeeded',
        'Metadata': {'ProcessingJob': {'Arn': 'arn:aws:sagemaker:us-east-1:          :processing-job/pipelines-1tb3dly187ct-proces
singstep-bqjtptpfbk'}}}]
```

Figure 9.28 – Results of execution.list_steps()

We can see in *Figure 9.28* that `ProcessingStep` and `TrainStep` have succeeded already.

22. Use `LineageTableVisualizer` to see the input and output artifacts that are connected to the execution steps in the workflow:

```
import time
from sagemaker.lineage.visualizer import
LineageTableVisualizer
session = sagemaker.session.Session()
viz = LineageTableVisualizer(session)
ess = reversed(execution.list_steps())

for execution_step in ess:
    print(execution_step)
    display(viz.show(
        pipeline_execution_step=execution_step
    ))
    time.sleep(3)
```

This should give us dictionaries and tables of values, similar to what is shown in the following screenshot:

Figure 9.29 – Using LineageTableVisualizer

Here, we can see the artifacts and resources associated with the execution steps in the workflow.

23. Navigate to the **Components and registries** tab and locate the pipeline we just created in this recipe:

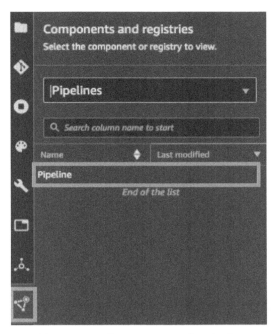

Figure 9.30 – Components and registries

To locate the pipeline we created in this recipe, use the **Pipelines** dropdown to list all the pipeline resources. Double-click the row that maps to the pipeline, similar to what is shown in the preceding screenshot.

24. Double-click the row corresponding to the execution we triggered in this recipe using `pipeline.start()`:

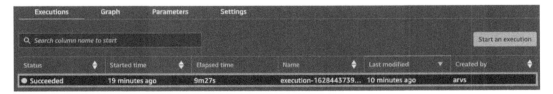

Figure 9.31 – Pipeline execution

Here, we can see that our pipeline execution has succeeded and that it took about 8 and a half minutes for the execution to complete.

25. We should see a graph of the execution, similar to what is shown in the following screenshot. Click the **TrainStep** node:

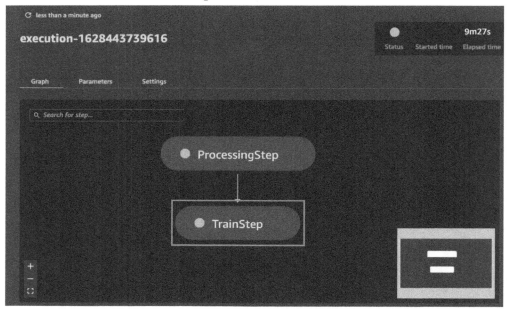

Figure 9.32 – Pipeline graph

Upon clicking the **TrainStep** node, we should see the following details:

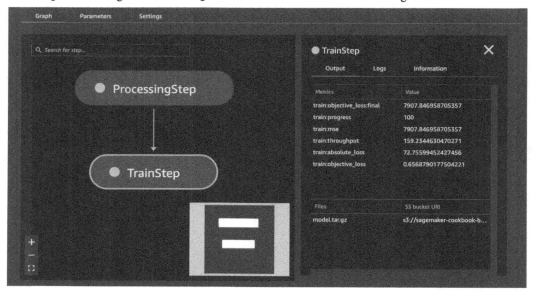

Figure 9.33 – TrainStep details

Here, we have the training job information, details, and results, including the values for `train:object_loss:final`, `train:mse`, `train:progress`, `train:throughput`, `train:absolute_loss`, and `train:objective_loss`.

> **Tip**
> You can also use `pprint(pipeline.describe())` to get more details.

26. Back in the notebook, use the `delete()` function to delete the pipeline:

```
pipeline.delete()
```

At this point, the pipeline should be deleted; however, the other resources that were created (for example, `model`) should not be deleted.

Now, let's see how this works!

How it works...

In this recipe, we have created and executed a pipeline using the different classes available in the **SageMaker Python SDK**. Similar to the **AWS Step Functions Data Science SDK**, there are several classes in the **SageMaker Python SDK** that map to the corresponding step or task in a machine learning workflow:

- `ProcessingStep`: This focuses on tasks such as automated data processing, feature engineering, data cleaning, and model evaluation.

- `TrainingStep`: This focuses on model training and fine-tuning tasks.

- `CreateModelStep`: This focuses on creating a SageMaker model from the output of a training step.

- `TransformStep`: This focuses on using an existing SageMaker model to perform batch transformation on a test dataset.

- `RegisterModelStep`: This focuses on creating a model package that packages the artifacts and model properties needed for inference.

- `ConditionStep`: This allows conditional execution support in the pipeline.

For a more complex example, we may decide to create a new project and use a template provided by SageMaker (for example, **MLOps template for model building, training, and deployment**). Here, we will get to create and use a relatively more complex version of the pipeline, which involves conditional expressions, model registration, automated model evaluation, and even a manual approval step:

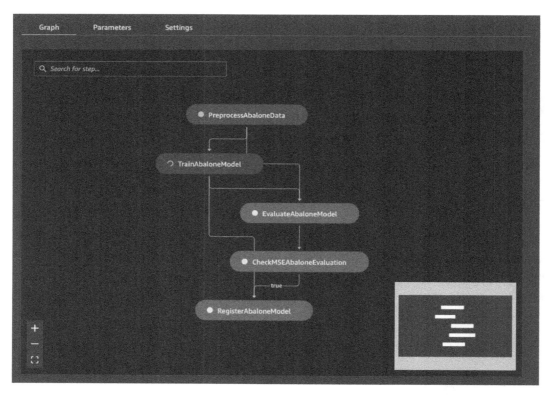

Figure 9.34 – Pipeline graph from the project template

We will not dive deep into the nitty-gritty details of using **SageMaker Pipelines** in this book, so feel free to check out the following link for more information on this topic: `https://docs.aws.amazon.com/sagemaker/latest/dg/pipelines.html`.

See also

If you are looking for examples of using **SageMaker Pipelines** on real datasets and more complex examples, feel free to check out some of the notebooks focusing on this topic here: `https://github.com/aws/amazon-sagemaker-examples/tree/master/sagemaker-pipelines/tabular`.

`Packt.com`

Subscribe to our online digital library for full access to over 7,000 books and videos, as well as industry leading tools to help you plan your personal development and advance your career. For more information, please visit our website.

Why subscribe?

- Spend less time learning and more time coding with practical eBooks and Videos from over 4,000 industry professionals

- Improve your learning with Skill Plans built especially for you

- Get a free eBook or video every month

- Fully searchable for easy access to vital information

- Copy and paste, print, and bookmark content

Did you know that Packt offers eBook versions of every book published, with PDF and ePub files available? You can upgrade to the eBook version at `packt.com` and as a print book customer, you are entitled to a discount on the eBook copy. Get in touch with us at `customercare@packtpub.com` for more details.

At `www.packt.com`, you can also read a collection of free technical articles, sign up for a range of free newsletters, and receive exclusive discounts and offers on Packt books and eBooks.

Other Books You May Enjoy

If you enjoyed this book, you may be interested in these other books by Packt:

Learn Amazon SageMaker

Julien Simon

ISBN: 978-1-80020-891-9

- Create and automate end-to-end machine learning workflows on Amazon Web Services (AWS)
- Become well-versed with data annotation and preparation techniques
- Use AutoML features to build and train machine learning models with AutoPilot
- Create models using built-in algorithms and frameworks and your own code
- Train computer vision and NLP models using real-world examples
- Cover training techniques for scaling, model optimization, model debugging, and cost optimization

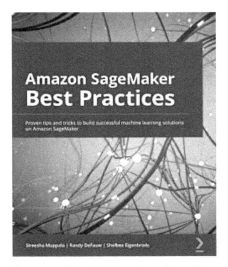

Amazon SageMaker Best Practices

Sireesha Muppala, Randy DeFauw, Shelbee Eigenbrode

ISBN: 978-1-80107-052-2

- Perform data bias detection with AWS Data Wrangler and SageMaker Clarify
- Speed up data processing with SageMaker Feature Store
- Overcome labeling bias with SageMaker Ground Truth
- Improve training time with the monitoring and profiling capabilities of SageMaker Debugger
- Address the challenge of model deployment automation with CI/CD with SageMaker Model registry
- Explore SageMaker Neo for model optimization
- Implement data and model quality monitoring with Amazon Model Monitor
- Reduce training time and costs with SageMaker Data and Model Parallelism

Packt is searching for authors like you

If you're interested in becoming an author for Packt, please visit `authors.packtpub.com` and apply today. We have worked with thousands of developers and tech professionals, just like you, to help them share their insight with the global tech community. You can make a general application, apply for a specifi c hot topic that we are recruiting an author for, or submit your own idea.

Share Your Thoughts

Now you've finished *Machine Learning with Amazon SageMaker Cookbook*, we'd love to hear your thoughts! Scan the QR code below to go straight to the Amazon review page for this book and share your feedback or leave a review on the site that you purchased it from.

https://packt.link/r/1800567030

Your review is important to us and the tech community and will help us make sure we're delivering excellent quality content.

Index

www.ingramcontent.com/pod-product-compliance
Lightning Source LLC
LaVergne TN
LVHW081326050326
832903LV00024B/1050